Chemical Engineering Process Simulation

Chemical Engineering Process Simulation

Second Edition

Edited by

Dominic C. Y. Foo
Department of Chemical and Environmental Engineering/Centre of Excellence for Green Technologies, University of Nottingham Malaysia, Semenyih, Selangor, Malaysia

ELSEVIER

IChemE ADVANCING CHEMICAL ENGINEERING WORLDWIDE

Elsevier
Radarweg 29, PO Box 211, 1000 AE Amsterdam, Netherlands
The Boulevard, Langford Lane, Kidlington, Oxford OX5 1GB, United Kingdom
50 Hampshire Street, 5th Floor, Cambridge, MA 02139, United States

Notices
Knowledge and best practice in this field are constantly changing. As new research and experience broaden our understanding, changes in research methods, professional practices, or medical treatment may become necessary.

Practitioners and researchers must always rely on their own experience and knowledge in evaluating and using any information, methods, compounds, or experiments described herein. In using such information or methods they should be mindful of their own safety and the safety of others, including parties for whom they have a professional responsibility.

To the fullest extent of the law, neither the Publisher nor the authors, contributors, or editors, assume any liability for any injury and/or damage to persons or property as a matter of products liability, negligence or otherwise, or from any use or operation of any methods, products, instructions, or ideas contained in the material herein.

ISBN: 978-0-323-90168-0

For information on all Elsevier publications visit our website at https://www.elsevier.com/books-and-journals

Publisher: Susan Dennis
Acquisitions Editor: Anita Koch
Editorial Project Manager: Judith Clarisse Punzalan
Production Project Manager: Paul Prasad Chandramohan
Cover Designer: Vicky Pearson Esser

Typeset by TNQ Technologies

Dedication

The editor and author Dominic Foo would like to dedicate this book to his wife Cecilia and their kids Irene, Jessica, and Helena. He would also like to dedicate this book to his students who join his process simulation classes at the University of Nottingham Malaysia.

Contents

3. Physical property estimation and phase behavior for process simulation

Rafil Elyas

4. Simulation of recycle streams

Dominic C.Y. Foo, Siewhui Chong and Nishanth Chemmangattuvalappil

Part III
Symmetry

Part IV
SuperPro designer

Part V
aspenONE engineering

Contributors

Abdul Rahim Norman, PETRONAS Group Technical Solutions, Process Simulation and Optimization, Kuala Lumpur, Malaysia

Alexandros Koulouris, Department of Food Science and Technology, International Hellenic University, Alexander Campus Sindos-Thessaloniki, Greece

Bor-Yih Yu, Department of Chemical Engineering, National Taiwan University, Taipei, Taiwan

Chang-Che Tsai, Department of Chemical Engineering, National Taiwan University, Taipei, Taiwan

Chien Hwa Chong, Department of Chemical and Environmental Engineering, University of Nottingham Malaysia, Semenyih, Selangor, Malaysia

Denny K.S. Ng, Heriot-Watt University Malaysia, Putrajaya, Malaysia

Dominic C.Y. Foo, Department of Chemical and Environmental Engineering/Centre of Excellence for Green Technologies, University of Nottingham Malaysia, Semenyih, Selangor, Malaysia

Jia Wen Chong, University of Nottingham Malaysia, Semenyih, Selangor, Malaysia

Jie Yi Goo, Heriot-Watt University Malaysia, Putrajaya, Wilayah Persekutuan Putrajaya, Malaysia

John Frederick D. Tapia, De La Salle University, Manila, Philippines

Lik Yin Ng, Department of Chemical and Petroleum Engineering, Faculty of Engineering, Technology and Built Environment, UCSI University, UCSI Heights, Cheras, Kuala Lumpur, Malaysia

Mijndert Van der Spek, Heiort-Watt University, Edinburgh, United Kingdom

Nishanth Chemmangattuvalappil, University of Nottingham Malaysia, Semenyih, Selangor, Malaysia

Noorhidayah Bt Hussein, PETRONAS Group Technical Solutions, Process Simulation and Optimization, Kuala Lumpur, Malaysia

Nurain Shakina Roslizam, PETRONAS Group Technical Solutions, Process Simulation and Optimization, Kuala Lumpur, Malaysia

Pei-Jhen Wu, Department of Chemical Engineering, National Taiwan University of Science and Technology, Taipei, Taiwan

Pitchaimuthu Diban, Pand.ai Pte Ltd, Singapore

Rafil Elyas, East One-Zero-One Sdn Bhd, Shah Alam, Selangor, Malaysia

Raymond E.H. Ooi, University of Nottingham Malaysia, Semenyih, Selangor, Malaysia

Rebecca Lim, Heriot-Watt University Dubai, Dubai Knowledge Park, Dubai, United Arab Emirates

Rui Ma, Department of Chemical and Environmental Engineering, University of Nottingham Malaysia, Semenyih, Selangor, Malaysia

Shahrul Azman Abidin, PETRONAS Group Technical Solutions, Process Simulation and Optimization, Kuala Lumpur, Malaysia

Shiang-Tai Lin, Department of Chemical Engineering, National Taiwan University, Taipei, Taiwan

Siewhui Chong, University of Nottingham Malaysia, Semenyih, Selangor, Malaysia; Xodus group, Perth, Australia

Siti Nurfaqihah Azhari, PETRONAS Group Technical Solutions, Process Simulation and Optimization, Kuala Lumpur, Malaysia

Vincentius Surya Kurnia Adi, Department of Chemical Engineering, National Chung Hsing University, Taichung, Taiwan

Zulfan Adi Putra, PETRONAS Group Technical Solutions, Process Simulation and Optimization, Kuala Lumpur, Malaysia

Acknowledgments

Dominic Foo would also like to acknowledge his former advisors Professor Zainuddin Abdul Manan and Professor Ramlan Abdul Aziz for their support in developing process simulation skillsets during his Ph.D. studies.

How to use this book

This book presents some of the most popular commercial steady-state process simulation software in the market. It is believed that this will be a good self-learning guide for students and working professionals in learning process simulation software, as well as for university instructors who conduct lectures on process simulation. Even though the chapters are interconnected, they are mostly written independently; this allows readers to read each chapter without having to read the preceding chapters. Note, however, that the four chapters in Part 1 of the book serve as important guide on the various basic principles behind all commercial simulation software, e.g., physical properties estimation (thermodynamic models), new component registration, recycle stream simulation, etc. Readers are encouraged to read these chapters before proceeding to other parts of the book (see Fig. 1). This is particularly important for novice in process simulation.

Parts 2–5 of the book cover four different families of commercial software in the market, i.e., UniSim Design, Symmetry, SuperPro Designer, and AspenONE Engineering. Each of these software has an introductory chapter (with step-by-step guide) to allow new users in mastering the usage of the software, before going into advanced topics. Table 1 shows the level of difficulties for all chapters in this book, in which the readers may refer to in selecting the topic for reading.

Apart from basic simulation knowledge and the use of simulation tools in designing various production processes (e.g., *n*-Octane, bio-jet fuel intermediate, citric acid, diethyl carbonate, etc.), several advanced topics on process optimization (e.g., heating medium systems for offshore platform, natural gas

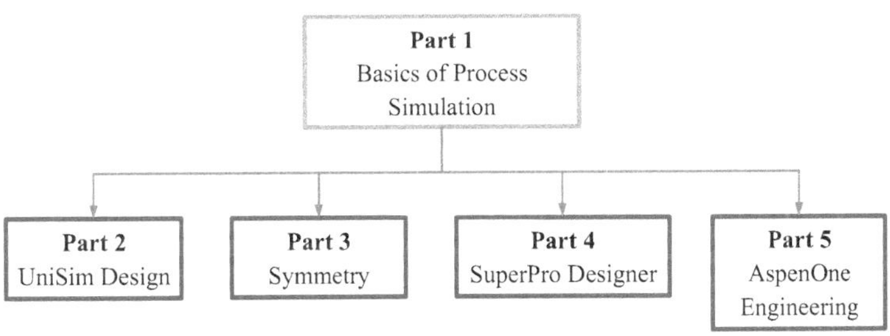

FIGURE 1 Suggested flow in reading this book.

TABLE 1 Level of difficulty for each chapter.

	Chapters	Level
Part 1—Basics of process simulation		
1	Introduction to process simulation	Basic
2	Registration of new components	Basic
3	Physical properties estimation for process simulation	Basic
4	Simulation of recycle streams	Basic
Part 2—UniSim design		
5	Basics of process simulation with UniSim Design	Basic
6	Design and simulation of distillation processes	Advanced
7	Modeling and optimization of separation and heating medium systems for offshore platform	Advanced
Part 3—Symmetry		
8	Basics of process simulation with Symmetry	Basic
9	Process modeling and analysis of a natural gas dehydration process using triethylene glycol (TEG) via Symmetry	Advanced
Part 4—SuperPro designer		
10	Basics of batch process simulation with SuperPro Designer	Basic
11	Modeling of citric acid production using SuperPro Designer	Advanced
12	Design and optimization of wastewater treatment plant (WWTP) for the poultry industry	Advanced
Part 5—AspenONE engineering		
13	Basics of process simulation with Aspen HYSYS	Basic
14	Process simulation and design of acetaldehyde production	Advanced
15	Dynamic simulation for process control with Aspen HYSYS	Advanced
16	Basics of process simulation with Aspen Plus	Basic
17	Design and Evaluation of alternative processes for the manufacturing of bio-jet fuel (BJF) intermediate	Advanced

TABLE 1 Level of difficulty for each chapter.—cont'd

	Chapters	Level
18	Production of diethyl carbonate from direct CO_2 conversion	Advanced
19	Multiplatform optimization on unit operation and process designs	Advanced
20	Flexible design strategy for process controllability	Advanced

dehydration) are also included in this book. Apart from steady-state modeling, chapters on dynamic simulation and process controllability are also covered.

To assist readers in better understanding, process simulation files of all chapters are made available on author support website, which is found in the following URL: https://www.elsevier.com/books-and-journals/book-companion/9780323901680. For university instructors, the solutions for various exercises in each chapter are also made available in the password-protected author support website: https://educate.elsevier.com/9780323901680, in which permission to access will be granted to university instructors who adopt the book for their lecture.

It is hoped that this book will serve as a useful guide for a good learning experience in process simulation knowledge. Have fun in your simulation exercises!

Dominic C.Y. Foo
updated for second edition on May 2022

Part I

Basics of process simulation

Chapter 1

Introduction to process simulation*

Dominic C.Y. Foo[1] and Rafil Elyas[2]

[1]*University of Nottingham Malaysia, Semenyih, Selangor, Malaysia;* [2]*East One-Zero-One Sdn Bhd, Shah Alam, Selangor, Malaysia*

Chapter outline

Process simulation is the representation of a chemical process by a mathematical model, which is then solved to obtain information about the performance of the chemical process (Motard et al., 1975). It is also known as *process flowsheeting*. Westerberg et al. (1979) also defined flowsheeting as *the use of computer aids to perform steady-state heat and mass balancing, sizing, and costing calculations for a chemical process*. In this chapter, some basic information about simulation will be presented. This includes the historical developments, basic architectures, and solving algorithms. Besides, 10 good habits of process simulation are also provided at the end of the chapter, to guide readers in nurturing some good practices in using process simulation software.

*Simulation and other electronic files of this chapter are made available on a companion website; URL: https://www.elsevier.com/books-and-journals/book-companion/9780323901680.

Chemical Engineering Process Simulation. https://doi.org/10.1016/B978-0-323-90168-0.00008-1

1.1 Process design and simulation

Many regard process simulation being equivalent to process design, which is indeed a misleading understanding. In fact, process simulation and *process synthesis* are two important and interrelated elements in chemical process design, which may be used to achieve optimum process design. The aim for process simulation is to predict how a defined process would actually behave under a given set of operating conditions. In other words, we aim to predict the outputs of the process when the process flowsheet and its inputs are given (Fig. 1.1). In the modern days, commercial process simulation software packages are often used for such exercises.

On the other hand, when an unknown process flowsheet is to be created for given process input and output streams, this entails the exercise on process synthesis (Fig. 1.2). Process synthesis has been an active area of research in the past 5 decades, with some significant achievements in specific applications, e.g., heat recovery system, material recovery system, and reaction network (El-Halwagi and Foo, 2014). Process synthesis and process simulation supplement each other well. In most cases, once a process flowsheet is synthesized, its detailed characteristics (e.g., temperature, pressure, and flowrates)

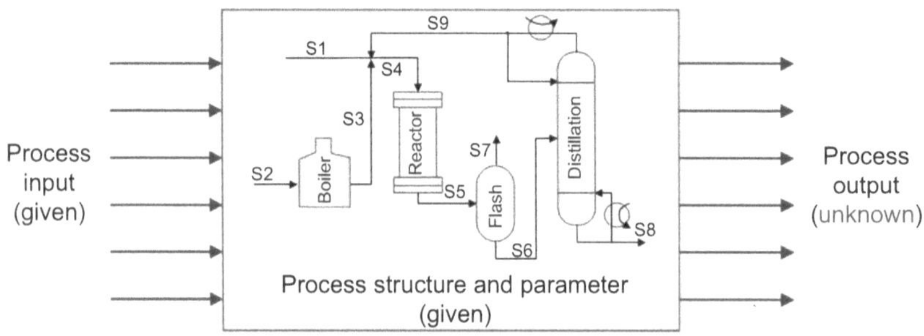

FIGURE 1.1 A process analysis problem (El-Halwagi, 2006; Foo, 2012).

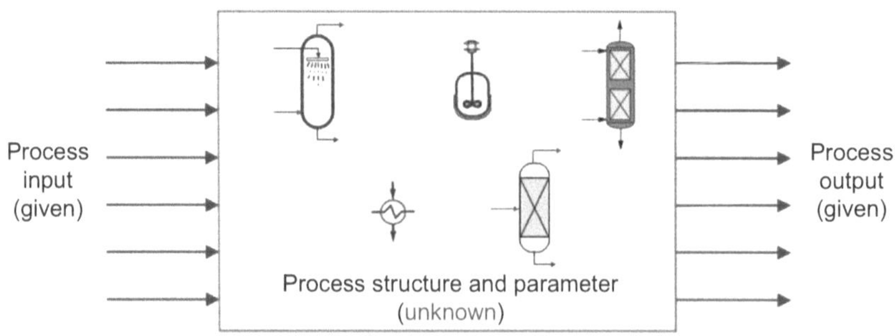

FIGURE 1.2 A process synthesis problem (El-Halwagi, 2006; Foo, 2012).

may be predicted using various process simulation tools, so that an optimum flowsheet may be developed. In other words, process simulation tools are useful in guiding process synthesis exercises (Foo et al., 2005; Lott, 1988). One may also explore the use of process simulation tools for other related activities, such as waste minimization (Sowa, 1994; Hilaly and Sikdar, 1996), debottlenecking (Koulouris et al., 2000), etc.

Within process synthesis, one of the important models to guide flowsheet synthesis is the *onion model* first reported in Linnhoff et al. (1982). As shown in Fig. 1.3, process design exercise begins from the core of the process and moves outward. In the center of the onion, the reactor system is first designed. The reactor design influences the separation and recycle structures at the second layer of the onion. Next, the reactor and separator structures dictate the overall heating and cooling requirement of the process. Hence, the heat recovery system is designed next, in the third layer. A utility system at the fourth layer is next designed, to provide additional heating and cooling requirements, which cannot be satisfied through heat recovery system. At the final layer, the waste treatment system is designed to handle various emissions/effluents from the process, prior to final environmental discharge.

In a later section of this chapter, the onion model will be used to guide the simulation of the integrated flowsheet of chemical processes.

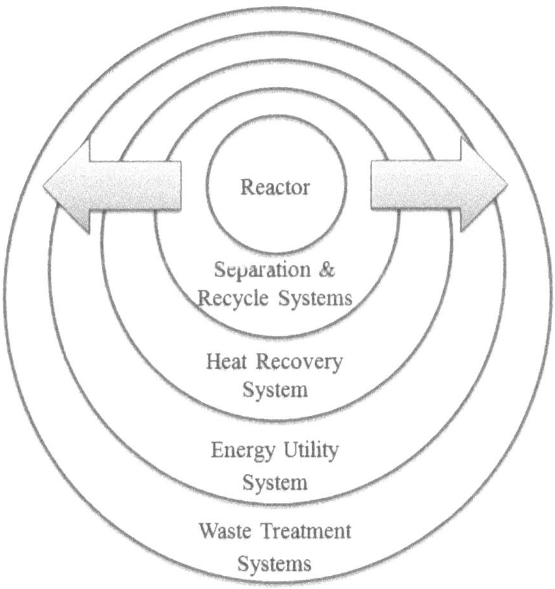

FIGURE 1.3 The onion model.

1.2 Historical perspective for process simulation

With the introduction of computers in the 1950s, we see the start of commercial process simulation software a decade later. The first generic process simulation software known as PROCESS was launched by Simulation Science based at Los Angeles (US) in 1966, for the simulation of distillation columns (Dimian et al., 2014). This software evolved into PRO/II and is marketed by AVEVA in recent years (AVEVA, 2022). Another commercial software for gas and oil applications, known as DESIGN, was launched in 1969 by ChemShare Corporation based at Houston (US) (Dimian et al., 2014). This software is marketed as DESIGN II for Windows by WinSim Inc. since 1995 (WinSim, 2022).

Stepping into the 1970s, which was generally known as the "golden age" of scientific computing, several important historical milestones mark the active developments of process simulation tools. First, FORTRAN programming language (introduced by IBM between 1954 and 1957) became the de facto standard among scientists and engineers (Evans et al., 1977). Two important books on process simulation (Crowe et al., 1971; Westerberg et al., 1979) described some important developments in the 1970s. Of particular importance is the formal introduction of *sequential modular* (SM) approach by Westerberg et al. (1979), which is commonly utilized in most software. Next, the first oil crisis in 1973 simulated the development of simulation tools that can be used for solid handling, i.e., power generation with coal. Following this was the important ASPEN (Advanced System for Process ENgineering) project at Massachusetts Institute of Technology (MIT) between years 1976 and 1979, sponsored by the US Department of Energy (Evans et al., 1979; Gallier et al., 1980). Aspen Technology Inc. (AspenTech) was then formed in 1981 to commercialize the technology, with Aspen Plus software being released in 1982 (AspenTech, 2022). Another important achievement in 1970s is the development of software based on *equation-oriented* (EO) approach. Important EO-based software includes SPEEDUP developed at Imperial College, London (UK) (Hernandez and Sargent, 1979; Perkins et al., 1982), which was later succeeded by gPROMS (Siemens PSE, 2022). Note that in the 1970s, simulation was mainly executed on fast but expensive mainframe systems, where user was connected via a remote terminal.

With the arrival of personal computer (PC) in the 1980s, several other important software packages such as ChemCAD (developed by ChemStations) and HYSYS (originally by Hyprotech) were launched. These software packages no longer operate on the mainframe systems but are PC based. Late 1980s also saw the needs of developing simulation software for biochemical processes (Petrides et al., 1989). This leads to the introduction of BioPro Designer (Petrides, 1994), which later evolved into SuperPro Designer marketed by Intelligen Inc. in the 1990s (Intelligen, 2022). It is worth mentioning that up to the 1980s, most basic developments of process simulation architectural are quite established, with review papers outlining their state-of-the-art techniques (Evans, 1981; Rosen, 1980). It is also worth mentioning the introduction of spreadsheet software during the 1980s, which allows quick solving of equation sets that follow a sequence (Seader, 1985). Spreadsheet software such as MS Excel[1] are still highly welcomed by industrial practitioners in solving day-to-day computational tasks to date.

In the 1990s, with the domination of Microsoft Windows, most software packages were migrated from their previous mainframe/keyword input versions into the more attractive graphical user interface (GUI). Another important milestone happened in the early 21st century. In 2002, Hyprotech was acquired by AspenTech, which resulted with the ownership of HYSYS software. However, the US Federal Trade Commission judged that acquisition of Hyprotech was anticompetitive and ruled AspenTech to divest its software to the approved buyer—Honeywell (Federal Trade Commission, 2003). This later leads to the introduction of UniSim Design (Honeywell, 2017), which shares the same GUI as HYSYS (of Hyprotech), while AspenTech continues to market its Aspen HYSYS software (with different GUI in year 2016).

Some of the commonly used process simulation software packages are listed in Table 1.1.

1.3 Basic architectures for commercial software

Fig. 1.4 shows the basic structure of a process simulation software and sequential steps in performing the simulation task (Turton et al., 2013). As shown at the upper side of the figure, a typical commercial simulation software includes the following components, i.e., component database, thermodynamic model database, flowsheet builder, unit operation model database, and flowsheet solver. Note that some other elements, e.g., subflowsheet, financial analysis tools, and engineering units option, are software dependent and hence are excluded in this figure.

The bottom side of Fig. 1.4 presents a list of sequential steps in solving a simulation problem. In step 1, the basic information for a simulation problem is first provided. This includes chemical components and thermodynamic model selection, which can be done easily through their associated databases of the software. Note that it is advisable to select all components needed for the flowsheet, even though some components will only be used at the later part of the flowsheet. The selection of thermodynamic model is a crucial step, as different thermodynamic models will lead to very different mass and energy balances for some processes. Next in step 2, the process flowsheet is constructed using the flowsheet builder. This involves the selection of appropriate

1. See Example 1.2 in this chapter for the use of MS Excel in solving a basic flowsheeting problem.

TABLE 1.1 Commercial process simulation software.

Corporations	Software	Websites
AspenTech	AspenONE Engineering (consists of Aspen Plus, Aspen HYSYS, Aspen Economic Analyzer, Aspen Energy Analyzer, etc.)	www.aspentech.com
Honeywell	UniSim Design	www.honeywellforge.ai/
AVEVA	AVEVA PRO/II	www.aveva.com
Chemstations	ChemCAD	www.chemstations.com
WinSim	DESIGN II for Windows	www.winsim.com
Intelligen	SuperPro Designer, SchedulePro	www.intelligen.com
Bryan Research and Engineering	ProMax	www.bre.com
Siemens Process Systems Enterprise	gPROMS	www.psenterprise.com
Schlumberger	Symmetry	www.software.slb.com/products/symmetry

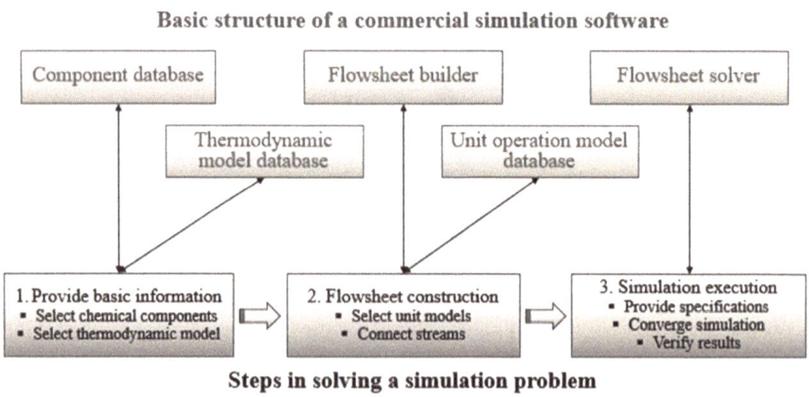

FIGURE 1.4 Basic structure of a commercial simulation software and sequence in solving a simulation model. *Adapted from Turton et al. (2013).*

unit operation models (from the unit model database) and the connections among them with the process streams (some software may need to have energy streams connected too). In step 3, specifications are to be provided for the unit models as well as important inlet streams (e.g., flowrate, temperature, pressure) to execute the simulation. Note that in all modern simulation software, users may choose to display the simulation results in various forms. Finally, it is important to cross-check the simulation results, either through some empirical model, mass, and energy balances or through reported plant/experimental data. Doing so will increase the confident level of the simulation model.

1.4 Basic algorithms for process simulation

Two main classical techniques used in solving process simulation models are SM and EO approaches. Most commercial simulation software packages in the market, e.g., Aspen Plus, ChemCAD, and PRO/II, are using SM approach and hence will be discussed more in depth in the following sections.

1.4.1 Sequential modular approach

The term "sequential modular" was formally introduced in the late 1970s (although commercial software packages were found in the market prior to that) by Westerberg et al. (1979). The concept of SM may be explained using Fig. 1.5. Each of the unit modules contains some algorithms that are utilized to solve a set of process models, provided that the inlet stream information and unit specifications are given. Once a module is solved to convergence, it will generate the results for the outlet stream(s). The latter is then connected as a feed stream for the following unit module, which is then solved for convergence (Turton et al., 2013). The same process is repeated until all process units

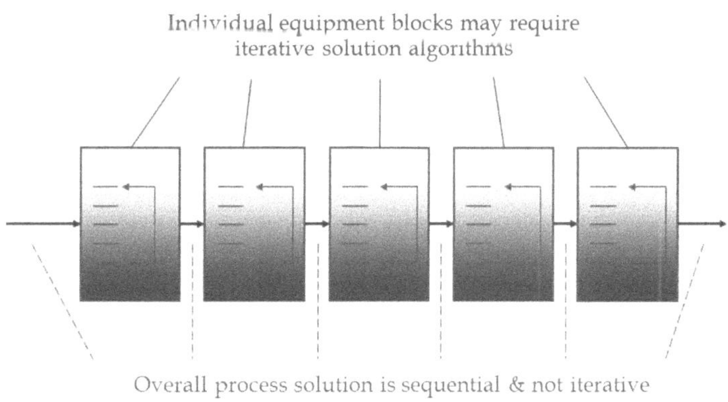

FIGURE 1.5 Concept of sequential modular approach (Turton et al., 2013).

in the flowsheet are solved and converged. Note that certain unit modules may require iterative solution algorithm to achieve convergence; the overall process is, however, sequential in nature, i.e., no iteration is required (Turton et al., 2013).

For process flowsheet that contains recycle stream(s), *tear stream* strategy is commonly used with SM approach to converge the recycle stream. As shown in Fig. 1.6A, the flowsheet consists of six operations, i.e., units A–F and a recycle stream that connects units C and F. If SM approach in Fig. 1.5 is adopted for the simulation exercise, one will start to simulate and converge unit A. This is followed by unit B and then unit C. However, because of the existence of the recycle stream, unit C can only be simulated once the recycle stream contains the necessary properties (e.g., pressure, temperature, and flowrates) after unit F is converged. However, unit F cannot be simulated without first converging unit C. In other words, the convergence of units C and F involves iterative steps.

To cater the iterative procedure, a tear-stream strategy is used. As shown in Fig. 1.6A, the recycle stream is virtually "torn" into two parts—r_1 (inlet for unit C) and r_2 (outlet from unit F). Some estimated data (e.g., temperature, pressure, and flowrate) are provided for r_1 to simulate unit C. Once unit C is converged, simulation then proceeds to unit D, E, and finally unit F. Once unit F is converged, the simulated results from outlet stream r_2 are then compared to those estimated data given to r_1 earlier on. If their values agree to the specified convergence tolerance (typically given in terms of percentage

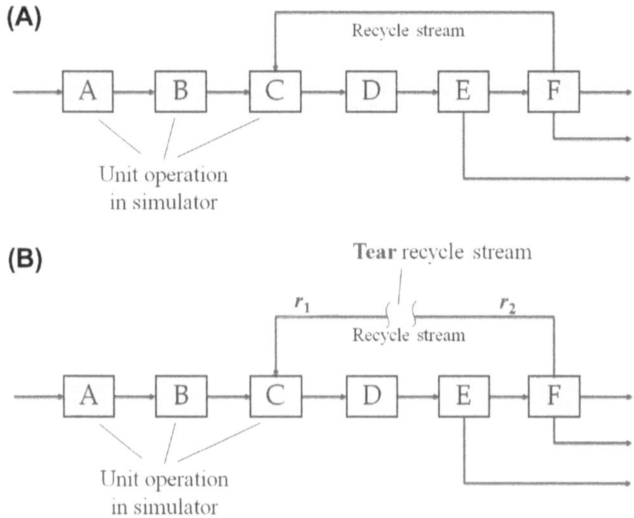

FIGURE 1.6 (A) A process consists of a recycle stream; (B) concept of tear stream (Turton et al., 2013).

difference), the simulation is converged, or else the simulated results from r_2 will substitute the estimated data in r_1 and the iterative procedure is carried on.[2]

The main advantage of SM is that it is intuitive and easy to understand. It allows the interactions of users as the model develops. However, large problems (with many recycle streams) may be difficult to converge.

1.4.2 Equation-oriented approach

In EO approach, a set of equations are solved simultaneously for a simulation problem. For instance, for a problem with n design variables, p equality constraints, and q inequality constraints, the problem is formulated as follows (Smith, 2016):

$$\text{Solve } h_i(x_1, x_2, x_3, \ldots, x_n) = 0 \ (i = 1, \ldots, p) \tag{1.1}$$

$$\text{subject to } gj \ (x_1, x_2, x_3, \ldots, x_n) \leq 0 \ (i = 1, \ldots, q) \tag{1.2}$$

The main advantage of EO approach is its ability to be formulated as an optimization problem. However, complex EO problems are difficult to solve and diagnose. It is also not as robust as SM approach (Smith, 2016). Hence, it has not been favored among commercial simulation software in the past few decades, until recent years where it is embedded in solving complex models in SM-based software (e.g., gPROMS).

One of the convenient ways of solving problem of linear equation set is through matrix inversion, which has the general structure as in Eq. (1.3):

$$\mathbf{A}\,\mathbf{X} = \mathbf{B} \tag{1.3}$$

where **A** is the coefficient matrix, **X** is variable matrix or an unknown vector ($i \times 1$), and **B** is a constant matrix or a known solution vector ($i \times 1$). The variables in variable matrix **X** can be determined by the product of inverse matrix **A** (i.e., $\mathbf{A}^{-1}$) with matrix **B**, given as in Eq. (1.4):

$$\mathbf{X} = \mathbf{A}^{-1}\mathbf{B} \tag{1.4}$$

The use of matrix inversion for equations solving is demonstrated with an example at later section (see Example 1.2).

1.5 Degrees of freedom analysis

DOF analysis is a useful tool to determine if a system has the sufficient information before it can be solved. Eq. (1.5) shows the basic equation of DOF

2. See Chapter 4 for a detailed discussion on handling recycle systems with sequential modular approach.

(n_{df}) of a system, which is given as the difference between its numbers of variables (n_{var}) and the numbers of independent equations (n_{eq}).

$$n_{df} = n_{var} - n_{eq} \tag{1.5}$$

For a system to be solvable, its n_{df} must be equal to zero. In other words, when numbers of independent equations (e.g., $n_{var} = 3$) are equal to its numbers of variables (e.g., $n_{eq} = 3$), the system can be solved readily.

On the other hand, when the numbers of variables (e.g., $n_{var} = 3$) are more than the number of independent equation (e.g., $n_{eq} = 2$), this system has a positive DOF (e.g., $n_{df} = 1$). For such a case, *design variables* are to be specified before other *state variables* can be calculated. There are also *over-defined* cases where the number of variables is less than the numbers of independent equations (e.g., $n_{df} \leq 1$). For such a case, the redundancies are to be identified and removed, before the system can be solved.

A more complete form of DOF analysis is given in Eq. (1.6), to account for chemical reaction, molecular balances, and additional equation relations (Felder and Rousseau, 2005).

$$n_{df} = n_{var} - n_{rxt} - n_{mol} - n_{rel} \tag{1.6}$$

where n_{rxt} is the number of chemical reactions, n_{mol} is the number of independent molecular species balances, and n_{rel} is the number of other equations relating the unknown variables.

Many commercial process simulation software (e.g., Aspen HYSYS, UniSim Design, etc.) have embedded DOF analysis into their unit operations and solution engines. Fig. 1.7 shows the DOF setting for some common unit operation models in commercial simulation software.

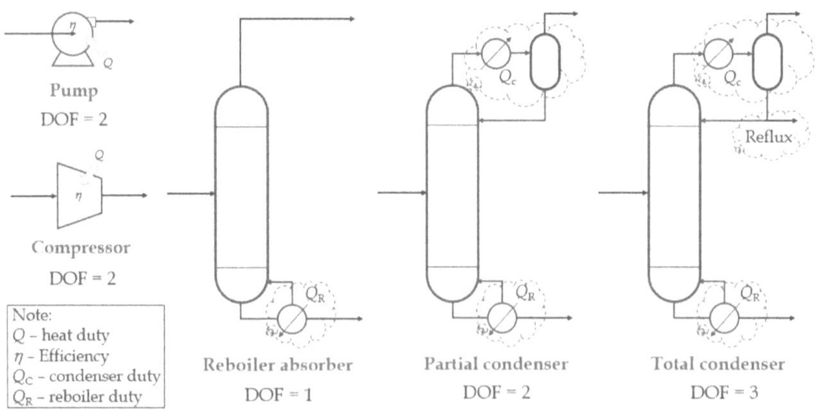

FIGURE 1.7 DOF for some common unit operation models.

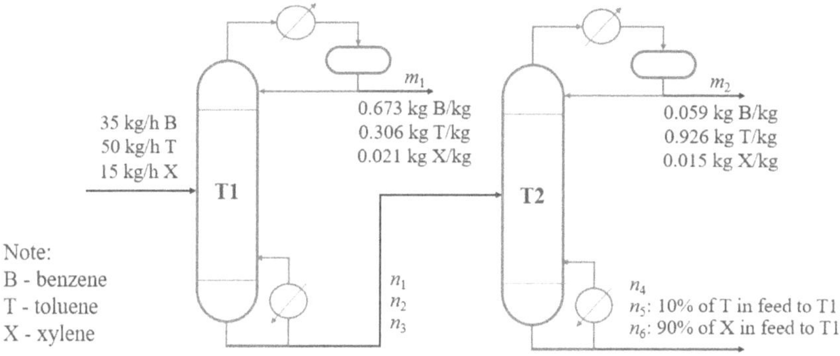

FIGURE 1.8 Two columns for the separation of BTX mixture (Felder and Rousseau, 2005).

Example 1.1. DOF

Fig. 1.8 shows two distillation columns, i.e., T1 and T2 that are used for the separation of a process stream containing a mixture of benzene, toluene, and xylene of known flowrates (Felder and Rousseau, 2005). The bottoms product of column T1 is fed to column T2. The overhead product flowrates of both columns (m_1 and m_2) are unknown, with known compositions. Furthermore, it is given that 10.0% of toluene and 90.0% of the xylene feeding to column T1 are recovered as bottom products of column T2. All unknown variables, i.e., overhead product flowrates (m_1 and m_2) and bottom product component flowrates ($n_1 - n_6$) are labeled in Fig. 1.8.

Determine the DOF for solving the above mass balance problems when the system boundary is set for the following unit(s):

1. Column T1
2. Column T2
3. Overall flowsheet.

Solution:

1. Fig. 1.8 indicates that there are four unknown variables (n_{var}) among the inlet and outlet streams of column T1, i.e., m_1, n_1, n_2, n_3. Since molecular balance is to be performed for benzene, toluene, and xylene, three independent molecular species balances (n_{mol}) can be written. Hence, its n_{df} is determined using Eq. (1.6) as 1 (= 4 − 3). In other words, one among the unknown variables ($m_2, n_1, n_2,$ or n_3) should be specified (as design variables), before the other state variables can be determined.
2. There are seven unknown variables (n_{var}) among the inlet and outlet streams of column T2, i.e., $m_2, n_1 - n_6$. As there are three independent molecular balances (n_{mol}) to be performed (for benzene, toluene, and xylene), its n_{df} is calculated with Eq. (1.6) as 4 (= 7 − 3). In other words,

one would have to specify up to four design variables before the other state variables can be determined.

3. When the system boundary is set for the overall flowsheet, one may ignore the variables of T1 bottom product, since it is not an outlet stream for the overall flowsheet. Hence, there are five unknown variables n_{var}, i.e., m_2, $n_4 - n_6$, and three independent molecular balances (n_{mol}) can be written. Besides, two equations can be written to relate the unknown variables of n_5 and n_6, i.e., $n_{rel} = 2$. Hence, the n_{df} is calculated as 0 (= 5 − 3 − 2), which means that the overall mass balance of the flowsheet can be solved without additional specification of variable. In other words, if one were to specify any design variable, the problem becomes overdefined.

Example 1.2. EO approach with MS Excel spreadsheet

For the distillation system in Example 1.1, solve for the following:

1. Derive equations that relate mass balance problem for the distillation system.
2. Use MS Excel spreadsheet to solve the mass balance problem.

Solution:

1. For column T1, molecular balances can be written for benzene, toluene, and xylene, given as in Eqs. (1.7)–(1.9), respectively.

$$35 = 0.673m_1 + n_1 \tag{1.7}$$

$$50 = 0.306m_1 + n_2 \tag{1.8}$$

$$15 = 0.021m_1 + n_3 \tag{1.9}$$

Similar balances are written for benzene, toluene, and xylene for column T2, given as in Eqs. (1.10)–(1.12), respectively. Next, Eqs. (1.13) and (1.14) may be written to relate variables n_5 and n_6 with their component feed flowrates to column T1.

$$n_1 = 0.059m_2 + n_4 \tag{1.10}$$

$$n_2 = 0.926m_2 + n_5 \tag{1.11}$$

$$n_3 = 0.015m_2 + n_6 \tag{1.12}$$

$$n_5 = 0.1(50) = 5.0 \tag{1.13}$$

$$n_6 = 0.9(15) = 13.5 \tag{1.14}$$

2. Since the mass balance problem has a total of eight equations and eight variables, the problem can be conveniently solved using matrix inversion technique. Eqs. (1.7)–(1.14) can be converted into its matrix form as in Eq. (1.3).

$$\begin{pmatrix} 0.673 & 0 & 1 & 0 & 0 & 0 & 0 & 0 \\ 0.306 & 0 & 0 & 1 & 0 & 0 & 0 & 0 \\ 0.021 & 0 & 0 & 0 & 1 & 0 & 0 & 0 \\ 0 & 0.059 & -1 & 0 & 0 & 1 & 0 & 0 \\ 0 & 0.926 & 0 & -1 & 0 & 0 & 1 & 0 \\ 0 & 0.015 & 0 & 0 & -1 & 0 & 0 & 1 \\ 0 & 0 & 0 & 0 & 0 & 0 & 1 & 0 \\ 0 & 0 & 0 & 0 & 0 & 0 & 0 & 1 \end{pmatrix} \begin{pmatrix} m_1 \\ m_2 \\ n_1 \\ n_2 \\ n_3 \\ n_4 \\ n_5 \\ n_6 \end{pmatrix} = \begin{pmatrix} 35 \\ 50 \\ 15 \\ 0 \\ 0 \\ 0 \\ 5 \\ 13.5 \end{pmatrix} \quad (1.15)$$

Hence, the unknown variables $(m_1, m_1, n_1 - n_6)$ in the variable matrix can be calculated using the inverse matrix formula in Eq. (1.4):

$$\begin{pmatrix} m_1 \\ m_2 \\ n_1 \\ n_2 \\ n_3 \\ n_4 \\ n_5 \\ n_6 \end{pmatrix} = \begin{pmatrix} 48.061 \\ 32.714 \\ 2.655 \\ 35.293 \\ 13.991 \\ 0.725 \\ 5 \\ 13.5 \end{pmatrix} \quad (1.16)$$

The detailed steps to solve this matrix inversion using MS Excel spreadsheet are shown in Fig. 1.9.

1.6 Incorporation of process synthesis model and sequential modular approach

In simulating an integrated flowsheet, when many units are involved, it is always good to break down the complex flowsheet into small systems that are manageable. A useful way of doing so is to make use of the onion model to guide the simulation tasks (Foo et al., 2005). As discussed in Section 1.1, a

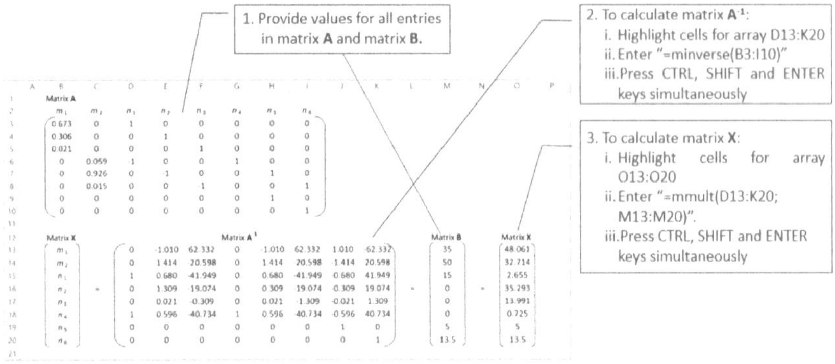

FIGURE 1.9 Detailed steps to solve mass balance problem of distillation system using matrix inversion in MS Excel.

chemical process is designed from the core of the onion and moves outward. Typically, at each of the layer, decision is made by the designer after detailed analysis. Hence, SM approach can be incorporated with the onion model in guiding the design process. This is particularly useful in evaluating new process pathways or generating alternatives for new process development (Foo et al., 2005).

An example on the production of *n*-octane is next demonstrated to illustrate this concept.

Example 1.3. *n*-Octane Production Example

Fig. 1.10 shows the process flow diagram for the production of *n*-octane (C_8H_{18}) (Foo et al., 2005). The fresh feed stream, containing ethylene (C_2H_6), *i*-butane (*i*-C_4H_{10}), and some trace amounts of impurities [i.e., nitrogen and

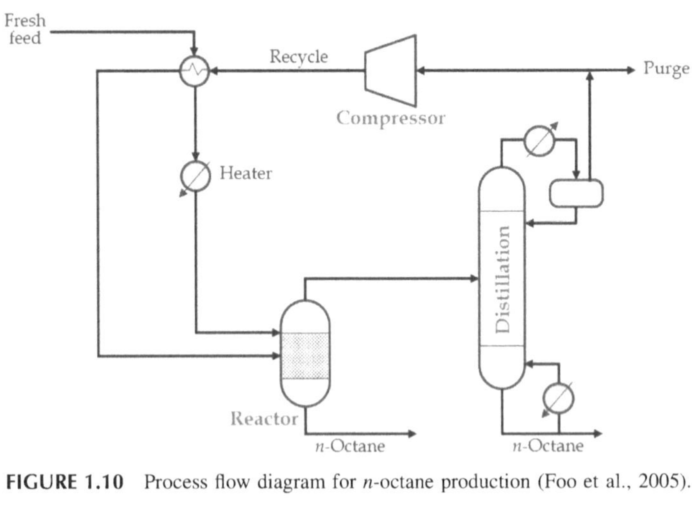

FIGURE 1.10 Process flow diagram for *n*-octane production (Foo et al., 2005).

n-butane (n-C_4H_{10})], is preheated before being fed to the reactor, along with the recycle stream. The reactor operates isothermally with a high conversion rate (98%) following the reaction stoichiometry in Eq. (1.16).

$$2C_2H_4 + C_4H_{10} \rightarrow C_8H_{18}$$

A big portion of n-octane product exits from the reactor bottom stream. The vapor effluent of reactor is then fed into a distillation column, where more n-octane product is recovered as bottom stream, while the unconverted reactant from the distillation top stream is being recycled to the reactor. The recycle stream passes through a compressor where its pressure is adjusted to match with that of the reactor and then exchanges its heat to preheat the fresh feed stream.

Following the concept of the onion model, one shall first simulate the reactor, which is the core of the process (see Fig. 1.3). Once the reactor model is converged, simulation is moved on to the distillation column and next the recycle systems; both of these are located at the second layer of the onion model. This is illustrated in Fig. 1.11.

Note that from the simulation perspective, recycle may be further classified as *material* and *heat recycle systems*. For the case of n-octane production, these subsystems for the complete flowsheet are shown in Fig. 1.12. The main challenge for this case is that these subsystems are interrelated, as they are connected by the process-to-process heat exchanger. The latter has two inlet streams, i.e., fresh feed and the outlet stream from the compressor. However, only the fresh feed stream contains the data (e.g., pressure, temperature, component, and flowrates) necessary for simulation to take place. The outlet stream from the compressor will have no result until the compressor is

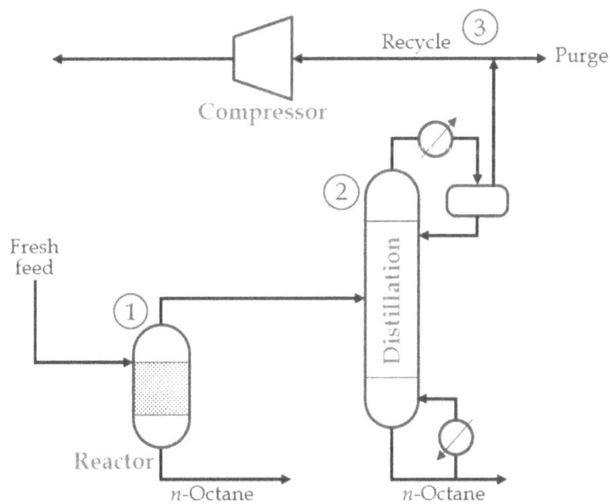

FIGURE 1.11 Sequence in simulating the n-octane case guided by onion model (numbers indicate the sequence of simulation).

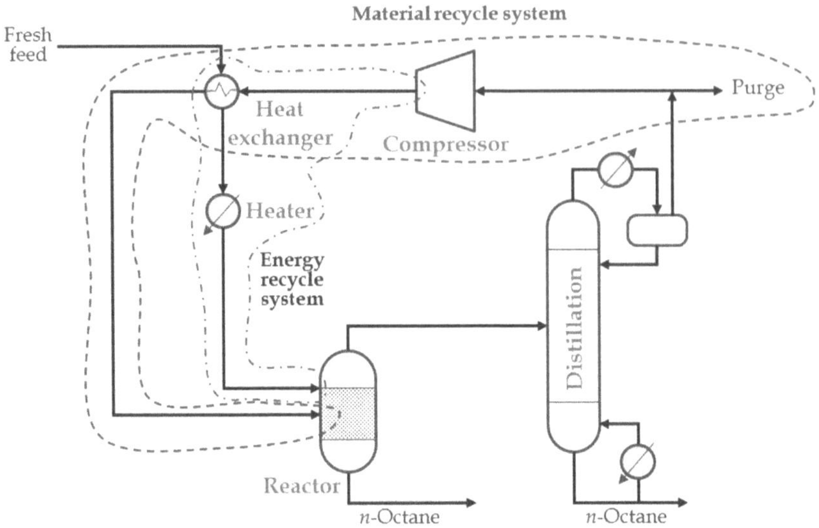

FIGURE 1.12 The material and energy recycle subsystems.

converged. Without complete data for its inlet streams, the heat exchanger simulation cannot be carried out, which means that its two outlet streams will have no results. This also leads to the nonconvergence of the subsequent units, i.e., reactor, distillation, etc.

A more convenient way to simulate material and heat recycle systems is to decouple them. This can be done by replacing the process-to-process heat exchanger with other equivalent heating and cooling units.[3] As shown in the revised flowsheet in Fig. 1.13, the material recycle stream is cooled using a utility cooler before being fed into the reactor, while the fresh feed stream is heated by the utility heater prior being sent to the reactor. In other words, energy recovery (and hence the energy recycle system) is not being considered at this stage. One then can make use of the tear stream concept to converge the material recycle stream.

Upon the convergence of the material recycle system, we then move on to the third layer of the onion model, where heat recovery system (i.e., the heat recycle system) is considered. In the converged flowsheet in Fig. 1.13, the heating duty from the utility heater, as well as the cooling duty of utility cooler, may be extracted. One can make use of the well-established *heat pinch analysis* technique to design the heat recovery system. For instance, for the flowsheet in Fig. 1.13, cooling and heating duties of the cooler and heater are extracted to plot a temperature versus enthalpy diagram (or termed as "*heat transfer composite curve*" in pinch analysis) in Fig. 1.14. With this diagram,

3. See detailed discussion for various strategies in converging recycle systems in Chapter 4.

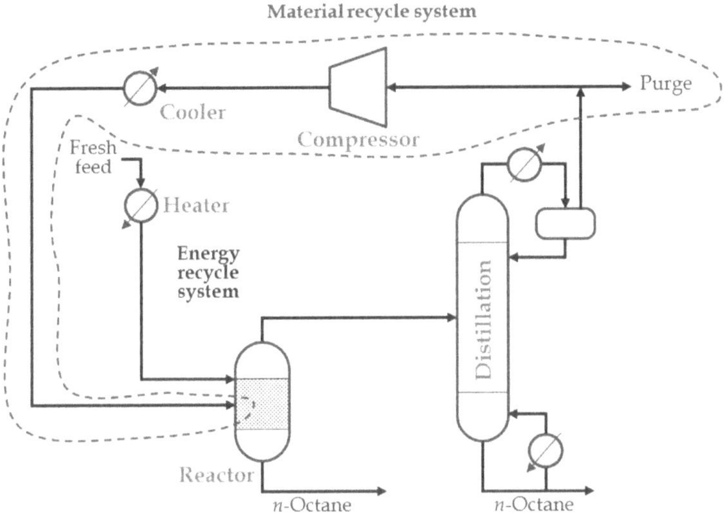

FIGURE 1.13 A decoupled material and energy recycle subsystems.

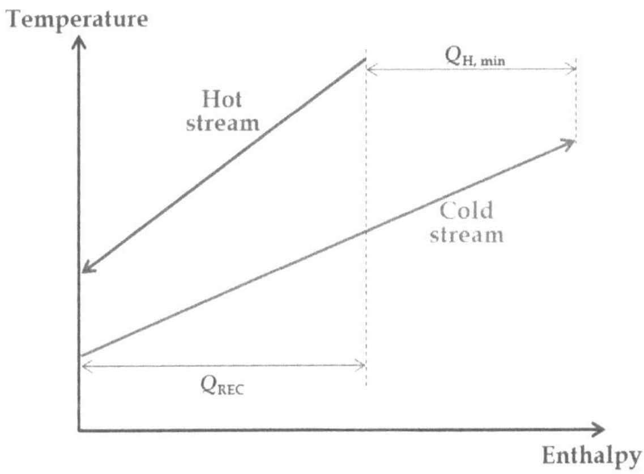

FIGURE 1.14 Temperature versus enthalpy diagram for *n*-octane case.

the minimum amount of hot ($Q_{H,min}$) and cold ($Q_{C,min}$) utility targets may be determined. Doing this also determines how much energy are recovered (Q_{REC}) between the *hot* (compressor outlet) and *cold* (fresh feed) streams. For the case in Fig. 1.14, the duty of the hot stream is completely recovered to the cold stream, resulting in a *threshold* case where no cold utility (i.e., $Q_{C,min} = 0$) is needed.

With the heat duties ($Q_{H,min}$ and Q_{REC}) identified, we then replace the cooler in the material recycle system with the process-to-process heat exchanger, while the heater is kept to provide additional heating duty ($Q_{H,min}$)

required by the fresh feed stream. This concludes the simulation of the *n*-octane case study. Note that the last two layers of the onion model are not applicable in this case. The *n*-octane case is utilized as work examples throughout this book (see Chapters 5, 8, 13 and 16).

1.6.1 Ten good habits for process simulation

In this section, we present ten good habits that should be practiced by all users while using any commercial simulation software. It is advisable that all process simulation software users to practice the following habits while performing their process simulation work.

Habit 1: You build a simulation model to meet an objective

As a process designer, the objective of building a simulation model should be understood well. In other words, the purpose of running a simulation should be clearly defined upfront. For instance, if one were to perform mass and energy balances for a preliminary flowsheet, it may be acceptable to use simplified unit operation models available in the simulation software. A good example is the use of a shortcut distillation model that uses the Fenske–Underwood–Gilliland method to provide a first-pass distillation model before constructing a rigorous tray-by-tray distillation model. The latter would require some detailed information, such as number of trays, top and bottom temperature. On the other hand, a shortcut model would normally require the definition of light and heavy keys, top, and bottom column pressures, along with reflux ratio to converge the column model. The shortcut distillation computation will provide useful information needed for building a rigorous tray-by-tray distillation model.[4]

Please also note that no single mathematical model can represent all fluids and processes. Hence, the simulation model must be purpose-built.

Habit 2: Identify the system or process and draw an envelope around it

It is important to identify the system that we wish to simulate. In some cases, not the entire flowsheet will be simulated. This could be due to several reasons. One typical example is the limitation of the simulation software. An example is shown in Fig. 1.15, where biomass is utilized as feedstock for a biorefinery. In the pretreatment section, biomass will go through some physical treatments for size reduction and moisture removal, before the biomass is fed into the gasification reactor (and other downstream separation system). If one were to utilize commercial software dedicated for the hydrocarbon-based industry (e.g., Aspen HYSYS and UniSim Design) for the simulation task, it would be best to leave out the pretreatment section, which is a very much

4. See Chapter 6 on the use of shortcut and rigorous distillation models in process simulation software.

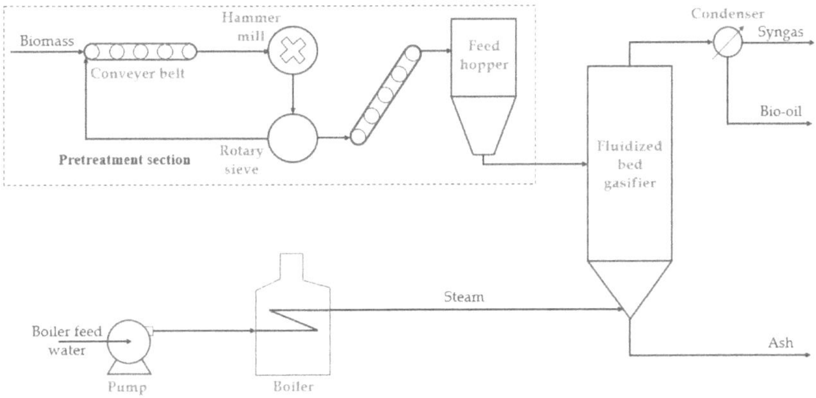

FIGURE 1.15 A biorefinery problem where pretreatment section is excluded from process simulation.

solid-based operation. In other words, the simulation task will focus on the gasification reactor and other downstream separation systems, but exclude the pretreatment section.

Another example is shown in Fig. 1.16, where a commercial process simulation software is to be used for the modeling of a multiphase pump. The latter is normally not found in the standard model library in most commercial software; hence, approximation is necessary if one were to determine the energy requirement of the multiphase pump. A possible option is shown in Fig. 1.16. As shown, the liquid and gas portions of the stream are first separated with a flash unit. The gas product from the flash unit is fed to a compressor, while the liquid portion is sent to a pump model. The compressor and pump models of the software will allow the energy requirement to be calculated (i.e. Q_{COMP} for compressor and Q_{PUMP} for pump), which can then be added for the energy requirement of the multiphase pump (Q_{MPP}). Such approximation is very commonly done whenever a unit is not found in a commercial software in use.

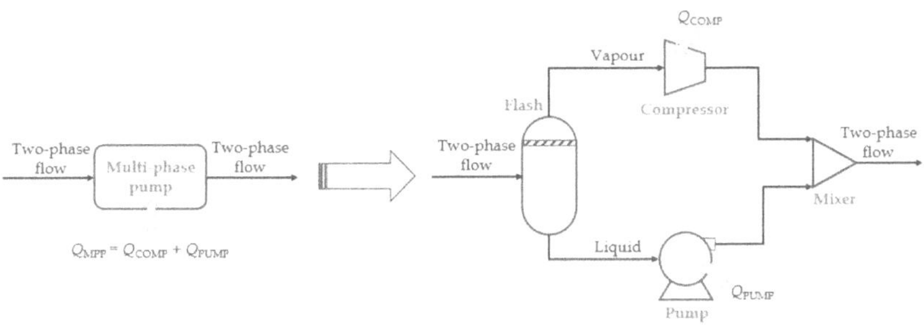

FIGURE 1.16 An approximation to model multi-phase fluid with commercial simulation software.

Habit 3: Imagine what is going on physically

The engineers who perform a simulation tasks should have good imagination. For instance, the engineer should imagine the state and flow pattern/regime for an inlet stream heading to a reactor/flash unit or an effluent stream emitted from a reactor. For the latter case, if the reactor effluent stream contains compressed liquid with light gases, an adiabatic separator (i.e., flash unit) may be added once the reactor is converged, to separate the light gases from the liquid components.

Habit 4: Translate the physical model to a mathematical model

Engineers should use their basic knowledge in reaction engineering, thermodynamics, separation processes, etc., to translate a physical process in the plant into an equivalent mathematical model. He/she then needs to pick the right process simulation software to perform the simulation tasks. One should note that most commercial simulation software packages are dedicated for continuous processes. Hence, having to simulate batch processes (e.g., biofermentation and polymerization) using a commercial software that is meant for continuous processes is indeed a challenging task. Even if one may be able to approximate the model to represent the process behavior, there is no way to model the time-related elements (e.g., production scheduling). Besides, many commercial software packages (e.g., Aspen HYSYS and UniSim Design) are more applicable to hydrocarbon-based chemical process industry, e.g., oil and gas and petrochemical, with component and thermodynamic databases that are suited for those industrial sectors. If one were to use these software packages for biochemical process modeling (e.g., production of yeast, vitamin, antibody, food, and beverage), one will have to customize the component as well as its associated thermodynamic models, which is time consuming and yet inaccurate. For the modeling of biochemical processes, it would be wiser to consider the use of dedicated software packages such as SuperPro Designer (www.intelligen.com) or Aspen Batch Process Developer (www.aspentech.com).

A similar situation also applies for environmental applications, where a high degree of accuracy and very small numbers need to be represented in stream composition (e.g., ppm level). If one were to use the hydrocarbon-based software for wastewater treatment plant design, the stream compositions will hardly be traced. Another example is the use of process simulation software for the modeling of momentum balance and mass transfer. Most commercial software (e.g., UniSim Design, SuperPro Designer, Aspen Plus, Aspen HYSYS) only address material and energy balances and heat transfer at equilibrium stage. They provide limited representation of momentum effects (only in pipe objects) and no representation of mass transfer. If one were to perform momentum balance or mass transfer computations, other dedicated software are to be used. Hence, the advice is to use the right software for the right applications.

Habit 5: Know your components

It is important to know the chemicals that are present in the system you are working on. Furthermore, it is important to understand the type of intermolecular interactions that may exist in that system. Understanding the components and their interactions will enable the engineer to choose the right property estimation methods.[5] In some cases, there may be some components that are not represented in the simulator's component database. In that case, it may be necessary to either find an equivalent component or create a user-defined or "hypothetical" component.[6]

One will also have to be aware of the presence of water in a predominantly hydrocarbon process stream. Some thermodynamic models may lump the organic and aqueous phases into a single liquid phase, even though water is immiscible with the hydrocarbon. Hence, the selection of thermodynamic model is crucial for this kind of system.

In some cases, azeotropic mixtures introduce challenges in distillation process. The presence of binary or tertiary azeotrope points leads to the existence of a distillation boundary that limits the degree of separation. An example is given in Fig. 1.17, which shows the temperature–composition (T-xy) plot for a binary mixture containing isopentane (iC5) and methanol. The T-xy plot shows that a minimum boiling azeotrope occurs for this binary mixture, which prevents the separation of high-purity products with normal

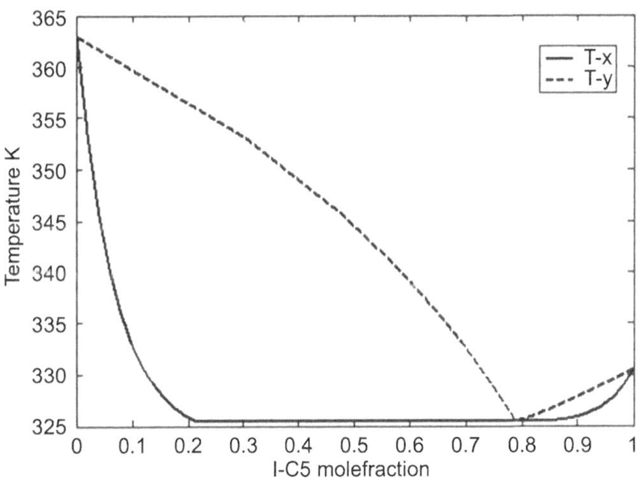

FIGURE 1.17 A binary mixture with minimum boiling azeotrope.

5. Refer to Chapter 3 for details on physical properties estimation.
6. Refer to Chapter 2 for new component registration.

distillation. One will have to utilize the enhanced distillation techniques to perform such separation tasks.

Habit 6: Know the context of your feed streams

It is important to know the characteristics of the feed stream, e.g., its origin, composition, sampling conditions, and the presence of impurities.

In some cases, a composition for a gas stream that comes from a three-phase (gas, oil, and water) separator may not contain any water. This is because the stream composition has been determined using gas chromatography (GC). In a GC analysis, a carrier gas, typically helium, transports the gas sample into a length of tubing (called a "column"), which is packed with a polymer or solid support. Water will damage this polymer and is removed before the gas enters the column.

In this case, it is necessary to saturate the gas with water prior to using it for any computations; otherwise, the contribution of water in the gas, particularly the heat of vaporization, will not be accounted for and will result in an inaccurate heat and material balance.

Habit 7: Know your components boiling points

In performing simulation for a separation system (e.g., flash and distillation), it is important to know the boiling points, and hence, the relative volatility of the chemicals involved. It is good practice to have the chemicals arranged in the ascending order of their boiling points (i.e., volatilities). An example is shown in Table 1.2. As shown, a process stream consists of five hydrocarbon components, arranged in the ascending order of their boiling points. If these components were to be used for a distillation computation, it would be easy to draw a line between the heavy and light keys and perform a quick material balance to determine distillate and bottom product flowrates.

Besides, we should also watch out for polar molecules and those chemicals that have hydrogen bonds. This may give rise to azeotropic systems where high boiling point components boil and vaporize before low boiling point components.

TABLE 1.2 Chemicals arranged in the ascending order of their boiling points.

Components	Boiling points (°C)
Ethane (C_2H_6)	−88.6
Propane (C_3H_8)	−42.1
n-Butane (C_4H_{10})	−0.5
n-Pentane (C_5H_{12})	36.2
n-Hexane (C_6H_{14})	68.8

Habit 8: Keep track of the units of measure in all calculations

Errors on units of measurement are very serious and the easiest errors to avoid with some simple discipline. The best practice is always to keep track of the units of measurement in all computation or simulation exercises. All modern commercial simulation software packages are equipped with unit conversion functions for all unit operations and stream parameters. These simulation packages allow for global units of measure profiles to be defined and saved. Local definitions of units are also possible to allow for flexibility, but when possible it is a good idea to preprocess any data to ensure consistency.

Prior to conducting any engineering analysis or project, it is necessary to first define what units of measure are to be used for the analysis and reporting purposes and ensure that this is adopted by all personnel working on the analysis or project.

Habit 9: Always do a simple material and energy balance first

An important good habit is to perform some manual calculations (or "hand calculations"). For example, performing a quick material and energy balances prior to executing a process simulation exercise enables the engineer to have a better "feel" for the orders of magnitude in the numbers that may be encountered. In addition to that, these hand calculations can also be used as initial guesses for more complex types of computations. An example is shown in Fig. 1.18, where a depropanizer column is to be modeled. If one were to set overhead mole fraction of propane to be 0.95, the simulation cannot be converged certainly. We can perform a quick material balance by assuming all propane (ideal separation) is recovered at the overhead vapor stream of the column and heavier components are recovered at the bottom stream. It is then determined that the mole fraction of propane in the overhead stream is determined to be approximately 35% (= 37/(33 + 36 + 37) kmol/h), since the lighter components of methane and ethane are also recovered to the overhead

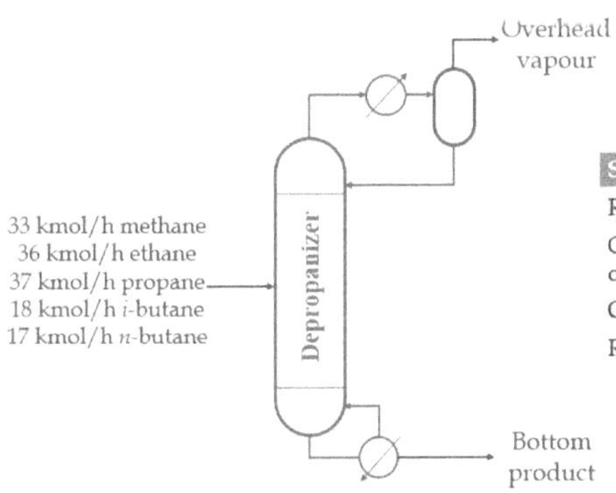

Specifications	Value	Activation
Reflux ratio	1.0	Yes
Overhead mole fraction of propane	0.95	Yes
Overhead mole flowrate	100	No
Reflux rate	Empty	No

FIGURE 1.18 Simulation of a depropanizer column that does not converge.

stream. In other words, the propane overhead mole fraction of 0.95 is an unrealistic setting. This kind of manual calculations is useful to provide an estimate for distillation simulation and can be used as an initial guess for a rigorous tray-by-tray distillation model.[7] Doing manual calculations can also avoid potential mistakes caused by bugs in the process simulation software (see a reported case by Le et al., 2000).

Performing mass and energy balances for some unit operations may be tedious, especially for those that involve nonlinear equations. Hence, some linearized models are always useful for preliminary flowsheet, especially for conceptual design stage. Linear models for some commonly used unit operations (distillation, absorption, etc.) may be found in Biegler et al. (1997).

Habit 10: Plot the phase envelope for important streams

It is extremely important to know in which state the fluid is in, e.g., is it a subcooled or saturated liquid? For a process stream with a subcooled liquid, temperature rise is expected when sensible heat is added; the latter is the product of mass flowrate, heat capacity, and temperature rise. On the other hand, no temperature rise will be reported when a saturated liquid is heated, as latent heat is involved. Also, it is necessary to check if the fluid is near the critical point or is it in the supercritical phase?

It is also important to know how close the process stream is to the dew point (either hydrocarbon or water). When adsorption beds or fuel gas systems were designed, liquids formed at the dew point may damage the adsorption bed or combustion chamber. Another important practice is to identify the retrograde region of the fluid. In the retrograde region, compressing a fluid may result in its vaporization instead of liquefaction.[8]

Exercises

Fig. E1.1 shows a five-stage process to extract sodium carbonate (Na_2CO_3) from black ash using water as solvent (Seader, 1985). Na_2CO_3 is partially

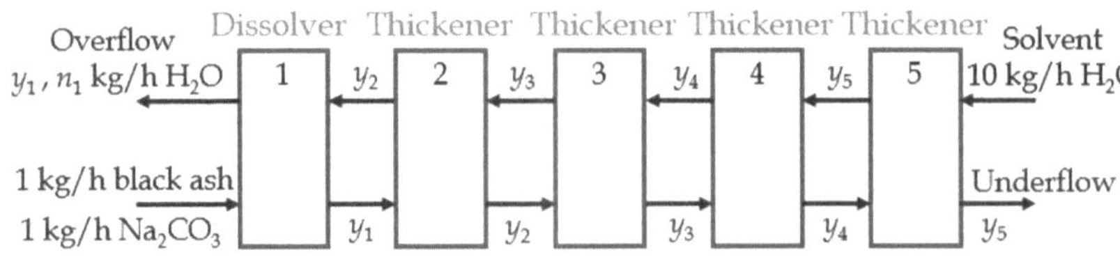

FIGURE E1.1 A five-stage extraction process to extract sodium carbonate.

7. See Chapter 6 for discussion of shortcut models in estimating values for rigorous distillation model.
8. See detailed discussion in Chapter 3.

dissolved in the first stage, while sludge (other component in black ash) is completely insoluble in water. The following assumptions are made:

i. The liquid overflow (top stream) from each stage is assumed to be free of solids.
ii. Underflow (bottom stream) is a slurry stream containing 3 kg of water (Na_2CO_3-free basis) per kg of insoluble sludge.
iii. Equilibrium is reached for each extraction stage, and hence, the mass fractions of Na_2CO_3 (y_i, kg Na_2CO_3/kg H_2O) in the overflow and underflow streams leaving each stage i are the same.

Calculate the following:

1. Solve the overall mass balance to determine the overflow in Stage 1 (n_1).
2. Determine the mass flowrates of H_2O in all stages.
3. Determine the mass fraction of Na_2CO_3 in the overflow in each stage i, i.e., y_i.

References

AspenTech, 2022. A History of Innovation. www.aspentech.com.

AVEVA, 2022. AVEVA™ PRO/II™ Simulation. www.aveva.com (accessed June 2022).

Biegler, L., Grossmann, I.E., Westerberg, A.W., 1997. Systematic Methods of Chemical Process Design. Prentice-Hall.

Crowe, C.M., Hamielec, A.E., Hoffman, T.W., Johnson, A.I., Woods, D.R., Shannon, P.T., 1971. Chemical Plant Simulation. Prentice Hall, Englewood Cliffs.

Dimian, A., Bildea, C., Kiss, A., 2014. Integrated Design and Simulation of Chemical Processes, second ed. Elsevier Science, Amsterdam.

El-Halwagi, M.M., 2006. Process Integration. Elsevier, San Diego.

El-Halwagi, M.M., Foo, D.C.Y., 2014. Process synthesis and integration. In: Seidel, A., Bickford, M. (Eds.), Kirk-Othmer Encyclopedia of Chemical Technology. John Wiley & Sons.

Evans, L.B., 1981. Advances in process flowsheeting systems. In: Mah, R.S., Seider, W.D. (Eds.), Foundations of Computer-Aided Chemical Process Design. Engineering Foundation, New York.

Evans, L.B., Boston, J.F., Britt, H.I., Gallier, P.W., Gupta, P.K., Joseph, B., Mahalec, V., Ng, E., Seider, W.D., Yagi, H., 1979. Computers & Chemical Engineering 3 (1–4), 319–327.

Evans, L.B., Joseph, B., Seider, W.D., 1977. System structures for process simulation. AIChE Journal 23 (5), 658–666.

Federal Trade Commission, 2003. FTC Charges Aspen Technologys Acquisition of Hyprotech. LTD. Was Anticompetitive. www.ftc.gov.

Felder, R.M., Rousseau, R.W., 2005. Elementary Principles of Chemical Processes, 3rd ed. John Wiley, New York, US.

Foo, D.C.Y., 2012. Process Integration for Resource Conservation. CRC Press, Boca Raton, Florida, US.

Foo, D.C.Y., Manan, Z.A., Selvan, M., McGuire, M.L., October, 2005. Integrate process simulation and process synthesis. Chemical Engineering Progress 101 (10), 25–29.

Gallier, P.W., Evans, L.B., Boston, J.F., Britt, H.I., Boston, J.F., Guupta, P.K., 1980. ASPEN: advanced capabilities for modeling and simulation of industrial processes. In: Squires, R.G., Hilaly, A.K., Sikdar, S.K. (Eds.), Process Simulation Tools for Pollution Prevention. Chemical Engineering, pp. 98–105. February 1996.

Hernandez, R., Sargent, R.W.H., 1979. A new algorithm for process flowsheeting. Computers & Chemical Engineering 3 (1–4), 363–371.

Hilaly, A.K., Sikdar, S.K., 1996. Process Simulation tools for pollution prevention. Chemical Engineering 103, 98–105.

Honeywell, 2022. UniSim Design R491 Suite. https://www.honeywellforge.ai (accessed June 2022).

Intelligen, 2022. Company information. www.intelligen.com (accessed June 2022).

Koulouris, A., Calandranis, J., Petrides, D., 2000. Throughput analysis and debottlenecking of integrated batch chemical processes. Computers & Chemical Engineering 24, 1387–1394.

Le, N.D., Sel, B.J., Edwards, V.H., 2000. Doublecheck your process simulations. Chemical Engineering Progress 51–52.

Linnhoff, B., Townsend, D.W., Boland, D., Hewitt, G.F., Thomas, B.E.A., Guy, A.R., Marshall, R.H., 1982. A User Guide on Process Integration for the Efficient Use of Energy. IChemE, Rugby.

Lott, D.H., 1988. Simulation software as an aid to process synthesis. In: Crump, P.R., Greenwood, D.V., Smith, R. (Eds.), Understanding Process Integration II. IChemE, Rugby.

Motard, R.L., Shacham, M., Rosen, E.M., 1975. Steady state chemical process simulation. AIChE Journal 21 (3), 417–436.

Perkins, J.D., Sargent, R.W.H., Thomas, S., 1982. SPEEDUP: a computer program for steady-state and dynamic simulation of chemical processes. IChemE Symposium Series 73, H78–H86.

Petrides, D., 1994. BioPro designer: An advanced computing environment for modeling and design of integrated biochemical processes. Computers & Chemical Engineering S18, 621–625.

Petrides, D., Cooney, C.L., Evans, L.B., 1989. Bioprocess simulation: an integrated approach to process development. Computers & Chemical Engineering 13 (4–5), 553–561.

Rosen, E.M., 1980. Steady state chemical process simulation: a state-of-the-art review. Computer applications to chemical engineering. ACS Symposium Series 124, 3–36.

Seader, J.D., 1985. Computer Modeling of Chemical Processes. In: AIChE Monograph Series No. 15. American Institute of Chemical Engineers, New York.

Siemens Process System Enterprise (PSE), 2022. Next-generation modelling tools across the process lifecycle. https://www.psenterprise.com/products/gproms (accessed June 2022).

Smith, R., 2016. Chemical Process Design and Integration, second ed. John Wiley & Sons.

Sowa, C.J., 1994. Explore waste minimisation via process simulation. Chemical Engineering Progress 1994, 40–42.

Turton, R., Bailie, R.C., Whiting, W.B., Shaeiwitz, J.A., 2013. Analysis, Synthesis and Design of Chemical Processes, fourth ed. Prentice Hall, New Jersey.

Westerberg, A.W., Hutchison, H.P., Motard, R.L., Winter, P., 1979. Process Flowsheeting. Cambridge University Press, Cambridge.

WinSim, I., 2022. About WinSim. www.winsim.com (accessed June 2022).

Further reading

Reklaitis, G.V. (Eds.), Computer applications to chemical engineering. ACS Symposium Series, vol. 124. American Chemical Society, Washington, DC, pp. 293–308.

Chapter 2

Registration of new components*

Denny K.S. Ng[1], Chien Hwa Chong[2] and Nishanth Chemmangattuvalappil[2]
[1]*Heriot-Watt University Malaysia, Putrajaya, Malaysia;* [2]*University of Nottingham Malaysia, Semenyih, Selangor, Malaysia*

Chapter outline

As mentioned in Chapter 1, the first step of setting up a process simulation is to define chemical components that involve in the entire process. However, it is noted that not all components are available in the database of the simulation software. Hence, one will have to define/register the chemical component(s) before one can make use of them in the simulation flowsheet. This may involve various types of hypothetical and oil components. In this chapter, the registration of components in several important process simulation software packages, i.e., Aspen HYSYS and PRO/II, is illustrated.

2.1 Registration of hypothetical components

Components that are not available in simulator library can be defined in the simulation software as "hypothetical" components based on the properties of the chemicals. Different simulation software packages treat this step differently. For example, in Aspen HYSYS, PRO/II and UniSim Design, it is possible to define such components in the library based on their properties. On the other hand, components are defined based on the molecular structure in

*Simulation and other electronic files of this chapter are made available on a companion website; URL: https://www.elsevier.com/books-and-journals/book-companion/9780323901680.

Chemical Engineering Process Simulation. https://doi.org/10.1016/B978-0-323-90168-0.00006-8

Aspen Plus. In this section, registration of hypothetical components is demonstrated with Aspen HYSYS and PRO/II.

2.1.1 Hypothetical component registration with Aspen HYSYS

To define the "hypothetical" components in Aspen HYSYS library, important properties of the chemicals such as molecular weight and boiling point are first defined in the simulator. Aspen HYSYS will then estimate the other missing properties based on UNIFAC group contribution models (AspenTech, 2015). However, it is necessary to provide the boiling point and molecular weight to estimate the rest of the properties. If more properties are made available, the accuracy of the prediction of the other properties would be much higher. Steps involved in building hypothetical components are illustrated in Example 2.1.

Example 2.1. This example involves the registration of a hydrocarbon component as a "hypothetical" component in Aspen HYSYS. For this case, the hydrocarbon has a molecular weight of 86 and boiling point of 64°C. Fig. 2.1 shows the detailed steps for registering this component in the software. Based on the boiling point and molecular weight, the other physical properties are then estimated, as shown in Fig. 2.2.

2.1.2 Hypothetical component registration with PRO/II

Two methods can be used to estimate components in PRO/II (Schneider Electric, 2015). A user can create a new component using the user-defined library and generate chemical structure using the UNIFAC structure and finally fill options to generate all the component properties of user-defined components. The second option is defining the component using the PROPRED property prediction methodology and then estimating with and without measured normal boiling point (NBP). Example 2.2 shows the first method used in defining a component.

Example 2.2. Estimating a component in PRO/II: In this case, gallic acid is selected to be developed as a new user-defined component using PRO/II. To

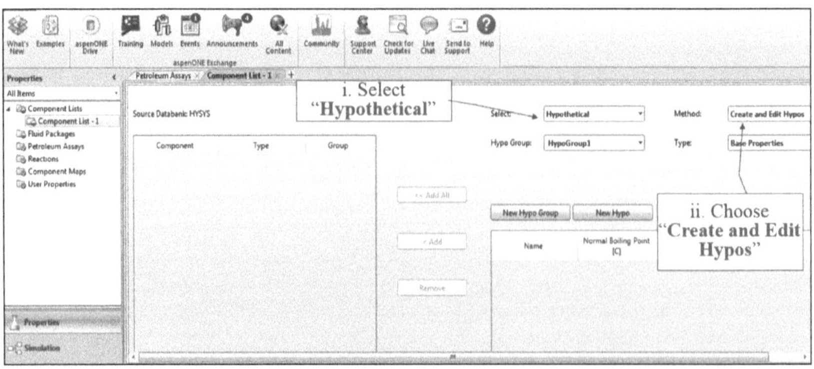

FIGURE 2.1 Hypo selection in Aspen HYSYS.

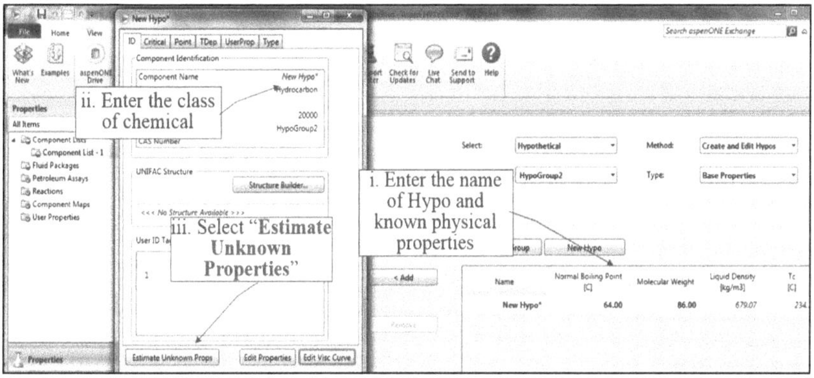

FIGURE 2.2 Property estimation of hypo components in Aspen HYSYS.

create a new component, the user needs to create a user-defined component name from the component selection section. The steps are shown in Figs. 2.3 and 2.4.

Next, the user is required to define UNIFAC structure. In this case, acids, aromatics, and alcohol category structures are selected for the gallic acid component. Detailed steps for creating the structures are shown in Fig. 2.5.

Molecular weight, critical temperature, critical pressure, NBP, and miscellaneous properties data can be modified in "fixed" thermophysical properties section. In this case, the user can modify the NBP value based on steps in Fig. 2.6.

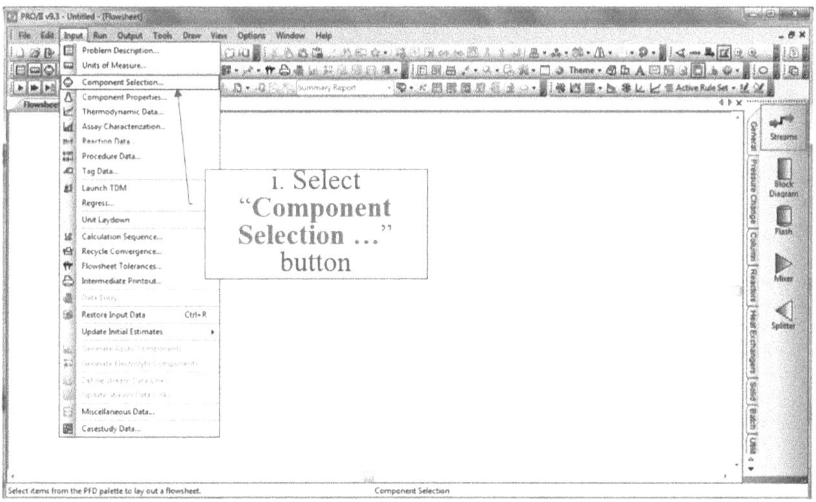

FIGURE 2.3 Start with component selection in PRO/II.

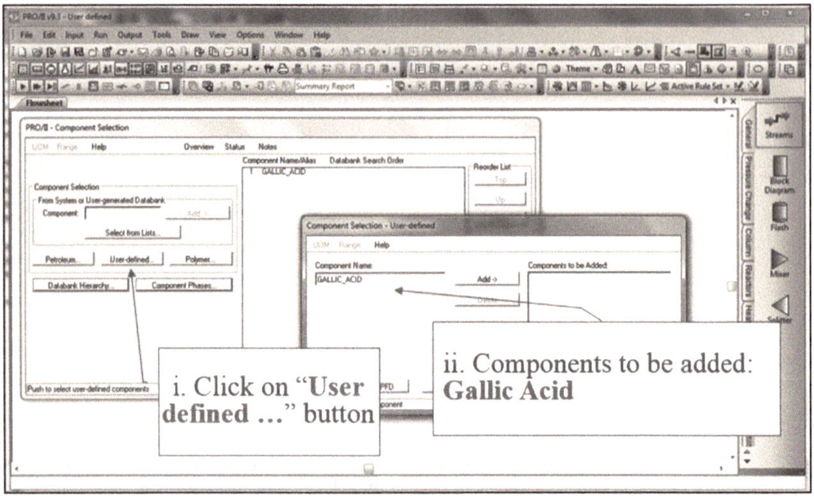

FIGURE 2.4 Define the name of a new component.

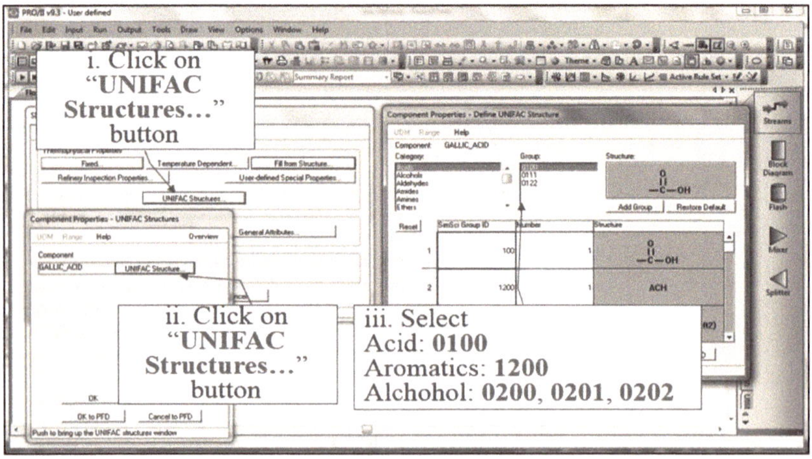

FIGURE 2.5 Using the UNIFAC structures to generate the new component in PRO/II.

To conduct a simulation, a heat exchanger can be added to the stream following detailed steps in Fig. 2.7.

2.2 Registration of crude oil

Crude oil is a naturally occurring, unrefined petroleum product composed of hydrocarbon deposits and other organic materials, which can be refined to

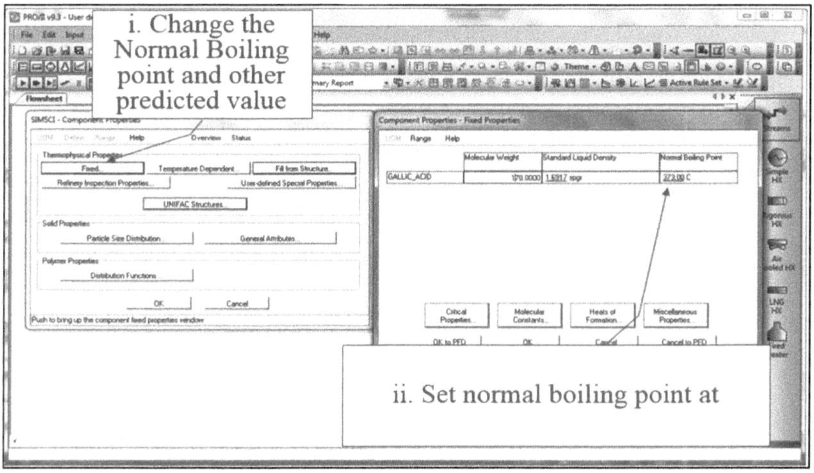

FIGURE 2.6 Specifying normal boiling point and other properties.

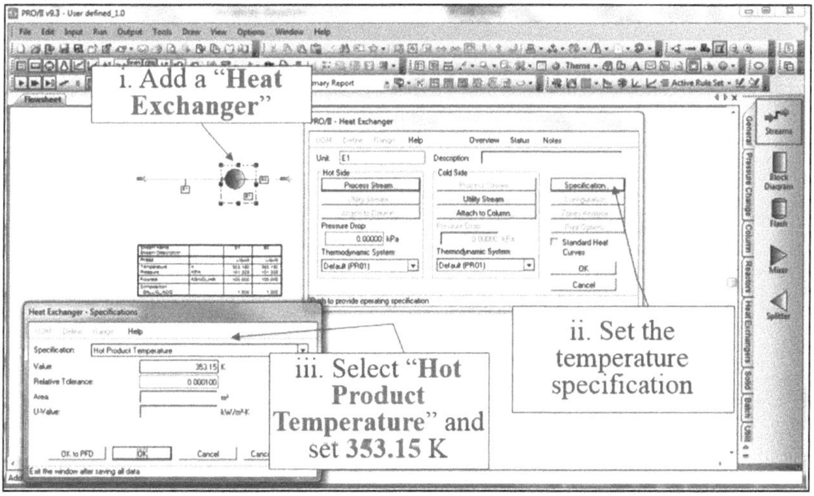

FIGURE 2.7 Performing a simulation by adding a heat exchanger as a unit operation.

produce other useable products such as gasoline, diesel, and various forms of petrochemicals. As crude oil is a mixture of multiple components, it is difficult to identify individual components present in the oil. In industry practice, characteristics of a crude oil sample are normally determined via laboratory distillation tests (see list in Table 2.1). Similarly, rather than defining individual oil components in process simulation, it is commonly defined based on the results of common laboratory distillation tests.

TABLE 2.1 Common laboratory distillation tests.

Test name	Reference	Main applicability
ASTM (atmospheric)	ASTM D86	Petroleum fractions (products do not decompose when vaporized at 1 atm)
ASTM (vacuum, 1.3 kPa)	ASTM D1160	Heavy petroleum fractions or products that tend to decompose in ASTM D86 test but can be partially or completely vaporized at a maximum liquid temperature of 400°C at 0.13 kPa
TBP (atmospheric or 1.3 kPa)	Nelson, ASTM 2892	Crude oil petroleum fractions
Simulated TBP (gas chromatography)	ASTM D2887	Crude oil petroleum fractions

ASTMs, American society for testing and materials; *TBP*, true boiling point.

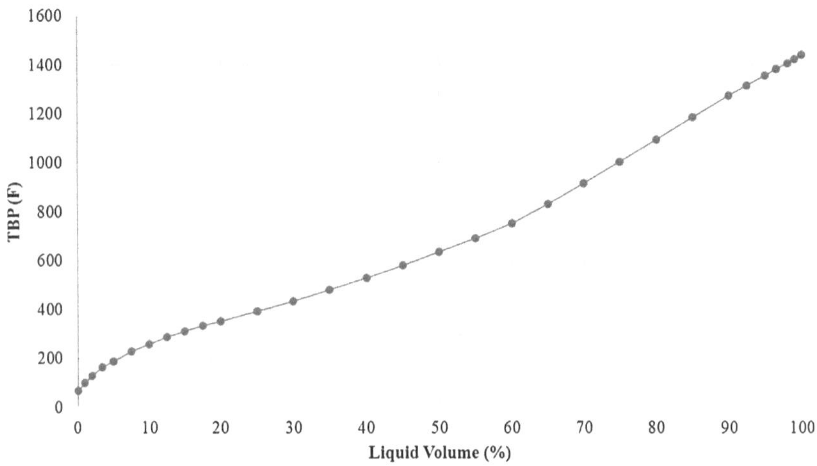

FIGURE 2.8 Sample of true boiling point (TBP) of a crude oil.

Fig. 2.8 shows the sample of true boiling point (TBP) of a crude oil. The TBP distillation (ASTM D2892) uses a 15-theoretical plate column, and 5:1 flux ratio is a classical method to obtain the distillation curve for the crude oil sample. The distillation fractionates the crude oil into a number of narrow fractions up to 400°C atmospheric equivalent temperature.

Example 2.3. Crude oil registration with Aspen HYSYS

To model crude oil in Aspen HYSYS, the oil needs to be defined based on its characteristics via Oil Manager. Based on the laboratory results, the characteristics of the crude oil are included in Aspen HYSYS library and hypothetical components. This involves the following three main steps:

- Characterization of crude assay
- Generate pseudocomponents—create, cut, and blend
- Install the oil in the flowsheet

In this example, 477,000 kg/h of crude oil is required to heat up to 482°F (523.15 K) under ambient conditions, 86.18°F and 14.7 psia (303.15 K and 101.325 kPa). To illustrate the proposed procedure in detail, the oil properties in Table 2.2 will be installed in Aspen HYSYS. Thermodynamic package to be used is Peng–Robinson, with "Auto Cut" as the option.

Step 1. Characterization of Crude Assay

Before defining the oil, the components that exist in simulator databank "HYSYS" are first defined. Next, the thermodynamic package (also known as fluid package in Aspen HYSYS) is also defined. Based on the information given, i-Butane, n-Butane, i-Pentane, and n-Pentane are installed via HYSYS databank and Peng–Robinson is installed as the fluid package in the simulator, as shown in Figs. 2.9 and 2.10.

Next, the oil properties are defined via "Oil Manager" in the simulator. As discussed, the characteristics of oil are determined based on the laboratory test. Therefore, the result of the test will be used as input information for the simulator. Detailed steps for characterizing the crude assay are shown in Figs. 2.11, 2.12, 2.13, 2.14 and 2.15.

As mentioned previously, there are few types of laboratory distillation tests; therefore, the user needs to select the respective test in Oil Manager, as shown in Fig. 2.13. In Aspen HYSYS, few options of Assay Data Type are available, i.e., TBP, ASTM D86, ASTEM D1160, ASTM D86–D1160, ASTM D2887, Chromatograph, and EFV.

Next, the user should select "Edit Assay" to key in the assay information that is obtained from the laboratory results (see Figs. 2.14 and 2.15). Note that liquid volume percent distilled versus temperature is first entered in the table. Next, liquid volume percent distilled, which requires other properties, such as density and molecular curve, is also included in the table (as shown in Fig. 2.15).

Once the assay table is completed, information about other properties such as bulk properties, molecular weight curve, light ends, density curve, and viscosity curve can also be included. Fig. 2.16 shows the input of bulk properties into the simulator. Besides, based on the given information, light ends, molecular weight, API gravity, viscosity (cP) at 100°F, and viscosity (cP) at 210°F are available; therefore, information about these properties is included to estimate the crude oil properties accurately in the simulator. However, in the event where the information is absent, the user can select the option of "Not Used" in the assay window. The simulator will then estimate the properties based on the available information accordingly.

TABLE 2.2 Information of crude oil.

(a) Bulk properties		
Bulk crude properties	Reference	
Molecular weight	300	
API gravity	48.75	
(b) Light ends liquid		
Light ends liquid volume percent distilled		
i-Butane	0.19	
n-Butane	0.11	
i-Pentane	0.37	
n-Pentane	0.46	
(c) True boiling point distillation assay		
Liquid volume percent distilled	**Temperature (°F)**	**Molecular weight**
0.0	80.0	68.0
10.0	255.0	119.0
20.0	349.0	150.0
30.0	430.0	182.0
40.0	527.0	225.0
50.0	635.0	282.0
60.0	751.0	350.0
70.0	915.0	456.0
80.0	1095.0	585.0
90.0	1277.0	713.0
98.0	1410.0	838.0
(d) API gravity assay		
Liquid volume percent distilled	**API gravity**	
13.0	63.28	
33.0	54.86	
57.0	45.91	
74.0	38.21	
91.0	26.01	

TABLE 2.2 Information of crude oil.—cont'd

(e) Viscosity assay

Liquid volume percent distilled	Viscosity (cP) 100°F	Viscosity (cP) 210°F
10.0	0.20	0.10
30.0	0.75	0.30
50.0	4.20	0.80
70.0	39.00	7.50
90.0	600.00	122.30

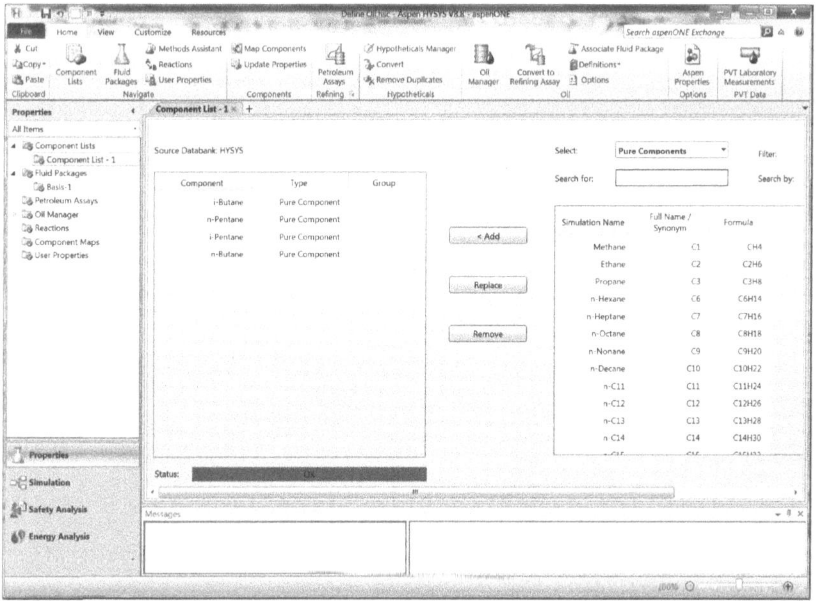

FIGURE 2.9 Define components in databank HYSYS.

Note that the following information is required to estimate the crude oil properties:

- Mass density
- Bulk viscosity at 100°F and at 210°F

Meanwhile, the other optional properties are recommended to be included. These include the following:

- Molecular weight curve
- Light ends
- Density curve
- Viscosity curve.

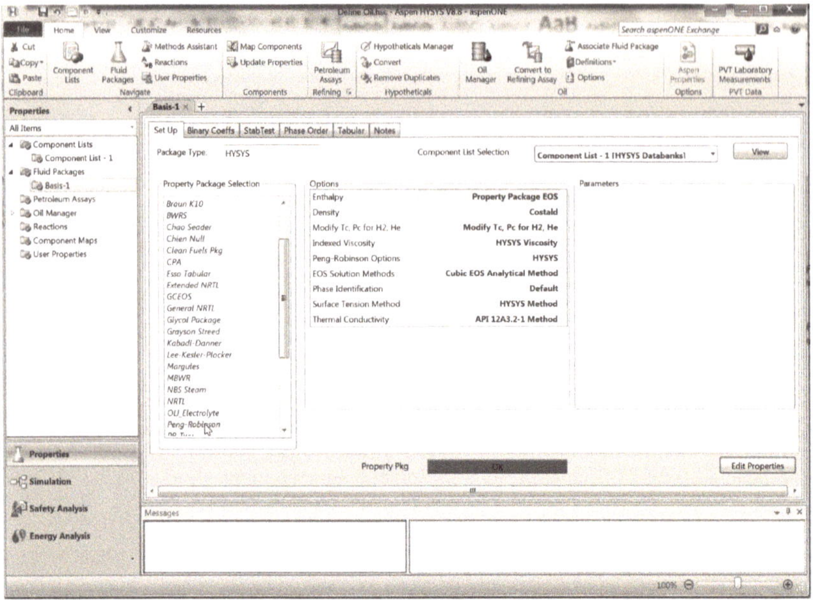

FIGURE 2.10 Define thermodynamic package.

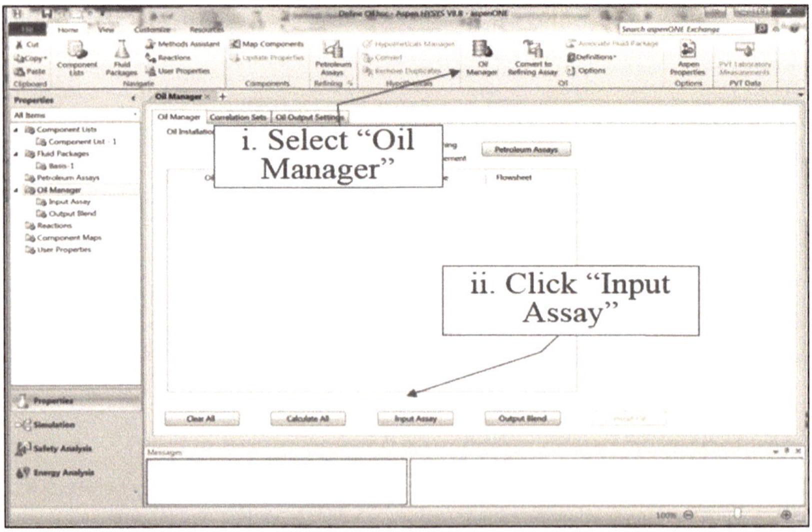

FIGURE 2.11 Defining crude oil in Oil Manager of Aspen HYSYS.

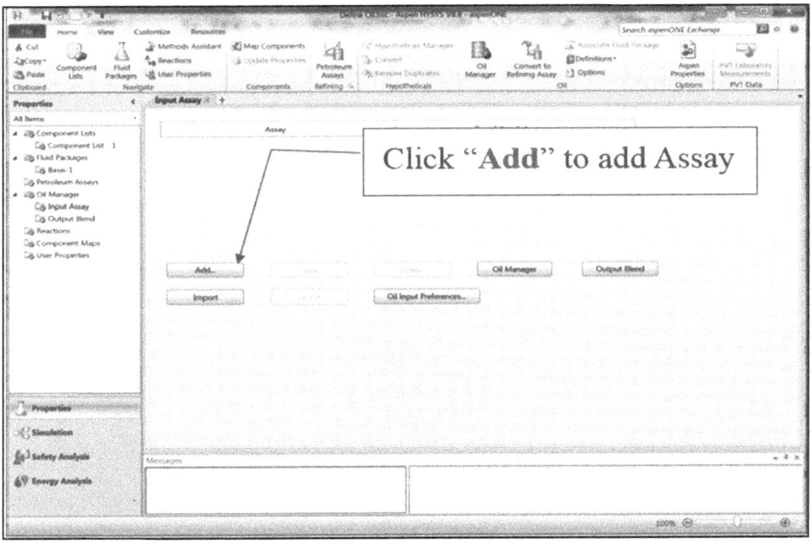

FIGURE 2.12 Add assay in Oil Manager.

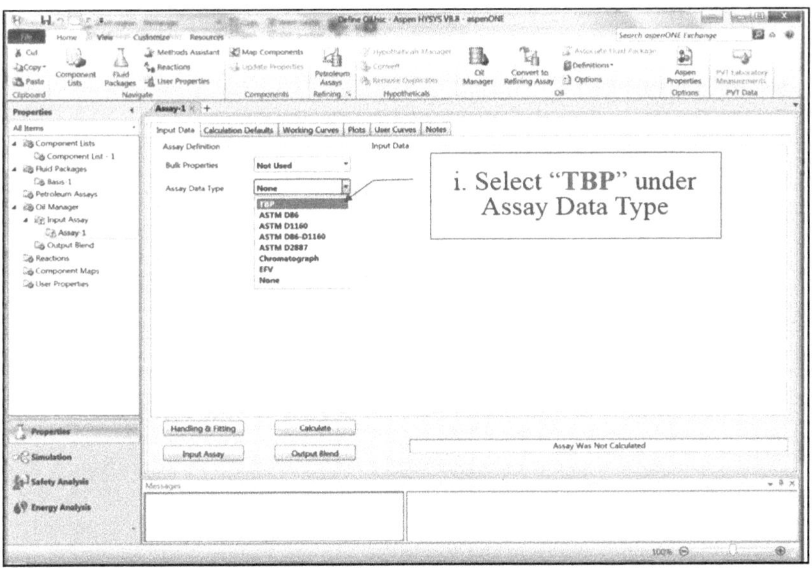

FIGURE 2.13 Selection of assay data type.

Figs. 2.16, 2.17, 2.18, 2.19 and 2.20 show the input of bulk properties, light ends, molecular weight, API gravity, viscosity (cP) at 100°F, and viscosity (cP) at 210°F.

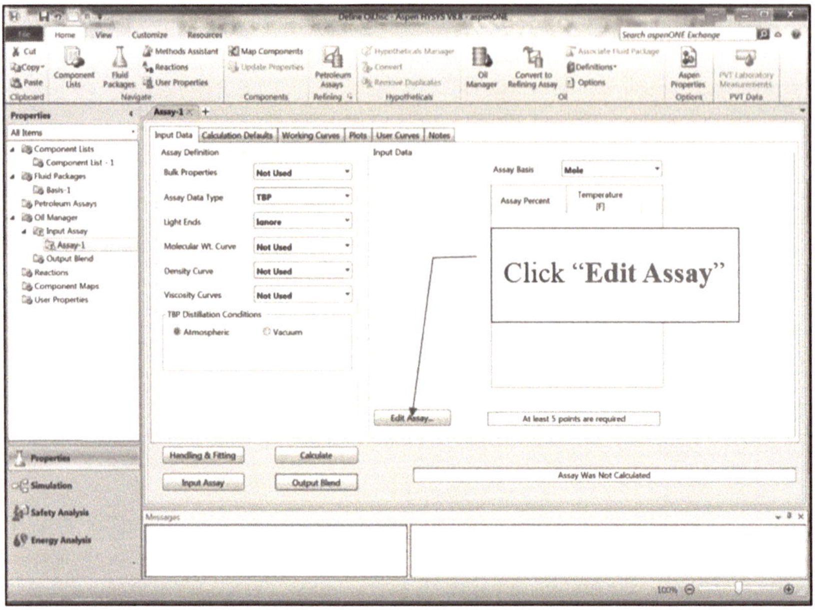

FIGURE 2.14 Editing assay and input of assay information.

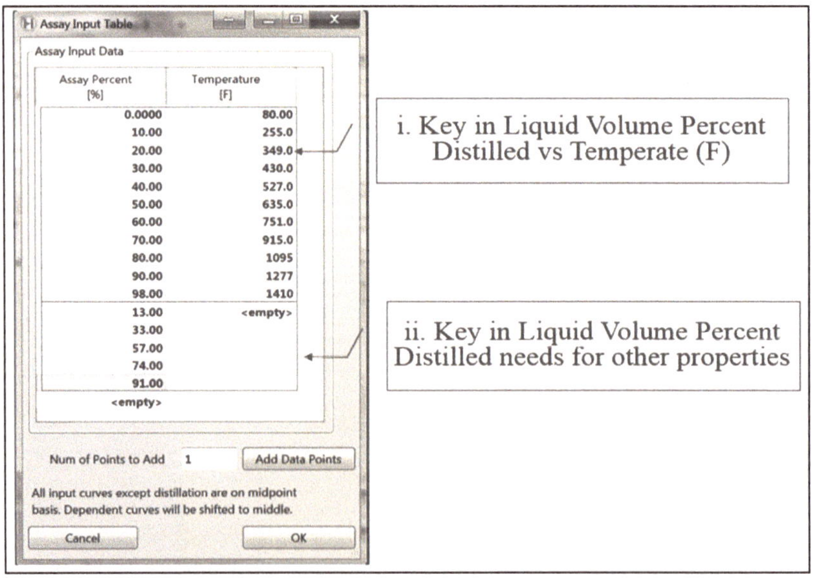

FIGURE 2.15 Input information of assay into assay table in Oil Manager.

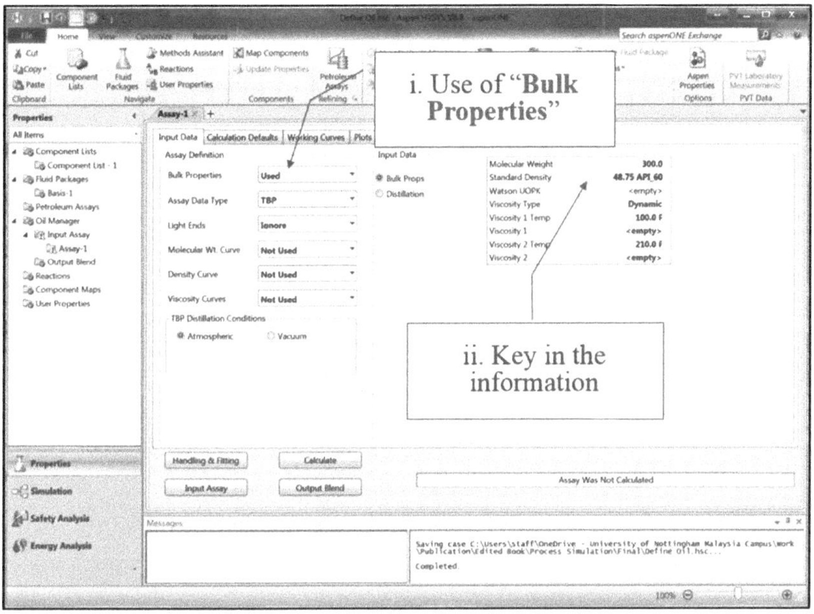

FIGURE 2.16 Input of bulk properties in Oil Manager.

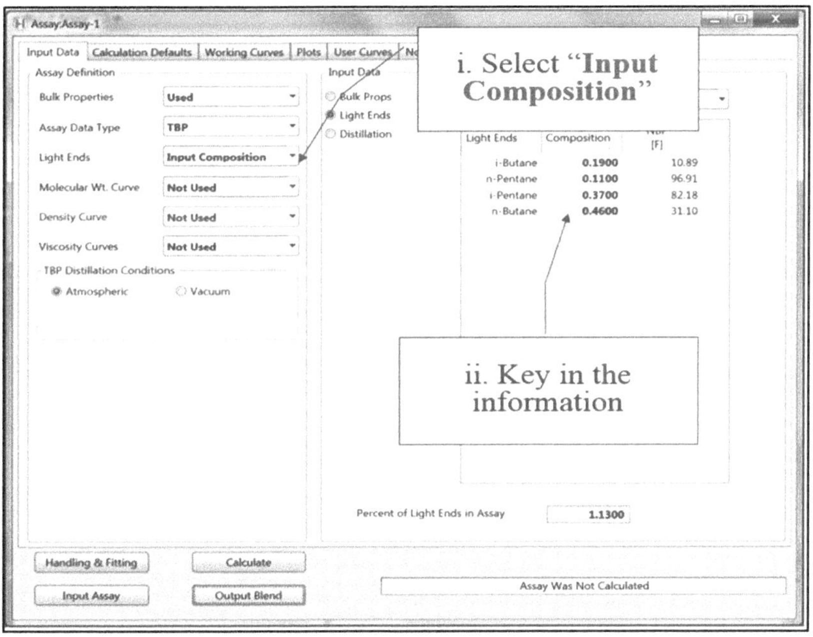

FIGURE 2.17 Input of light ends in Oil Manager.

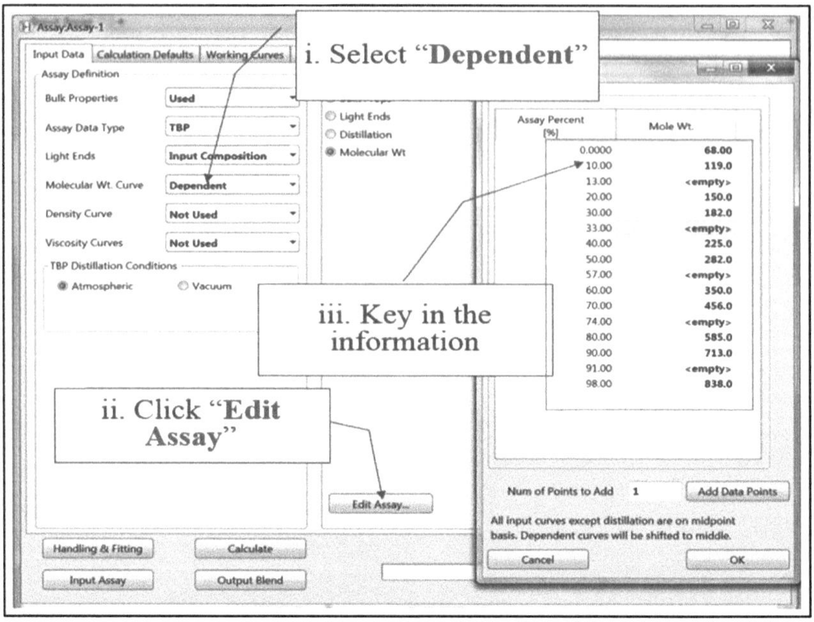

FIGURE 2.18 Input of molecular weight curve in Oil Manager.

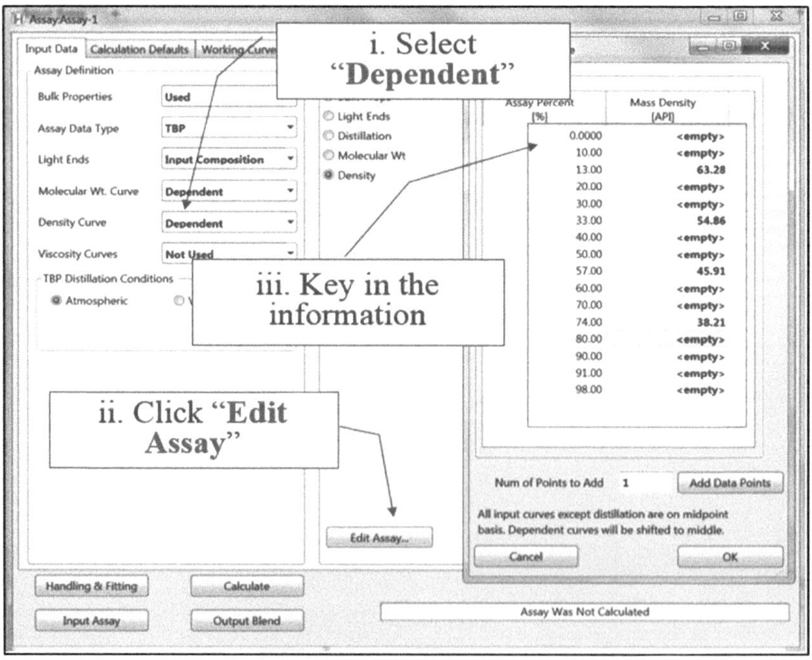

FIGURE 2.19 Input of density curve in Oil Manager.

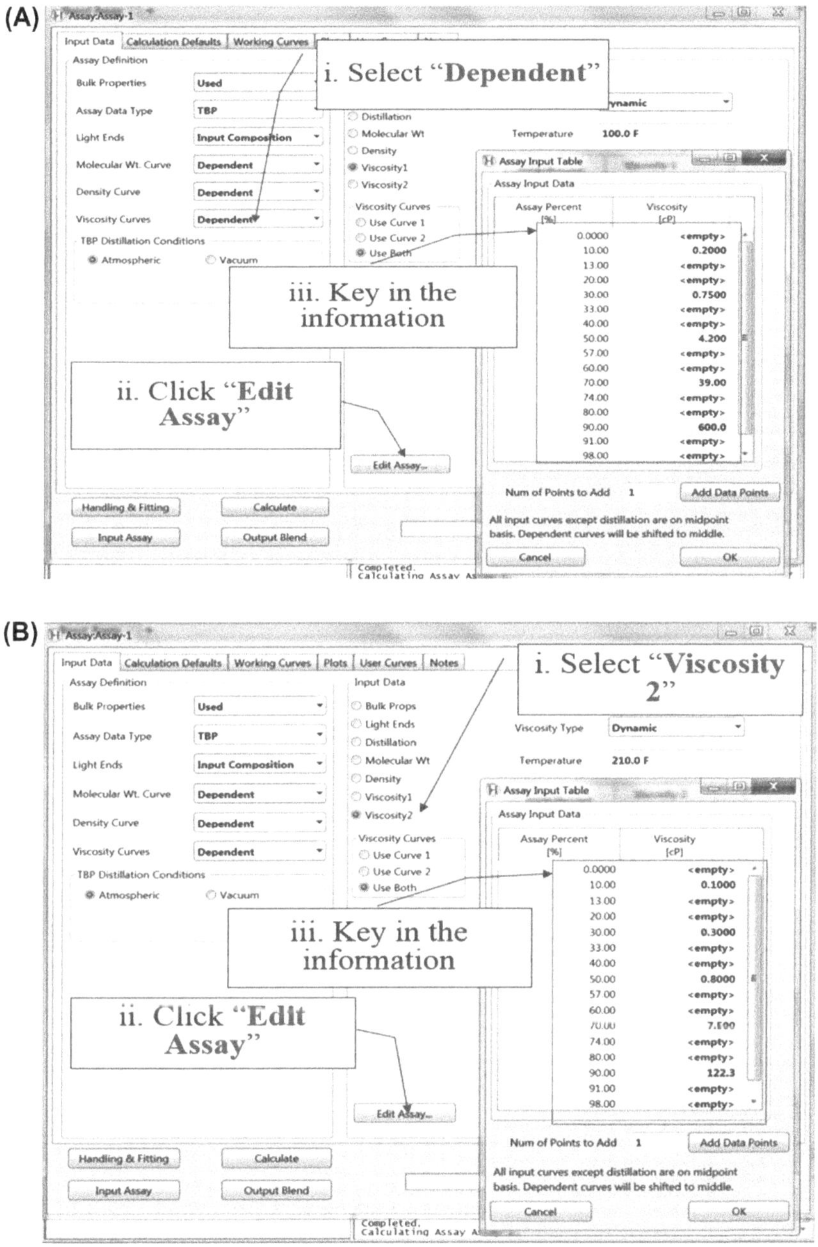

FIGURE 2.20 Input of viscosity curves (A) at 100°F (B) at 210°F in Oil Manager.

Once all the information is included into the simulator, the assay can be determined by a click on "Calculate" (see Fig. 2.21). This has shown that the assay is ready to generate pseudocomponents in the following step.

Step 2. Generate Pseudocomponents—Create, Cut and Blend

When more information is provided to the simulation software, more accurate properties of the crude oil assay can be generated. In Aspen HYSYS, blend and cut functions of the assay are required to generate the pseudocomponents. Fig. 2.21 shows the blend of assay to generate the general presentation of the whole curve. Next, the calculated blend is required to add into cut function (Fig. 2.21). After the blend is added (Figs. 2.22 and 2.23), the array is ready to perform cut function.

In Aspen HYSYS, three types of cut, which are "Auto Cut," "User Ranges," and "User Points" are provided (see Fig. 2.23). Note that "Auto Cut" option is based on the values specified internally to determine the cut; "User Ranges" specified boiling point ranges and then number of cuts per range; and "User Points" specified cut points that are proportioned based on internal weight scheme (Fig. 2.23).

Once the blend and cut are ready, the pseudocomponents can be determined. The pseudocomponents that represent the characteristics of the oil will be generated. To view the generated oil information, click at tabs "Tables,"

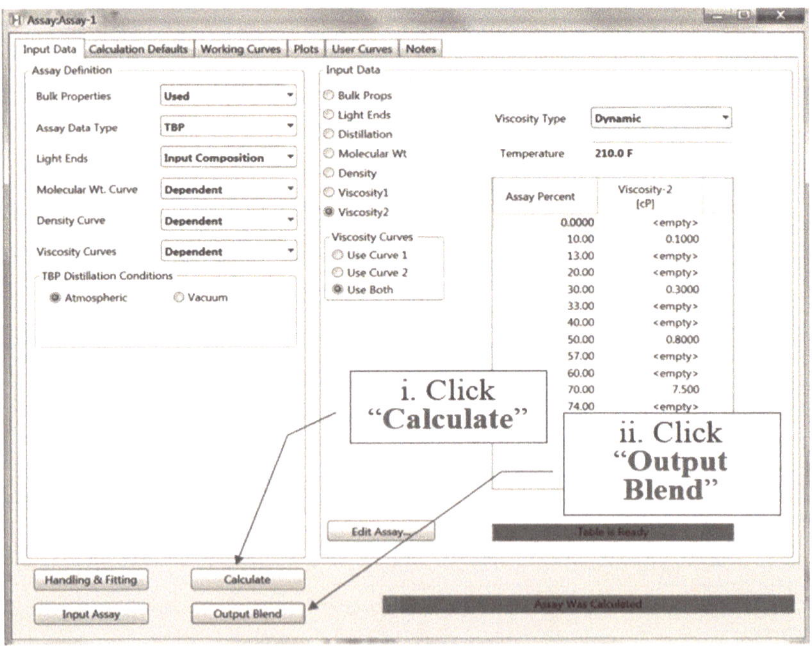

FIGURE 2.21 Generating the blend based on the input assay.

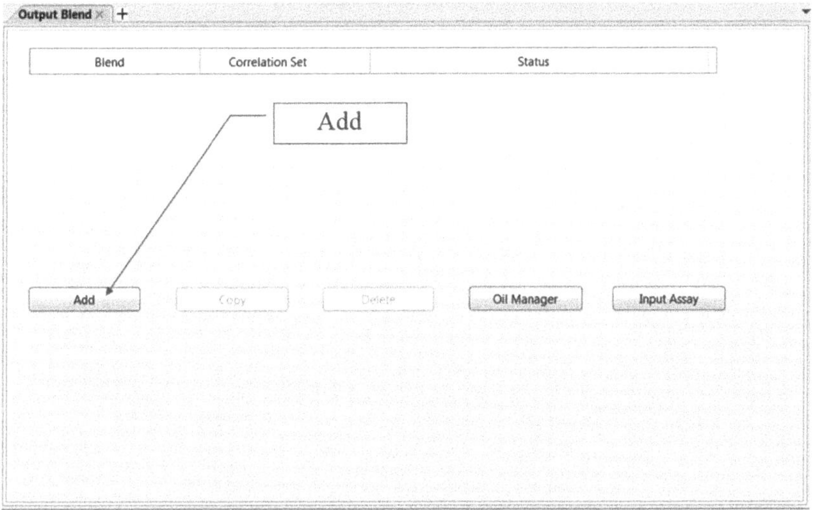

FIGURE 2.22 Add calculated blend into the simulator.

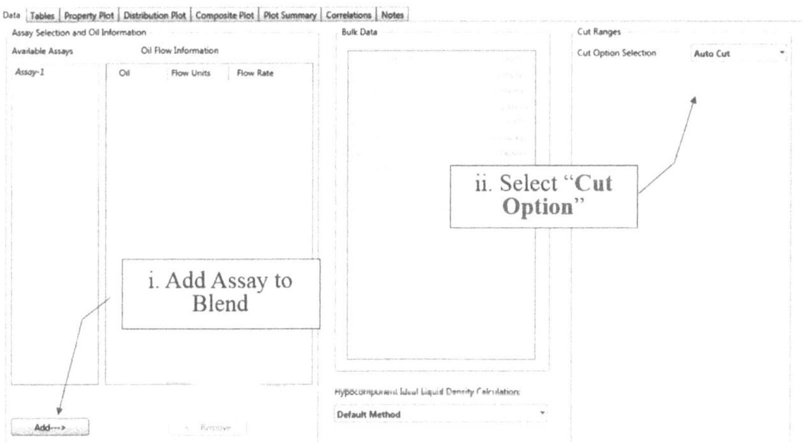

FIGURE 2.23 Selection of option for cuts.

"Composite Plot," and others as well. Figs. 2.24 and 2.25 show the information of oil. Note that in Fig. 2.25, both information of calculated oil (red) and input of user (green) can be viewed clearly. Note that both curves should be superimposed with each other when the calculated oil is very similar with the input oil information.

Step 3. Install the Oil in the Flowsheet

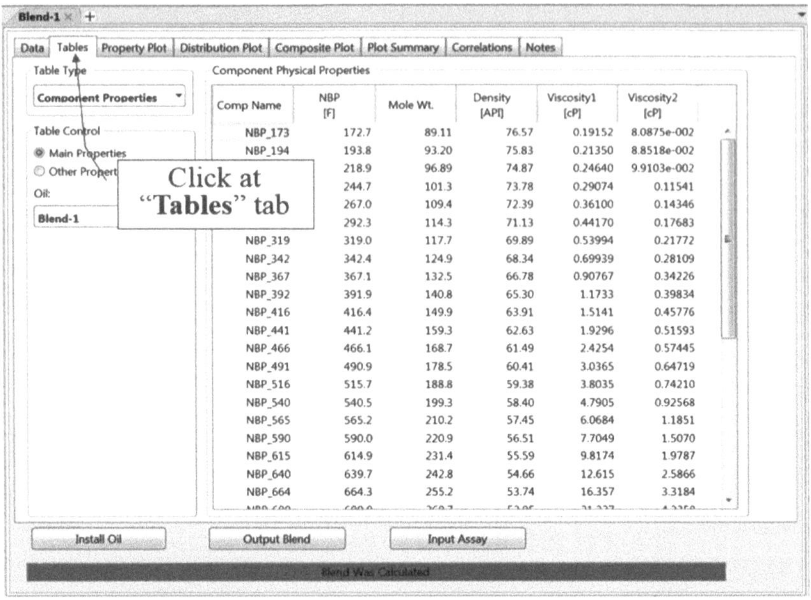

FIGURE 2.24 Component physical properties of calculated oil.

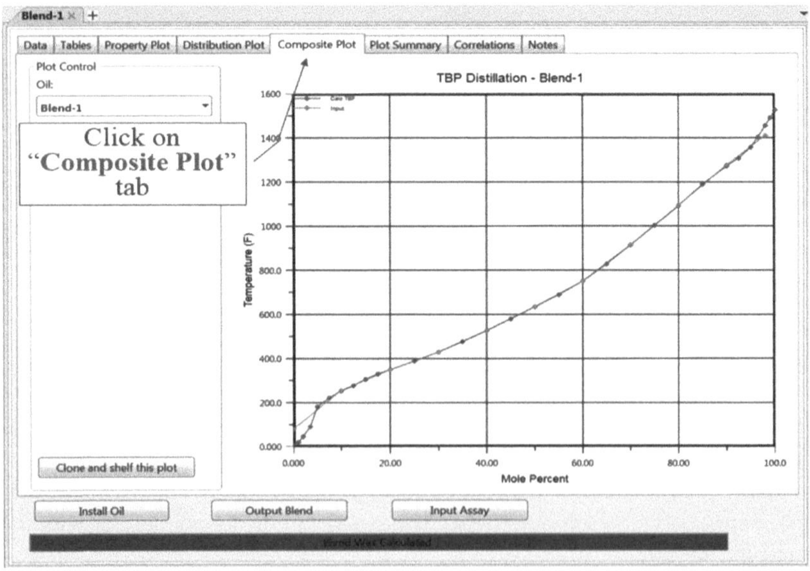

FIGURE 2.25 Composite plot of calculated oil versus the input oil information [calculated oil (*red*) and input of user (*green*)]. (For interpretation of the references to colour in this figure legend, the reader is referred to the web version of this article.)

At this stage, the oil is well defined. For the oil to be used in the simulation, it has to be installed with the fluid package (Fig. 2.26). The pseudocomponent information can then be included into the flowsheet via create stream with a defined composition. If the oil information is not installed in the fluid package, the oil composition will not exist in the flowsheet. Alternatively, the oil can be added as hypothetical components as discussed in the previous section.

Next, to view the information in simulation environment (Fig. 2.27), click on the "Crude Oil," which is defined previously. Note that the composition of the hypothetical components will be included in the stream. In the event where the oil composition is not included in the "Crude Oil," the stream should be redefined based on the previous steps.

When "Crude Oil" stream is ready, stream properties (e.g., temperature, pressure, and flowrate) are included as shown in Fig. 2.28. Next, the crude oil is heated up via heater in simulator. Setting of heater and heated crude oil is shown in Fig. 2.29.

Example 2.4. Crude oil registration in PRO/II

To model crude oil in PRO/II, the oil needs to be defined based on its characteristics via assay characterization. Based on the laboratory results, the characteristics of the crude oil are included in PRO/II library and hypothetical components. This involves the similar steps shown in Example 2.3.

An example of registration of crude oil is created based on information provided in Table 2.2, but the thermodynamic method selected is

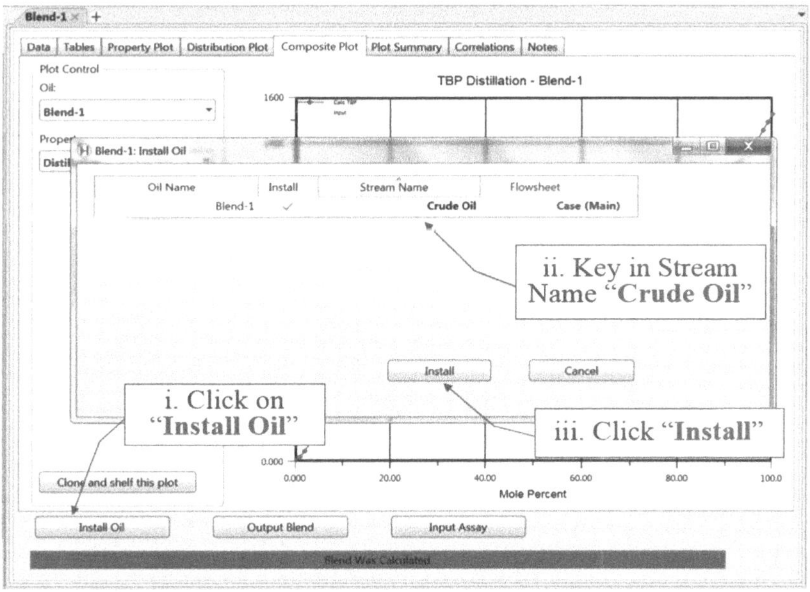

FIGURE 2.26 Install oil in fluid package and stream name.

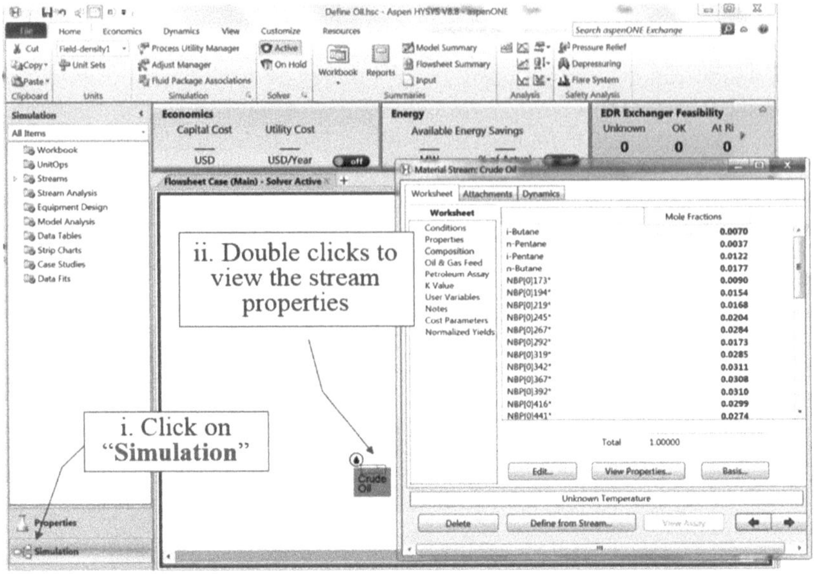

FIGURE 2.27 Install oil in the flowsheet and view the stream composition.

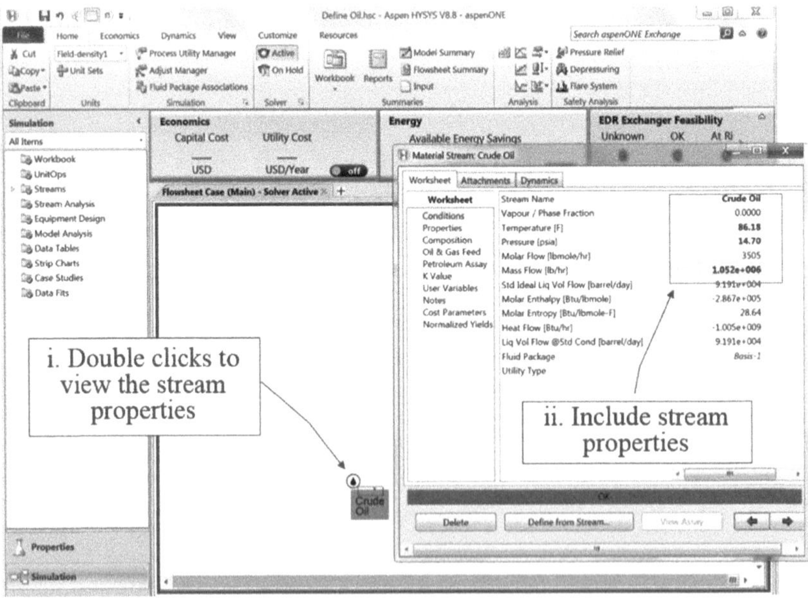

FIGURE 2.28 Define stream.

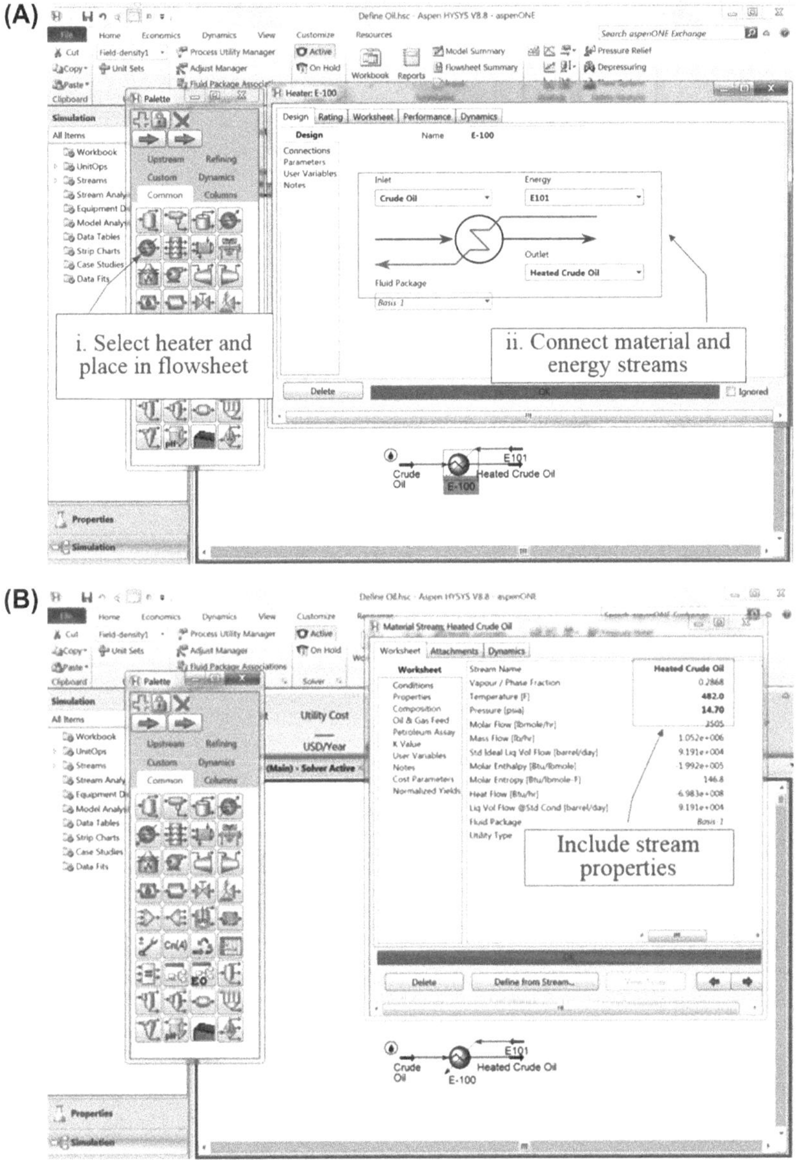

FIGURE 2.29 Heat up crude oil via heater: (A) setting of heater and (B) setting of stream properties of heated crude oil.

Soave–Redlich–Kwong. In PRO/II, the user is required to define light ends components first prior to modify the assay characterization data. Fig. 2.30 shows components required for this example. In terms of thermodynamic method, Soave–Redlich–Kwong is selected (Fig. 2.31).

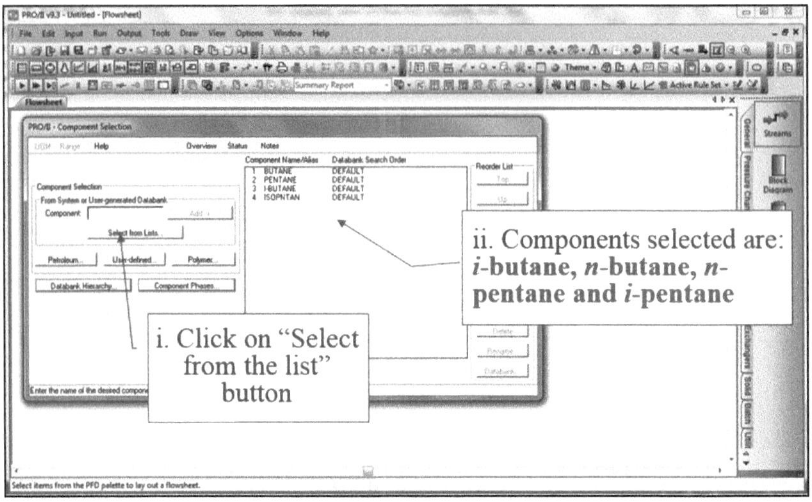

FIGURE 2.30 Define components for the light ends.

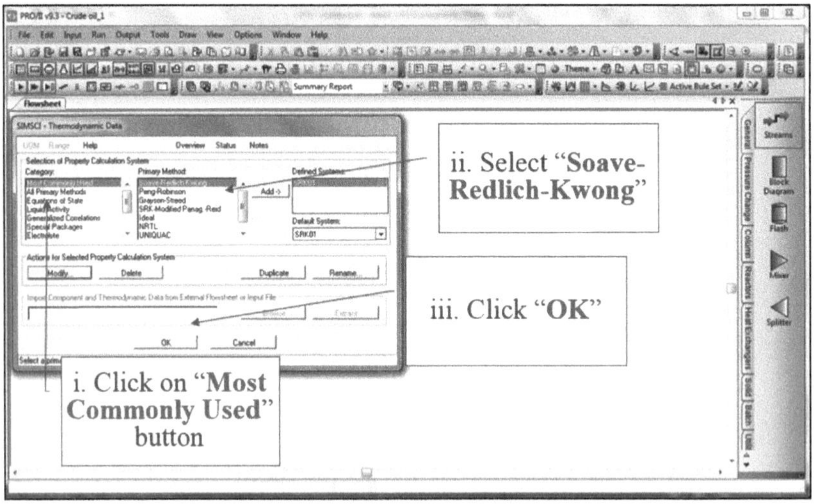

FIGURE 2.31 Thermodynamic properties for crude assay.

Next, modify the assay Primary TBP Cutpoints Definition based on the feed data (Fig. 2.32). Make sure the minimum temperature for the first interval is below the temperature of the initial point of distillation boundaries value. Referring to the feed data, initial and end point boundaries value are set at 5% and 98%, respectively (Fig. 2.33). Besides, the user is required to define assay and light ends data including distillation test, percentage distilled, temperature,

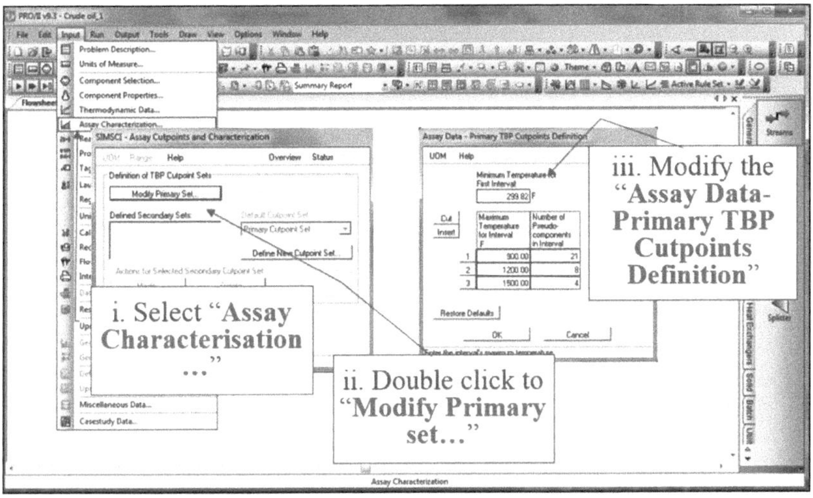

FIGURE 2.32 Modify the assay characterization data.

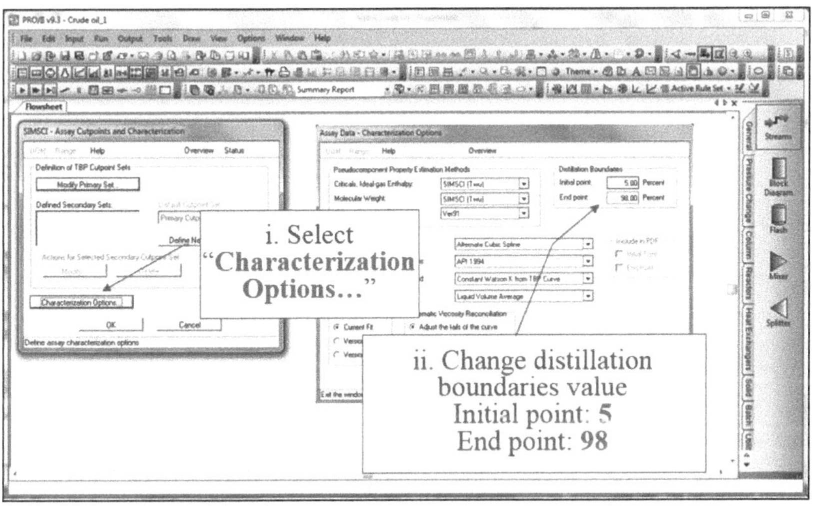

FIGURE 2.33 Specifying distillation initial and end boundaries value.

composition, and gravity data. Detailed steps for defining assay, flowrate, and assay light ends data for crude oil are shown in Figs. 2.34 and 2.35.

Finally, generate the assay component output report and distillation curve to view the profile of crude oil assay. Detailed steps are shown in Figs. 2.36 and 2.37.

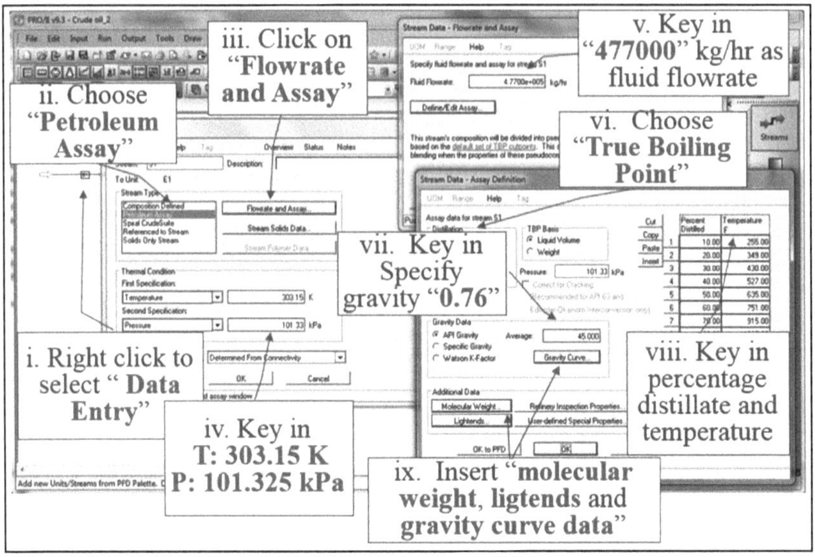

FIGURE 2.34 Defining assay and flowrate for crude oil.

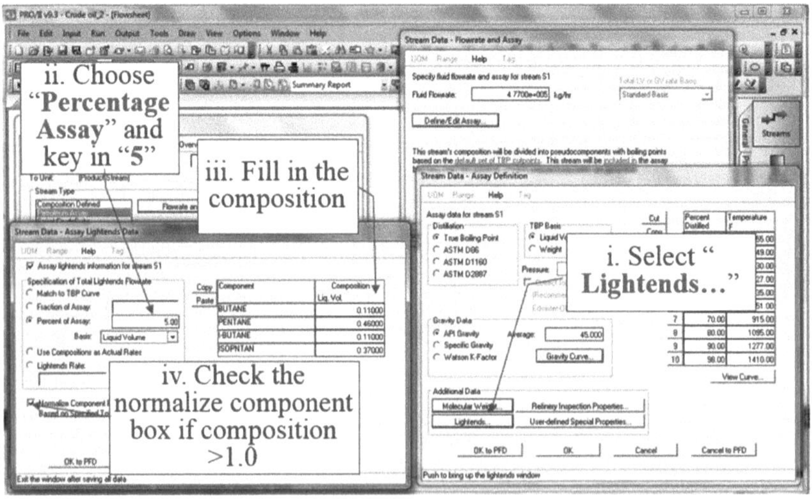

FIGURE 2.35 Specification of total light ends flowrate and compositions of components.

A heat exchanger unit is added to the feed stream to preheat the registered crude oil to 523.15 K. Detailed steps of adding a heat exchanger are shown in Fig. 2.38.

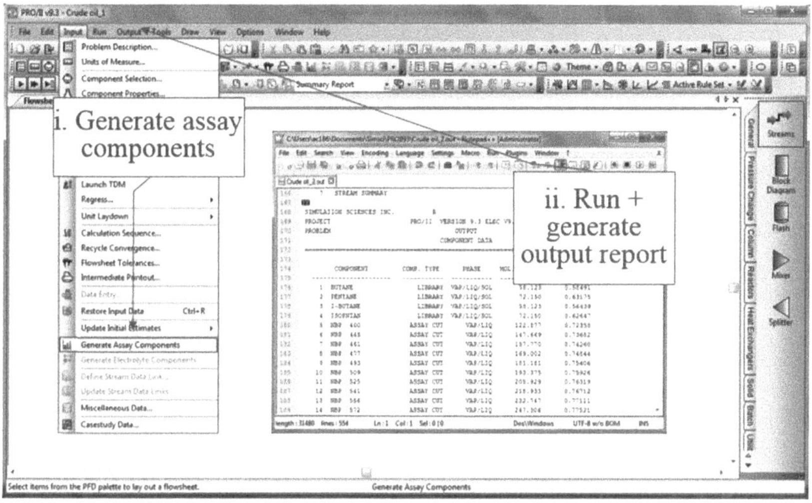

FIGURE 2.36 Generate assay component output report.

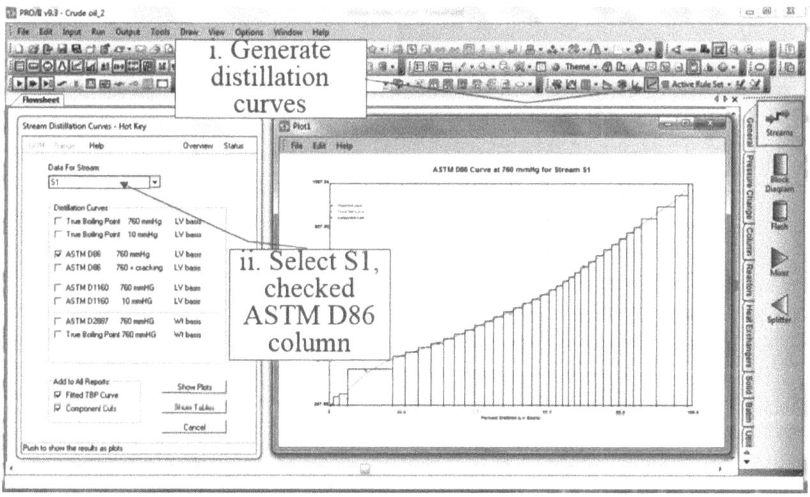

FIGURE 2.37 Distillation curve for defined crude oil assay.

Exercise

Define the oil stream based on the properties given in Table E.1. The oil is to be heated to 100°C. Perform the simulation task using the following:

1. Aspen HYSYS
2. PRO/II

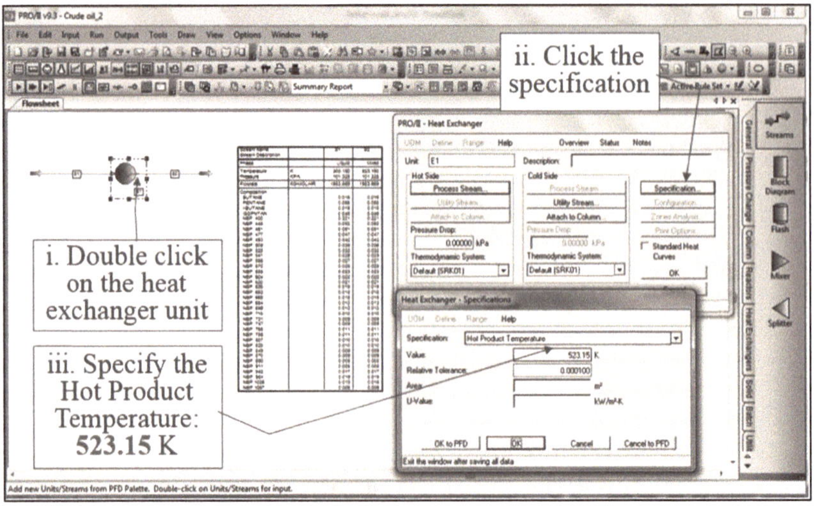

FIGURE 2.38 Heating of registered crude oil.

TABLE E.1 ASTM D86 and properties of distillate oil.

Properties	Unit	Distillate oil
Mass flowrate	kg/h	3.925×10^5
Temperature	°C	15
Pressure	kPa	500
Density at 15°C	kg/m^3	828
Assay basis		Liquid volume
ASTM D86 0%	°C	−5
ASTM D86 5%	°C	166
ASTM D86 10%	°C	226
ASTM D86 30%	°C	265
ASTM D86 50%	°C	283
ASTM D86 70%	°C	301
ASTM D86 90%	°C	330
ASTM D86 95%	°C	341
ASTM D86 100%	°C	359
Light ends		Auto calculate
		i-Butane, *i*-pentane, *n*-butane, *n*-pentane, thiophene, *m*-mercaptan
Cut ranges		Auto cut
Thermodynamic properties		SRK

SRK, Soave–Redlich–Kwong.

References

AspenTech, 2015. Aspen HYSYS user guide.
Schneider Electric, 2015. SimSci PRO/II v9.3.2 reference manual.

Chapter 3

Physical property estimation and phase behavior for process simulation*

Rafil Elyas

East One-Zero-One Sdn Bhd, Shah Alam, Selangor, Malaysia

Chapter outline

*Simulation and other electronic files of this chapter are made available on a companion website; URL: https://www.elsevier.com/books-and-journals/book-companion/9780323901680.

Chemical Engineering Process Simulation. https://doi.org/10.1016/B978-0-323-90168-0.00005-6

Like the foundation of a building, the methods used for physical property estimation determine the integrity of a chemical engineering computation. These days, most engineers rely on commercial simulators to perform their computations, and all commercial simulators these days come with a myriad of property packages, where various *property estimation methods* have been combined into *property packages* such as Peng–Robinson, Soave–Redlich–Kwong (SRK), BWRS, Grayson–Streed, Braun-K10, NRTL, UNIQUAC, and the list goes on. It is critical to know which property package would be applicable for one's computation. The objective of this chapter is to provide some insight into the workings of those property packages and enable the reader to make the correct selection.

3.1 Chemical engineering processes

One is often "advised" by "seniors" to use the Peng–Robinson property package in commercial simulators such as UniSIM Design or Aspen HYSYS, or the SRK property package in PRO/II. It is a common misconception that the property package is a single equation used to calculate "everything." Each property package consists of *several* computational methods that are used to estimate *thermophysical* and *transport properties of interest*. What sorts of thermophysical and transport properties are of interest?

If we take a look at most processes, a chemical engineer works with, we find that a majority of these processes involve vaporizing, condensing, separating, compressing, and moving fluids. Let us take a look at a typical process, the accompanying computational requirements, and the corresponding properties required. To simplify things, we shall look at a typical upstream oil and gas production facility in Fig. 3.1. In this facility, oil, gas, and water are separated in the three-phase separator. Gas stream is then compressed, while the oil is pumped and stabilized. Let us analyze some equipment in this facility and identify what properties are required to execute some typical computations.

3.1.1 Inlet separator

Starting with the gas wells, the first process that is encountered is an inlet separator. In the design and sizing of this inlet separator, it is necessary to understand *how much vapor and liquid are produced* at their operating pressure and temperature and *how the components partition themselves between vapor and liquid phases* and to establish the volume–mass relationships of the phases. Hence, the required thermophysical properties include *vapor and liquid densities, enthalpies, and vapor pressures.*

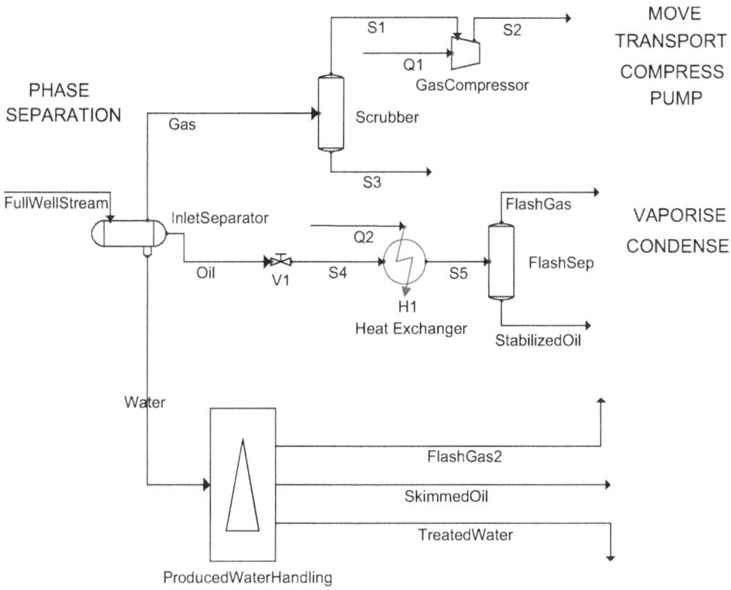

FIGURE 3.1 A typical upstream oil and gas production facility.

3.1.2 Heat exchanger

Heat exchangers allow fluids to either be heated or cooled. In this example, it is necessary to stabilize the oil by removing the lighter or more volatile components, to allow the oil to be stored in atmospheric tanks or transported in single-phase liquid pipelines. To represent the heating or cooling process, the following thermophysical properties are required: *vapor and liquid heat capacities and densities, liquid vapor pressure, and heat of vaporization.*

3.1.3 Gas compressor

A gas compressor moves a fluid by adding work to it. There are positive displacement compressors that simply displace the volume using a piston, or centrifugal compressors that transfer momentum to a gas, which increases its velocity and converts it to pressure by reducing its velocity through a diffuser. Compression computations require the properties of *gas density, compressibility, and heat capacities (Cp/Cv).*

In summary, the basic thermophysical properties required for engineering computations of these oil and gas production facilities are *density, vapor pressure, and energy characteristics*. In addition, it is also necessary to estimate the transport properties of the fluids because the fluids will obviously have to flow across these processes. Some of the typical transport properties would be *viscosity and surface tension.*

3.2 Thermodynamic processes

One of the most basic requirements when processing fluids is to understand how a state of a fluid changes with respect to pressure and temperature. This is largely addressed by thermodynamics. Most chemical engineering processes such as those in Fig. 3.1 are expressed as thermodynamic processes in commercial process simulators. Processes are represented by changes in state of a fluid. Gas compression is used to illustrate this. As discussed earlier, the following *properties* were identified for gas compression: density, compressibility, and *Cp*/*Cv*; these properties are used to determine the *thermodynamic state variables* of the fluid.

The thermodynamic path is described in the phase diagram in Fig. 3.2.

An actual compression is an irreversible process. Most simulators employ a two-step procedure for this computation. First, a reversible compression, A-B, is used along isentrope *S*1. $H_2 - H_1$ is the reversible work for this process. This work is then multiplied (or divided) by an efficiency to give the actual work (point C). Total work imparted to the fluid is given by the difference of H_1 and H_3. Hence, the difference between H_2 and H_3 may be regarded as the "lost work." The adiabatic efficiency for this process is hence defined as the ratio of these two differences, i.e.,

$$\eta_{ad} = \frac{H_2 - H_1}{H_3 - H_1} \tag{3.1}$$

where η_{ad} is the adiabatic efficiency, $H_2 - H_1$ is the reversible work for the process, and $H_3 - H_1$ is the total work.

For this process, the thermodynamic state variables required in this computation are *temperature* (*T*), *pressure* (*P*), *volume* (*V*), *enthalpy* (*H*), *and entropy* (*S*). In addition to the above, other state variables are also required to represent other processes; for example, Gibbs (*G*) and Helmholtz (*A*) free energies are used to describe thermodynamic potential, which in turn is used

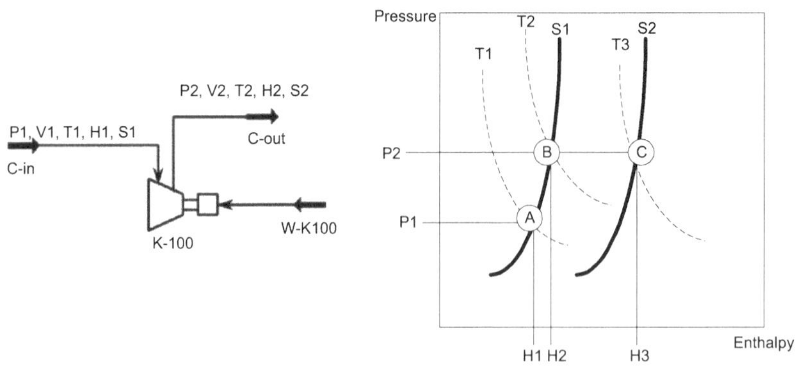

FIGURE 3.2 Gas compression.

for phase equilibrium computations. These state variables are conveniently related to each other from characteristic thermodynamic definitions and the Maxwell relationships. These are discussed in the following subsections.

3.2.1 Characteristic thermodynamic relationships (Smith et al.)

3.2.1.1 Internal energy (U)

The internal energy of a substance can be seen as the energy possessed by the molecules or atoms making up the substance. Most physical chemistry models describe atoms and molecules in ceaseless vibrational, rotational, or translational modes. In addition to that, there are energies that hold the atoms and molecules together in the form of bonds.

It is impossible to know what the absolute value of internal energy of a substance is, but this is not a problem as most thermodynamic analysis deals with changes in energy (ΔU). For example, adding heat (Q) to the system increases this molecular activity; work (W) is extracted when this activity is made to interact with external forces. Hence, this can be described by Eq. (3.1).

$$\Delta U = Q - W \tag{3.2}$$

Note that Eq. (3.2) is also the mathematical statement of the first law of thermodynamics, i.e., the conservation of energy.

3.2.1.2 Enthalpy (H)

Enthalpy is defined as the summation of the internal energy system and a work term, which is given by the product of pressure (p) and volume (V).

$$H = U + pV \tag{3.3}$$

3.2.1.3 Entropy (S)

In simple terms, entropy is used to quantify "lost work." For example, one first puts in 100 W of work to compress a gas from V_1 to V_2. Expanding the gas from V_2 to V_1, however, does not produce 100 W of work. Some of the work is lost because of intermolecular "friction" or other phenomena. The change in entropy provides an indication of the amount of "lost work." In Fig. 3.2 the amount of lost work (difference between H_2 and H_3) corresponds to an increase in entropy from $S1$ to $S2$.

The mathematical definition of entropy is given by

$$dS = \frac{dQ}{T} \tag{3.4}$$

or

$$\Delta S = \frac{\int dQ}{T} \tag{3.5}$$

Eq. (3.4) is also a statement of the second law of thermodynamics, which states that total entropy of an isolated system increases with time. It only remains constant in a system that remains in steady state or that undergoes a reversible process.

3.2.1.4 Gibbs free energy (G)

The Gibbs free energy (G) is defined as follows:

$$G = H - TS \tag{3.6}$$

From this definition, it can be seen as the "available" energy a fluid may have. It can be used to describe the maximum or net amount of extractable work from a fluid (TS can be seen as the work that will be lost). However, Gibbs free energy is commonly used in chemistry and chemical engineering to determine the chemical potential of a system.

3.2.1.5 Helmholtz free energy (A)

The Helmholtz free energy is defined as follows:

$$A = U - TS \tag{3.7}$$

Similar to the Gibbs free energy, the Helmholtz free energy, too, can be used to describe the maximum or net amount of extractable work from a fluid (TS can be seen as the work that will be lost), and it can also be used to determine the chemical potential of a system.

3.2.2 Maxwell relationships

Now that the characteristic thermodynamic relationships have been developed, it is necessary to relate them to each other. This is made possible using the mathematical framework of the Maxwell relationships.

$$dU = TdS - PdV \rightarrow \left(\frac{\partial T}{\partial V}\right)_S = -\left(\frac{\partial P}{\partial S}\right)_V \tag{3.8}$$

$$dA = -SdT - PdV \rightarrow \left(\frac{\partial S}{\partial V}\right)_T = -\left(\frac{\partial P}{\partial T}\right)_V \tag{3.9}$$

$$dH = TdS + VdP \rightarrow \left(\frac{\partial T}{\partial P}\right)_S = \left(\frac{\partial V}{\partial S}\right)_P \tag{3.10}$$

$$dG = -SdT + VdP \rightarrow -\left(\frac{\partial S}{\partial P}\right)_T = \left(\frac{\partial V}{\partial T}\right)_P \tag{3.11}$$

The equations in this section were not intended to induce thermodynamic class nightmares but to highlight a very important point; that is, the characteristic energy functions that are essential for the definition of thermodynamic processes [i.e., internal energy (U), Helmholtz free energy (A), enthalpy (H), Gibbs free energy (G), and entropy (S)] may be related to temperature, pressure, and volume. Temperature, pressure, and volume in turn are related by the equations of state.

3.3 Equations of state

3.3.1 The ideal gas law (c.1834)

One of the most basic relationships to describe the state of a gas with respect to pressure and temperature, or an equation of state, is the ideal gas law.

$$PV = nRT \tag{3.12}$$

where P is the absolute pressure of the gas, V is the volume of the gas, n is the amount of substance of the gas, usually measured in moles, R is the gas constant, and T is the absolute temperature.

This is sometimes rewritten in terms of molar volume as follows:

$$Pv = RT \tag{3.13}$$

where v is the molar volume, i.e., V/n.

The ideal gas law provides reasonable estimates for gasses at low pressure and high temperature, for at these conditions the distance between the gas molecules is large enough and the kinetic energy is sufficiently high to eliminate any effects of the size of the molecules or any possible molecular interactions.

Pushing the ideal gas law to low-temperature and high-pressure limits yields the following impossible results:

$$\frac{1}{T\rightarrow 0}\; v = \frac{RT}{P} = 0 \tag{3.14}$$

$$\frac{1}{P\rightarrow \infty}\; v = \frac{RT}{P} = 0 \tag{3.15}$$

Observing Eqs. (3.13) and (3.14) leads to a potential mistake. Because matter occupies space and it is highly unlikely that materials shrink to nothingness at absolute zero or under high pressure! This leads to the corrections needed for the ideal gas law, which is discussed next.

3.3.2 Corrections to the ideal gas law (cubic equations of state)

The objective of this section is to provide a basic overview of the ideal gas law corrections to provide the reader a basis for understanding how the equation of state is implemented in commercial process simulators. This section is by no means a detailed analysis of the various equations of states; there are many thermodynamic texts available that provide descriptions of much greater rigor.

The first set of corrections came in 1873 from Johannes Diderik van der Waals.

The first correction addresses the fact that molecules have volume. The effect of molecular volume would be realized at lower temperatures and higher pressures when molecules were "closer" together and "less energetic."

To compensate for this, a volume offset term, b, was added to the ideal gas equation.

$$v = \frac{RT}{P} + b \tag{3.16}$$

Eq. (3.15) may be rewritten as follows:

$$P = \frac{RT}{v - b} \tag{3.17}$$

Now, the problem of material vanishing at low temperatures and high pressures has been addressed.

The next correction takes into account cases in which molecules are attracted to one another at certain distances. This effect becomes observable when a liquid is vaporized. To vaporize a liquid, it is necessary to add energy to it. Therefore, there must be some sort of attraction force among the molecules.

Attractive forces are believed to be proportional to concentration or inverse molar volume. The attraction term, a, is appended to Eq. (3.17) and yields the familiar van der Waal's equation:

$$P = \frac{RT}{v - b} - \frac{a}{v^2} \tag{3.18}$$

Most work following van der Waals continued to address the attraction term component, and the general form of a two-parameter cubic equation of state can be expressed as follows:

$$P = \frac{RT}{(v - b)} - \frac{a}{v^2 + ubv + wb^2} \tag{3.19}$$

From this structure, the following common equations of state can be defined:

van der Waals: $u = 0$, $w = 0$ (c.1873)
Redlich–Kwong: $u = 1$, $w = 0$ (c.1949)
Peng–Robinson: $u = 2$, $w = -1$ (c.1976).

The attraction (a) and molecular volume (b) terms are expressed as functions of pure component critical pressure and temperature. In the case of the Peng–Robinson equation, the attraction term (a) is multiplied by a temperature-dependent $\alpha(T)$.

3.3.2.1 Van der Waals

$$a = \frac{27R^2T_c^2}{64P_c}; b = \frac{RT_c}{8P_c} \tag{3.20}$$

3.3.2.2 Redlich–Kwong

$$a = \frac{0.42748R^2T_c^{2.5}}{P_c}; b = \frac{0.08664RT_c}{P_c} \tag{3.21}$$

3.3.2.3 Peng–Robinson

$$a = \frac{0.45724R^2T_c^2}{P_c}\alpha(T); b = \frac{0.07780RT_c}{P_c} \tag{3.22}$$

$$\alpha(T) = \left(1 + \left(0.37464 + 1.5422\omega - 0.2699\omega^2\right)\left(1 - \sqrt{\frac{T}{T_c}}\right)\right)^2$$

However, fluids consist of mixtures of components in most applications. It is then necessary to use mixing rules for the a and b terms. These rules are essentially mole-weighted averages a and b terms of the pure component, which can be expressed as follows.

The mixture b term is a simple mole-weighted average of the pure component b terms.

$$b = \sum_i y_i b_i \tag{3.23}$$

where y_i and b_i are the mole fraction and b term of the ith component, respectively.

The mixture a term is addressed differently as it represents the attraction between molecules. It is an average of the attraction terms between all possible component pairs in the mixture.

$$a = \sum_i \sum_j y_i y_j a_{ij} \tag{3.24}$$

where a_{ij} is referred to as the cross-parameter between components i and j in the mixture. The cross-parameter is simply the root mean square of the pure components a_i and a_j with a tunable "gain" term k_{ij}.

$$a_{ij} = \sqrt{a_i a_j}\left(1 - k_{ij}\right) \tag{3.25}$$

The k_{ij} term in Eq. (3.24) is referred to as the *binary interaction parameter* (BIP) in most commercial simulators and is allowed to change by users.

Here is a simple way to understand how the value of BIP affects the molar volume. We know that the a term accounts for intermolecular attraction. When the attraction force is large, the molecules get "closer" together and the overall molar volume decreases. On the other hand, when the attraction force is small, the molecules get "farther" from each other and the overall molar volume increases. This is next demonstrated with a simple example.

3.3.2.4 Reducing the "attractive force"

Starting with a k_{ij} of zero for a fixed pair of components (with fixed a_i and a_j) and increasing the number to approach 1, one can see that a_{ij} decreases as k_{ij} starts to approach 1. This means that the attraction term is weakened as k_{ij} is increased, and the molar volume of the mixture increases.

3.3.2.5 Increasing the "attractive force"

One can also go the other way around and reduce the attractive force by starting with a k_{ij} of zero and reducing the k_{ij} to -1. This has the effect of "strengthening" the attraction between components i and j, resulting in a smaller molar volume of the mixture.

Example 3.1. A gas mixture consisting of equimolar methane and ethane at 25°C and 10 barg. Its molar volume is to be determined using the Peng–Robinson equation of state, with values of the BIP ranging from −0.5 to 0.5. The default BIP provided in the simulator for this mixture was 0.00224. The results for this computation are presented in Table 3.1.

It should be noted that the values of the BIP were *arbitrarily spanned* from −0.5 to 0.5 to demonstrate the effect this tuning parameter has on the molar volume of the gas mixture. In reality, the tuning of the cubic equation BIPs requires experimental data for binary systems (and in some cases ternary systems if available).

TABLE 3.1 Comparison of molar volumes for a range of binary interaction parameters.

K_{ij}	V	Δ	% Δ
−0.5	2.0913	−0.0352	−1.66
−0.1	2.1226	−0.0039	−0.18
0.00224	**2.1265**	**0.0000**	**0.00**
0.1	2.1380	0.0115	0.54
0.5	2.1680	0.0415	1.95

Most commercial simulators have done this work for some common component pairs. However, it is impossible to cover every single component pair that may ever be encountered in industry. If no experimental data were used to estimate the BIP for a component pair, the BIP value is either set to zero (for the case of SimSci PRO/II) or estimated using extrapolation, or group contribution method (for the case of Aspen HYSYS and UniSim Design). Note that some process simulators provide tuning functions or packages that allow the users to calculate BIPs from their own data sets of P-XY,[1] T-XY,[2] bubble, and dew points. Note also that in Peng–Robinson and SRK equations, the k_{ij} is symmetric, $k_{ij} = k_{ji}$. However, there are "advanced" implementations of these equations of state (e.g., Peng–Robinson Stryjek Vera); the BIPs can be asymmetric, $k_{ij} \neq k_{ji}$ essentially doubling the number of tuning parameters.

3.4 Liquid volumes (Walas, 1985)

The cubic equations of state described in Section 3.3.2 are not applicable for determining liquid densities and must be avoided as they will result in grossly incorrect estimates of the liquid density. Remember that all these equations of state come from one common "ancestor," i.e., the ideal *gas* law.

The liquid phase is far more complicated than the gas phase because molecules are in "closer quarters" and intermolecular interactions begin to dominate. This makes it difficult to predict how these molecules may "arrange" themselves in a liquid continuum, particularly if there are different functional groups present, which may give rise to complexes or exhibit hydrogen bonding.

It should be noted that many users select the Peng–Robinson property package when using software such as Aspen HYSYS or UniSim Design, thinking that the Peng–Robinson equation is used to determine gas and liquid densities. This is (fortunately) incorrect. As mentioned earlier in this chapter, a property package is not a single equation but a collection of methods used to estimate various properties.

Some of the typical methods used to estimate liquid densities in commercial simulators are as follows:

1. Volume translated equation of state methods, where the calculated volume by the equation of state is "corrected." An example of this is the PRF method by Peneloux et al. This is used in PVTSim.[3]

1. Pressure–liquid composition and gas composition at constant temperature.
2. Temperature–liquid composition and gas composition at constant pressure.
3. PVTSim Help Calsep A/S 2014.

2. Correlations or stand-alone equations. Examples would be the Rackett equation and corresponding state liquid density (COSTALD[4]) method by Hankinson–Brobst–Thomson.

In the case of Aspen HYSYS and UniSim Design, Peng–Robinson COSTALD is the default method for liquid density estimation. There are other methods for estimating liquid volumes. The COSTALD equation is an empirical equation that has the following form for a pure component:

$$\frac{v_s}{v} = v_r^o\left(1 - \omega_{SRK} v_r^\delta\right) \tag{3.26}$$

$$v_r^o = 1 + a\frac{(1-T_R)^1}{3} + b\frac{(1-T_R)^2}{3} + c(1-T_R) + d\frac{(1-T_R)^4}{3} \tag{3.27}$$

$$v_r^\delta = \frac{e + fT_R + gT_R^2 + hT_R^3}{T_R - 1.00001} \tag{3.28}$$

where v_s is the molar volume of saturated liquid at some temperature; v_r^o is a characteristic of spherical molecules and v_r^δ is a correction factor, both correlated in terms of the reduced temperature; ω_{SRK} is the acentric factor determined by the reduced pressure and temperature of the system; T_R is the reduced temperature; a, b, c, d, e, f, g, and h are fixed equation parameters; and v is an empirical "characteristic volume"; this has been determined for various substances. It has also been correlated in terms of the SRK acentric factor, ω_{SRK}, for nine substance groups.

$$v = \frac{RT_c}{P_c}\left(k_1 + k_2\omega_{SRK} + k_3\omega_{SRK}^2\right) \tag{3.29}$$

where T_c is the critical temperature and P_c is the critical pressure.

Empirical values for k_1, k_2, and k_3 are provided for nine substance groups:

1. Paraffins
2. Olefins and diolefins
3. Cycloparaffins
4. Aromatics
5. Other hydrocarbons
6. Sulfur compounds
7. Fluorocarbons
8. Cryogenic liquids
9. Condensable gasses

The mixing rules for mixtures are highly complicated and are not shown for brevity. The purpose of this section was to highlight that liquid densities

4. An Improved Correlation for Densities of Compressed Liquids and Liquid Mixtures, G.H. Thomson, K.R. Brobst, and R.W. Hankinson. AIChE Journal (Vol. 28, No. 4) July 1982.

and corresponding volumes are generally determined using a completely different description than gas densities and volumes.

3.5 Viscosity and other properties

Equations of state cannot be used to calculate transport properties such as viscosity. The latter is highly dependent on the chemical nature of the substance, fluid mechanics, and surface chemistry. For example, paraffinic change entanglement and wax formation in oils or colloidal/emulsion formation in water and oil mixtures. The methodologies to estimate viscosities are generally empirical in nature. Some examples of these methods are summarized in Table 3.2.

3.6 Phase equilibria

When phases are said to be in thermodynamic equilibrium, then their potential or driving forces to transfer material from one phase to the other are equal. To illustrate this concept, consider a two-component (*A* and *B*), two-phase (gas and liquid) system as per Fig. 3.3.

TABLE 3.2 Examples for viscosity estimation.

Phase/fluid	Method
Gas	Lucas, Jossi–Stiel–Thodos, Ely–Hanley
Liquid	Lucas, Letsou–Stiel, Twu
Oil/water emulsion	Woelflin

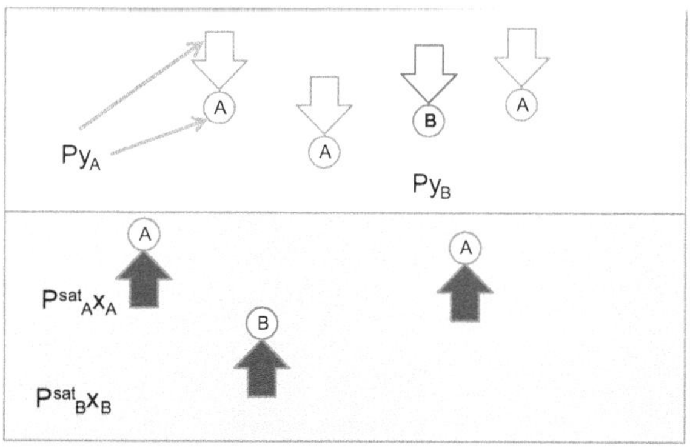

FIGURE 3.3 Two-phase system with two components, A and B.

The gas-side driving force to move component A to the liquid phase is referred to as the gas phase fugacity of component A and for now can be expressed simply as its partial pressure Py_A. The liquid driving force exerted on component A to move it to the liquid phase is the liquid phase fugacity and for now can be considered its prorated vapor pressure $P_A^{sat}x_A$.

At equilibrium, gas and liquid fugacities are equal:

$$f_A^V = f_A^L \tag{3.30}$$

or in this example:

$$Py_A = P_A^{sat}x_A \tag{3.31}$$

Eq. (3.31) is the familiar Raoult's law, in which A is the more volatile component, x_A and y_A are the mole fractions of component A in the liquid and gas phases, respectively, P is the total system pressure, P_A^{sat} is the vapor pressure of component A, and is a function of temperature.

The vapor pressure of a component is a function of its temperature. The higher the temperature, the more energy a molecule has and the higher its tendency to "escape" to the vapor phase. This can be seen in the equilibrium plot of methane and ethane mixture at 20 bara in Fig. 3.4. As the temperature increases, the composition of methane in the gas phase increases monotonically.

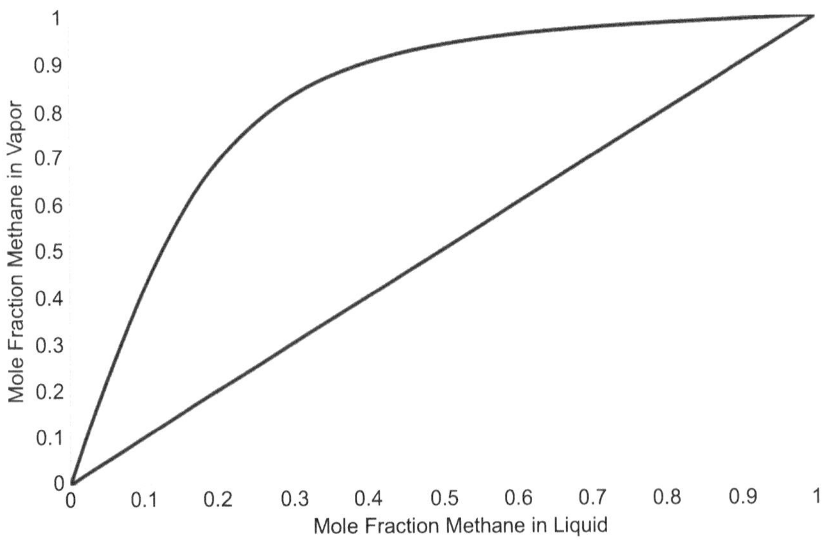

FIGURE 3.4 Methane and ethane equilibrium plot at 20 bara.

3.6.1 Vapor phase correction

The gas phase driving force in Eq. (3.31) is expressed as the partial pressure. The partial pressure of a component in a mixture is simply a mole fraction prorated pressure. This is derived from Dalton's law:

$$P = \sum_{i}^{n} Py_i \tag{3.32}$$

where Py_i is the partial pressure of the ith component and n is the total number of components in the mixture.

Assuming the gas phase driving force or fugacity of a component to be its mole fraction prorated pressure is a rather ideal representation. Consequently, it is often necessary to correct the gas phase fugacity by implementing the gas phase fugacity coefficient.

$$\varnothing_i P y_i = P_i^{sat} x_i \tag{3.33}$$

where $\varnothing_i$ is the gas phase fugacity coefficient of component i.

The gas phase fugacity coefficient's purpose is to correct for nonidealities in the gas phase. To illustrate this, the definition of the fugacity coefficient of a pure substance will be analyzed.

The fugacity coefficient for a pure substance is defined as follows:

$$\ln(\varnothing) = \frac{-1}{RT} \int_0^P \left(\frac{RT}{P} - V_d \right) dP \tag{3.34}$$

where $\varnothing$ is the pure component fugacity, V_d is either the volume determined by an equation of state or an experimental value.

From Eq. (3.34), it can be seen that if:

1. The pressure approaches 0 (vacuum), $\ln\varnothing$ approaches 0, and $\varnothing$ approaches 1.
2. If the ideal gas law is used as the equation of state, the fugacity coefficient also reduces to 1.
3. At higher pressures using a nonideal equation of state will yield a fugacity coefficient less than or greater than 1.

A fugacity coefficient greater than 1 results when $\frac{RT}{P} - V_d$ is negative (and the entire term on the right side of Eq. 3.34 is positive). This happens when the volume of the ideal gas volume is less than that of the real or nonideal gas. This means that the molecules in the real gas may be "repelling" each other. Applying this in Dalton's law will result in a higher system pressure for a given volume.

Analogously, a fugacity coefficient less than 1 results when $\frac{RT}{P} - V_d$ is positive (and the entire term on the right side of Eq. 3.34 is negative). This happens when the volume of the ideal gas is greater than that of the real or nonideal gas. This means that the molecules in the real gas may have a strong attractive force. Applying this in Dalton's law will result in a lower system pressure for a given volume.

For mixtures, the fugacity coefficient is expressed in the following manner:

$$\ln\varnothing_i = \int\left(\left(\frac{\partial P}{\partial n_i}\right)_{T,V,n_j} - \frac{RT}{V}\right)dV - \ln\left(\frac{PV}{RT}\right) \tag{3.35}$$

where $\left(\frac{\partial P}{\partial n}\right)_{T,V,n_j}$ is the partial derivative of pressure with respect to the ith component and is computed using the equation of state.

From here, we see that in addition to being used to determine gas densities and thermodynamic energy state functions, the equation of state is also used to calculate the gas phase correction, or gas phase fugacity coefficient.

3.6.2 Liquid phase corrections

In an ideal liquid, with components A and B, the interactions between $A-A$, $B-B$ and $A-B$ are considered the same. From Raoult's law, the vapor pressure contribution of component A was simply defined as its mole fraction prorated vapor pressure, $P_A^{sat}x_A$. The total vapor pressure exerted by components in an ideal multicomponent mixture is as follows:

$$P = \sum_i^n x_i P_i^{sat} \tag{3.36}$$

If the real solution vapor pressure is greater than the ideal pressure, then the fluid is said to exhibit a positive deviation from Raoult's law, the different molecules can be seen to be "repelling" each other. And vice versa, if the real solution vapor pressure is less than the ideal pressure, then the fluid is said to exhibit a negative deviation from Raoult's law, the different molecules can be seen to be "attracting" each other. In a situation like this, the equilibrium line would no longer be monotonic.

Fig. 3.5 shows a mixture of ethyl acetate and water. It can be seen that initially the more volatile ethyl acetate (normal boiling point 77.1°C) boils off. However, at a concentration of around 0.6 mol fraction of ethyl acetate in the vapor, water (normal boiling point 100°C) begins boiling off and diluting ethyl acetate in the vapor phase. This is an example of an azeotropic mixture that results from hydrogen bonding in the liquid phase.

In cases like this, it is necessary to add a correction factor to the vapor pressure term to account for this behavior, the activity coefficient, γ_i.

$$P = \sum_i^n \gamma_i x_i P_i^{sat} \tag{3.37}$$

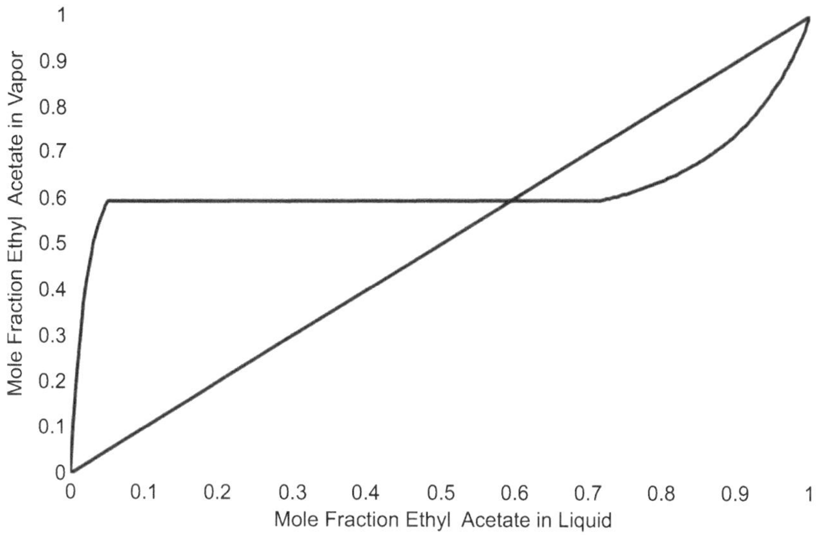

FIGURE 3.5 Water and ethyl acetate at 5 bara.

When implemented in the equilibrium equation:

$$Py_i = \gamma_i P_i^{sat} x_i \tag{3.38}$$

The activity coefficient can be related to the excess Gibbs energies in the following manner:

$$\ln\gamma_i = \frac{G^{ex}}{RT} - \sum_{k \neq i} x_k \left(\frac{\left(\frac{\partial G^{ex}}{RT}\right)}{\partial x_k} \right)_{T,P,x_j \neq i,k} \tag{3.39}$$

Commercial simulators typically have a large selection of activity coefficient methods. The most commonly used activity coefficient would be the nonrandom two liquid (NRTL), which has the following form:

$$\ln(\gamma_j) = \frac{\sum_j \tau_{ji} G_{ji} x_j}{\sum_k G_{ki} x_k} + \frac{\sum_j xjGij}{\sum_k G_{kj} x_k}\left[\tau_{ij} - \frac{\sum_k x_k \tau_{kj} G_{kj}}{\sum_k G_{kj} x_k}\right]$$

$$\tau_{ij} = a_{ij} + \frac{b_{ij}}{T} + \frac{c_{ij}}{T^2} \tag{3.40}$$

$$Gij = \exp(\alpha_{ij}\tau_{ij})$$

$$\alpha_{ji} = \alpha_{ji}^1 + \beta_i^1 T$$

The tuning parameters are $a_{ij/ji}$, $b_{ij/ji}$, and $c_{ij=ji}$. This equation can be used for strongly nonideal and polar mixtures, as well as partially miscible systems. The concept of NRTL is based on the concept of a "local" composition, where the local concentration around a molecule can be different from the bulk concentration.

In addition to NRTL, other activity coefficient models are also made available in commercial simulators. These include Wilson, van Laar, and the UNIQUAC method.

It should be noted that in commercial simulators, unlike the tuning parameters (BIPs) for the equations of state, which are available for almost all of the library component combinations, activity coefficient tuning parameters can be somewhat scarce. Hence it is necessary to either

1. acquire binary T-XY, P-XY, and vapor pressure data to tune the activity coefficient models or
2. estimate the activity coefficient tuning parameters computationally. Most commercial simulators have a UNIFAC group contribution method tuning parameter estimation algorithm. This algorithm generally estimates the tuning parameters for a user-specified temperature at atmospheric pressure.

Because most engineers or companies do not have access to binary data or expertise for tuning the activity coefficient models, it may be necessary to outsource this to specialist organizations. In a majority of cases, option 2 using the built-in tuning parameter generator is typically used. Caution should be exercised and results should be checked for vapor pressure accuracy.

3.6.3 Bringing it all together

In Section 3.6.1, vapor phase corrections were addressed, and in Section 3.6.2, liquid phase corrections were discussed. Implementing both vapor and liquid phase corrections to the equilibrium equation yields the following:

$$\varnothing_i P y_i = \gamma_i P_i^{sat} x_i \tag{3.41}$$

This can also be written as follows:

$$y_i = K_i x_i \tag{3.42}$$

K_i is referred to as the K-value or the distribution coefficient for component i.

$$K_i = \frac{\gamma_i P_i^{sat}}{\varnothing_i P} \tag{3.43}$$

From Eq. (3.43), one can see that implementing *both* vapor and liquid phase corrections (a dual model) would lead to a very complex endeavor. There would be two groups of tuning parameters that would have to be determined. In most cases, details will be put into the liquid activity coefficient model, while the gas phase fugacity coefficient is assumed ideal, or estimated using an untuned equation of

TABLE 3.3 Common practices for vapor–liquid equilibrium method selection.

Application/system	Gas phase fugacity coefficient method	Liquid activity coefficient method
Oil and gas processing. Estimation of bulk flow rates of oil, gas, and water. Accurate compositional tracking and partitioning of CO_2, H_2S, and water not critical	Peng–Robinson Soave–Redlich–Kwong Kabadi Danner Lee–Kessler–Plocker	Ideal ($\gamma_i = 1$)
Gas dehydration or hydrate suppression with glycol or alcohols	Peng–Robinson Soave–Redlich–Kwong Specialty glycol or alcohol packages	Ideal ($\gamma_i = 1$)
Refining (traditional computations)	Empirical vapor pressure models, Grayson–Streed, Braun-K10	
Refining (recent developments)	Peng–Robinson Soave–Redlich–Kwong	Ideal ($\gamma_i = 1$)
Sour gas. H_2S partitioning across vapor, hydrocarbon, and aqueous predictions critical	Specialty packages. Typically enhanced cubic equations of state	
Gas sweetening using amines/solvents	Specialty packages	
Nonideal liquids, highly polar, self-associating	Ideal Peng–Robinson Soave–Redlich–Kwong	NRTL UNIQUAC
Steam systems (only water)	ASME or NBS steam tables	

state. In addition to that solving for both the fugacity and liquid activity coefficients would place a greater demand on computational requirements particularly if the simulation were large, or when dynamic simulation were to be performed. Table 3.3 summarizes the methods used in industry for typical vapor-liquid equilibrium and vapor–liquid–liquid equilibrium computations.

3.7 Flash calculations (Smith and Van Ness, 1975)

All commercial simulators perform vapor–liquid and vapor–liquid–liquid (not addressed in this chapter) equilibria computations for fluid systems. This is generally referred to as a flash calculation. This is a common computation that is used to predict phase splits, bubble and dew points, and distribution/partitioning of components across the phases. The calculations involved are described in the following subsections.

3.7.1 "MESH" equations

The flash calculation involves the solving of four main sets of equations, which are known by the "MESH" equations:

1. **M**aterial Balance
2. **E**quilibrium
3. **S**ummation
4. **H**eat Balance

3.7.1.1 Material balance

For a simple flash unit in Fig. 3.6, the material balance consists of the overall balance and component balance.

Overall balance:

$$F = V + L \tag{3.44}$$

Component balance:

$$Fz_i = V_i y_i + L_i x_i \tag{3.45}$$

where F is the total molar flow rate of material, V is the vapor molar flow rate, L is the liquid molar flow rate, z_i is the mole fraction of component i in the combined phases, y_i is the mole fraction of component i in the gas phase, and x_i is the mole fraction of component i in the gas phase.

3.7.1.2 Equilibrium

Equilibrium and component distribution across phases are determined by the equilibrium equation introduced in Section 3.6.3:

$$y_i = K_i x_i$$

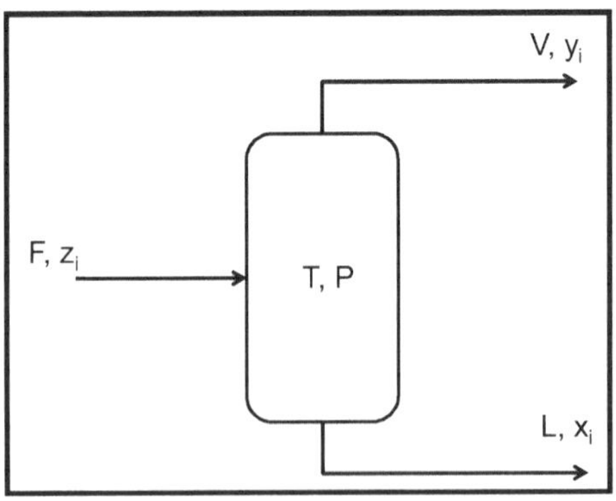

FIGURE 3.6 Flash calculation at a given temperature and pressure.

3.7.1.3 Summation

The sum of all mole fractions must equal 1

$$\sum_i z_i = 1; \sum_i x_i = 1; \sum_i y_i = 1; \tag{3.46}$$

3.7.1.4 Heat balance

$$Fh_f = Vh_{vap} + Lh_{liq} \tag{3.47}$$

where h_f is the molar enthalpy of the combined phases; h_{vap} and h_{liq} are the molar enthalpies of the vapor and liquid phases, respectively.

3.7.2 Bubble point flash

The bubble point or saturated liquid flash is typically performed to determine the bubble point temperature *or* pressure. Here, the composition of the liquid is the composition of the entire fluid.

3.7.2.1 Methodology

$$\sum_i K_i x_i = \sum_i y_i = 1$$

- Assume a temperature for the known pressure or vice versa.
- Find K_i at the pressure and temperature known and assumed.
- Check that the summation approaches your desired tolerance (close to zero).
- If not, repeat until satisfactory tolerance has been achieved.

In commercial simulators such as Aspen HYSYS or UniSim Design, this would be performed by setting the vapor fraction equal to zero, and specifying a temperature (to calculate a bubble point pressure) *or* pressure (to calculate a bubble point temperature).

3.7.3 Dew point flash

The dew point or saturated vapor flash is typically performed to determine the dew point temperature *or* pressure. Here, the composition of the vapor is the composition of the entire fluid. Procedure of dew point flash calculation consists of the following steps:

- Set $\frac{\sum_i y_i}{K_i} = \sum_i x_i = 1$.

- Assume a temperature for the known pressure or vice versa.
- Find K_i at the pressure and temperature (known or assumed).
- Check that the summation approaches your desired tolerance (close to zero).
- If not, repeat until satisfactory tolerance has been achieved.

In Aspen HYSYS or UniSim Design, this would be performed by setting the vapor fraction equal to 1, and specifying a temperature (to calculate a dew point pressure) *or* pressure (to calculate a dew point temperature).

3.7.4 Two-phase pressure–temperature flash

The pressure–temperature (PT) flash is used to determine the amount of liquid and vapor and their compositions at a given pressure and temperature.

Equations:

1. $F = V + L$
2. $Fz_i = Vy_i + Lx_i$
3. Take a basis of $F = 1$ mol
4. $z_i = Vy_i + Lx_i$
5. $y_i = K_i x_i$
6. $x_i = z_i/(L + VK_i)$
7. $y_i = z_i/(V + L/K_i)$
8. Summation criteria:

$$\sum x_i = \sum (z_i / (L + VK_i)) = 1$$

$$\sum y_i = \sum (z_i / (V + LK_i)) = 1$$

Procedure:

- Find K at T and P
- Assume V or L
- Solve step 8
- Iterate until step 8:

$$\sum x_i - \sum y_i < \text{Tolerance}$$

3.7.5 Other flash routines

In addition to the dew point, bubble point, and PT flashes described in this chapter, most commercial simulators include the following flash routines:

1. Pressure enthalpy (used in the valve and mixer computations)

2. Temperature enthalpy (used in heat exchanger computations)
3. Pressure entropy and temperature entropy (used for rotating equipment, compressors, and pumps)

3.8 Phase diagrams

Phase envelopes are one of the most useful tools a chemical engineer has at his or her disposal. The phase envelope provides a graphical overview of the possible phases a fluid may have for a range of pressures and temperatures. All commercial simulators come with phase envelope utilities. The previous section addressed flash routines that are commonly built into commercial simulators. We next discuss the phase envelopes generated by running these flash routines for a given pair of variables.

3.8.1 Pressure–temperature diagrams of pure components and mixtures

A PT phase diagram is generated by running the bubble point and dew point flash routines over a range of temperatures and pressures. When the routine is run for a single component, a phase diagram or a saturation line is generated. PT diagrams for pure methane are given in Fig. 3.7, while those for pure ethane are given in Fig. 3.8. As shown, methane and ethane have different boiling and critical points.

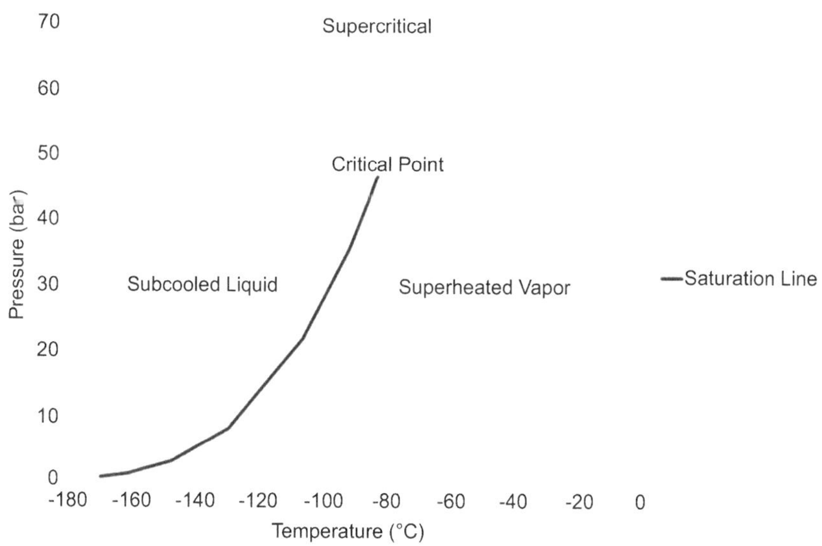

FIGURE 3.7 Phase diagrams/saturation lines (PTs) for pure methane.

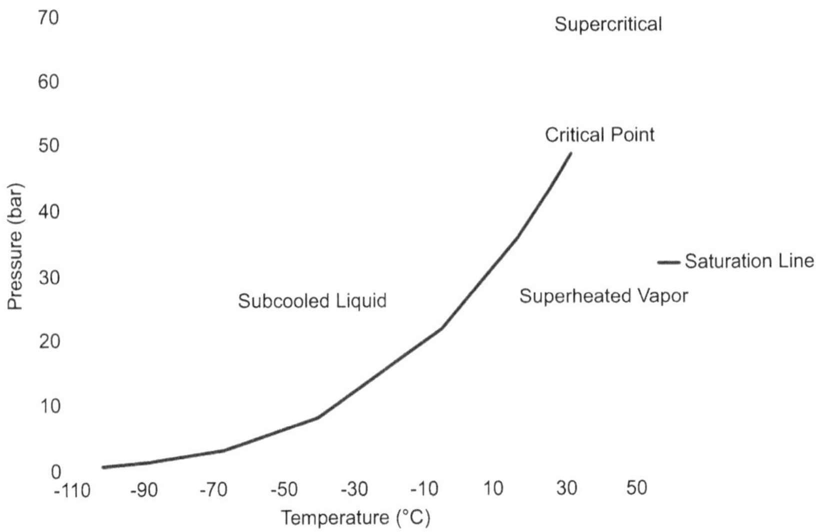

FIGURE 3.8 Phase diagrams/saturation lines (PTs) for pure ethane.

The PT diagram for pure component can be used to determine if the fluid is a subcooled liquid, superheated gas, or supercritical. It does not, however, give an indication that the fluid is a saturated liquid, saturated vapor, or two-phase as all these phases fall within the one-dimensional saturation line. To determine this, it is necessary to look at the pressure–enthalpy or temperature–enthalpy diagrams, such as those in Fig. 3.9. The maximum temperature and

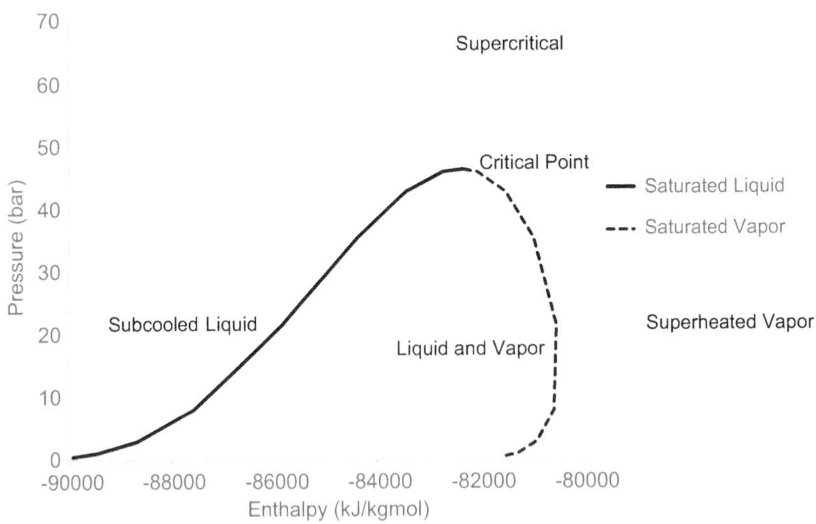

FIGURE 3.9 Pressure–enthalpy diagram for pure methane.

pressure where the fluid can exist as vapor or liquid is the critical point and the terminus of the saturation line.

For an equimolar mixture of methane and ethane, its phase diagram looks quite different from the pure component phase diagrams (e.g., Figs. 3.7 and 3.8). In this case, we see a phase envelope, such as that in Fig. 3.10.

Methane is the more volatile component with a normal boiling point of −161.5°C. The normal boiling point of ethane is significantly higher at −89°C. Instead of a single saturation line like those pure component cases in Figs. 3.7 and 3.8, two lines are observed for Fig. 3.10. The solid line is the saturated liquid or bubble point line, while the dashed line is the saturated vapor or dew point line. Both these lines meet at the critical point.

Because of the difference in boiling points in this mixture, the saturation line has "spread out." Starting from a subcooled liquid at point *A* and increasing the temperature of the mixture moves the operating point to point B, the bubble point line. Here, a bubble begins to nucleate; the composition of that bubble is predominantly the more volatile component, methane. As the temperature of the mixture is increased, both the methane and ethane vaporize. At point C, the mixture consists of a vapor and a liquid phase. Point *D* is the dew point where the last drop of liquid vaporizes. The composition of this last drop of liquid is mainly the heavier component, ethane. As the temperature is increased, the operating point moves to the superheated gas region.

If all three diagrams (Figs. 3.7, 3.8 and 3.10) are overlaid in one plot, we can gain some insight into the phase envelope of the mixtures, such as that in Fig. 3.11.

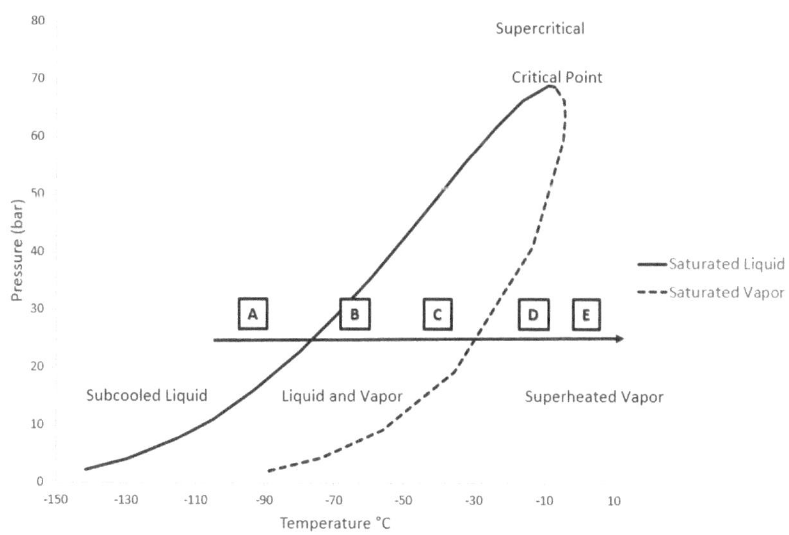

FIGURE 3.10 Phase diagram (PT) for an equimolar mixture of methane and ethane.

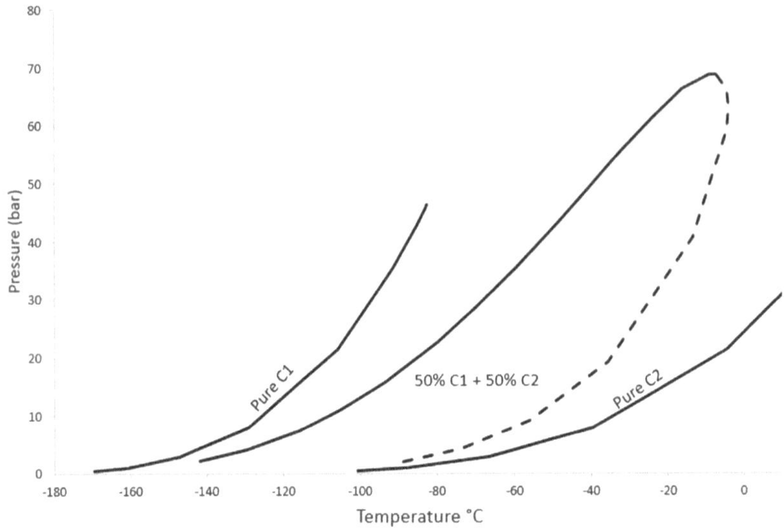

FIGURE 3.11 Phase diagrams for pure methane, pure ethane, and an equimolar mixture of methane and ethane.

The following observations can be made. The bubble point and dew point lines are bound by the saturation lines of the lightest and heaviest components. The mixture critical point is higher than the individual pure component critical points because of the intermolecular interactions between the components.

Phase diagrams for various methane and ethane ratios are overlaid in one plot (Fig. 3.12). The dashed line drawn tangent to the critical points of the

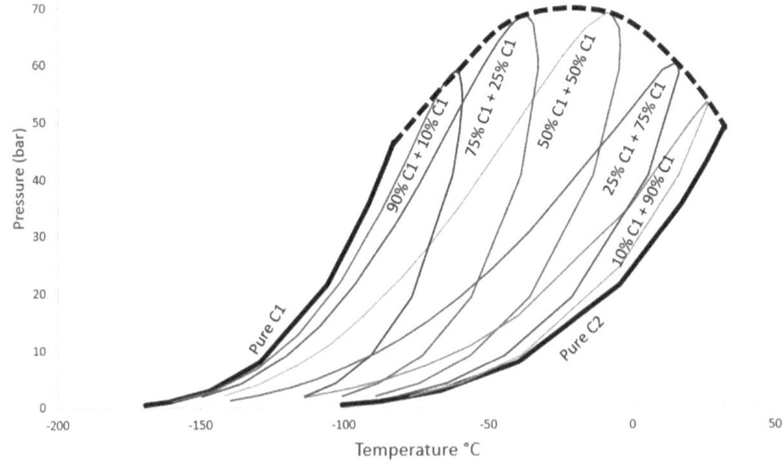

FIGURE 3.12 The critical locus for a binary mixture of methane and ethane.

phase envelopes for compositions at the critical point at each curve is referred to as the critical locus (Campbell, 2004).

3.8.2 Retrograde behavior

In Fig. 3.9, the phase envelope for a 50:50 mol mixture of methane and ethane exhibits a single critical point, the maximum temperature and pressure at which vapor and liquid may exist.

Fig. 3.13 is a typical phase diagram for multicomponent natural gas mixture.

In this case, there is no single point defining the temperature and pressure maxima; instead there are three points:

1. The cricondenbar, the maximum temperature where liquid and vapor may exist.
2. The cricondentherm, the maximum temperature where liquid and vapor may coexist in equilibrium.
3. The retrograde region, the shaded area in the phase envelope where liquid condensation can occur by increasing temperature or lowering pressure (which is counterintuitive).

This demonstrates the importance of plotting the phase envelope when performing any computation involving a multicomponent mixture to avoid "unexpected" phase behavior.

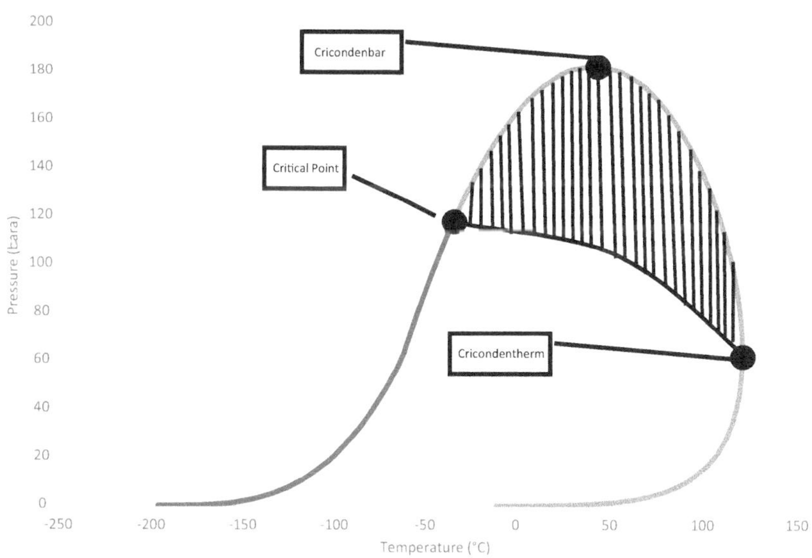

FIGURE 3.13 Typical phase diagram for a multicomponent natural gas mixture.

3.9 Conclusions

A good chemical engineer must have a strong foundation in thermodynamics!

A significant number of chemical engineering processes are characterized using thermodynamic processes, which are in turn defined by the thermodynamic state functions such as internal energy, enthalpy, entropy, and Gibbs and Helmholtz free energies. These thermodynamic state functions are derived from pressure, volume, and temperature using equations of state.

For nonpolar systems, where liquid phase activity can be neglected, phase equilibrium and component phase (and corresponding distribution or partitioning) may be estimated using equations of state. However, in cases where there are strong molecular interactions in the liquid phase, it would be necessary to use liquid activity coefficient models.

Caution must be exercised when using liquid activity models because in many cases, tuning parameters may not be available in most commercial simulations. They would need to be estimated from group contribution methods or experimentally determined. Always check the vapor pressure prediction and compare that to experimental or field observations.

When working with pure component systems, exercise caution when using PT flashes. It is not possible to determine the phase of a saturated pure component fluid by knowing the pressure and temperature alone. It is necessary to look at the pressure–volume, pressure–enthalpy, and temperature–volume diagrams to determine whether a saturated fluid is a saturated liquid, saturated vapor, or multiphase. It is absolutely critical to plot the phase envelope of the fluid under study. The phase envelope will provide a convenient visual indication of whether a fluid is a subcooled liquid, superheated gas, multiphase, critical, or supercritical. In particular, mixtures with wide boiling point ranges may exhibit retrograde behavior, where the phase behavior is counterintuitive. Having a graphical representation will allow one to flag potential issues like these.

While all commercial simulation software packages conveniently support most of the equations of state commonly used for engineering computations, it is important to remember that these equations of state have been derived from gas phase relationships. Hence, these process simulation packages will have additional corrections or empirical correlations to describe liquid densities and transport properties, such as viscosities.

Exercises

Use your process simulator and favorite equation of state–based property package to perform the following tasks:

- Generate PT, PV, and PH plots for pure $C1$ and pure $C2$. Determine the critical temperatures and pressures for $C1$ and $C2$.

- On one PT diagram, plot (overlay the following curves) (hint: you would have to copy the data from your simulator into a spreadsheet program):
  - pure *C1*
  - pure *C2*
  - 60% *C1* + 40% *C2*
  - 40% *C1* + 60% *C2*
  - 20% *C1* + 80% *C2*

  For these phase envelopes, what can you say about the

1. critical points and
2. bubble and dew point lines?

Create fluids with the condition given in Table E.1 using an equation of state property package.

TABLE E.1 Stream condition.

Component	Mole%
H_2O	0.00
Nitrogen	0.01
H_2S	4.78
CO_2	0.71
Methane	78.15
Ethane	3.41
Propane	5.92
i-Butane	1.82
n-Butane	0.94
I-Pentane	0.61
n-Pentane	0.77
n-Hexane	1.23
*C7+**	1.64
Molecular weight (kg/kg mol)	111.00
Normal boiling point (°C)	110.00
Ideal liquid density (kg/m^3)	745.38

*C7+** is a hypothetical component (refer to Chapter 2 for details on registration of new components.) with the following properties.

Perform flash calculations and determine the following pressures and temperatures of the stream:

1. Bubble point temperature at 6000 kPa
2. Dew point temperature at 6000 kPa
3. Bubble point temperature at 14,000 kPa
4. Dew point temperature at 14,000 kPa
5. Dew point pressure at 60°C

References

Campbell, J.M., 2004. Gas conditioning and processing, 8th ed., Vol. 1. BBS.

Smith, J.M., Van Ness, H.C., 1975. Introduction to chemical engineering thermodynamics, 3rd ed. McGraw-Hill.

Walas, S.M., 1985. Phase equilibria in chemical engineering. Butterworth.

Chapter 4

Simulation of recycle streams*

Dominic C.Y. Foo[1], Siewhui Chong[1,2] and Nishanth Chemmangattuvalappil[1]

[1]*University of Nottingham Malaysia, Semenyih, Selangor, Malaysia;* [2]*Current affiliation: Xodus group, Perth, Australia*

Chapter outline

Recycle system is commonly found in a process flowsheet. However, it is one of the systems that are difficult to achieve convergence, especially for novice in process simulation. In this chapter, strategies to converge recycle systems will be covered. Emphasis is placed on sequential modular (SM) approach, which is commonly used in commercial process simulation software. Chapter 1 discussed how SM approach is used to simulate recycle stream simulation briefly. More tips on recycle stream convergence are given in this chapter.

4.1 Types of recycle streams

From process simulation perspective, recycle system may be generally categorized as material or heat recycle. The former normally involves the recovery of material (e.g., unconverted feedstock) to certain process units (e.g., reactors) for further processing. One such example is shown in Fig. 4.1. When sequential modular (SM) approach is adopted in solving this system, the recycle stream will

*Simulation and other electronic files of this chapter are made available on a companion website; URL: https://www.elsevier.com/books-and-journals/book-companion/9780323901680.

Chemical Engineering Process Simulation. https://doi.org/10.1016/B978-0-323-90168-0.00004-4

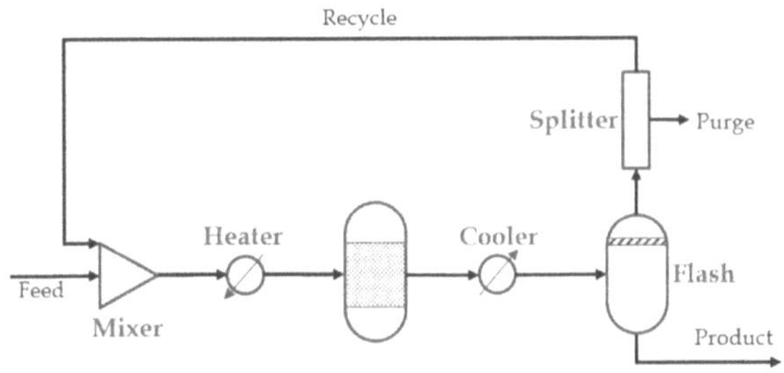

FIGURE 4.1 A material recycle system.

FIGURE 4.2 A heat recycle system.

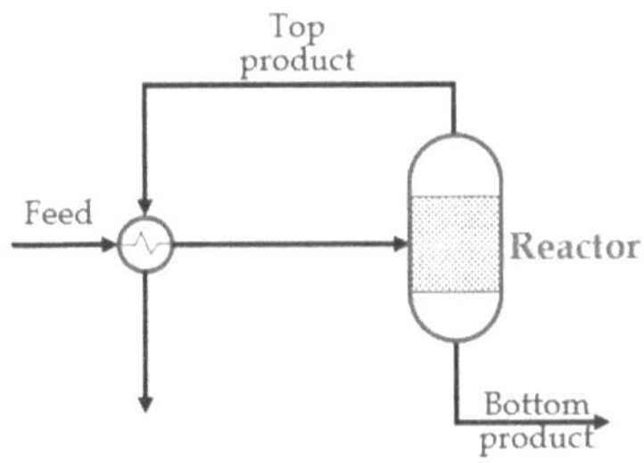

have no data for the reactor simulation to converge. As discussed in Chapter 1, *tear-stream* strategy can be used to converge this material recycle stream.

On the other hand, heat recycle system involves the recovery of energy between the process streams. Fig. 4.2 shows such a case, where the top product stream from a reactor is used to preheat the feed stream to the reactor. In this case, the top product stream contains no data to simulate the heat exchanger unit. Similarly, tear-stream concept can be used to handle this heat recycle system.

4.2 Tips in handling recycle streams

To converge recycle streams, some of the following strategies (WinSim, 2017) are very useful.

4.2.1 Analyze the flowsheet

Not all flowsheet contains a recycle system. Fig. 4.3 shows a flowsheet that has no recycle stream, even though it looks like having one. Hence, it is important to analyze the flowsheet carefully before attempting to solve a flowsheet with tear-stream strategy directly.

For the process in Fig. 4.4, if one were to simulate the process with conventional sequence, i.e., from feed stream to product stream, one will have to employ the tear-stream concept for both the recycle streams. However, if one were to employ the tear-stream concept on a commonly shared stream on both recycle systems, i.e., outlet stream of Mixer 2 (i.e., inlet for Splitter 2), only one tear stream needs to be solved to converge the entire process. This greatly enhances the speed of convergence in this simulation exercise.

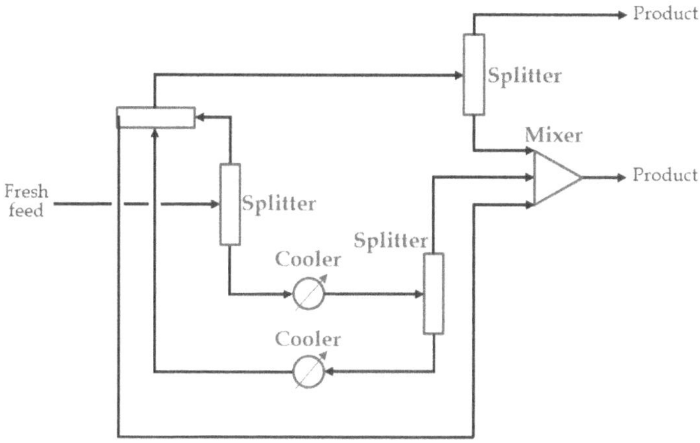

FIGURE 4.3 A flowsheet without recycle stream (WinSim, 2017).

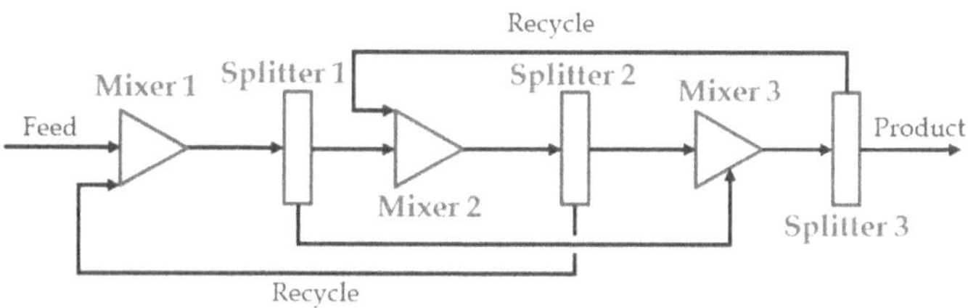

FIGURE 4.4 A flowsheet with two recycle streams (WinSim, 2017).

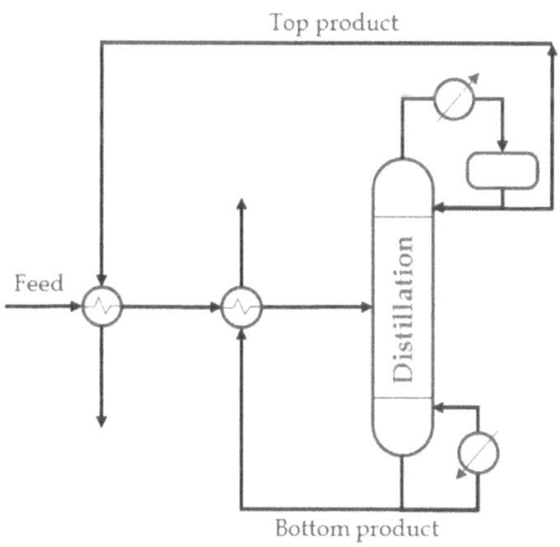

FIGURE 4.5 A distillation column with feed preheat (WinSim, 2017).

The same strategy is applicable for the process in Fig. 4.5, where the top and bottom product streams are used to preheat the feed stream of the distillation. Instead of applying the tear-stream concept on both of these energy recycle streams, one can apply the tear-stream concept on the outlet stream of the second heat exchanger (i.e., inlet to distillation), as it is the commonly shared stream between both the energy recycle streams.

4.2.2 Provide estimates for recycle streams

When applying the tear-stream concept on a stream, it is always good to supply some known data for the stream, to assist its convergence. For instance, for the distillation example in Fig. 4.5, the stream flowrate and compositions of the outlet stream from second heat exchanger (i.e., inlet stream to distillation) should stay the same as those of the feed stream. With modern process simulation software, it is often possible to copy the properties of this stream from the fresh feed stream. Besides, one may also set the pressure drop of the heat exchanger, which leaves the outlet stream with an unknown temperature. Specifying these data for the distillation inlet stream will greatly enhance the convergence of simulation.

Another example is the example in Fig. 4.6. Instead of specifying the recycle stream as tear stream, one may also specify the reactor inlet stream if its properties are known. In both examples, one notices that the tear-stream concept needs not be necessarily applied on the recycle stream; it can be applied to any stream within the recycle loop to promote faster convergence.

4.2.3 Simplify the flowsheet

Before simulating a complex unit, it is always a good practice to perform simulation on a simplified unit. A good example is the modeling of rigorous distillation column. In most simulation software, shortcut distillation model is made available to determine the basic parameters needed for a rigorous column. Hence, it is encouraged to converge a shortcut distillation model to

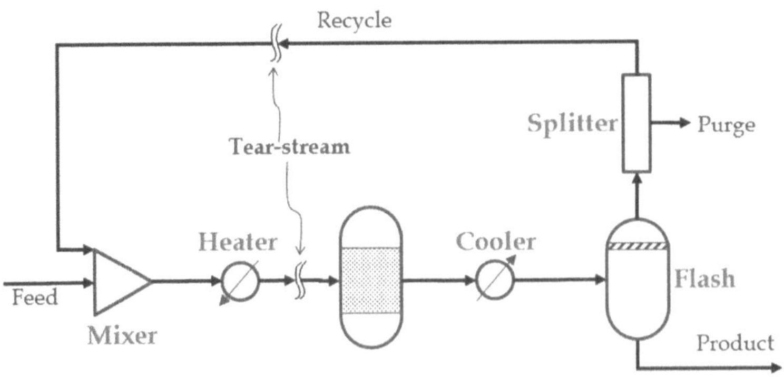

FIGURE 4.6 Specifying the tear stream. *Reproduced from Fig. 4.1.*

provide some initial guess for the rigorous model.[1] Besides, it is also a good practice to converge the rigorous distillation model as a stand-alone unit, before it is connected to a complete flowsheet (especially for cases where distillation is part of the recycle systems).

For flowsheets that contain both material and energy recycle systems that are interconnected, it is easier to converge the individual recycle system when they are decoupled. The *n*-octane case study in Example 1.1 provides a good illustration for this strategy.

It is important to remember the objective of a simulation exercise. For instance, for preliminary mass and energy balances at conceptual design stage, it is acceptable to simplify a flowsheet to achieve faster convergence. An example is shown in Fig. 4.7A, where a flash unit contains a reboiler and an internal recycle system. This unit will definitely require some iterative calculation before the unit can converge. This system can be simplified such as that in Fig. 4.7B for easier convergence.

4.2.4 Avoid overspecifying mass balance

When a purge stream is found in a material recycle system (e.g., Fig. 4.1), it is best to set the recycle fraction of the purge stream, rather than the exact material flowrate of the recycle stream. Setting an exact flowrate may prevent the convergence of the recycle stream.

For a distillation train that contains multiple columns (e.g., see Fig. 4.8), specifying the top or bottom product flowrates may be overconstraining the mass balances of the system. For these cases, apart from converging the shortcut distillation model prior to rigorous model (see earlier tip), it will be useful to specify the reflux ratio and the top/bottom product rate (WinSim, 2017).

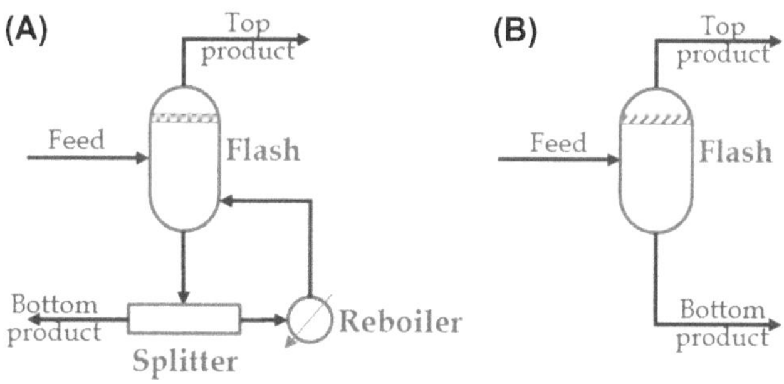

FIGURE 4.7 (A) A flash unit with reboiler and internal recycle system; (B) a simplified system without reboiler (WinSim, 2017).

1. Refer to VCM production example in Chapter 12 for better understanding.

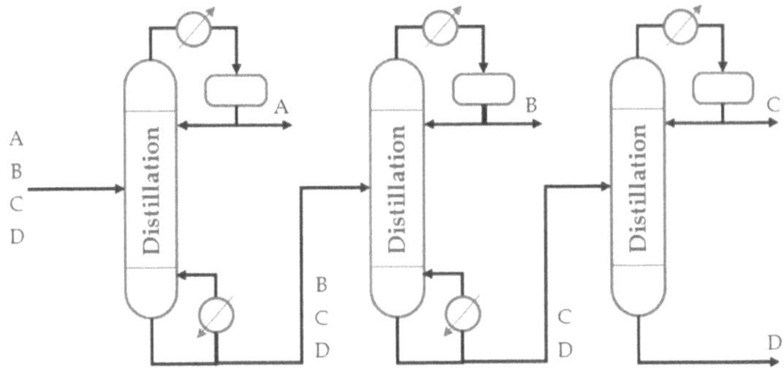

FIGURE 4.8 Distillation train with three columns.

4.2.5 Check for trapped material

When a recycle system does not converge, it is useful to check if there exists any unnecessary material buildup in the system. In general, component with medium boiling range is easily trapped. For instance, for the system in Fig. 4.9, water is condensed by the cooler, which is then sent to the heater. The heater vaporizes water into steam, which is then sent back to the cooler. This causes the water component to be trapped in the recycle loop.

In checking an unconverged recycle loop, it is necessary to check the material balance summary to see which components have the largest error. When the component flowrate leaving the recycle loop is less than that in the entering stream, it is likely that the component is being trapped in the recycle loop.

4.2.6 Increase number of iterations

Most flowsheets will converge easily within a few iterations. However, when a recycle loop is unconverged after some iterations, it will be good to increase its iterations. When the recycle loop is approaching convergence, properties of the stream (where tear-stream concept is applied) such as pressure,

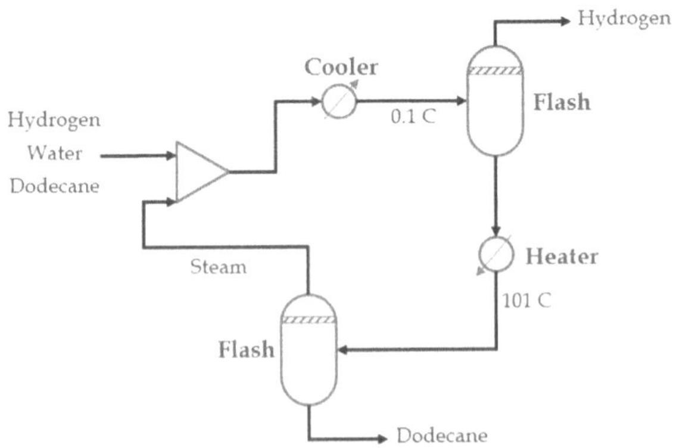

FIGURE 4.9 A recycle system where water is trapped in the recycle loop (WinSim, 2017).

temperature, flowrate, and composition may be updated to provide better guess for the recycle loop to converge (WinSim, 2017).

4.3 Recycle convergence and acceleration techniques

Chapter 1 describes how SM approach may be used to converge a recycle loop. In most cases, a tear stream is to be chosen at which a convergence solver is to be placed. The convergence solver computes the difference between the calculated and estimated values of the tear stream and then updates the estimated value with the calculated one. The simplest method to converge the recycle calculation is the *direct substitution* method (or *successive substitution*) (Smith, 2016), in which the calculated value of the variable [$G(x)$] simply updates the estimated value of the variable (x). The substitution will stop when the convergence criteria are met, which is given by Eq. (4.1) (known as the scaled residue; Smith, 2016):

$$-Tolerance \leq \frac{G(x) - x}{x} \leq Tolerance \tag{4.1}$$

There are several techniques that can be used to accelerate the convergence of a recycle calculation; the most commonly used one is called the *Wegstein acceleration* method. This method will be illustrated using Example 4.1. For highly nonlinear and interdependent equations, other acceleration methods such as dominant eigenvalue, Newton–Raphson, and Broyden's quasi-Newton methods may also be used (Smith, 2016).

Example 4.1. An isomerization process (Smith, 2016) is used to illustrate the concept of recycle convergence using SM approach. In an isomerization process, component A is converted to component B. As shown in Fig. 4.10A, the mixture from the reactor is then separated into relatively pure A, which is then recycled back to the reactor system, and relatively pure B, which is the product. Fig. 4.10B shows the block structure of the SM approach where the convergence solver is placed for the selected tear stream—the recycle stream, $S5$. The following assumptions are made (Smith, 2016):

- No by-products are formed.
- The reactor performance can be characterized by its conversion.
- The performance of the separator is characterized by the recovery of A (r_A) and B (r_B) to the recycle stream.

Material balances for components A and B are conducted for the isomerization process, which consists of eight equations and 13 variables.

At the mixer:

$$\dot{m}_{A,S2} = \dot{m}_{A,S1} + \dot{m}_{A,S5} \tag{4.2}$$

$$\dot{m}_{B,S2} = \dot{m}_{B,S1} + \dot{m}_{B,S5} \tag{4.3}$$

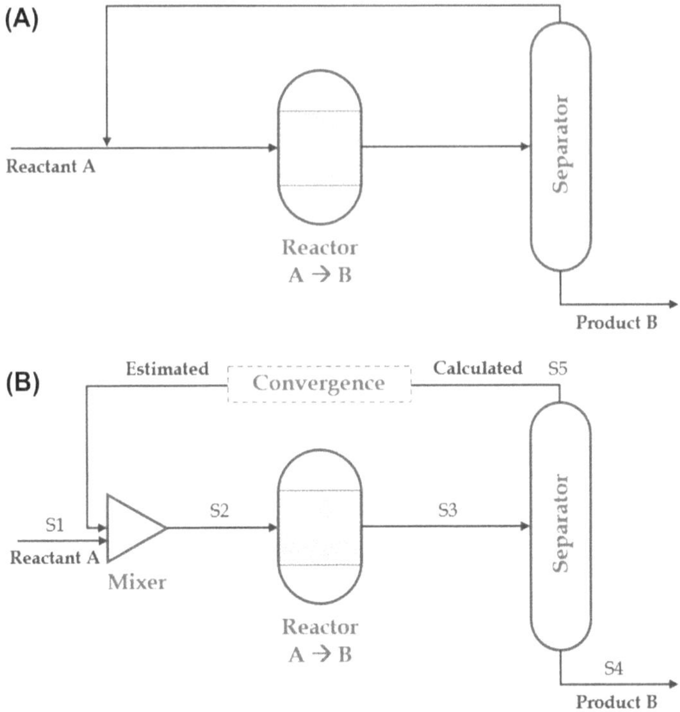

FIGURE 4.10 An isomerization process in: (A) process flowsheet; (B) block structure of sequential modular approach.

At the reactor:

$$\dot{m}_{A,S3} = \dot{m}_{A,S2}(1 - X) \tag{4.4}$$

$$\dot{m}_{B,S3} = \dot{m}_{B,S2} + \dot{m}_{A,S2}X \tag{4.5}$$

At the separator:

$$\dot{m}_{A,S4} = \dot{m}_{A,S3}(1 - r_A) \tag{4.6}$$

$$\dot{m}_{A,S5} = \dot{m}_{A,S3}(r_A) \tag{4.7}$$

$$\dot{m}_{B,S4} = \dot{m}_{B,S3}(1 - r_B) \tag{4.8}$$

$$\dot{m}_{B,S5} = \dot{m}_{B,S3}(r_B) \tag{4.9}$$

where $\overset{i,j}{\dot{m}}$ = molar flowrate of component i in stream j; X = reactor conversion of A; and r_i = fractional recovery of component i to the separator top stream.

An example of converging the recycle stream using direct substitution is demonstrated in Table 4.1. A spreadsheet is created for the calculations of

TABLE 4.1 Construction of spreadsheet for calculation of material balances using direct substitution method.

(A) Material balance for component *A*

Iteration	$\dot{m}_{A,S1}$ (fixed)	$\dot{m}_{A,S2}$ (Eq. 4.2)	$\dot{m}_{A,S3}$ (Eq. 4.4)	$\dot{m}_{A,S4}$ (Eq. 4.6)	$\dot{m}_{A,S5}$ Assumed	$\dot{m}_{A,S5}$ Calculated with (Eq. 4.7)	Scaled residue (Eq. 4.1)
1	100	150	45	2.25	50	42.75	−0.14500
2	100	142.750	42.825	2.141	42.750	40.684	−0.04833
3	100	140.684	42.205	2.110	40.684	40.095	−0.01447
4	100	140.095	42.028	2.101	40.095	39.927	−0.00419
5	100	139.927	41.978	2.099	39.927	39.879	−0.00120
6	100	139.879	41.964	2.098	39.879	39.866	−0.00034
7	100	139.866	41.960	2.098	39.866	39.862	−0.00010
8	100	139.862	41.959	2.098	39.862	39.861	−0.00003
9	**100**	**139.861**	**41.958**	**2.098**	**39.861**	**39.860**	−0.00001 (converged)

Continued

TABLE 4.1 Construction of spreadsheet for calculation of material balances using direct substitution method.—cont'd

(B) Material balance for component *B*

Iteration	$\dot{m}_{B,S1}$ (fixed)	$\dot{m}_{B,S2}$ (Eq. 4.2)	$\dot{m}_{B,S3}$ (Eq. 4.4)	$\dot{m}_{B,S4}$ (Eq. 4.6)	$\dot{m}_{B,S5}$ Assumed	$\dot{m}_{B,S5}$ Calculated with (Eq. 4.9)	Scaled residue (Eq. 4.1)
1	0	5	110.000	104.500	5	5.500	0.10000
2	0	5.500	105.425	100.154	5.500	5.271	−0.04159
3	0	5.271	103.750	98.562	5.271	5.187	−0.01589
4	0	5.187	103.254	98.091	5.187	5.163	−0.00478
5	0	5.163	103.112	97.956	5.163	5.156	−0.00138
6	0	5.156	103.071	97.917	5.156	5.154	−0.00039
7	0	5.154	103.059	97.906	5.154	5.153	−0.00011
8	0	5.153	103.056	97.903	5.153	5.153	−0.00003
9	0	**5.153**	**103.055**	**97.902**	**5.153**	**5.153**	−0.00001 (converged)

material balances for components A and B. The initial conditions $-\dot{m}_{A,S1}$, $\dot{m}_{B,S1}$, X, r_A, and r_B are known and are given as follows:

$$\dot{m}_{A,S1} = 100\ \text{kmol/h}$$

$$\dot{m}_{B,S1} = 0\ \text{kmol/h}$$

$$X = 0.7$$

$$r_A = 0.95;\ r_B = 0.05$$

Using initial estimated values of $\dot{m}_{A,S5} = 50\ \text{kmol/h}$ and $\dot{m}_{B,S5} = 5\ \text{kmol/h}$, the scaled residue is computed for the estimated and the calculated values. Iterations follow the profile as shown in Fig. 4.11A until the specified tolerance is met, for instance, 0.00001 in this case.

The major limitation of this method is the requirement of many iteration steps and that some values may fail to converge to the required tolerance (Smith, 2016). If the direct substitution method is linearized:

$$G(x) = ax + b, \tag{4.10}$$

with the slope:

$$a = \frac{G(x_k) - G(x_{k-1})}{x_k - x_{k-1}}, \tag{4.11}$$

where x_k and x_{k-1} are the calculated variable values for iterations k and $k-1$, respectively. For iteration k, the intercept can be calculated by the following:

$$b = G(x_k) - ax_k \tag{4.12}$$

Substituting Eq. (4.12) in Eq. (4.10):

$$G(x_{k+1}) = ax_{k+1} + [G(x_k) - ax_k] \tag{4.13}$$

With the required intersection:

$$G(x_{k+1}) = x_{k+1} \tag{4.14}$$

Eq. (4.13) is then updated to the following:

$$x_{k+1} = ax_{k+1} + [G(x_k) - ax_k]$$

Or

$$x_{k+1} = \left(\frac{a}{a-1}\right)x_k - \left(\frac{1}{a-1}\right)G(x_k) \tag{4.15}$$

If we define

$$q = \frac{a}{a-1}, \tag{4.16}$$

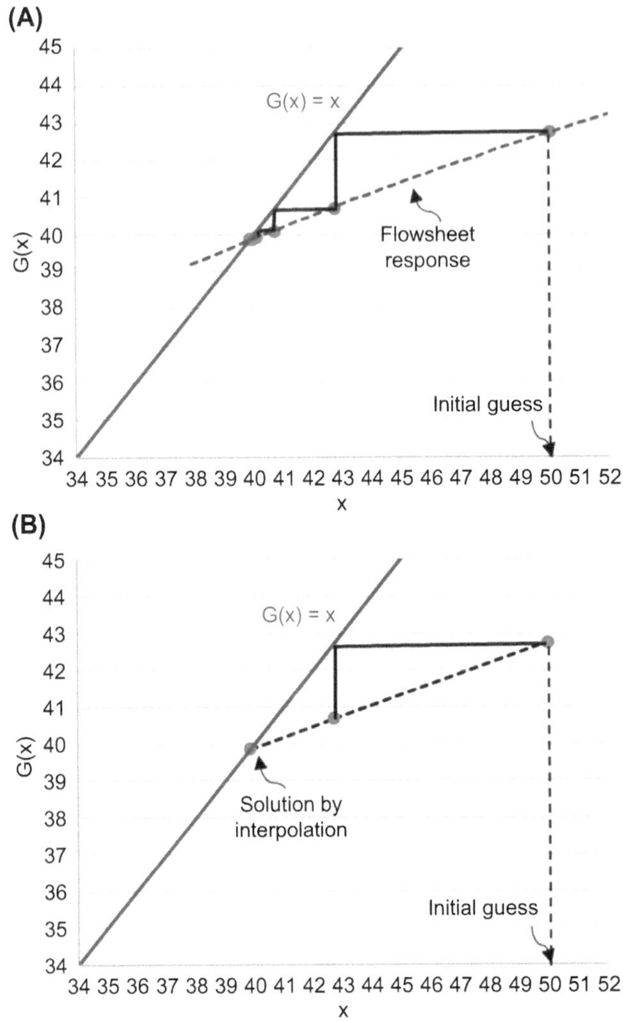

FIGURE 4.11 Convergence of recycle loops using the sequential modular approach via (A) direct substitution or (B) Wegstein method.

the final expression becomes the following:

$$x_{k+1} = x_k + (1 - q)[G(x_k) - x_k] \tag{4.17}$$

Eq. (4.17) is known as the Wegstein method (Wegstein, 1958), as interpreted in Fig. 4.11B, which can accelerate the convergence. If $q = 0$, Eq. (4.17) becomes direct substitution; if $q < 0$, acceleration of convergence of the iteration processes occurs, and if $0 < q < 1$, damping occurs. Typically, q is bound between -20 and 0 to ensure stability and reasonable rate of convergence (Seider et al., 2009). If we apply Wegstein method for the second iteration process in Table 4.1:

$$a = \frac{G(x_k) - G(x_{k-1})}{x_k - x_{k-1}}$$

$$\frac{42.75 - 40.684}{50 - 42.750}$$

$$0.2850$$

$$q = \frac{a}{a-1}$$

$$\frac{0.2850}{0.2850 - 1}$$

$$-0.3986$$

Substituting in Eq. (4.17), the new estimated value becomes the following:

$$x_{k+1} = 42.750 + (1 + 0.3986)[40.684 - 42.750] = 39.860\ \text{kmol/h}$$

which is equivalent to the ninth direct substitution iteration process in Table 4.1. Compared to direct substitution, the Wegstein method has accelerated the convergence of solutions.

Exercises

Redo the process in Example 4.1, assuming components *A* and *B* are entering at the flowrates of 150 and 10 kmol/h, respectively. Further assume that the conversion of *A* is 60% in the reactor and the fractional recovery in the separator is 0.9 for both components. With this information, perform the following tasks:

1. Perform component mass balances for ***A*** and ***B*** for each unit operation in the process.
2. By assuming molar flowrate of the mixture outlet (stream *S*2) is 400 kmol/h with 80% mole fraction of component *A* for the first iteration, use MS Excel to find the molar flowrate of components ***A*** and ***B*** in product stream (stream *S*4) and recycled stream (stream *S*5). Iteration stops when the scaled residue is smaller than 1×10^{-5}.
3. Apply Wegstein acceleration after the third iteration and calculate flowrates of the streams. Show the complete calculations and values of *q*, *a*, and the flowrates. Comment on the acceleration by comparing the results of direct substitution.
4. Determine the purity of the product stream. If the fractional recovery for component ***A*** and component ***B*** is changed to 0.8 and 0.2, respectively, what would be the effect to the purity of the product stream?

5. Assume that a by-product C is formed during the conversion of A to B and the selectivity of B over C is 4. Component C can be separated from A and B in a unit before the separator. Let the recovery of C be 90% and the remaining C will be leaving along with product B stream. With this information, reestimate the flowrates and purity of the product stream.

References

Seider, W.D., Seader, J.D., Lewin, D.R., 2009. Product & Process Design Principles: Synthesis, Analysis and Evaluation (With CD). John Wiley & Sons.

Smith, R., 2016. Chemical Process Design and Integration, second ed. John Wiley, West Sussex, England.

Wegstein, J.H., 1958. Accelerating convergence of iterative processes. Communications of the ACM 1, 9–13.

WinSim, I., 2017. DESIGN II for windows training guide [Online]. Available: www.winsim.com.

Further reading

Schad, R.C., 1994. Don't let recycle streams stymie your simulation. Chemical Engineering Progress 90 (12), 68–76.

Part II

UniSim design

Chapter 5

Basics of process simulation with UniSim design*

Dominic C.Y. Foo

University of Nottingham Malaysia Campus, Semenyih, Selangor, Malaysia

Chapter outline

This chapter aims to provide a step-by-step guide in simulating an integrated process flowsheet using UniSim design (www.honeywellprocess.com). The concept of simulation is based on *sequential modular* approach and follows the onion model for flowsheet synthesis (see Chapter 1 for details). A simple example involving the production of *n*-octane (C_8H_{18}) is demonstrated, with detailed descriptions given in Example 1.1.

5.1 Example on *n*-octane production

In simulating the integrated flowsheet of *n*-octane production, it will be good to follow the concept of onion model,[1] where simulation starts at the center of the onion and moves outward. In general, the simulation can be carried out in the following stages:

1. Basic simulation setup—This stage involves the setting up of basic information needed for simulation, which includes registration of components, thermodynamic model, and reaction stoichiometry.

*Simulation and other electronic files of this chapter are made available on a companion website; URL: https://www.elsevier.com/books-and-journals/book-companion/9780323901680.

1. See detailed discussion in Chapter 1.

Chemical Engineering Process Simulation. https://doi.org/10.1016/B978-0-323-90168-0.00007-X

2. Reactor system—This involves the setting up of the basic structure of simulation, as well as the selected reactor model.
3. Separation units—This stage involves the simulation of a distillation column.
4. Modeling of recycle system—This involves the simulation of purge stream, compressor, and heat recovery system in the recycle loop.

The basic simulation setup is to be carried out in the *Basis Environments* of UniSim design, while the other stages are carried out in the *Simulation Environments*. The individual stages are discussed in the following subsections.

Tips: Component selection with SimName allows faster identification of necessary components.

5.2 Stage 1: basic simulation setup

First, it is necessary to select the components and thermodynamic model for the process. For UniSim design, these steps were performed using the Simulation Basis Manager in the Basis Environments (Fig. 5.1). All components involved in the simulation, including the reactants, inerts, and the product (see Table 5.1), are to be selected from the component database (see detailed steps for component selection in Fig. 5.1).

Similarly, thermodynamic model (or "Fluid Package" in UniSim design terminology) for the flowsheet is also selected from the database using the

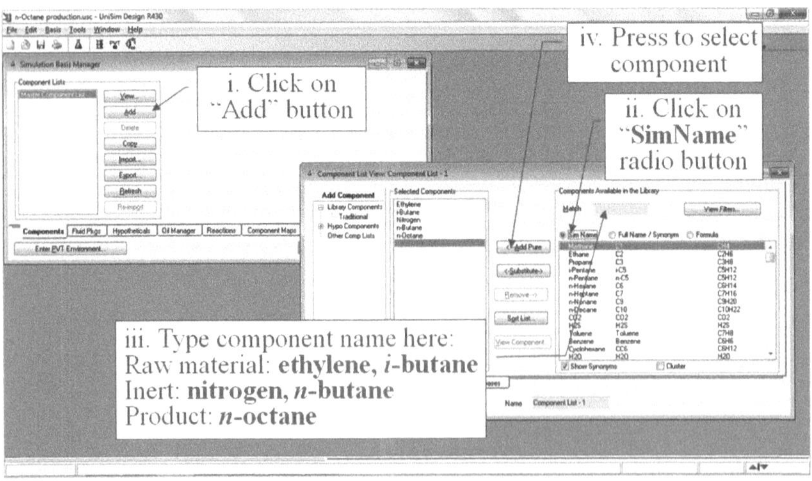

FIGURE 5.1 Component selection using Simulation Basis Manager (basis environments).

TABLE 5.1 Components needed for simulation flowsheet.

Components	Role
Ethylene (C_2H_6)	Reactant
i-Butane (*i*-C_4H_{10})	Reactant
n-Octane (*n*-C_8H_{18})	Product
Nitrogen (N_2)	Inert
n-Butane (*n*-C_4H_{10})	Inert

Simulation Basis Manager, however, in the "Fluid Pkgs" tab (see detailed steps in Fig. 5.2). For this process, Peng–Robinson model is to be used.

Next, we move to define the reaction specification for the simulation flowsheet (details given in Table 5.2). This is also carried out using the Simulation Basis Manager, at the "Reaction" tab (see Fig. 5.3). For this example, a simple conversion reaction is modeled. Note that conversion reactor model may be used for preliminary design, when the main purpose of the simulation is to perform basic mass and energy balances. For more detailed modeling, other types of reaction model should be used.[2] Hence, the basic

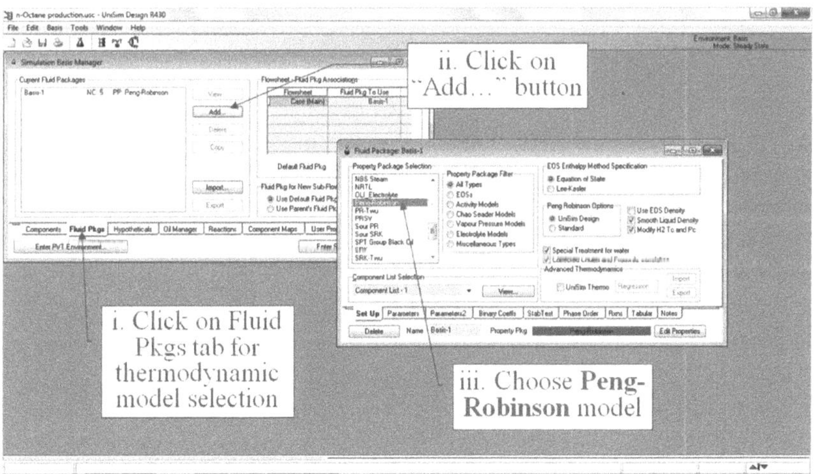

FIGURE 5.2 Thermodynamic model selection using Simulation Basis Manager.

2. See the use of continuous stirred tank reactor (CSTR) and plug flow reactor (PFR) models in Chapter 12.

TABLE 5.2 Specifications for conversion reaction.

Specifications	Details
Reaction type	Conversion
Conversion	98%
Limiting reactant	Ethylene

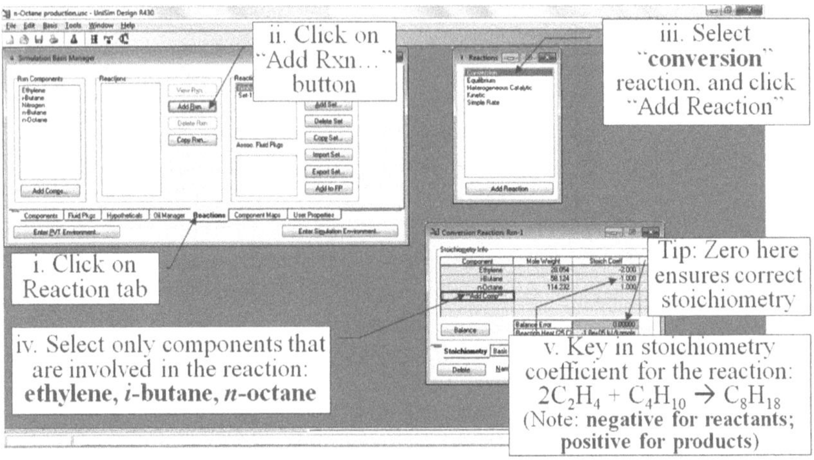

FIGURE 5.3 Specifying reaction stoichiometry ("Stoichiometry" tab).

information for reaction stoichiometry and conversion is needed. Note that reaction stoichiometry has to be specified prior to the specification of conversion rate. Fig. 5.3 shows the detailed step in specifying the reaction stoichiometry in the "Stoichiometry" tab. Next, the limiting reactant (i.e., ethylene for this case) and its conversion rate (98%) are defined in the "Basis" tab, following the detailed steps in Fig. 5.4.

Tips: It is always useful to select all components needed for the entire simulation flowsheet, before a thermodynamic model is selected. Doing this avoids the situation where the earlier selected thermodynamic model does not suit a component that would be selected at a later stage.

Once the reaction details are specified, we proceed to assign a thermodynamic package (or fluid package—FP) for this reaction. For this case, the earlier specified thermodynamic model for the flowsheet, i.e., Peng–Robinson, is to be

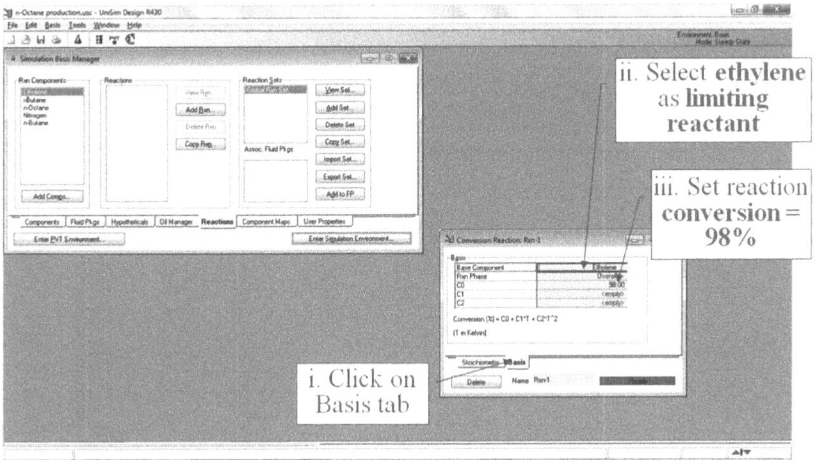

FIGURE 5.4 Specifying limiting reactant and its conversion rate ("Basis" tab).

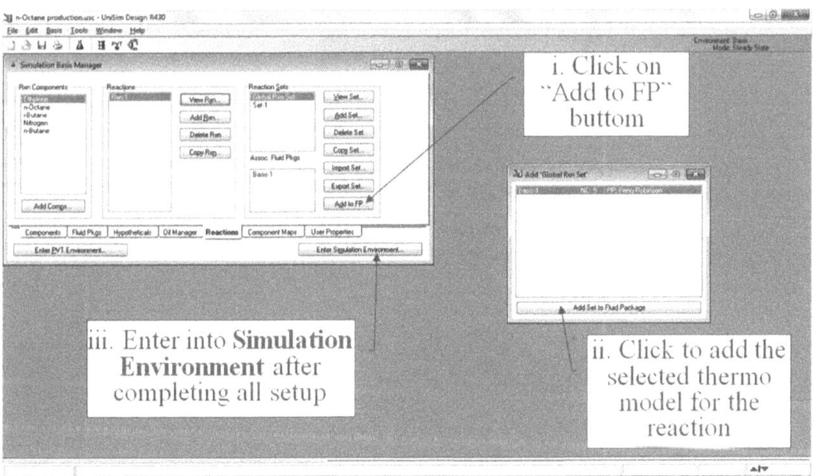

FIGURE 5.5 Associating thermodynamic model for reaction set.

used. Hence, it is associated with this reaction (see detailed steps in Fig. 5.5). In other words, the thermodynamic model will estimate the process condition of all the components during the reaction.

Tips: A good practice in using any computer software is to save the file regularly.

Once the basic information for the simulation flowsheet is specified, we may proceed to the Simulation Environments to perform modeling of the unit operations. To enter the Simulation Environments, we shall press the "Enter Simulation Environment" button on the Simulation Basis Manager (Fig. 5.5).

Tips: To edit basic information of the flowsheet, one can switch back to the Basis Environments by pressing the shake flask icon on the toolbar.

5.3 Stage 2: modeling of reactor

The Simulation Environments of UniSim consist of main flowsheet, subflowsheet, and column subflowsheet environments. For the *n*-octane production example, only the main flowsheet is used. We next move on to configure the simulation flowsheet for *n*-octane production. The topology of the flowsheet is to be built on the simulation interface called the "PFD" in UniSim design. Different types of reactor models exist within UniSim design, e.g., continuous stirred tank reactor, plug flow reactor, and other general reactor models (e.g., Gibbs, equilibrium, conversion reactors, etc.). For the *n*-octane production example, we shall utilize the "Conversion Reactor" model. Note that this reaction model should only be used for preliminary flowsheet development, where basic mass and energy balances are to be sorted. Other rigorous reactor models should be utilized for equipment modeling and design. Detailed steps to draw the flowsheet on the PFD are shown in Fig. 5.6, where the conversion reactor consists of two inlet (Streams 1 and Q-101) and two

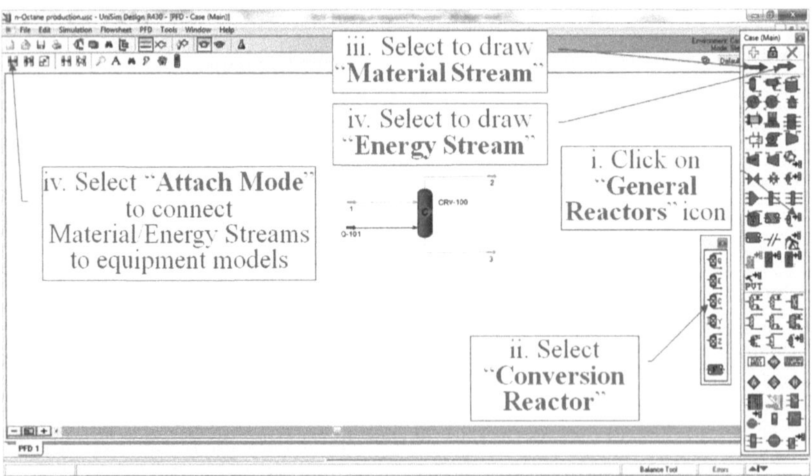

FIGURE 5.6 Construction of flowsheet topology on PFD (Simulation Environments).

outlet streams (Streams 2 and 3). Note that Streams 1, 2, and 3 are the actual process streams that consist of material (termed as Material Stream in UniSim design), while Stream Q-101 is actually virtual stream that is used for performing heat balances.

Tips: If you do not see the Object Palette on the PFD (where the unit operation models are found), press the F4 button on the keyboard to launch it.

Tips: You may edit the background color of the PFD, model icons, or text font by visiting Tools/Preferences ("Resources" tab).

We then proceed to define reactor feed stream properties. Component flowrates, temperature (T), and pressure (P) of the stream are given in Table 5.3. Detailed steps for defining stream properties are shown in Fig. 5.7.

TABLE 5.3 Specifications for process feed stream (Foo et al., 2005).

Components	Flowrate (kg mol/h)	Condition
Ethylene (C_2H_6)	20	
i-Butane (*i*-C_4H_{10})	10	T: 93°C
Nitrogen (N_2)	0.1	P: 20 psia
n-Butane (*n*-C_4H_{10})	0.5	

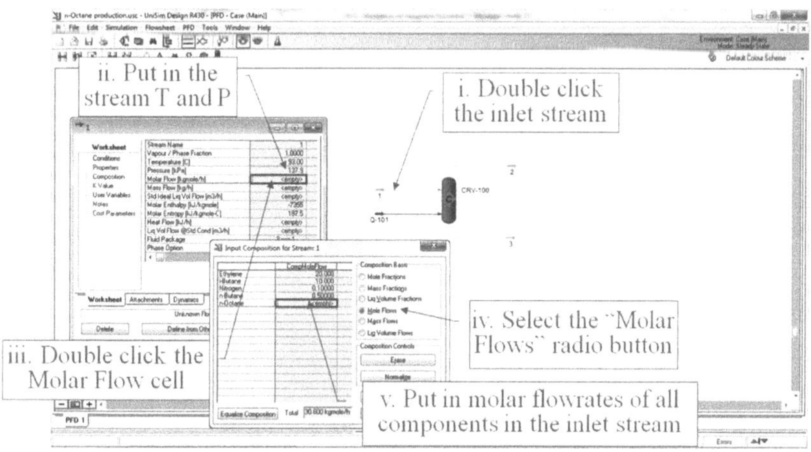

FIGURE 5.7 Defining feed stream properties.

TABLE 5.4 Specifications for reactor.

Equipment	Specifications
Reactor	Delta P: 5 psi
	Operational mode: isothermal

Next, we proceed to provide specifications for the reactor model (given in Table 5.4). To specify pressure drop (known as "Delta P" in UniSim design terminology) and reaction set for UniSim design, one would make use of the "Design" and "Reactions" tabs of the reactor model interface (see detailed steps in Fig. 5.8). The isothermal mode of the reactor is specified by setting the temperature for one of the outlet streams of the reactor (Stream 3 for this case), in the "Worksheet" tab (see Fig. 5.9). Once this is specified, the reactor model changes its color into black, indicating that the simulation model has been converged. Note that UniSim design software is configured to perform simulation once the necessary data are sufficient. For instance, in Fig. 5.9, the flowsheet is converged once all necessary data for the Conversion Reactor model are complete (i.e., when Stream 3 temperature is specified).

Tips: Red icon and light blue streams indicate that the flowsheet has not been converged, while black icon and royal blue streams indicate a converged flowsheet.

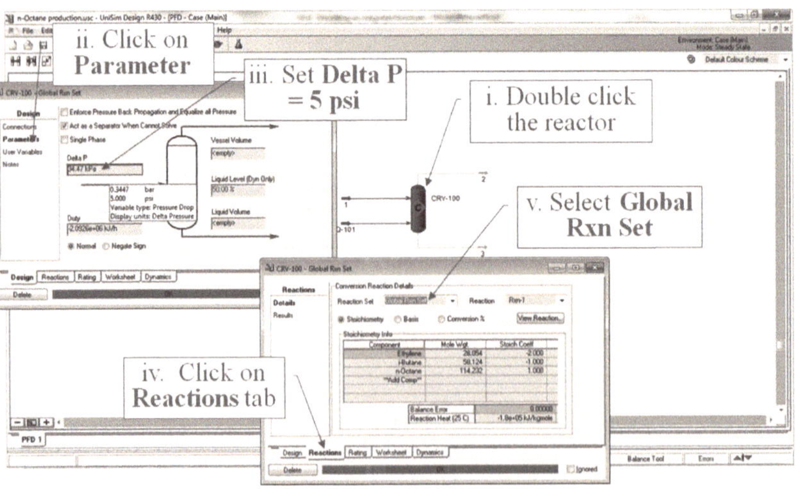

FIGURE 5.8 Specifying pressure drop and reaction set for conversion reactor.

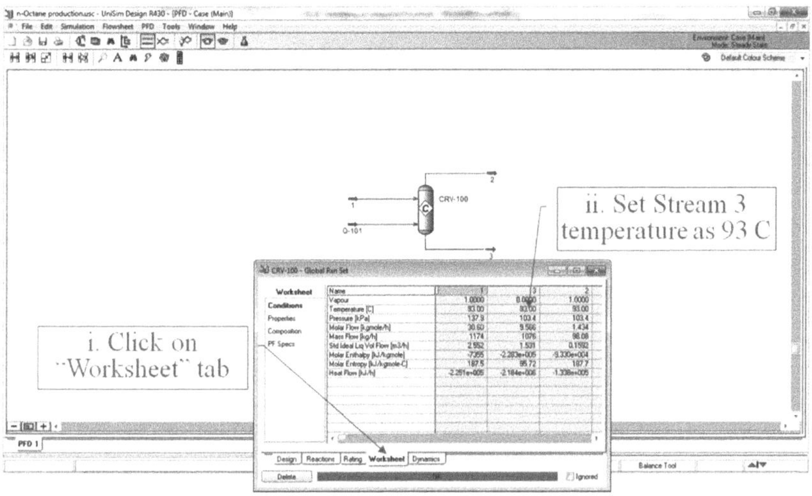

FIGURE 5.9 A converged conversion reactor model, after Stream 3 temperature is specified.

We then move on to display the simulation results. A convenient way of displaying the simulation results in UniSim design is via the *Workbook*. Fig. 5.10 shows the detailed steps in displaying molar flowrates of all components on the Workbook. One may also insert the *Workbook Table* within the PFD. This is illustrated with Fig. 5.11, where material and energy streams are displayed (one may also choose to display the stream compositions).

Tips: Most commercial simulation software requires users to perform simulation model solving manually.

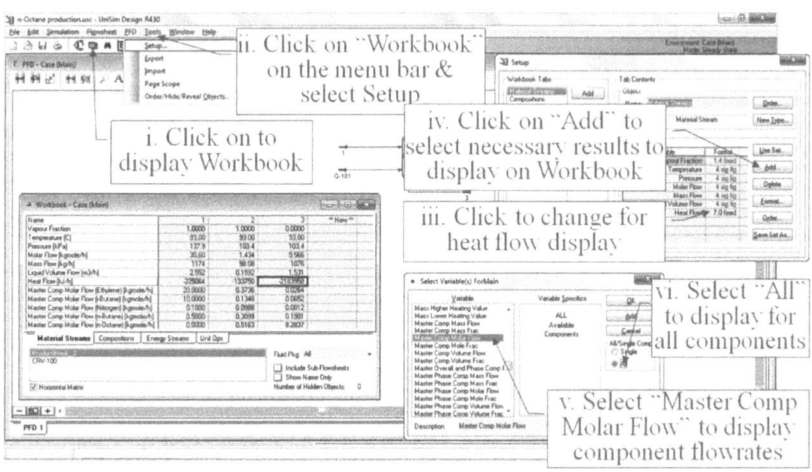

FIGURE 5.10 Displaying results with workbook and steps to display component molar flowrates.

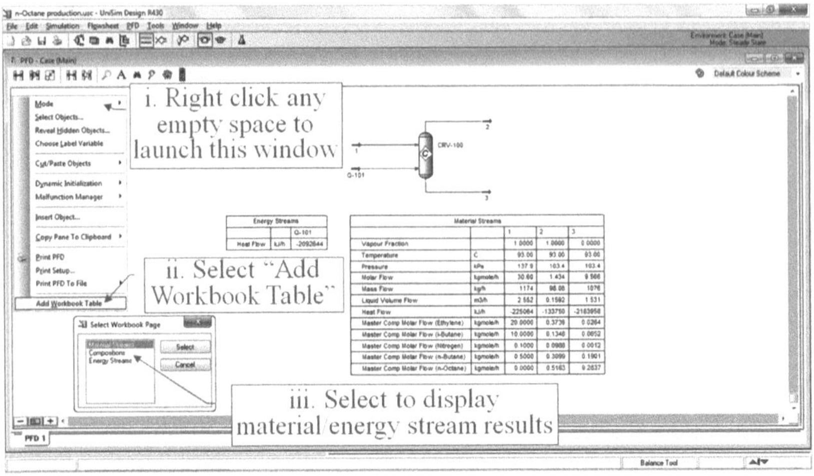

FIGURE 5.11 Adding workbook table to the PFD.

The energy stream table in Fig. 5.11 indicates that the energy stream of the reactor (Q-101) has an enthalpy of -2.09×10^6 kJ/h, indicating that the conversion of *n*-octane is an exothermic reaction. In other words, cooling utility is needed for heat removal so to operate this reactor in isothermal condition.

5.4 Stage 3: modeling of separation unit

In this stage, we shall model the only separation unit of the flowsheet, i.e., the distillation column. The simulation task is performed using the "shortcut distillation" model. Note that this distillation model is based on the Fenske–Underwood–Gilliland model, which is useful for preliminary flowsheet development. The parameters obtained from shortcut distillation model can be used as initial estimates in the rigorous distillation model, which performs stage-by-stage calculations.[3] To construct the topology of the flowsheet, one may refer to Fig. 5.12 for the alternative steps in connecting the material and energy streams. Fig. 5.12 also shows that the column is set to operate with partial condenser, where the column top stream exists in vapor form.

Tips: You may use the following hot keys to display the important stream properties at the PFD: pressure (SHIFT P), temperature (SHIFT T), mass flowrates (SHIFT M), and molar flowrates (SHIFT F). See other hot keys under the Help menu of the main screen.

---

3. See Chapter 12 for an example on the use of shortcut distillation model for generating data for rigorous distillation model on VCM production.

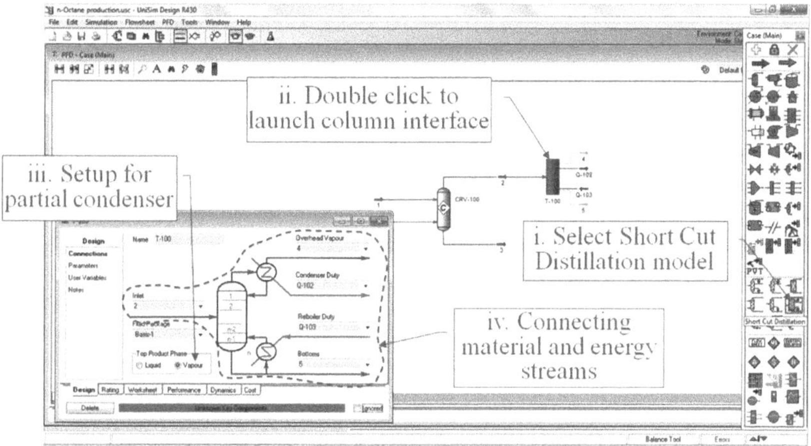

FIGURE 5.12 Adding shortcut distillation model to the PFD.

TABLE 5.5 Specifications for distillation model.

Equipment	Specifications
Distillation	Condenser type: partial condenser
	Light key in bottom stream: Ethylene (0.0001 mol fraction)
	Heavy key in distillate stream: *n*-octane (0.0500 mol fraction)
	Top P: 15 psia
	Bottom P: 25 psia
	External reflux ratio: 1

Once the distillation is added to the flowsheet, with its streams connected, we move on to provide specifications for the distillation model (given in Table 5.5). Detailed steps to provide specifications for the distillation model are shown in Fig. 5.13. The latter also shows the converged distillation model once all specifications are provided. We next move on to display the simulation results using the Workbook Table (Fig. 5.13).

5.5 Stage 4: modeling of recycle system

As discussed in Chapters 1 and 4, recycle simulation may be further classified as *material* and *energy recycle systems*. For the case of *n*-octane production,

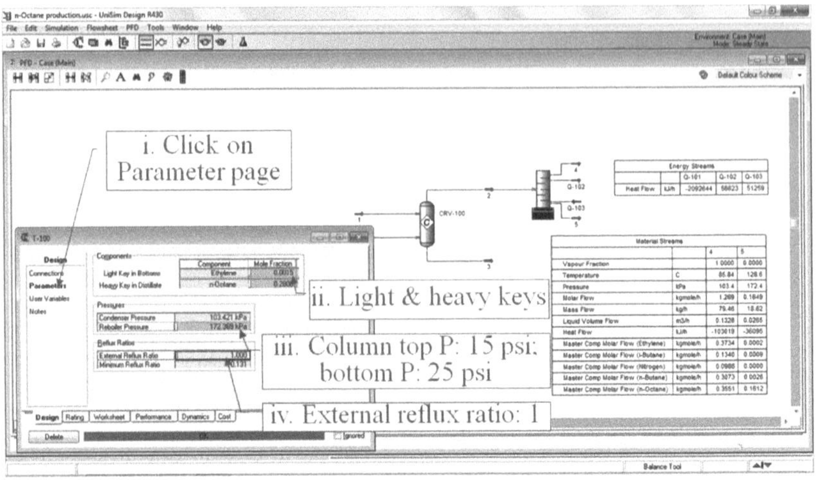

FIGURE 5.13 Specifications for the shortcut distillation model.

material recycle corresponds to the recycling of the unconverted raw material to the reactor (Fig. 5.13 indicates that the distillate stream contains some unconverted ethylene and *i*-butane), while the energy recycle stream corresponds to the heat recovery system. Example 1.1 details out the use of sequential modular approach to converge these recycle loops. Because both of these material and energy recycle systems are connected, it is easier to decouple them for the ease of convergence. This is done by replacing the process-to-process heat exchanger in the heat recovery system with equivalent heater and cooler, so that the cooling requirement of the material recycle stream is taken care by the cooler, while the heating of the fresh feed stream is provided by a heater. Once the material recycle system is converged, the cooler is replaced by the .process-to-process heat exchanger to converge the energy recycle system. Doing this allows the flowsheet to be converged easily. This is demonstrated in the following subsections.

5.5.1 Material recycle system

As described in Example 1.1, the material recycle system involves purge stream unit, compressor, and cooler (see Fig. 1.10). At the purge stream unit, 10% of the distillation top stream is purged, while the remaining is recycled. The compressor and a cooler are then used to adjust the pressure and temperature of the recycle stream to match those of the reactor. Specifications for these units are given in Table 5.6.

To model the purge unit, a stream splitting model (called the "Tee" in UniSim design) is utilized. Detailed steps to connect the Tee model (to distillation top stream) and to provide its model specifications are shown in

TABLE 5.6 Specifications for units in the material recycle system.

Equipment	Specifications
Purge unit	Flow ratio for recycle stream: 0.9
Compressor	Outlet P: 22 psia
Cooler	Outlet T: 93°C
	Delta P: 2 psi

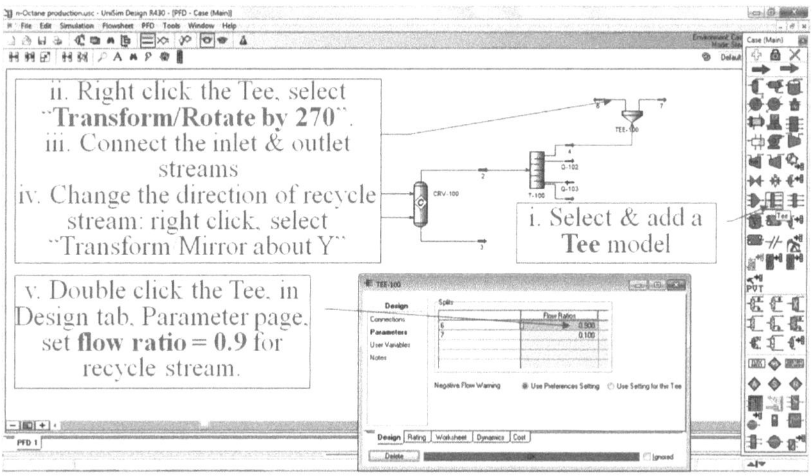

FIGURE 5.14 Steps to connect the purge stream unit and its specifications.

Fig. 5.14. As mentioned, 10% of the distillation top stream is purged. Hence, the flow ratio for recycle stream (corresponds to Stream 6 in Fig. 5.14) is set to 0.9.

Because the recycle stream has a pressure of 15 psia (verify this from your flowsheet converged earlier), which is lower than the operating pressure of the reactor, a compressor is added to raise its pressure to 22 psia. Detailed steps to connect the compressor model and to provide its specifications are shown in Fig. 5.15.

Outlet stream of the compressor has a temperature of 98.9°C (see Fig. 5.15), which is higher than the reactor operating temperature. A cooler unit is next added to reduce the temperature to 93°C. Fig. 5.16 shows the detailed steps to connect the cooler model and to provide its specifications.

With both pressure and temperature of the recycle stream adjusted to match those of the reactor, the recycle stream can now be connected to the reactor. To

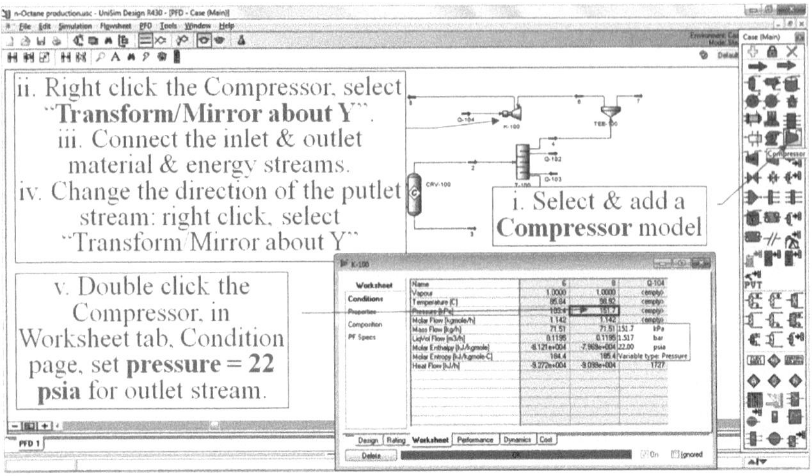

FIGURE 5.15 Steps to connect the compressor unit and its specifications.

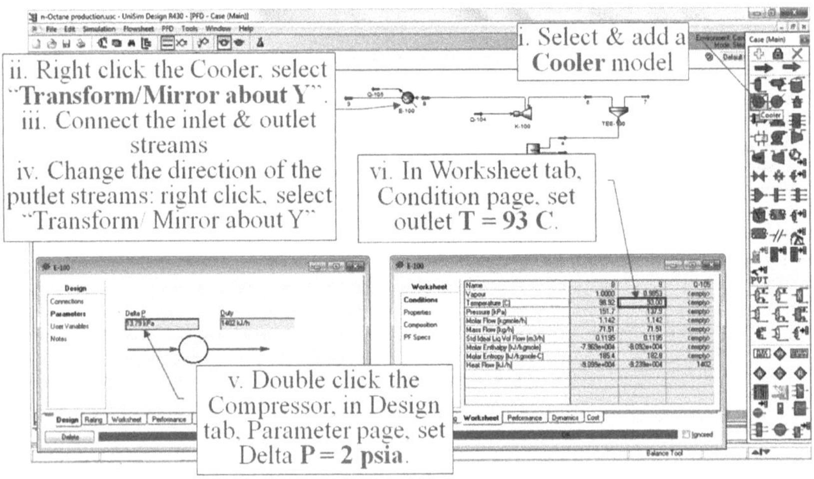

FIGURE 5.16 Steps to connect the cooler unit and its specifications.

converge this material recycle stream, we can make use of a useful model in UniSim design—the "Recycle" unit. The latter facilitates the convergence of a recycle stream following the "tear stream" concept.[4] Detailed steps in configuring the "Recycle" unit are shown in Fig. 5.17. Note that the Recycle unit shows a yellow outline when the recycle stream is first connected to the reactor. This means that some parameters are not converged after 20 rounds of

4. See Chapter 4 for detailed discussion on the convergence of recycle stream simulation.

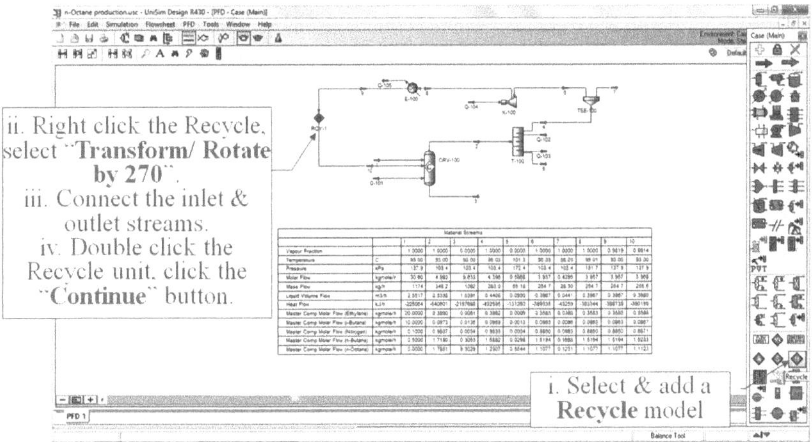

FIGURE 5.17 Steps to converge the material recycle stream with recycle unit.

iteration (default setting in UniSim design). Hence, more iterations are needed to ensure all parameters are converged completely[5]; this is done by pressing the "Continue" button in its Connections page. The simulation results are also displayed in the Workbook Table in Fig. 5.17.

5.5.2 Energy recycle system

As discussed in Example 1.1, the process-to-process heat exchanger of the *n*-octane case was replaced by a pair of heater and cooler to decouple the material and energy recycle systems, to facilitate flowsheet converge. After the material recycle system is converged, we next proceed to converge the energy recycle stream. The convergence of the energy recycle stream is to be done using the "tear stream" concept, i.e., without the use of the Recycle unit. Specifications for heat exchanger and heater in the energy recycle system are given in Table 5.7.

In earlier stage, it has been assumed that the fresh feed stream is available at 93°C (see Table 5.3). This assumption is now relaxed. A heater is added to raise the temperature of the fresh feed stream from 30°C. Detailed steps to do so are given in Fig. 5.18. The simulated results indicate that the heater requires a total heating duty of 131 MJ/h (indicated by energy stream Q-106), while 5.4 MJ/h of energy needs to be removed by the cooler (indicated by energy stream Q-105).

Fig. 5.19 shows a *temperature–enthalpy plot*[6] for the streams undergoing heating and cooling in the Heater and Cooler. As shown, the temperature

5. See Section 4.2 for tips in handling recycle streams.
6. See Example 1.1 for more details.

TABLE 5.7 Specifications for units in the energy recycle system.

Equipment	Specifications
Heat exchanger	Delta P: 2 psi (tube side)
	Delta P: 2 psi (shell side)
Heater	Outlet T: 93°C
	Delta P: 2 psi

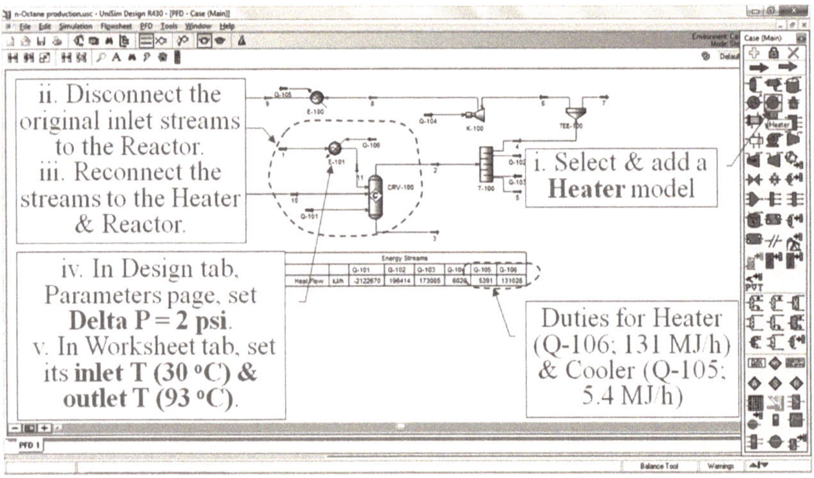

FIGURE 5.18 Determination of heating and cooling duties.

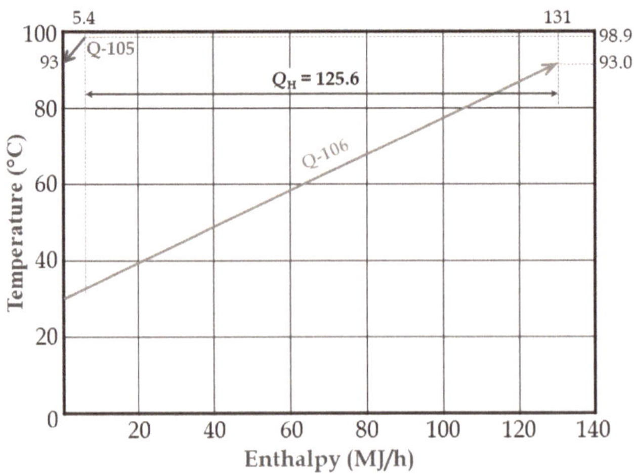

FIGURE 5.19 Temperature–enthalpy plot for heat recovery system.

profiles of the Cooler (Q-105—the material recycle stream) are higher than those of the Heater (Q-106—fresh feed). Hence, energy released from the heater can be completely recovered to the cold stream. In other words, part of the heating requirement of the Heater (5.4 MJ/h) is to be fulfilled by the cooling duty of the Cooler, through a process-to-process heat exchanger. The remaining heating duty of the cold stream (Q_H = 125.6 MJ/h) is to be supplied by the Heater, as shown in Fig. 5.19.

With the heating and cooling requirements identified, the process-to-process heat exchanger will then be included in the PFD. The "Heat Exchanger" model is utilized and added to replace the Cooler model. Because the Recycle model is not utilized in this case, we shall create a *tear stream* for the energy recycle system, for the stream connecting the Heat Exchanger and Heater. Detailed steps for doing so are shown in Fig. 5.20. Note that the pressure of the fresh feed stream has been revised to account for pressure drop across the Heat Exchanger and Heater. Note also that the flowsheet is unconverged at this stage.

We next proceed to provide the missing parameters to converge the flowsheet. These include pressure drop (Delta P) for the Heat Exchanger (see Fig. 5.21), as well as the estimated values for the tear stream. Because this stream is essentially the same fresh feed stream that enters the Heat Exchanger at the shell side, their parameters are very similar (except that with different temperature). A convenient way to define the tear stream condition is to refer that with a source stream,[7] with detailed procedure shown in Fig. 5.21. Once the tear stream is specified, the "open loop" flowsheet is converged.

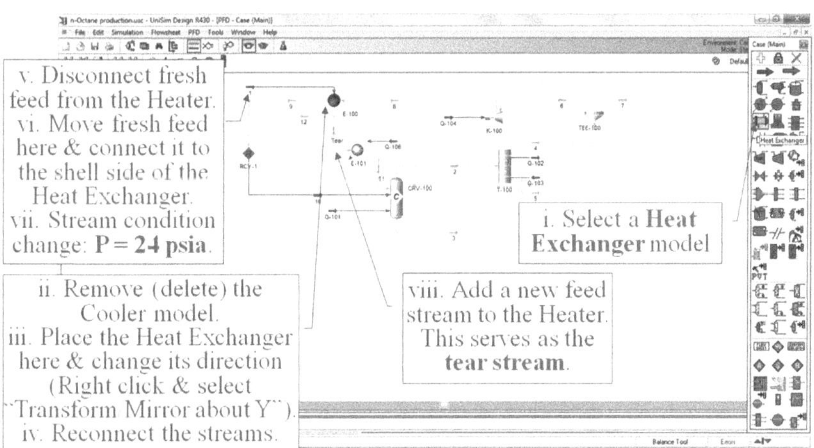

FIGURE 5.20 Introducing a tear stream for energy recycle system.

7. See Section 4.2 for tips in handling recycle streams.

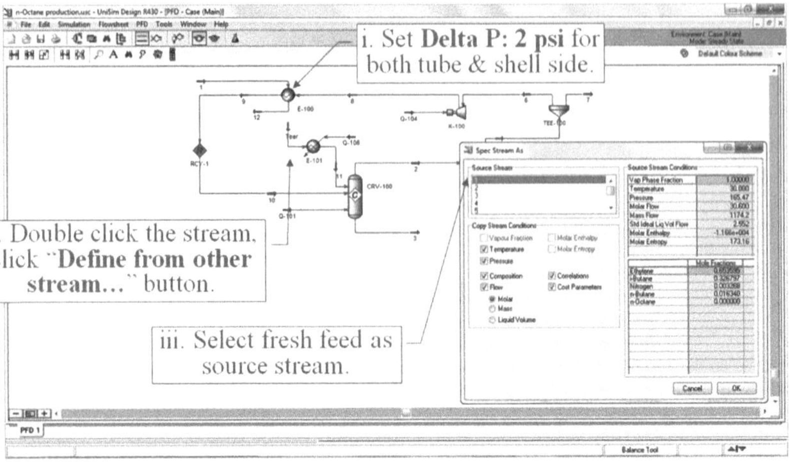

FIGURE 5.21 Specifying the tear stream to converge the "open loop" energy recycle system.

In the final step, the tear stream is removed and the outlet stream from the Heat Exchanger is connected to the Heater. A converged "close loop" flowsheet is resulted and is shown in Fig. 5.22.[8] The material stream conditions are shown using the Workbook Table in Fig. 5.22.

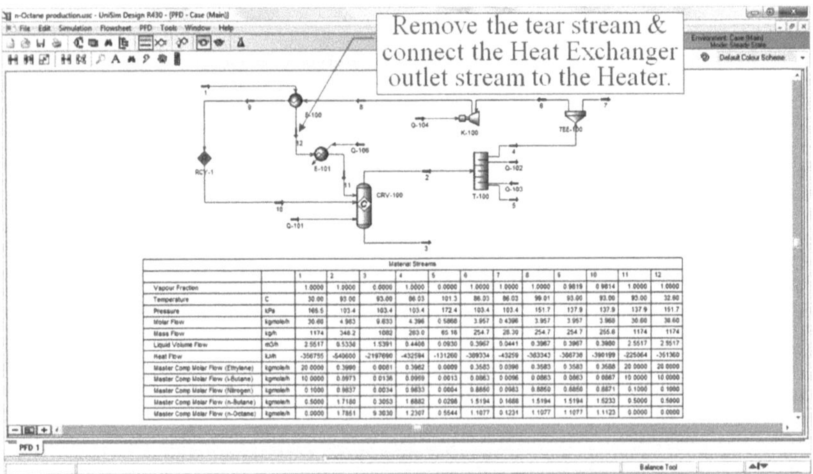

FIGURE 5.22 A converged flowsheet for *n*-octane production process.

8. Readers should verify the heat recovery targets have been materialized as in Fig. 5.19.

5.6 Conclusions

The *n*-Octane production in Example 1.1 is simulated using UniSim design, using the concept of sequential modular approach, guided by onion model. Material and energy recycle systems are handled efficiently when they are decoupled. Note that unit models used in this example are meant for preliminary design. Rigorous models should be used for more detailed engineering design.

Exercises

1. Fig. E.1 shows the "synthesis loop" for methanol (CH_3OH), in which a mixture of carbon dioxide (CO_2) and hydrogen (H_2) is reacted to form methanol product at high pressure. The reaction stoichiometry is given as follows:

$$CO_2 + 3H_2 \leftrightarrows CH_3OH + H_2O$$

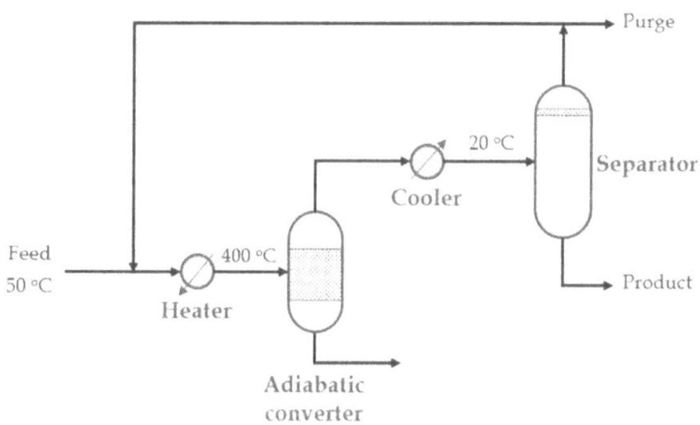

FIGURE E.1 Methanol synthesis loop (Seider et al., 2009).

The feed specification is given in Table E.1. As shown, the synthesis gas consists of mainly hydrogen and carbon dioxide, but with traces of inert gases.
Additional specifications for the process are given as follows:

- Thermodynamic model: SRK
- Pressure drops in all units are neglected
- The adiabatic converter can be approximated as a Conversion Reactor, operated adiabatically. Reactor conversion is set to 30%, with CO_2 being the limiting reactant.
- The separator may be approximated by a flash unit.
- Flow ratio of the purge stream is set to 0.02.

TABLE E.1 Feed stream specification (methanol synthesis loop).

Parameter	Value
Temperature (°C)	50
Pressure (MPa)	5, 7.5, 10
Flowrate (kg mol/h)	1000
Composition (mol %)	
Hydrogen	74.85
Carbon dioxide	24.95
Methane (inert)	0.1
Argon (inert)	0.1

Determine the following:

a. Molar flowrate and mole fraction of the product and purge streams.
b. The effect of purge ratio (range: 0.02–0.08) on the duties of Heater and Cooler.
c. In normal practice, an equilibrium reactor model is used to model a reversible reaction like this case. Justify why in the above case, the reaction can be approximated with an irreversible reaction, modeled by a conversion reactor model?

Fig. E.2 shows an ammonia production process, where a mixture of nitrogen (N_2) and hydrogen (H_2) is reacted to form the ammonia (NH_3) product at high pressure, with the following stoichiometry:

$$N_2 + 3H_2 \leftrightarrows 2NH_3$$

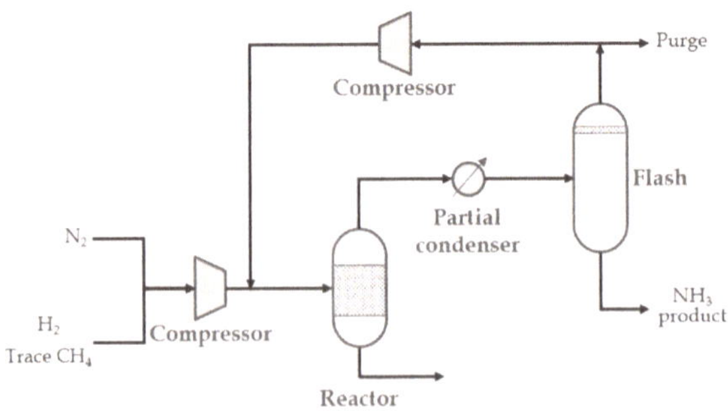

FIGURE E.2 Ammonia production (Seider et al., 2009).

The feed gas to the process is largely composed of N_2 and H_2, but with traces of inert CH_4, as shown in Table E.2.

Additional specifications for the process are given as follows:

- Thermodynamic model: Chao–Seader
- Both compressors are used to adjust stream pressure according to the reactor operating pressure.
- Conversion reactor model is used, to be operated isothermally at 500°C and 400 bar. Nitrogen is set as the limiting reactant, with a conversion of 30%. Pressure drop is ignored for the reactor.
- Partial condenser is approximated by a cooler, to cool the reactor effluent to −33°C; ignore its pressure drop.
- Model the flash vessel with a separator, with pressure drop (Delta P) of 250 bar.
- Flow ratio of the purge stream is set to 0.04.

Determine the following:

a. Molar flowrate and mole fraction of the product and purge streams.
b. The effect of purge ratio (range: 0.02–0.08) on the duties of the compressor.
c. In normal practice, an equilibrium reactor model is used to model a reversible reaction like this case. Justify why in the above case, the reaction can be approximated with an irreversible reaction, modeled by a conversion reactor model?

TABLE E.2 Feed stream specification (ammonia production).

Parameter	N_2 stream	H_2 stream
Temperature (°C)	25	25
Pressure (bar)	1	1
Flowrate (kg mol/h)		
Nitrogen	24.0	
Hydrogen		74.3
Methane (inert)		1.1

References

Foo, D.C.Y., Manan, Z.A., Selvan, M., McGuire, M.L., October, October 2005. Integrate process simulation and process synthesis. Chemical Engineering Progress 101 (10), 25–29.

Seider, W.D., Seader, J.D., Lewin, D.R., 2009. Product & Process Design Principles: Synthesis, Analysis and Evaluation. John Wiley & Sons.

Chapter 6

Design and simulation of distillation processes*

Nishanth Chemmangattuvalappil and Jia Wen Chong
University of Nottingham Malaysia, Semenyih, Selangor, Malaysia

Chapter outline

6.1 Fundamentals of distillation calculations

The simulation of a distillation column involves the solving of mass balance, energy balance, and equilibrium relationships between the components. During simulation exercises, each distillation stage is treated as 100% efficient, which means that the vapor and liquid streams are assumed to be in equilibrium with each other as they leave the stage. The equations associated with each stage are referred to as the "MESH" equations:[1]

1. **M**aterial balances
2. **E**quilibrium relationships
3. **S**ummation relationships
4. **H**eat (energy) balances

*Simulation and other electronic files of this chapter are made available on a companion website; URL: https://www.elsevier.com/books-and-journals/book-companion/9780323901680.

1. See Chapter 17 for the use of MESH equation for flash calculation.

Chemical Engineering Process Simulation. https://doi.org/10.1016/B978-0-323-90168-0.00011-1

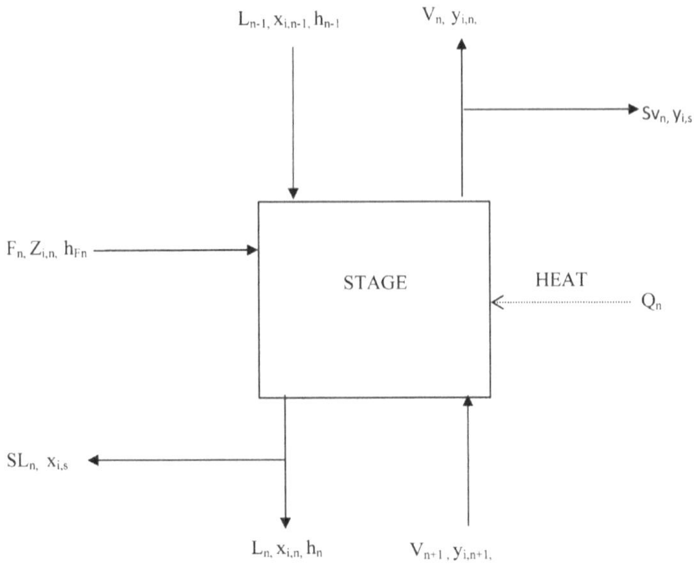

FIGURE 6.1 Equilibrium stage of a distillation column.

Fig. 6.1 shows an equilibrium stage of a distillation column, where liquid and vapor streams entering and leaving the stage are labeled. In addition, there are sidestream withdrawals (liquid and/or vapor), and external feed, and provision for the input or output of external heat. If the sidestreams are ignored, equations for the equilibrium stage can be represented using Eqs. (6.1)–(6.4) (Wankat, 2016):

1. Material balance (component flows):

$$l_{i,n-1} - l_{i,n} - v_{i,n} + v_{i,n+1} + f_{i,n} = 0 \tag{6.1}$$

where $l_{i,n} = Lx_{i,n}$, $v_{i,n} = V\,y_{i,n}$ and $f_{i,n} = F_n z_{i,n}$

2. Equilibrium relationship:

$$y_{i,n} = K_{i,n} x_{i,n} \tag{6.2}$$

3. Summation relationship on mole fractions:

$$\sum_i xi, n = 1.0 \tag{6.3a}$$

$$\sum_i yi, n = 1.0 \tag{6.3b}$$

4. Energy balance around stage n:

$$L_{n-1}h_{n-1} - L_n h_n - V_n H_n + V_{n+1}H_{n+1} + F_n h_{Fn} + Q_n = 0 \tag{6.4}$$

where L and V are the liquid and vapor flowrates; K is the vapor liquid equilibrium ratio (K value); x and y are the mole fractions of liquid and vapor phases of component i; h and H represent the liquid and vapor phase enthalpies; and Q_n is the energy input to stage n.

6.2 Distillation column simulation

The design of distillation column involves the solving of MESH equations. Regardless of the solution techniques, the same equations are to be solved. However, solving all MESH equations simultaneously is challenging, due to the large number of equations involved in MESH calculation. Before the development of simulation packages, hand calculation methods such as Lewis-Matheson (Lewis & Matheson, 1932) method were used to solve MESH equations under several simplifying assumptions. While these assumptions were reasonable for near ideal systems, those were not appropriate in the systems where components have different heats of vaporizations and also when they form nonideal mixtures. The approaches for solving the stage equations has been changed over the decades as the level of sophistication of computational aids available to the process engineer has been developed in the form of hardware and software. With the help of process simulation software, large number of equations can be solved in very short computational time (Seader & Henley, 1998). In simulating a distillation column, the main aims are to identify the following.

(1) Estimate of condenser and reboiler pressure
(2) Reflux ratio, R
(3) Number of theoretical stages, N
(4) Feed tray location

If pressure and number of stages are kept as variables while solving the MESH equations, solving those equations will be challenging even with the computational power of the modern computers. Because of this, the optimal pressure of a column will be estimated first by considering the effect of pressure on relative volatility and utility selection. The selected pressure will be a good compromise for high relative volatility and utility cost. Once the pressure is set, shortcut distillation techniques based on Fenske-Underwood-Gilliland (FUG) method is used to get an estimate on the reflux ratio, number of theoretical stages and feed tray location. While conducting the rigorous simulation, the number of stages and feed inlet location identified from short cut design will be used. This is to reduce the number of variables needed to be solved at this stage. With these estimates, rigorous distillation method can calculate an accurate estimation of distillate and bottoms compositions. The design will be modified if the targeted compositions are not met after the rigorous distillation (Seider et al., 2016).

The applications of these methods are illustrated using an example in the following sections. Please note that even though the solution procedure is illustrated using UniSim Design (Honeywell, 2017), the procedure is generic and is applicable for any commercial simulation tools (though minor differences are expected for different simulation tools).

6.3 Debutanizer example

A simple illustrative example that involves the separation of butanes from a hydrocarbon mixture is used to demonstrate the distillation column design steps using UniSim Design. Based on the feed information in Table 6.1, a distillation column to separate butanes from other components is to be simulated in UniSim Design. It is expected to recover 90% of *n*-butane in the distillate and 90% of i-pentane at the bottoms.

6.3.1 Setting up the problem

The first step in building the simulation flowsheet is to select the components involved in the process and an appropriate thermodynamic model to describe the vapor liquid equilibrium in the system.[2] In this example, Peng-Robinson equation of state is selected because the feed stream contains hydrocarbon mixture at moderate pressure.

Once the components involved in the process are selected, a mole balance calculation needs to be conducted to estimate the approximate distillate and bottoms composition expected from the distillation column. The specifications on the target components can be used to perform the mole balance. At this stage, the key components are defined. In this case, *n*-butane is chosen as the light key as it is the heaviest component expected in the distillate. Similarly, i-pentane is chosen as the heavy key. In the preliminary stage, it is acceptable

TABLE 6.1 Conditions and composition of feed stream.

Component	Composition (mol%)	Feed condition
i-butane	10	Bubble point feed Pressure: 500 kPa Flowrate of 100 kmol/h
n-butane	30	
i-pentane	10	
n-pentane	15	
n-hexane	20	
n-heptane	15	

2. See Chapters 15–19 for details.

to assume that all the light non-key components (i-butane) are collected at the distillate and all the heavy non-key components (n-pentane, n-hexane, and n-heptane) are collected at the bottoms. The mass balance can be performed by a splitter unit as shown in Fig. 6.2.

In a distillation column, the vapor leaves the top of the column at its dew point and the liquid leaves the bottom at its bubble point. To adjust the temperatures, an energy stream needs to be included in the splitter unit. In order to obtain an initial estimation on pressure to be set inside the column, the condenser pressure is assumed to be the feed pressure as shown in Fig. 6.2. Since the top of the distillation column is vapor, the vapor fraction is entered as 1 so that the vapor temperature is now estimated at the set pressure. A constant pressure drop is assumed inside the column. To relate the bottom pressure to top pressure, the SET function in UniSim can be used. For that, the SET function is to be selected from the object palette and select pressure of the bottom stream (stream 3) as the target variable. In this example, a pressure drop of 50 kPa is assumed between the top and bottom of the distillation column. To define pressure drop, the top stream pressure can be selected as the source variable and the bottom stream pressure can be defined from the top stream using the SET function. The relationship between the pressures is shown in Eq. (6.5).

$$\text{Bottom pressure (KPa)} = \text{Top Pressure (KPa)} \times 1 + 50 \text{ (KPa)} \tag{6.5}$$

The splitter unit will now converge and bottom pressure is related to the top in terms of a constant pressure drop. At this point, appropriate pressure to operate the column needs to be selected.

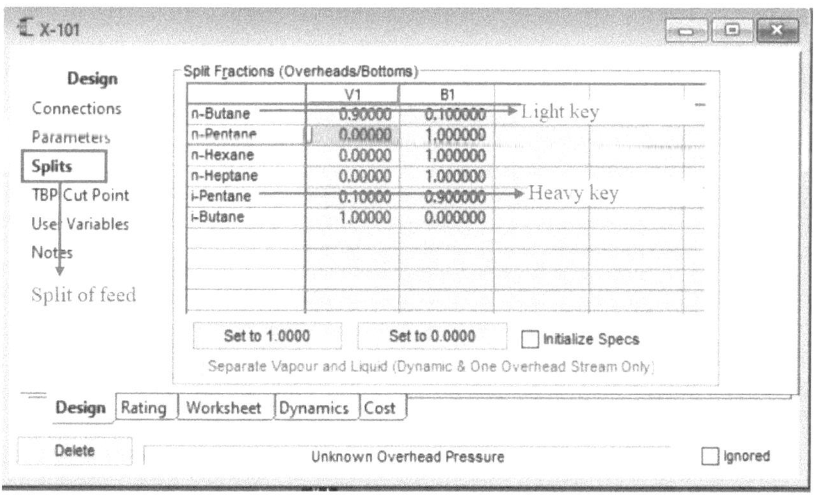

FIGURE 6.2 Splitter to perform the mass balance.

6.3.2 Operating pressure selection

The selection of pressure is based on the following considerations:

1. Effect of pressure in material selection
2. Overall topology of the process.
3. Effect of pressure on relative volatility
4. Effect of pressure on the selection of utilities

In general, high pressure or vacuum operation should be avoided (unless if it is necessary), as high pressure or vacuum operations require expensive materials and larger wall thickness to withstand the operating conditions. In addition, if the process requires further pressurizing of gas streams after a distillation column, it is recommended to continue the distillation at high pressure to reduce avoidable compressors. These decisions are usually taken during the flowsheet synthesis stage. The focus in this chapter is the effect of pressure on the relative volatility and the selection of utilities.

6.3.3 Effect of pressure on relative volatility

The following procedure can be used to study the effect of pressure on relative volatility (α). Since relative volatility is the ratio of K-values of the components, the latter needs to be tracked for different pressures to obtain the relative volatilities at those conditions. The α-values may be plotted against pressure to see the effect of pressure on α.

To estimate and track the α-values, "Spreadsheet" tool in UniSim can be used. In the spreadsheet, the K-values of key components in the feed are imported. The relative volatility can then be calculated by taking the ratio of K-values. The relative volatility estimated now is at the current pressure. To study the effect of pressure on alpha, the pressure is to be varied over a range and alpha is to be plotted against the pressure. To do that, the Databook in UniSim can be used where the dependent and independent variables can be defined. In this case, relative volatility is the dependent variable and pressure is the independent variable. Relative volatility is estimated in the spreadsheet whereas the pressure refers to the pressure of stream 2. The variables can be entered by choosing the relevant object (spreadsheet for alpha and stream 2 for pressure).

Once the variables to be studied are selected, a case study can be conducted to study the effect of pressure on alpha. The case study is initiated from the databook. In the next step, pressure is selected as the independent variable and alpha is selected as the dependent variable as our target is to see the effect of pressure on alpha. The range of pressure in which the column is to be operated is chosen. With these details, the case study can be initiated as shown in Fig. 6.3.

The result of the case study which gives the effect of pressure on relative volatility is shown in Fig. 6.4. It can be seen that the relative volatility decreases with increasing pressure. Therefore, lower pressure is recommended for the ease of separation.

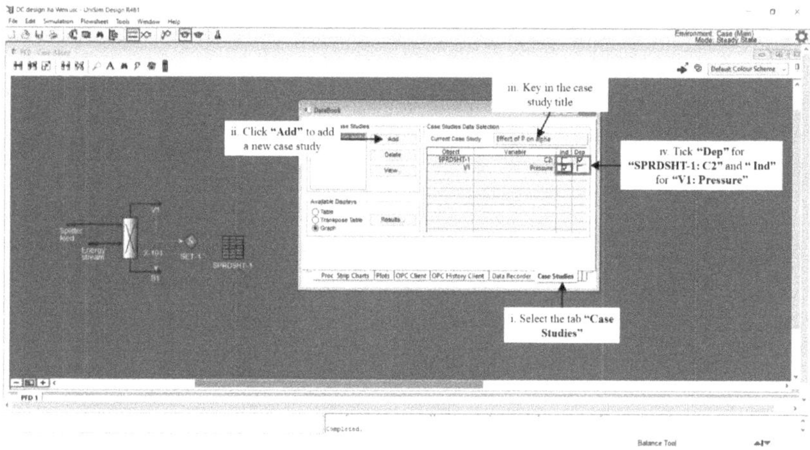

FIGURE 6.3 Initiating case study on the effect of pressure on relative volatility.

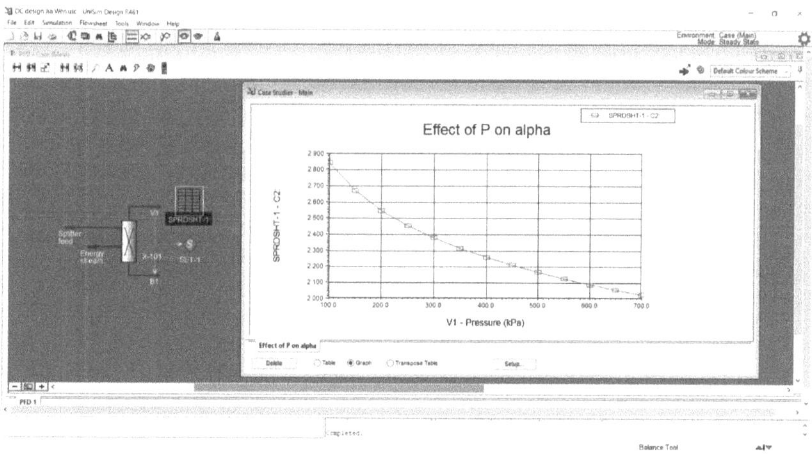

FIGURE 6.4 Effect of pressure on alpha.

6.3.4 Effect of pressure on utility selection

Since the pressure inside the column affects the temperature inside the distillation column, it is necessary to quantify this effect for the selection of the most appropriate utilities to be used in the condenser and the reboiler. The utility used in the condenser depends upon the top temperature while the utility used in the reboiler depends upon the bottom temperature of the distillation column. Therefore, the effect of pressure on these temperatures must be studied. To do that, another case study needs to be conducted by keeping pressure as the independent variable and top and bottom temperatures as the dependent variables.

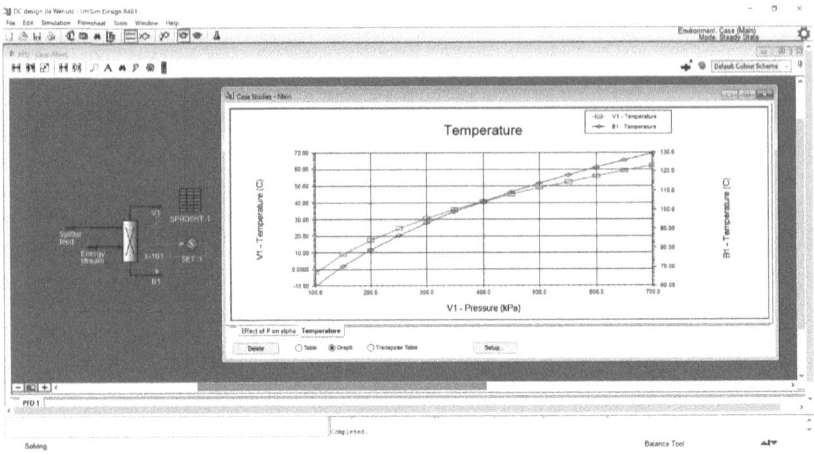

FIGURE 6.5 Effect of pressure on top and bottom temperatures.

The procedure in Section 6.3.3 has been repeated to generate the plot between top and bottom temperatures and the column pressure. Note that the top temperature should be set to at least 50°C so that cooling water can be used as cold utility in the condenser. The use of cooling water as utility in the condenser will minimize utility cost. Similarly, it is recommended to use MP stem as the reboiler utility for which the bottom temperature must be maintained below 180°C. From Fig. 6.5, it can be seen that the minimum pressure to attain both these conditions is 525 kPa. However, Fig. 6.4 indicates that lower pressure is preferred for a higher relative volatility. At the same time, the relative volatility at 525 kPa is not significantly lower than that at lower pressures. So, it is advisable to operate this column at 525 kPa so that a cheaper utility can be used without significant drop in the relative volatility value.

6.4 Preliminary design using short cut distillation

As described in Section 6.1, distillation column design involves solving the MESH equations. The rigorous distillation unit in UniSim can be used to solve the MESH equation. However, because of the large number of variables involved in solving the MESH equations, it is advisable to conduct short cut distillation before conducting the rigorous distillation. The purpose of performing shortcut distillations is to provide an estimate of variables such as number of stages, feed entry tray and reflux ratio to the rigorous distillation operation. To set up a shortcut distillation unit, the mole fractions of key components in the distillate and bottoms needs to be estimated. From the problem specifications, the mole fractions of light key (*n*-butane) and heavy key (i-pentane) at the bottoms are set to 0.048 and 0.026.

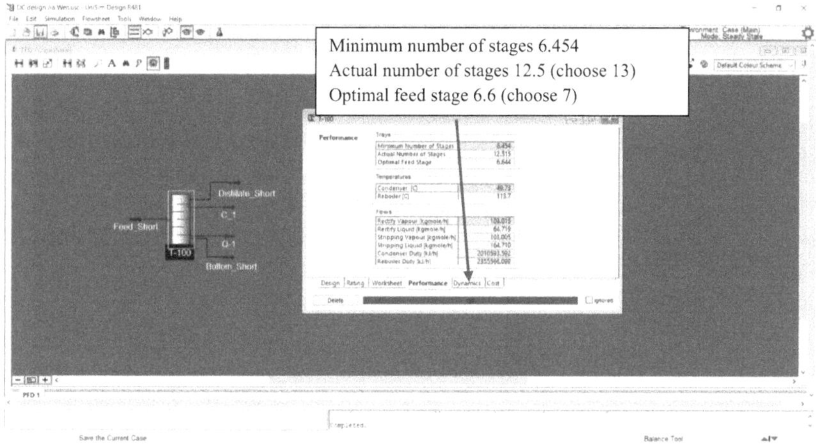

FIGURE 6.6 Short cut distillation results.

The shortcut distillation unit in UniSim makes use of Fenske and Underwood equations to find the minimum number of stages and minimum reflux based on the specification on key components. Based on the designer's decision, the actual reflux can be entered and Gilliland equation will be applied to find the actual number of stages and feed location. In this example, the actual reflux ratio is set to be 1.5 times the minimum reflux. Once the reflux is entered, the column will converge and the minimum number of stages, feed location and actual number of stages can be seen under the "Performance" tab in the shortcut unit as shown in Fig. 6.6. In this example, we can obtain the desired separation with 13 stages if a reflux ratio of 1.69 (1.5 times the minimum reflux ratio) is used.

6.5 Rigorous distillation column design

After setting up the pressure and getting a preliminary estimate on the number of stages, reflux and feed entry stage, the rigorous distillation column design can be performed. At this stage, tray-by-tray calculations have been performed using the rigorous distillation unit in UniSim to estimate the distillate and bottom compositions that can be obtained using the number of stages and reflux obtained from short cut distillation column design. Please note that the rigorous distillation simulation is with the inherent assumption that the efficiency is 100%. So, any correction for efficiency should be implemented after the simulation is completed.

In this example, the desired separation can be obtained in a column with 13 number of stages if a reflux ratio of 1.69 is used; both of these parameters were based on results from the simulation of short cut distillation performed in the previous section. The feed enters on the seventh stage from the top. A pressure of 525 kPa has been maintained on the top of the distillation column. To begin the simulation, the rigorous distillation column has been chosen from the object palette. To set up the simulation, the number of trays and feed entry stage needs to be defined. The type of condenser is chosen and the product streams and energy streams are entered. It is to be noted that, in the UniSim terminology, a partial condenser refers to a condenser with both liquid and vapor product streams where as full reflux refers to a condenser with only vapor product. The first specification page in setting up a rigorous distillation unit in UniSim is shown in Fig. 6.7. In the next steps (steps 2–4), the pressure of the column (obtained from Section 6.2), top and bottom temperature estimates (from short cut distillation) and reflux ratio (from short cut distillation) are entered as estimates. The parameters needed for these steps are shown in Table 6.2.

In the initial rigorous distillation column setup, the only specification entered in the estimated reflux ratio from the short cut distillation. It can be seen that the degree of freedom is shown to be 1. Hence, one additional specification is needed to run the simulation. Since there are two specifications of the recovery of key components, once of those can be used as shown in Fig. 6.8.

Note that in UniSim Design (Honeywell, 2017), there are several options to incorporate specifications depends on the design targets. However, it is advisable to converge column based on the results from the short cut design and see whether the design specifications are met. It may be hard to converge the column directly for several specific design targets. Because of that, it is recommended to keep reflux ratio as one of the specifications and include the recovery of light key component in the distillate. For that, "Column

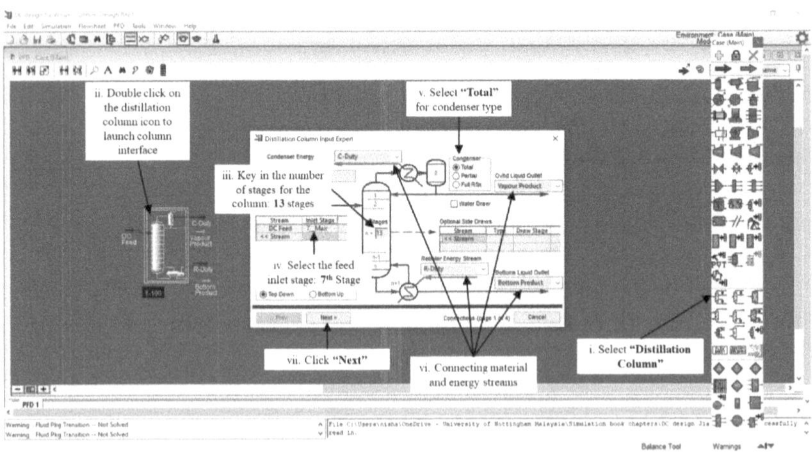

FIGURE 6.7 Setup for rigorous distillation column.

TABLE 6.2 Initial specifications for distillation column simulation.

Step	Parameter	Value
2	Condenser and reboiler pressure	525 and 575 kPa
3	Condenser and reboiler temperature estimate	50 and 114°C
4	Reflux ratio	1.69

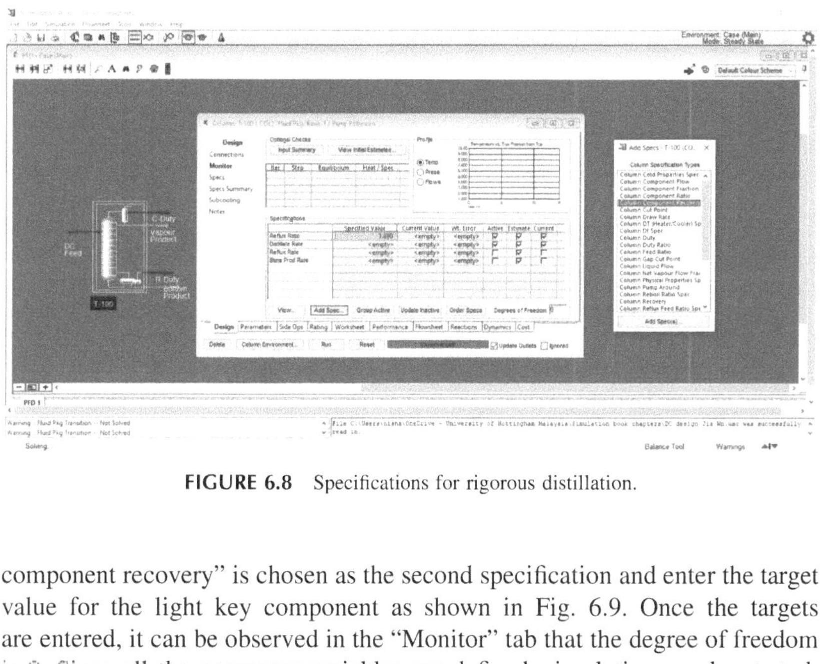

FIGURE 6.8 Specifications for rigorous distillation.

component recovery" is chosen as the second specification and enter the target value for the light key component as shown in Fig. 6.9. Once the targets are entered, it can be observed in the "Monitor" tab that the degree of freedom is 0. Since all the necessary variables are defined, simulation can be started. The simulation is now converged for the given targets.

Since the purpose of performing this operation is to obtain the specifications for distillate and bottoms, the simulation needs to be performed with the specifications on the product streams. To do that, the specification on reflux ratio needs to be replaced with the original design target on the recovery of heavy key component by following the similar procedure used for the light key component target as shown in Fig. 6.10. It is to be noted that the reflux ratio has been changed slightly from the value obtained in the short cut design when the specification is changed to the recovery of key components. However, the deviation is not significant and having a close initial starting point on reflux ratio has helped to converge the column first and the fine tuning to specific targets was done in the second stage.

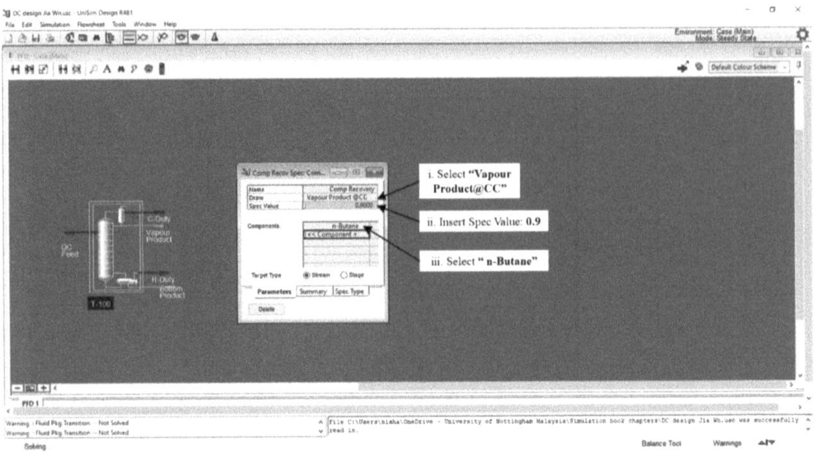

FIGURE 6.9 Defining additional specifications.

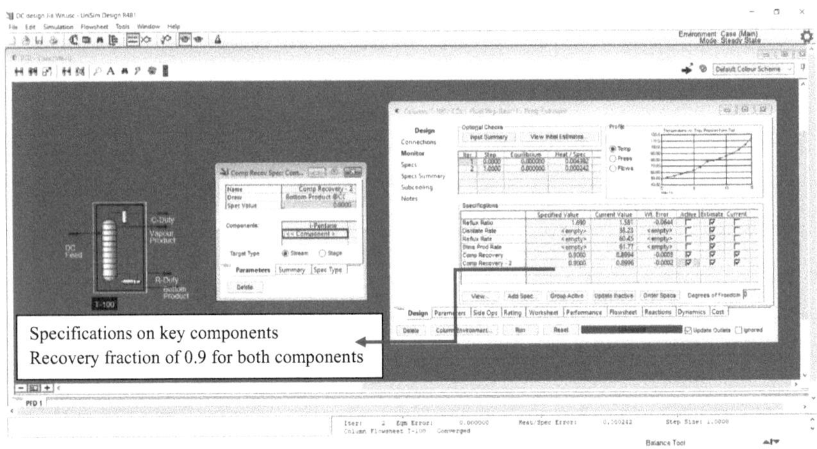

FIGURE 6.10 Converged distillation column simulation with the target specifications.

The final product specifications can be seen from the "performance" tab in the rigorous distillation unit as shown in Fig. 6.11. There are options available to analyze the results in different basis.

In the rigorous distillation units, there are several options to fine tune the design to meet several design targets. While it may be possible to specify these targets in the first step of simulation, it is recommended to get the column converged based on the specifications obtained from short cut distillation in the first step. On a converged flowsheet, it is easier to make changes to the specifications to obtain more accurate results.

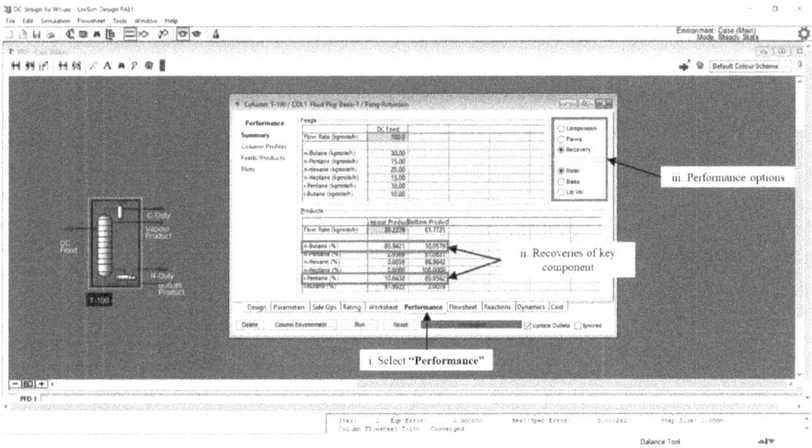

FIGURE 6.11 Final results from rigorous distillation column simulation.

6.6 Conclusions

In this chapter, step-by-step procedures needed for the simulation of distillation columns are covered. The simulation exercise started with the setting up of pressure based on its effect on the relative volatility between key components and utility selection. A compromise has to be made between these targets on several occasions. Based on the chosen pressure, a shortcut column design has been performed to get a preliminary design of the column. The column specifications obtained from the shortcut design has been used to conduct a rigorous simulation of the column using the rigorous distillation unit in UniSim. It is to be ensured that the target specifications have been attained with the design by checking the performance of the column.

Exercises

1. Perform the simulation of the column presented in the debutanizer example if the target is to recover 98% of the butanes with a purity of 95%. To perform this simulation, do a preliminary mass balance by selecting appropriate key components and specifications. Based on the results, simulation steps in the debutanizer example can be repeated. Once the column is converged, enter the specifications on the new recovery and purity of butanes.
2. A continuous distillation column is used to separate xylenes (ortho-, meta-, and para-xylenes) from a feed that contains benzene, toluene and xylenes (all the three types). It is required to recover 98% of the xylenes in the bottoms with a purity of 99%. The feed is entering at its bubble point temperature. The composition of the feed is shown in Table 6.3.

TABLE 6.3 Feed composition.

Component	Mol%
Benzene	40
Toluene	25
p-Xylene	15
m-Xylene	15
o-Xylene	5

a. Estimate the compositions of distillate and bottoms from this column. With this composition, estimate the optimal conditions at which this column needs to be operated. It can be assumed that the non-key components are completely collected at the top or at the bottom of the distillation column.
b. If the distillation column has 15 actual trays, estimate the reflux ratio required to achieve the desired separation using short cut distillation method. Assume a column efficiency of 100%.
c. Explain why the assumption on the separation of heavy non-key components may not be justifiable while that of light non-key component is acceptable. Perform rigorous distillation simulation to estimate the actual purity and recovery of xylenes we may get if we use the design in part (b).
d. Perform the simulation to meet the design specifications given in the problem.

References

Honeywell, 2017. In: www.process.honeywell.com.

Lewis, W.K., Matheson, G.L., 1932. Studies in distillation. Industrial and Engineering Chemistry 24 (5), 494–498.

Seader, J.D., Henley, E.J., 1998. Separation process Principles. John Wiley and Sons, New York.

Seider, W.D., Lewin, D.R., Seader, J.D., Widagdo, S., Gani, R., Ng, K.M., 2016. Product and process design Principles: Synthesis, Analysis and Evaluation, 4th ed. John Wiley and Sons, New York.

Wankat, P.C., 2016. Separation process engineering: Includes mass Transfer Analysis, 4th ed. Prentice Hall, Upper Saddle River, NJ.

Chapter 7

Modeling and optimization of separation and heating medium systems for offshore platform*

Dominic C.Y. Foo[1], Raymond E.H. Ooi[1] and Pitchaimuthu Diban[2]
[1]*University of Nottingham Malaysia, Semenyih, Selangor, Malaysia;* [2]*Pand.ai Pte Ltd, Singapore*

Chapter outline

7.1 Oil and gas processing facility for offshore platform

Fig. 7.1 shows the processes found on a typical offshore oil and gas platform (Voldsund et al., 2014), where the oil and gas products are processed before they are exported to the downstream clients, i.e., gas processing plants and oil refineries. As shown in Fig. 7.1, the oil and gas resources from reservoirs first enter the production manifolds, where they are mixed and depressurized. Crude oil from the manifolds is then sent for a series of separation processes where oil, gas, and water phases are separated. Produced gas is sent for pressure adjustment in the recompression section, prior to entering the gas treatment section, where gases underwent dehydration and conditioning prior to export. Condensate from the gas treatment and recompression sections (liquid recovered from aftercoolers) may be recycled to the separation

*Simulation and other electronic files of this chapter are made available on a companion website; URL: https://www.elsevier.com/books-and-journals/book-companion/9780323901680

Chemical Engineering Process Simulation. https://doi.org/10.1016/B978-0-323-90168-0.00014-7

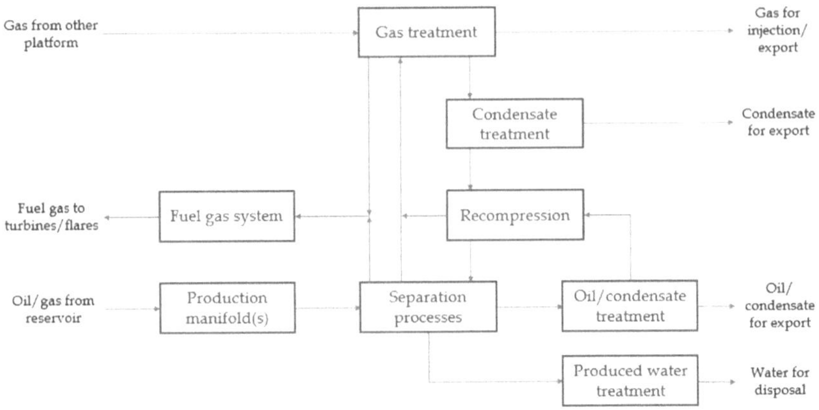

FIGURE 7.1 Processing facilities on an offshore oil and gas platform. *Adopted from Voldsund, M., Nguyen, T.-V., Elmegaard, B., Ertesva°g, I.S., Røsjorde, A., Jøssang, K., Kjelstrup, S., 2014. Exergy destruction and losses on four North Sea offshore platforms: A comparative study of the oil and gas processing plants. Energy 74, 45–58.*

processes, or undergo dehydration and pumping in the condensate treatment section for export. When the condensate is recycled to the separation processes, it is mixed with oil from the manifolds and undergo treatment processes (e.g., desalting) prior to export. The produced water from the separation processes will also be treated prior to final discharge. Note that different processing units may be installed on the offshore facilities depending on the fluid properties and its characterization, such as American Petroleum Institute (API) gravity, contaminant contents (e.g., mercury, H_2S, CO_2). Some of the major common designs were reported by Voldsund et al. (2014).

7.2 Modeling of oil and gas processing facilities

A case study by Diban et al. (2019) is used to illustrate the modeling of separation processes of the processing facilities in Fig. 7.1. The process model is constructed using UniSim Design R481 (Honeywell, 2021), given as in Fig. 7.2. Thermodynamic model used for this case is Peng-Robinson, with properties for the light and heavy oils given in Table 7.1. As shown, there are hypothetical components that need to be registered by the user, with their physical properties given in Table 7.2.

As shown in Fig. 7.2, different separation trains are used for light and heavy oil products. The latter refers to oil products that have API gravity of lower than 30°, while those with API gravity of 30° or higher are considered as light oil. The light and heavy oil products are first sent to their respective *slug catchers*, where light gases are vent to their gas scrubbers in which the condensate and water content are removed. The liquid products of the slug catchers are then heated and fed to the *production separators*, which are typically three-phase separators. In the latter, the carried-over gases, water, and oil are separated. The separated gas is sent to a series of compression stage

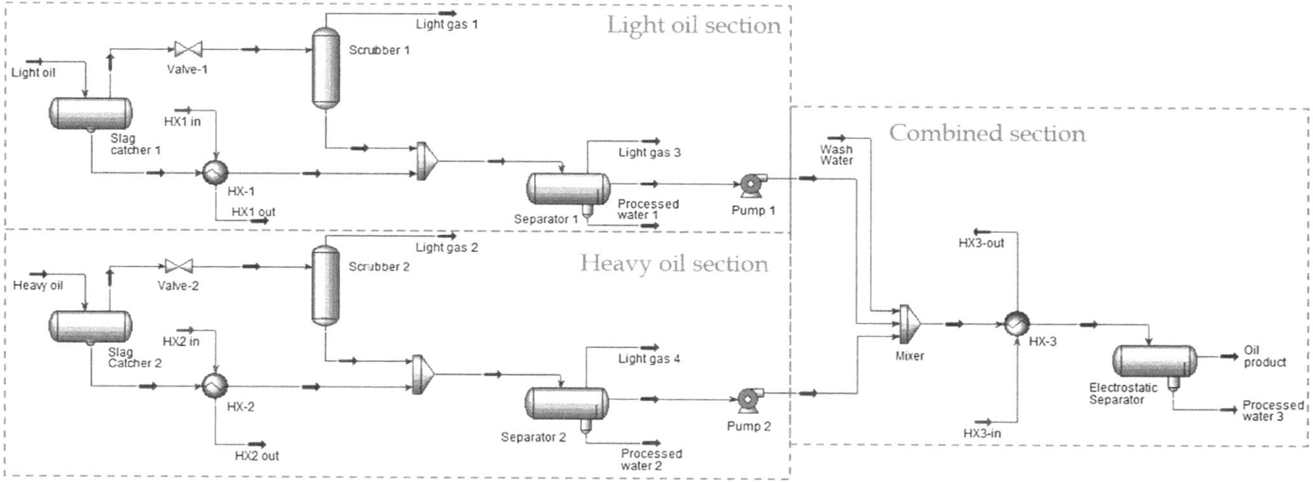

FIGURE 7.2 Simulation model for separation processes for oil and gas processing facilities. *Adapted from Diban, P., El-Halwagi, M.M., Foo, D.C.Y., 2019. A decomposition-based approach for the optimum integration of heating utility and phase separation systems in oil and gas platform. Industrial and Engineering Chemistry Research 58 (47), 21584–21601.*

TABLE 7.1 Properties for light and heavy oil streams (hypothetical component are identified with *).

	Light oil	Heavy oil
Temperature (C)	40	30
Pressure (bar)	20	20
Molar flowrate (k-mol/h)	20,000	4000
Composition		
Nitrogen	0.0005	0.0001
CO_2	0.0100	0.0100
H_2S	0.0000	0.0000
Methane	0.2000	0.2000
Ethane	0.0300	0.0150
Propane	0.0200	0.0040
i-Butane	0.0050	0.0020
n-Butane	0.0040	0.0020
i-Pentane	0.0020	0.0020
n-Pentane	0.0015	0.0020
n-Hexane	0.0015	0.0030
Mcyclopentan	0	0.0002
Benzene	0	0.0002
Cyclohexane	0	0.0010
Mcyclohexane	0	0.0010
Toluene	0	0.0005
E-Benzene	0	0.0005
m-Xylene	0	0.0005
o-Xylene	0	0.0005
C7*	0.0030	0.0050
C8*	0.0030	0.0085
C9*	0.0030	0.0090
C10*	0.0024	0.0130
C11*	0.0020	0.0130

TABLE 7.1 Properties for light and heavy oil streams (hypothetical component are identified with *).—cont'd

	Light oil	Heavy oil
C12*	0.0020	0.0140
C13*	0.0020	0.0160
C14*	0.0018	0.0160
C15*	0.0016	0.0170
C16*	0.0012	0.0180
C17*	0.0010	0.0180
C18*	0.0010	0.0180
C19*	0.0010	0.0200
C20+H*	0	0.2500
C20+L*	0.0120	0
H_2O	0.6885	0.3200

(depending on the export pressure requirement), while the processed water is sent for treatment prior to discharge.

The light and heavy crude oil from the production separator are pumped to a common header where they are mixed. The mixed oil stream usually contains some sediment and water content which needs to be reduced to the permissible content (typically between 0.1 and 3.0 wt%), in order to meet the sales specification. Besides, the salt content in the mixed oil is corrosive for pipelines and oil vessels. Hence, dehydration and desalting processes are carried out in an electrostatic separator in order to meet these requirements. Before entering the electrostatic separator, wash water (see specification in Table 7.3A) is added to the mixed oil stream in order to dilute its salt concentration. The mixture is then heated prior to entering the electrostatic separator. Produced water from the separator is sent for treatment prior to discharge, while the oil product is exported. Specification for simulation of all unit models, such as pressure drop (ΔP) and outlet temperature (T_{out}) are given in Table 7.4.

In Fig. 7.2, three heat exchangers (HX 1, HX 2, and HX 3) are found upstream of the separators, which are used to provide heating in order to lower the viscosity of the oil. Oil, water, as well as glycol and water mixture are commonly used as the heating medium in these heat exchangers. In general, oil is used for high-temperature heating, while the other two mediums are favored for low-temperature heating. For this case, hot water of 4 bar and 140°C is available for use (see Table 7.3B).

TABLE 7.2 Properties of hypothetical components.[a]

	C7*	C8*	C9*	C10*	C11*	C12*	C13*
MW	96	107	121	134	147	161	175
NBP (C)	94.81	118	144.93	168.32	185.75	201.88	216.35
Ideal liquid density (kg/m^3)	722	745	764	778	789	800	811
Critical temp (°C)	270.66	298.25	327.93	352.85	371.38	388.59	404.2
Critical pressure (kPa)	2923	2801	2579	2393	2275	2181	2108
Critical volume (m^3/kgmol)	0.4063	0.4407	0.4972	0.5504	0.5888	0.6257	0.6586
Acentricity	0.3102	0.342	0.3802	0.4236	0.4615	0.4854	0.5044

	C14*	C15*	C16*	C17*	C18*	C19*	C20+H*	C20+L*
MW	190	206	222	237	251	263	463.4	330
NBP (C)	231.77	252.01	270.8	287.9	302.87	315.44	520.08	390.92
Ideal liquid density (kg/m^3)	822	832	839	847	852	857	1004	930
Critical temp (°C)	419.8	433	452.51	473.05	486.52	497.95	689.58	576.21
Critical pressure (kPa)	2030	1919	1818	1769	1667	1591	1038	1381
Critical volume (m^3/kgmol)	0.6952	0.7455	0.7951	0.8363	0.8972	0.9491	1.6321	1.1728
Acentricity	0.5253	0.5614	0.5973	0.6126	0.6346	0.6527	1.0264	0.7423

[a]*Download spreadsheet file from book support website that contain the properties.*

TABLE 7.3 Properties for (A) wash water streams; (B) Hot water.

(A) Wash water stream	Values	(B) Hot water	Values
Temperature (°C)	20	Temperature (°C)	140
Pressure (bar)	32.5	Pressure (bar)	4
Mass flowrate (kg/h)	1100	Composition (water)	1
Composition (water)	1		

7.3 Process optimization of heating medium systems

In this section, an optimization approach based on *process integration* is used to minimize the overall flowrate of the heating medium. Doing this brings several advantages, e.g., lower capital and operating costs due to smaller pumps and waste heat recovery unit (for reheating of heating medium—see latter section). The general procedure for the graphical *pinch analysis* approach of *heat recovery pinch diagram* in minimizing heating medium flowrate is given as follows (Diban & Foo, 2018):

1. *Heating profiles* of the heating medium users (i.e., heat exchangers) are plotted on a temperature versus enthalpy diagram. To ease the graphical plot, it is more convenient to place the heating profile(s) with lower temperatures on the left.
2. Within each temperature intervals, vector addition is carried out to merge the individual segments of different heating profiles. The combined segment represents the total energy needed to raise the temperatures of the individual segments within the respective temperature intervals. The combined segments then form the *heating medium composite curve.* Horizontal span of the latter represents the total energy required to raise the temperature of all the heating medium users.
3. The *heating medium supply line* is then drawn from its supply temperature to cover the entire horizontal span of the heating medium composite curve; this forms the *heat recovery pinch diagram.* To keep the heating medium flowrate to its minimum while ensuring feasible heat transfer, its supply line should have the steepest slope, and yet stays entirely above the composite curve with a minimum approach temperature (ΔT_{min}). In other words, the supply temperature is used as a pivot point where the supply line is rotated to reach its steepest slope. The inverse slope of the supply line may be used to determine the minimum flowrate of the heating medium (F_{HM}) needed to fulfill the total energy requirement of the heating medium system (Q), which is given in Eq. (7.1).

TABLE 7.4 Specification for unit models: (A) light oil section; (B) heavy oil section; (C) combined section.

(A) Light oil section	ΔP (bar)	T_{out} (C)	(B) Heavy oil section	ΔP (bar)	T_{out} (C)	(C) Combined section	ΔP (bar)	T_{out} (C)
Slag catcher 1	0		Slag catcher 2	0		Electrostatic separator	0	
Valve 1	0.5		Valve 2	0.5		HX 3		
Scrubber 1	0		Scrubber 2	0		Tube side	0.5	90
Separator 1	0		Separator 2	0		Shell side	0.5	80
Pump 1	13		Pump 2	13				
HX 1			HX 2					
Tube side	0.5	65	Tube side	0.5	65			
Shell side	0.5	80	Shell side	0.5	80			

$$F_{\text{HM}} = \frac{Q}{C_p\left(T_{\text{HM, in}} - T_{\text{HM, out}}\right)} \tag{7.1}$$

where C_p is the heat capacity of the heating medium (i.e., hot water for this case), while $T_{\text{HM, in}}$ and $T_{\text{HM, out}}$ are its inlet and outlet temperatures.

The above procedure is then applied for the earlier case study. From the converged simulation model (Fig. 7.2), the heating duties of these heat exchangers are determined and summarized in Table 7.5. The total amount of hot water streams may also be determined from the simulation model as 122.5 kg/s, contributed by HX 1 (37.0 kg/s), HX 2 (44.6 kg/s) and HX 3 (40.9 kg/s) respectively, as all of them are supplied at 140°C.[1]

Following step 1 of the graphical pinch approach, the heating profiles of the three heaters (given in Table 7.5) are plotted on a temperature versus enthalpy diagram and shown in Fig. 7.3.

Step 2 is then followed to construct the heating medium composite curve. Within the temperature interval of 40–65°C, the individual segments of HXs 1 and 2 are merged into a single segment, with a total enthalpy of 18,320 kW (= (21.71 – 3.39) × 1000 kW), as shown in Fig. 7.4. Note that in temperature intervals 30–40 and 65–90°C, only a single profile is found; hence no vector addition is necessary. The resulted heating medium composite curve is shown in Fig. 7.4. Horizontal span of the latter represents the total heating requirement for the three heating medium users, i.e., 32,570 kW.

Following Step 3, the hot water supply line is drawn from 140°C, by rotated at the latter as a pivot point. As shown in the heat recovery pinch diagram in Fig. 7.5, the hot water supply line has a return temperature of 40°C, which stays above the composite curve by ΔT_{min} of 10°C. For the case in Fig. 7.5, heat capacity of the hot water stream is reported as 4.35 kJ/kg °C.[2] Hence, the hot water flowrate can be determined from Eq. (7.1) as 74.76 kg/s (= 32,520/4.35/(140–40) kg/s). Comparing with the hot water flowrate of 122.5 kg/s in Fig. 7.2, this represents a saving of close to 40%.

TABLE 7.5 Simulation results of heaters.

Unit	Heating duty (kW)	T_{in} (°C)	T_{out} (°C)
HX 1 (light oil section)	9850	40	65
HX 2 (heavy oil section)	11,860	30	65
HX 3 (combined section)	10,860	65	90
Total duty	32,570		

1. Readers are encourage to verify this from the simulation file (Fig. 7.2).
2. Readers are encourage to verify this from the outlet stream of heat exchanger in the simulation file.

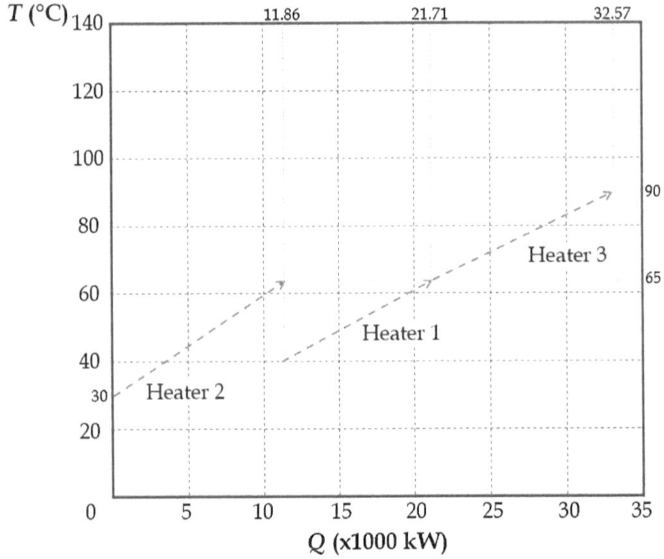

FIGURE 7.3 Heating profiles of three heaters are plotted on temperature versus enthalpy diagram.

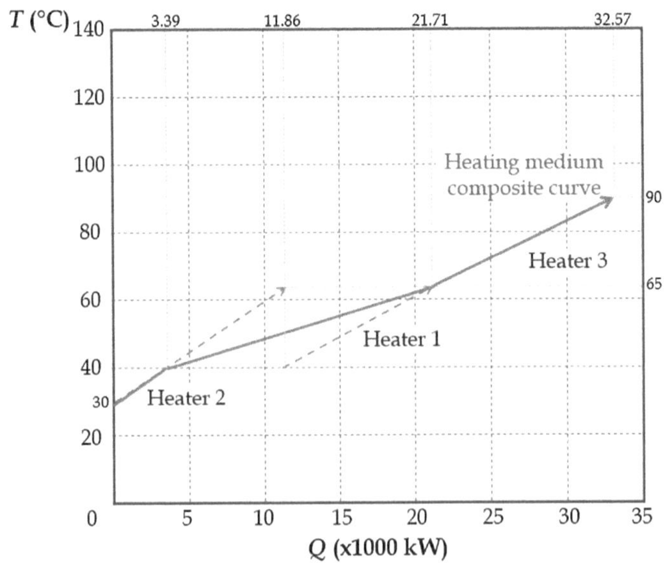

FIGURE 7.4 Combination of different heating profiles to form the heat medium composite curve.

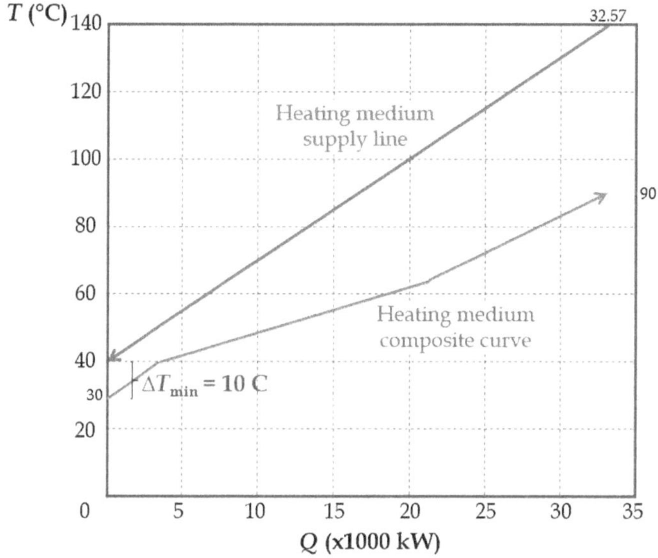

FIGURE 7.5 Heat recovery pinch diagram with hot water outlet temperature of 40°C.

Process simulation is next performed for the integrated heating medium system. Note that the sequence of the hot water for the heaters should follow the heating profiles that form the heat recovery pinch diagram. As shown in Fig. 7.6, the hot water should first enter HX 3, which is then followed by HX 1 and finally HX 2. In the simulation model, the three heat exchangers are connected in series. An additional pump and a waste heat recovery unit (WHRU) are added at the outlet of HX 2 so to adjust the water pressure and temperature before it enters HX 3 (see details in Fig. 7.7).

7.4 Heat exchanger design consideration

For this case, it is assumed that 1−2 shell-and-tube heat exchangers are to be used for the heating operation. In designing a heat exchanger, it is desired to have a good geometric correction factor, i.e., F_t value. An F_t value of lower than 0.75 is to be avoided (Smith, 2016). The converged simulation model (Fig. 7.7) indicates that F_t value of HX 2 is reported as 0.2,[3] which is far too low to be accepted. The main reason that leads to the low F_t value is due to its large temperature cross; the latter may be displayed using the simulation model (see Fig. 7.8 for detailed steps).

Several strategies may be explored to overcome the problem. For instance, one may increase the number of shells of HX 2, while maintaining its outlet

3. Readers are encourage to verify this from Performance page of HX 2 in the simulation model.

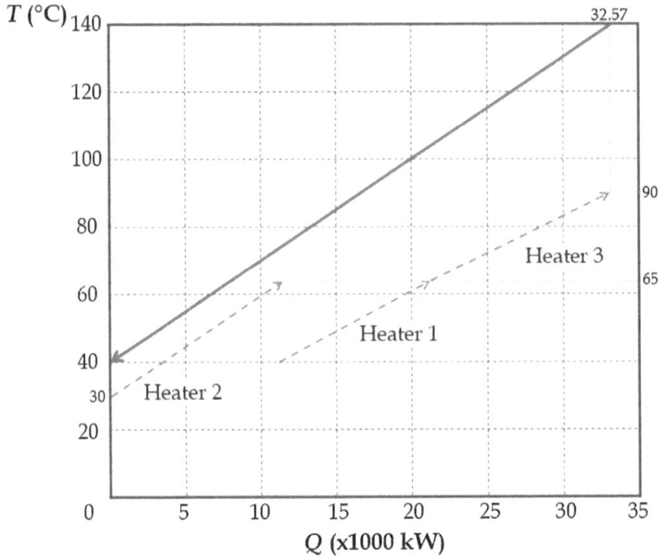

FIGURE 7.6 Sequence of heating profiles.

temperature at 40°C (Strategy 1). By having two shells for HX 2, its F_t value is raised to 1 (see steps in Fig. 7.9). Doing this however, lead to higher capital cost of HX 2 for the additional shell.

In Strategy 2, while the number of shell is kept to one, the hot water stream of HX 2 is set to have high flowrate (see detailed steps in Fig. 7.10). Note that doing this leads to higher return temperature at the outlet of HX 2. Results of this strategies are summarized in Table 7.6. As shown, for the F_t value of 0.75, the heating medium flowrate has to be increased to 95 kg/s, i.e., 27% higher than the minimum flowrate in strategy 1. Implementing this strategy entails higher operating (due to pumping) and capital costs. The latter is contributed by larger pump and pipeline, as well as larger area for the WHRU (due to lower heat transfer driving force).

Another strategy that is worth exploring is the use of plate heat exchanger for HX 2. However, plate heat exchanger has much higher capital cost, as compared to that of shell-and-tube heat exchanger. Note also that it should be avoided for crude oil with high slurry content (e.g., ≥70%), in which fouling can be severe. Detailed cost and technical analysis is needed before a decision can be made among the three possible strategies.

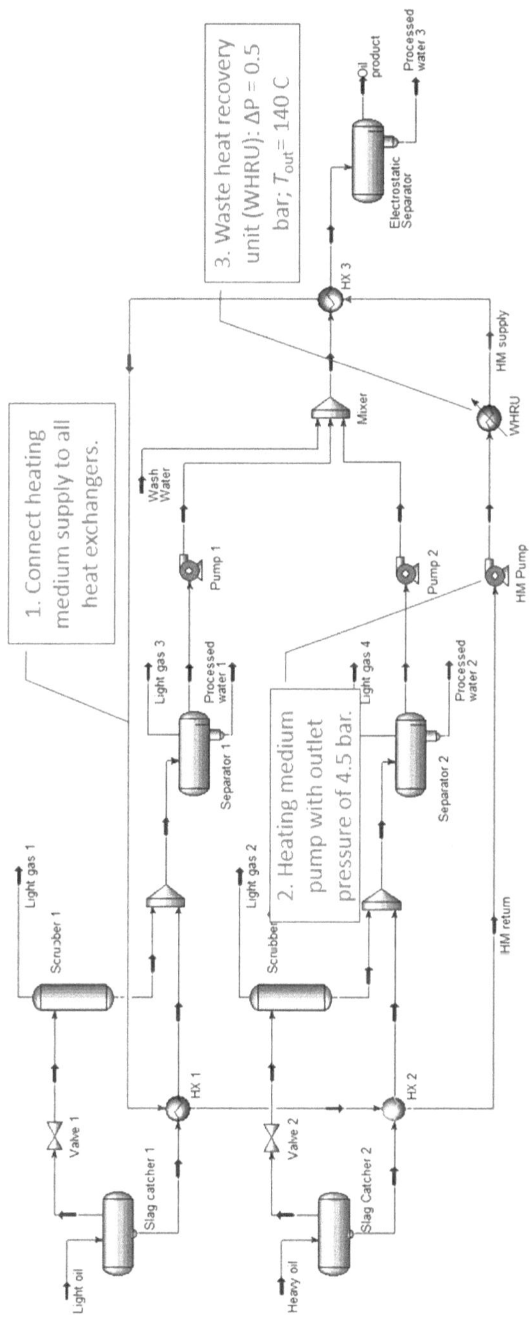

FIGURE 7.7 A converged simulation model with recycle loop of heating medium system.

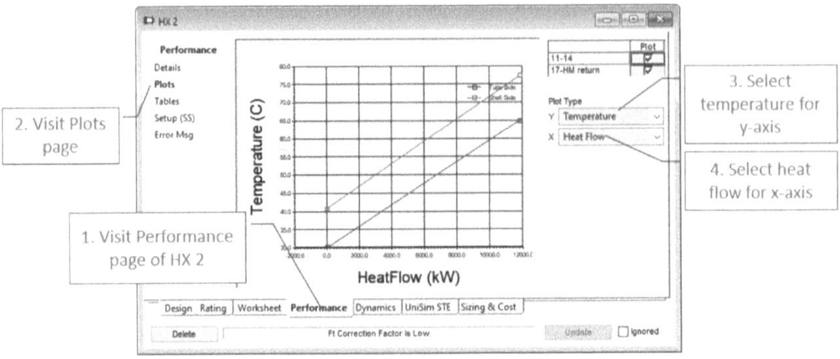

FIGURE 7.8 Steps to display temperature profile for HX 2.

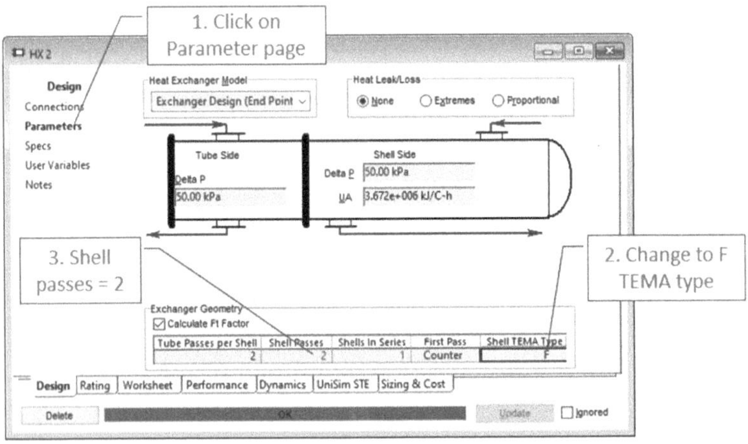

FIGURE 7.9 Steps for Strategy 1—two shell passes for HX 2 with F-type shell.

Exercises

If the hot water stream of 150°C (5 bar) is to be used as heating medium, solve for the following (remember to set the HM pump to raise the stream pressure to 5.5 bar):

1. Minimum flowrate for the heating medium, with minimum approach temperature of 10 °C.
2. Determine the minimum flowrate of heating medium that will achieve the F_t value of 0.75, when HX 2 has only one shell.

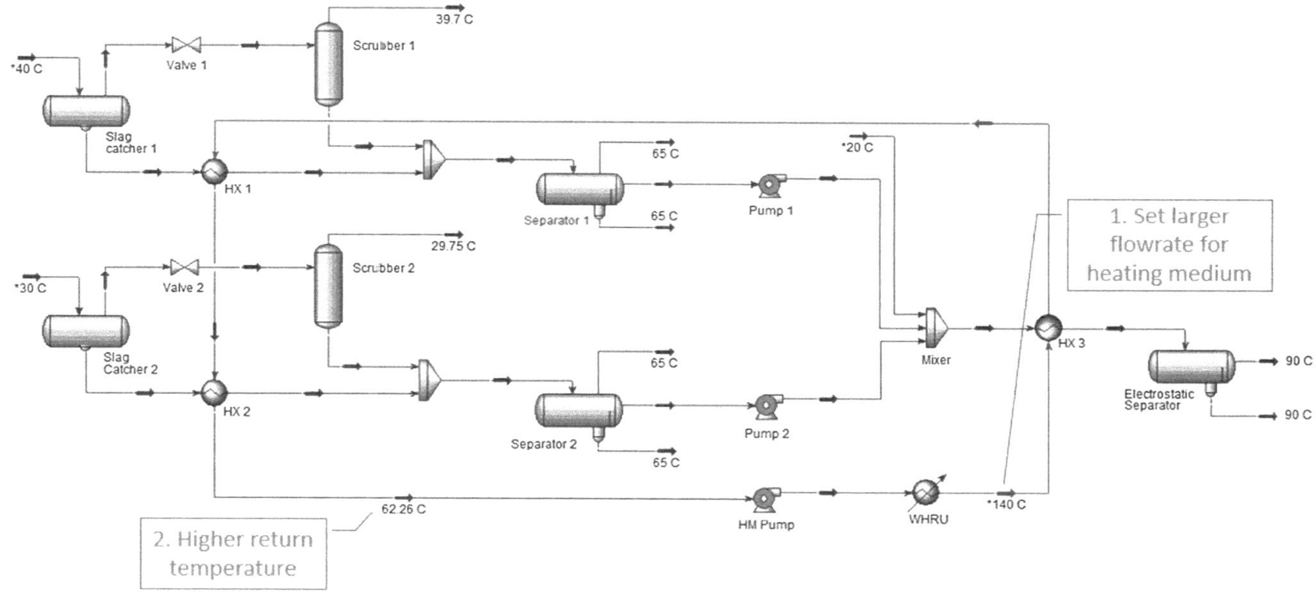

FIGURE 7.10 Strategy 2—higher flowrate for heating medium, leading to higher return temperature for HX 2.[4]

4. Tips: display stream temperatures with shortcut "Shift + T".

TABLE 7.6 Summary of Strategy 2—higher heating medium flowrate.

Flowrate (kg/s)	Return temperature (°C)	Ft value for HX-2
74.8	40.8	0.20
80	47.4	0.20
90	57.9	0.60
95	62.3	0.75
100	66.2	0.82

References

Diban, P., El-Halwagi, M.M., Foo, D.C.Y., 2019. A decomposition-based approach for the optimum integration of heating utility and phase separation systems in oil and gas platform. Industrial and Engineering Chemistry Research 58 (47), 21584–21601.

Diban, P., Foo, D.C.Y., 2018. Targeting and design for heating utility system for offshore platform. Energy 146, 98–111.

Honeywell, 2021. UniSim Design User Guide.

Smith, R., 2016. Chemical Process Design and Integration, 2nd ed. John Wiley & Sons, Inc, Chichester, West Sussex, United Kingdom.

Voldsund, M., Nguyen, T.-V., Elmegaard, B., Ertesvåg, I.S., Røsjorde, A., Jøssang, K., Kjelstrup, S., 2014. Exergy destruction and losses on four North Sea offshore platforms: A comparative study of the oil and gas processing plants. Energy 74, 45–58.

Part III

Symmetry

Chapter 8

Basics of process simulation with Symmetry*

Nurain Shakina Roslizam, Abdul Rahim Norman, Shahrul Azman Abidin and Zulfan Adi Putra

PETRONAS Group Technical Solutions, Process Simulation and Optimization, Kuala Lumpur, Malaysia

Chapter outline

8.1 Example on *n*-octane production

The modeling approach of the *n*-octane production process will be divided into two major stages. First, the thermodynamic model is configured in the *Thermodynamics Environment* of Symmetry. Next, process modeling is to be carried out in the *Process Flow Diagram Environment* of Symmetry. Detailed steps are discussed in the following sections.

8.2 Establishing the thermodynamic model

Regardless of process complexities, the quality of simulation results depends on the quality of the underlining thermodynamic method in estimating the thermodynamic properties such as vapor-liquid equilibrium, thermal properties, and transport properties.

*Simulation and other electronic files of this chapter are made available on a companion website; URL: https://www.elsevier.com/books-and-journals/book-companion/9780323901680.

Chemical Engineering Process Simulation. https://doi.org/10.1016/B978-0-323-90168-0.00002-0

First step in establishing the property package configuration is to choose the appropriate thermodynamic model. In Symmetry, thermodynamic models are divided into three categories: equation of state, activity coefficient models, and specialty packages. For the example of *n*-octane production, the equation of state of "Advanced Peng-Robinson" is chosen considering the hydrocarbon components involved in the process and its operating conditions. The detailed steps of thermodynamic model selection is shown in Fig. 8.1.

The second step involves the definition of all components involved in the *n*-octane production process. Pure components can be selected from the component database of Symmetry, as shown in Fig. 8.2.

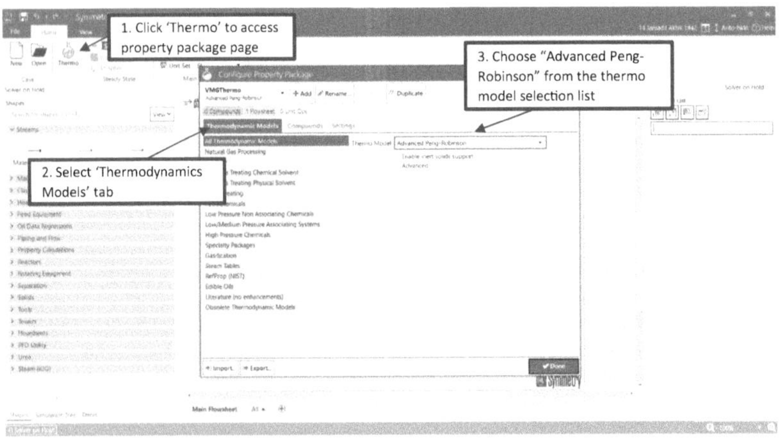

FIGURE 8.1 Steps to select thermodynamic model.

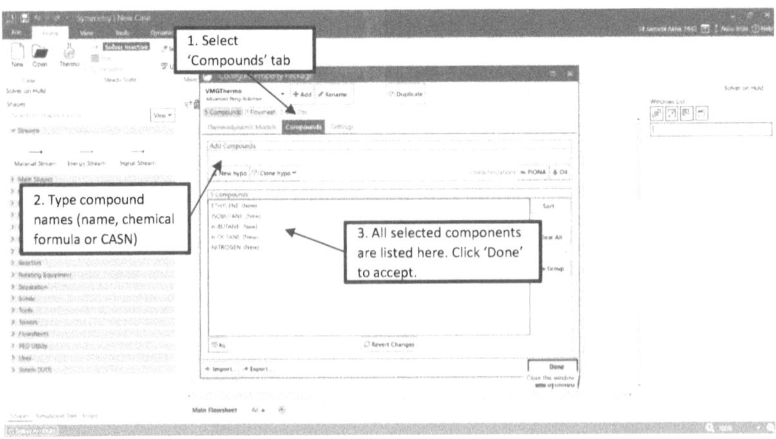

FIGURE 8.2 Steps to define components.

8.3 Process modeling

The process modeling will be carried out in the Process Flow Diagram Environment where the user is able to access all unit operations available in Symmetry. This Process Flow Diagram Environment uses MS Visio as its graphical user interface. User can "drag and drop" the unit operation of interest from the Shapes panel onto the PFD. It is important to conduct degree of freedom analysis[1] prior to process modeling to ensure that sufficient process parameters information are available while avoiding over-specification of the unit operations.

In Symmetry, the fundamental concept of ports is very useful for the users. Every unit operation will consist of at least two ports, which propagates the basic thermodynamic information from inlet to outlet of the equipment (iCON-Symmetry, 2021). For instance, in Fig. 8.3, a cooler unit has three ports: material inlet port, material outlet port, and energy port. The material inlet and outlet ports carry all information regarding the material flowing through the cooler, while the energy port carries the amount of energy removed from the cooler. The ports are very useful for degree of freedom analysis as Symmetry will automatically calculate the state of the outlet stream as soon as a specification is provided to the unit operation.

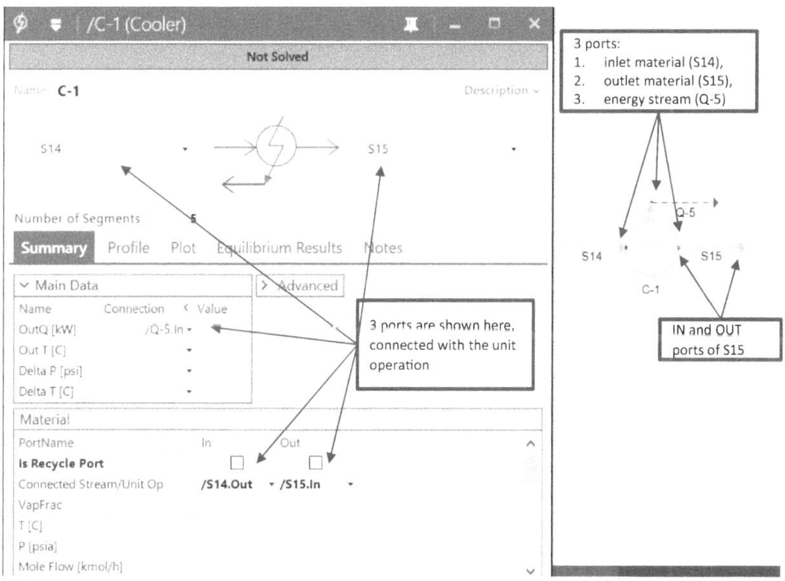

FIGURE 8.3 Ports in Symmetry.

1. Refer to Chapter 1 for better understanding.

Another important feature is to check the unit of measurements of the information available. A new unit of measurement can be created if it is not the same with the default unit sets such as SI, British, and Field.

8.3.1 Defining reactor inlet feed streams

Inlet stream for the *n*-octane production can be defined by selecting the "material stream" of the Process Flow Diagram Environment, by specifying its molar flowrate, pressure, and temperature, as shown in Table 8.1. The stream will converge (indicated by the green notification of "Solved" shown in Fig. 8.4) as the degrees of freedom are satisfied with information from Table 8.1.

TABLE 8.1 Feed stream data.

Component	Flowrate (kmol/h)	Condition
Nitrogen, N_2	0.1	T = 93°C
Ethylene, C_2H_4	20	P = 20 psia
n-Butane, C_2H_{10}	0.5	
i-Butane, C_4H_{10}	10	
n-octane, C_8H_{18}	0	

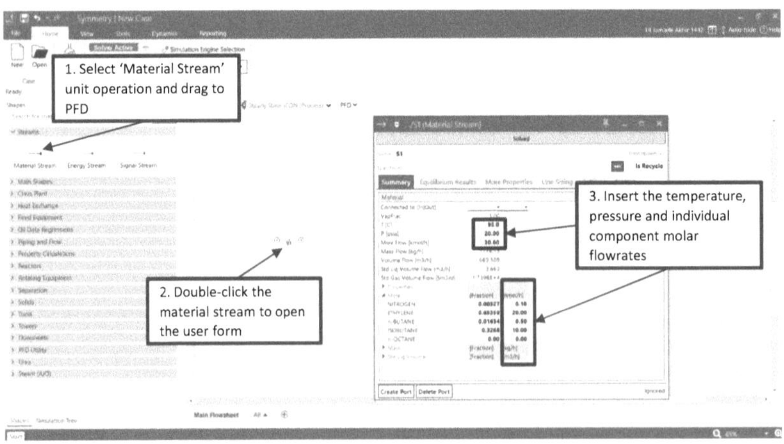

FIGURE 8.4 Specification of reactor feed stream in Process Flow Diagram Environment.

8.3.2 Modeling of reactor

Next step of simulation is to define the reaction in the *n*-octane reactor. There are several generic reactor models in Symmetry, e.g., conversion, plug flow, or equilibrium reactors, as well as specific reactors such as ethylene cracker and other refinery-type reactors. The selection of reactors will depend on the process technology of interest and the available information. In this example, a simple conversion reaction is used, as shown in Fig. 8.5. The chemical reaction is specified as shown in Table 8.2.

The reactants (ethylene and i-butane) are set with negative coefficients, while product (*n*-octane) is set with positive coefficient in the reaction data, according to the reaction stoichiometry. Inert components in the chemical reaction (nitrogen and n-butane) are specified as "0" in the reaction data. The balance of the reaction stoichiometry will be calculated as a reference to the user. If the reaction stoichiometry has been defined correctly, the "balance" will be shown as zero (see Fig. 8.6). A nonzero value indicates incorrect input

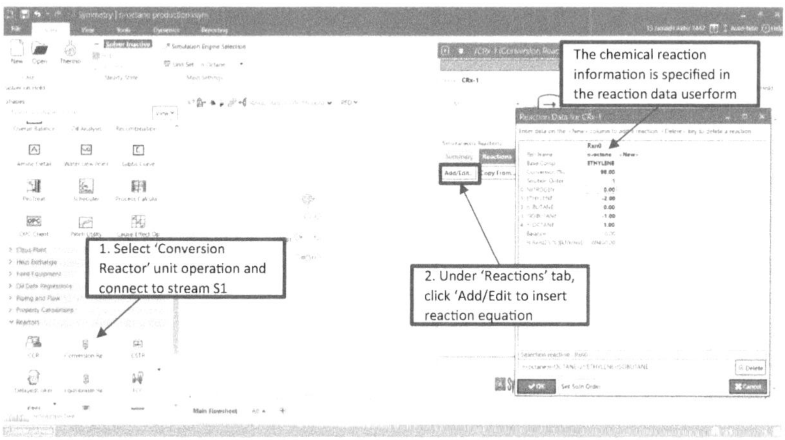

FIGURE 8.5 Steps to selecting the reactor model.

TABLE 8.2 Details of the reaction.

Specifications	Details
Reaction type	Conversion
Conversion rate	98%
Limiting reactant	Ethylene
Reaction stoichiometry	$2C_2H_4 + C_4H_{10} \rightarrow C_8H_{18}$

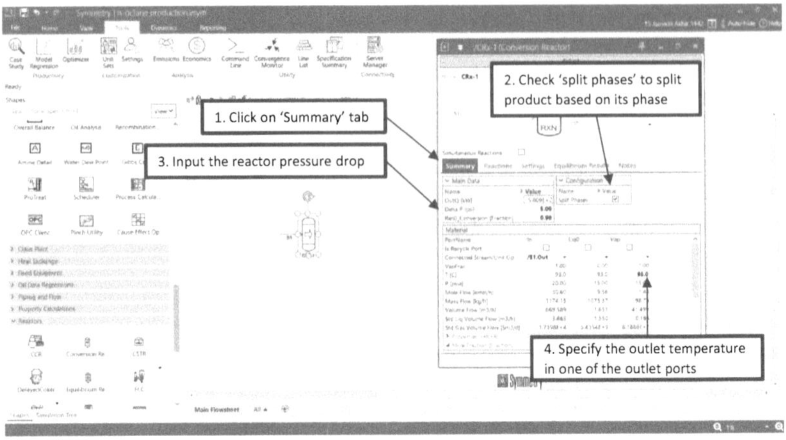

FIGURE 8.6 Overall reactor configuration.

of the coefficients. Symmetry also allows multiple chemical reactions in a single reactor. Solution orders can be set if there are chemical reactions occurring in sequence. The heat of reaction is calculated by the simulation, which is used to evaluate whether a chemical reaction is of endothermic or exothermic type.

After defining the reaction data, the overall reactor configuration must be set, as shown in Fig. 8.6. First, the product stream of the reactor will be split based on its phase. Only the vapor phase product will undergo further processing to recover the *n*-octane product. Note that the conversion reactor can also be configured to have a single outlet stream. Next, the operating conditions of the reactor has to be specified as shown in Table 8.3. The reactor in this example is an isothermal reactor where the temperature is kept constant. In this regard, the outlet temperature is set to be the same as the inlet temperature. The amount of heat input/output to achieve the desired outlet temperature will be calculated accordingly. The conversion reactor should now be fully converged. As mentioned earlier, Symmetry is set to solve the model once the necessary data are sufficient (DOF is satisfied).

The final step to complete the reactor model is to connect two material streams to the reactor outlets (S2 for vapor outlet, S3 for liquid outlet) and one energy stream. The latter is connected directly to the reactor's heat input/

TABLE 8.3 Specification required in the reactor.

Pressure drop	5 psi
Reactor mode	Isothermal

output and is used to perform energy balances. A convenient way to display the simulation results in Symmetry without having to access the unit operation's user form individually is to create a material balance table. The latter displays both material balances and specific parameters of interests. Figs. 8.7–8.10 show detailed steps in displaying the material balances and specific parameters of interest.

8.3.3 Modeling of separation units

In this model, a distillation column is used to simulate the vapor-liquid separation process. Design and operating parameters of the column are shown in Table 8.4.

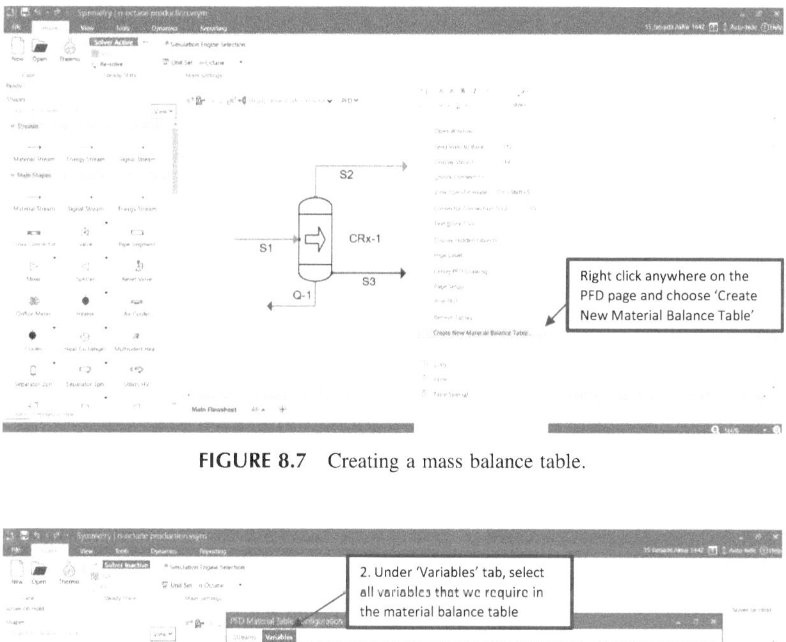

FIGURE 8.7 Creating a mass balance table.

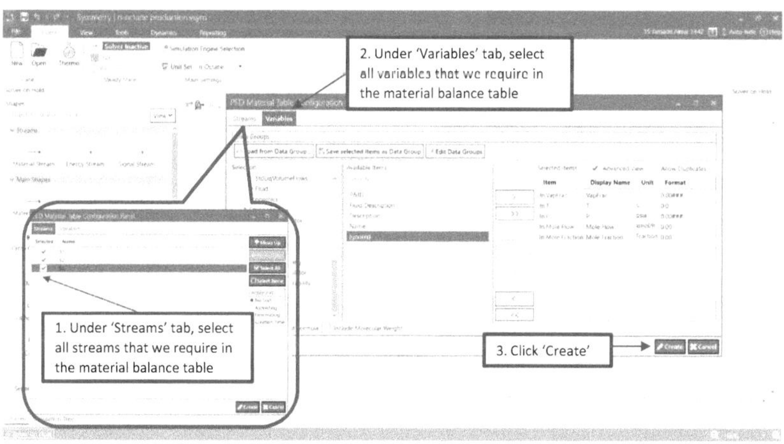

FIGURE 8.8 Selecting which stream to be showed in the mass balance table.

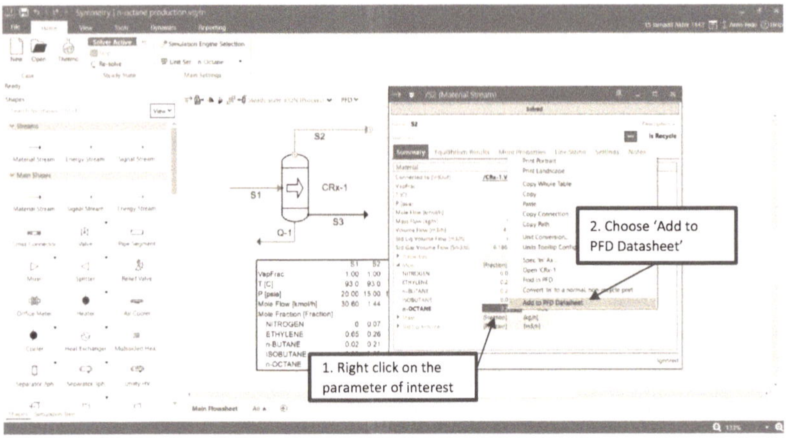

FIGURE 8.9 Steps to select specific information from a stream.

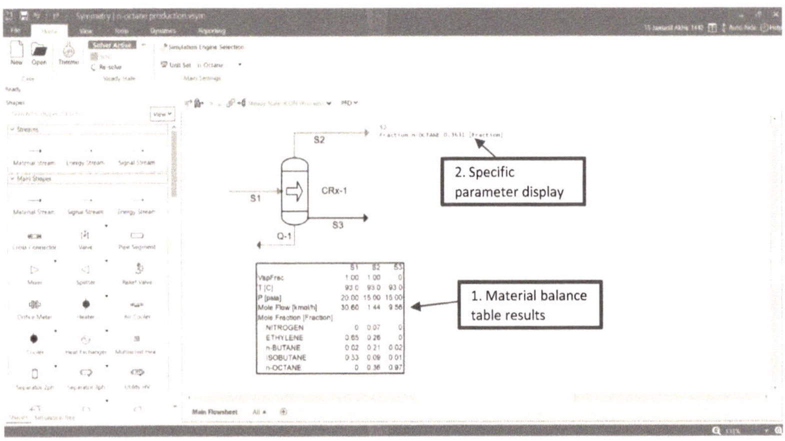

FIGURE 8.10 Steps to display mass balance table and specific parameters.

TABLE 8.4 Specifications used in the distillation column.

Specifications	Details
Condenser type	Partial condenser
Top pressure	10 psia
Bottom pressure	15 psia
Number of trays	10
Liquid distillate flow	0 kmol/h
Heavy key in vapor distillate (*n*-octane)	20 mol%
Light key in bottoms (ethylene)	0.01 mol%

Figs. 8.11–8.16 show the detailed steps in connecting the material streams to the distillation column and configuration of the column shows the result of a converged column simulation after it is connected to the upstream reactor.

Steps in Figs. 8.15 and 8.16 are repeated for reboiler with details shown in Table 8.5. For the liquid draw specification of the reboiler (Table 8.5), select the "reboiler.L" option of the Stage 10 to signify that it is the liquid part of the stage that is being drawn from the column. To converge the column, click the "Solve" tab after inserting all specifications in the column. Fig. 8.17 shows the mass balance result of the converged column simulation.

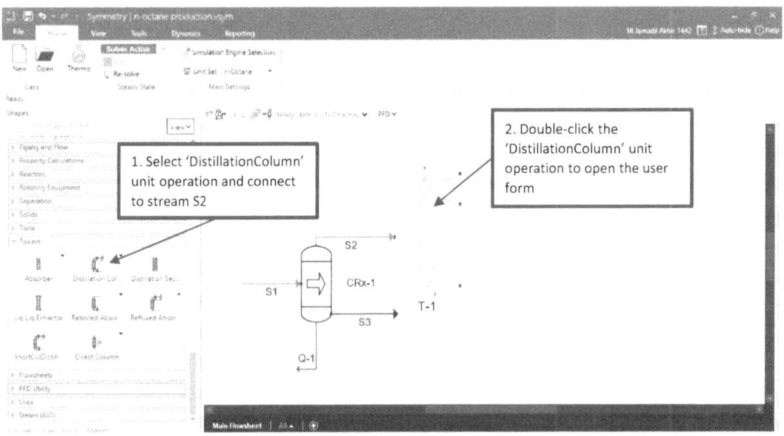

FIGURE 8.11 Steps to add a distillation column into the flowsheet.

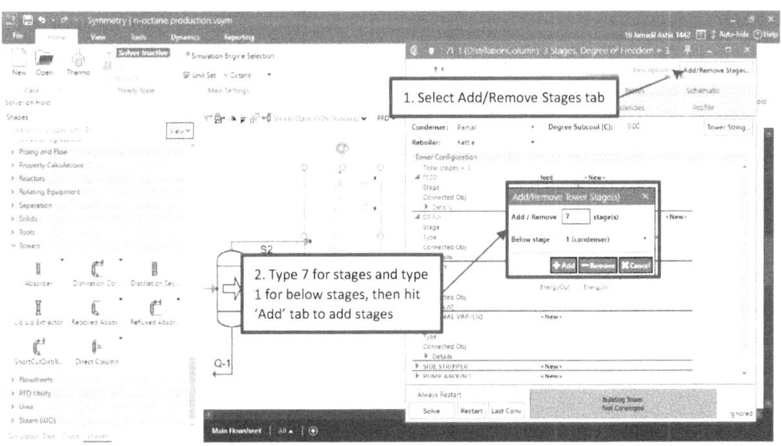

FIGURE 8.12 Steps for adding/removing stages in the distillation column.

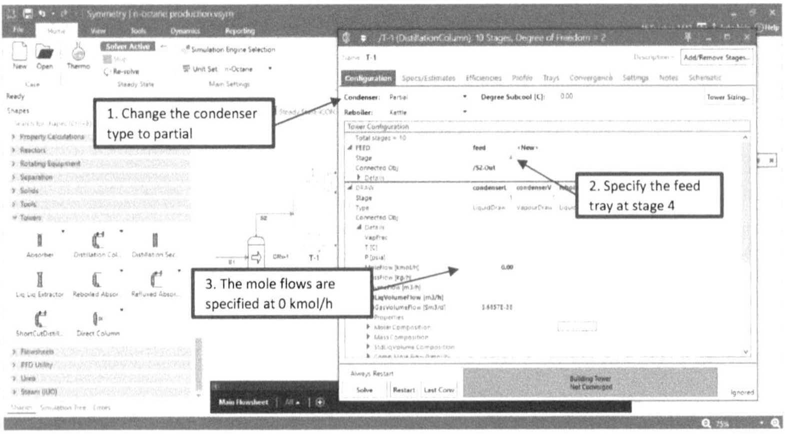

FIGURE 8.13 Steps to configure inlet and outlet streams of the distillation column.

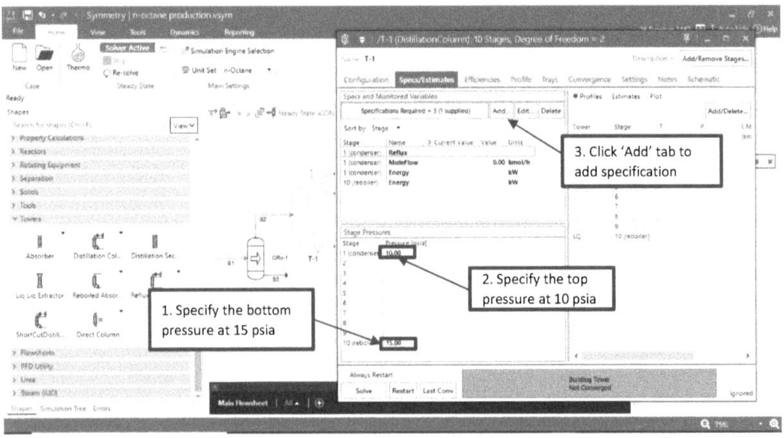

FIGURE 8.14 Steps for adding specifications for the distillation column.

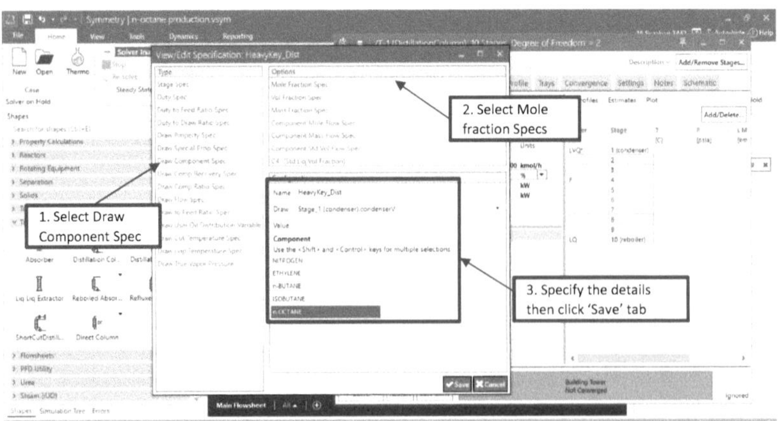

FIGURE 8.15 Steps for adding specifications for the distillation column (continued).

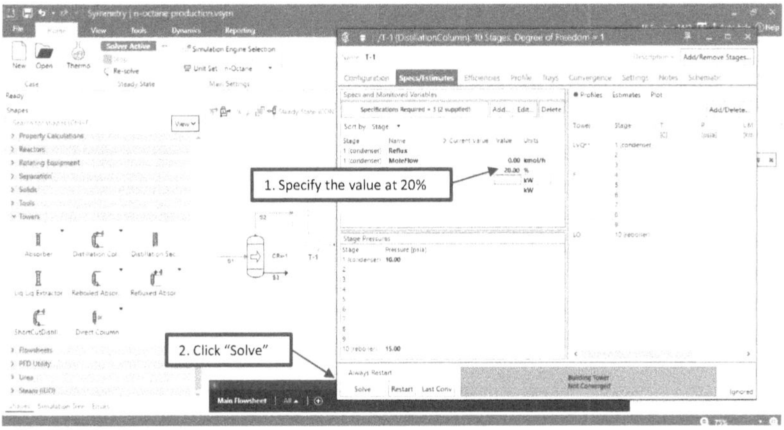

FIGURE 8.16 Steps for adding specifications for the distillation column.

TABLE 8.5 Specifications for the reboiler.

Specifications	Details
Name	Light Key_Bottom
Draw	Stage_10 (reboiler).reboilerL
Component	Ethylene
Value	0.01%

DRAW	condenserL	condenserV	reboilerL	<New>
Stage	1	1	10	
Type	LiquidDraw	VapourDraw	LiquidDraw	
Connected Obj	/S5.In	/S4.In	/S6.In	
Details				
VapFrac	0.00	1.00	5.78E-08	
T [C]	66.5	66.5	109.0	
P [psia]	10.00	10.00	15.00	
MoleFlow [kmol/h]	**0.00**	3.38	0.89	
MassFlow [kg/h]	0.00	199.44	100.40	
VolumeFlow [m3/h]	0.000	136.959	0.157	
StdLiqVolumeFlow [m3/h]	0.000	0.380	0.143	
StdGasVolumeFlow [Sm3/d]	5.6857E-38	.9232E+3	5.0683E+2	
Properties				
Molar Composition				
NITROGEN	0.00026	0.24424	0.00	
ETHYLENE	0.00079	0.1007	0.0001	
n-BUTANE	0.04484	0.43097	0.02712	
ISOBUTANE	0.00184	0.02409	0.00113	
n-OCTANE	0.95227	0.20	0.97164	
Mass Composition				
StdLiqVolume Composition				

FIGURE 8.17 Simulation results of converged distillation model.

8.3.4 Modeling of recycle systems

As shown in Fig. 8.18, stream S4 is a vapor stream exiting the distillation column. This stream contains about 10 mol% of unconverted ethylene. Hence, this stream is to be recycled to the reactor for higher conversion of unreacted raw materials. In this example, apart from unreacted material, stream S4 also contains inert component such as nitrogen. To avoid its accumulation, 10% of S4 is to be purged via a splitter (see steps in Fig. 8.18). For the latter, the fraction of recycled stream is set as 0.9 (see Fig. 8.19).

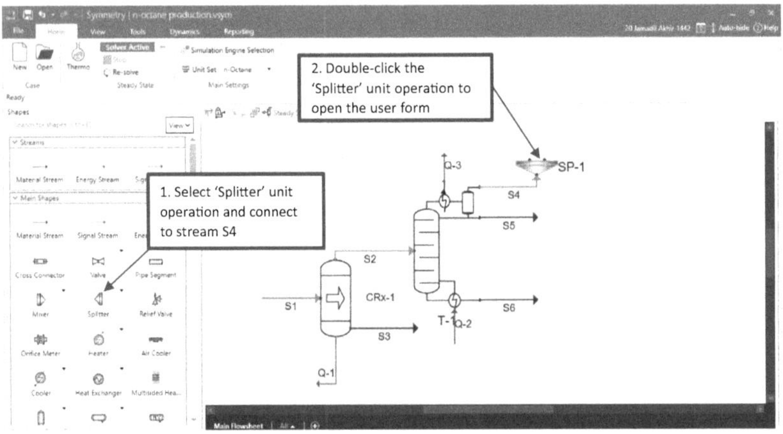

FIGURE 8.18 Steps to configure the splitter.

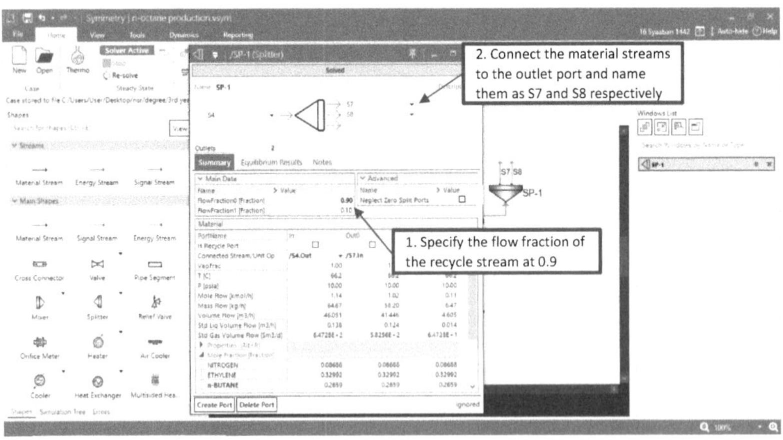

FIGURE 8.19 Step to add specifications for the splitter.

The recycled stream S7 has a pressure of 10 psia while the reactor is operated at 20 psia. Therefore, a compressor is used to increase its pressure of stream S7 to 22 psi (take into account 2 psi pressure drop of subsequent cooler) to fulfill the required pressure in the reactor. The detailed steps are shown in Figs. 8.20 and 8.21.

To complete the compressor model, a material stream (S9) and an energy stream (Q-4) are attached to its outlets. Simulation indicates that the temperature of its outlet stream S9 is higher than the required temperature, due to the specified efficiency of the compressor. Hence, a cooler is needed to reduce its temperature to 93°C. Figs. 8.22 and 8.23 show the steps to connect the

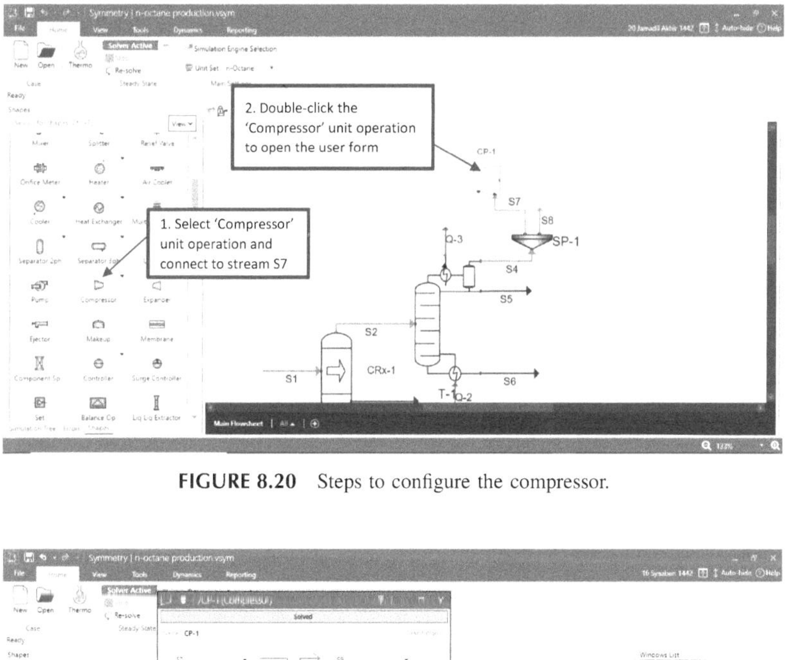

FIGURE 8.20 Steps to configure the compressor.

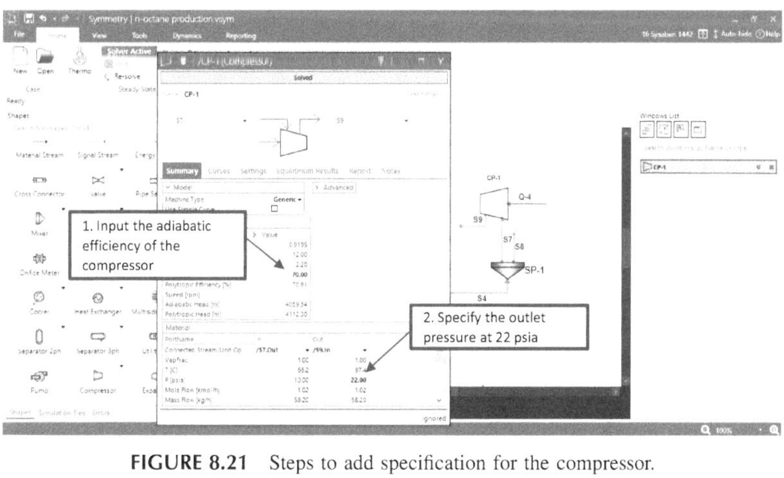

FIGURE 8.21 Steps to add specification for the compressor.

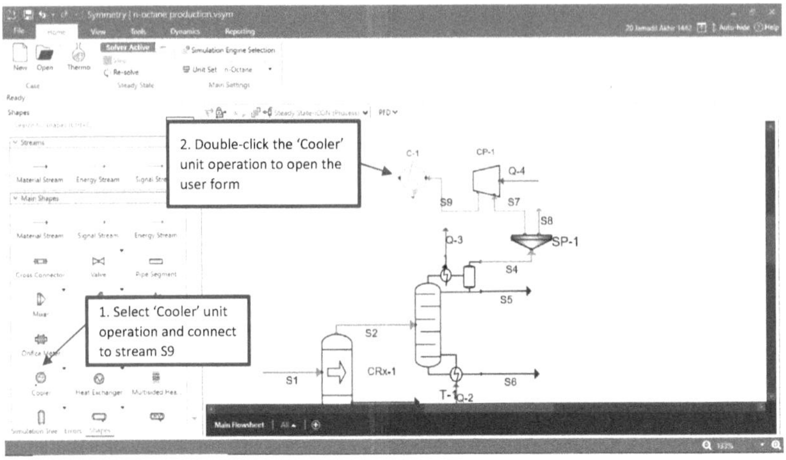

FIGURE 8.22 Step to add cooler in the flowsheet.

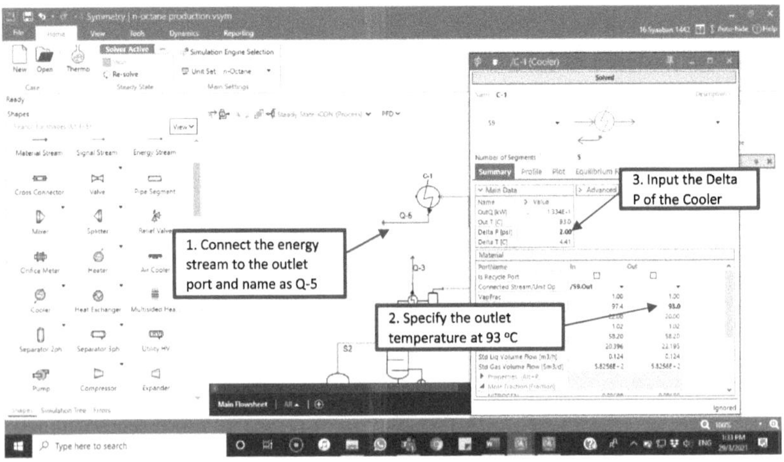

FIGURE 8.23 Steps to add specifications for the cooler.

cooler model to this stream. Table 8.6 summarizes the required specifications for the splitter, compressor, and cooler.

In Symmetry, a recycle stream must be specified prior to connecting it to a unit operation (i.e., a closed loop cycle). This specification is done by ticking "Is Recycle" box, as shown in Fig. 8.24. As shown, the recycle stream is marked as ~S10, indicating it being a recycle stream.

The recycle stream is next connected to the outlet of cooler C-1. Prior to that, the feed stream S1 to reactor (a mixed stream between fresh feed stream

TABLE 8.6 Specification of other unit operations.

Equipment	Specifications	Details
Splitter	Flow fraction for recycle stream	0.9
Compressor	Outlet pressure	22 psia
	Adiabatic efficiency	70%
Cooler	Outlet temperature	93°C
	Delta P	2 psi

and recycle stream ~S10) contains unreacted raw material. Selection of the mixer and steps to configuring it is shown in Fig. 8.25. Stream S1 is then reconnected to become one of the inlet streams of the mixer, as shown in Fig. 8.26.

Once these streams are connected to the reactor, the recycle stream ~S10 is then connected to the outlet of the cooler, as shown in Fig. 8.27. In this example, the solver is made nonactive while converging the recycle stream (see Fig. 8.25). Nonetheless, the solver can also be kept active during this process, in which Symmetry will perform iterative calculations until the recycle error in stream ~S10 becomes less than the allowable error tolerance. This latter is found in the recycle stream (see Fig. 8.28). A complete process flowsheet is shown in Fig. 8.29.

In previous step, it has been assumed that the fresh feed stream S1 is accessible at 93°C. This assumption is now relaxed. A heater model is added to increase the temperature of the fresh feed stream from 30 to 93°C. Figs. 8.30 and 8.31 show the detailed steps to install a heater (H-1), with specification given in Table 8.7.

Table 8.8 shows all calculated energy streams for the flowsheet. Heater H-1 for the feed stream requires a heating duty of 36.58 kW (indicated by stream Q-6), while the energy removed from the recycle stream via cooler C-1 is determined as 0.43 kW (stream Q-5). It is beneficial to recover this removed heat to increase the feed stream temperature by using a process-to-process heat exchanger. Doing this lead to reduced overall energy consumption of the process. Figs. 8.32 and 8.33 show the detailed steps to add a heat exchanger for heat recovery purpose, with specifications tabulated in Table 8.9.

Fig. 8.34 shows the converged flowsheet after feed stream S1 is integrated with recycle stream S10 for heat recovery. Material stream table is shown in Table 8.10.

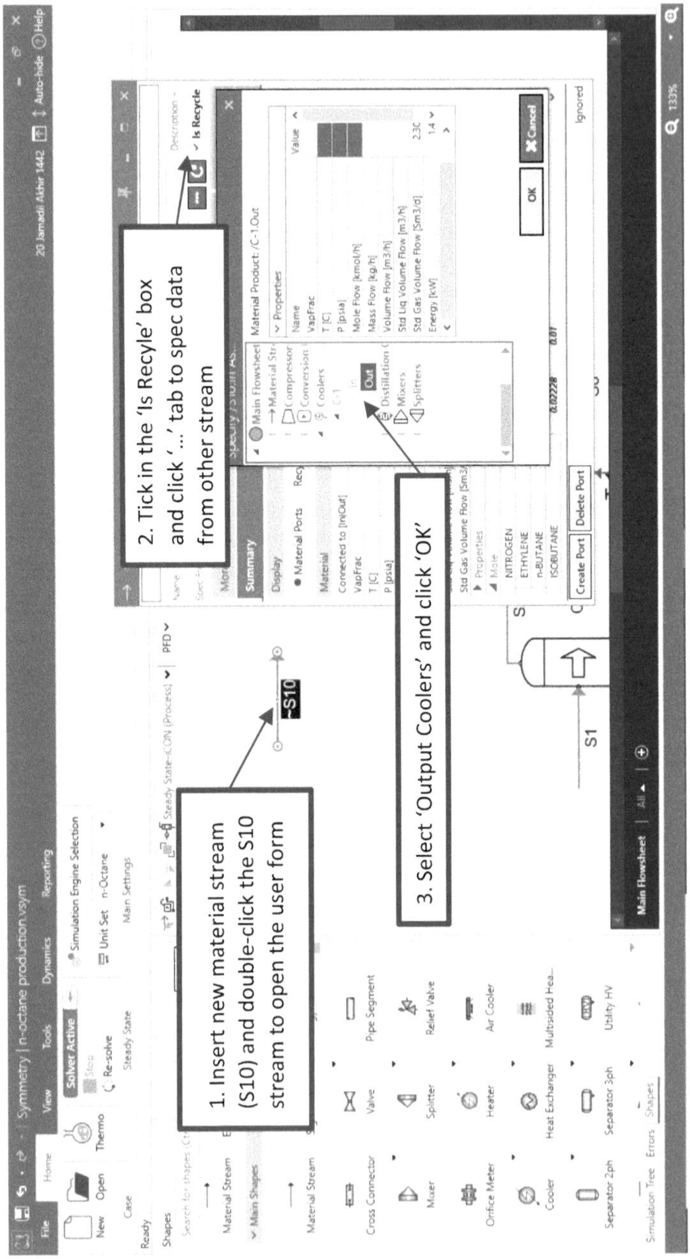

FIGURE 8.24 Steps to create a recycle stream.

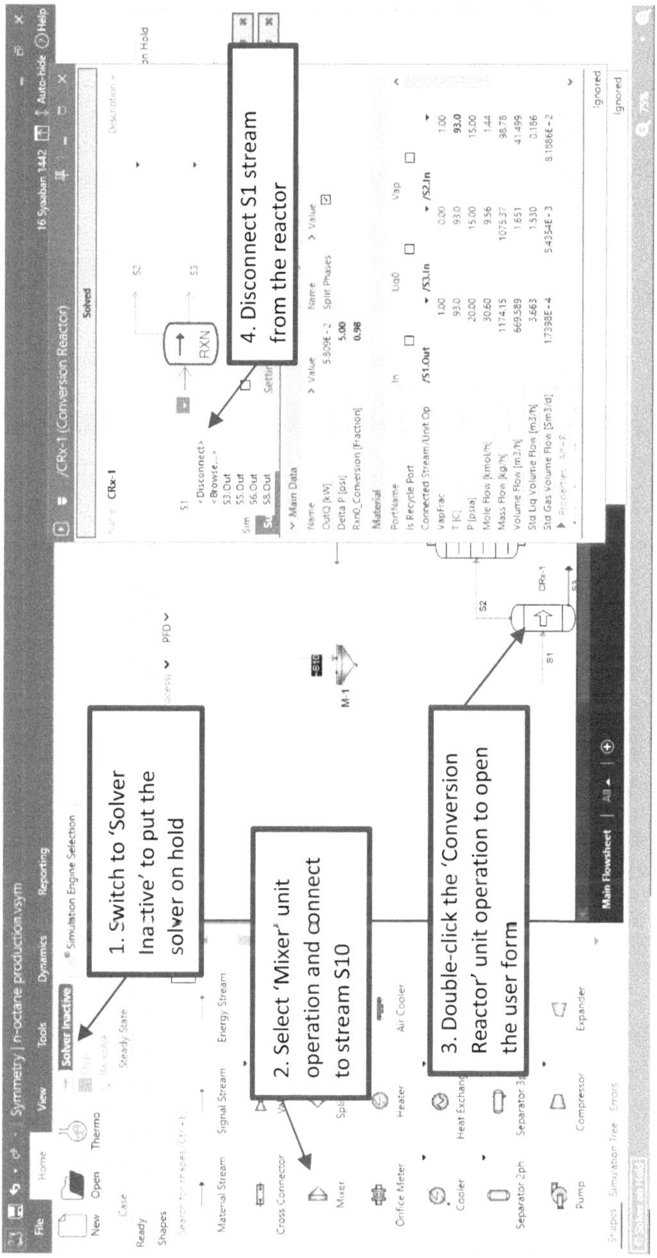

FIGURE 8.25 Steps to configuring the mixer.

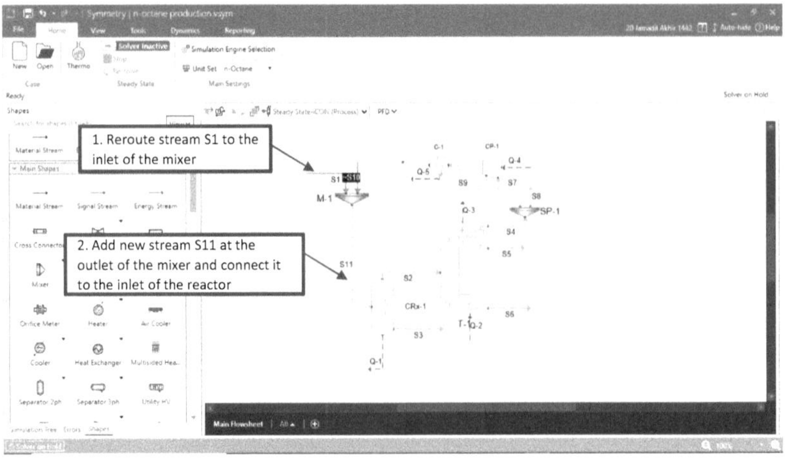

FIGURE 8.26 Reroute the inlet stream.

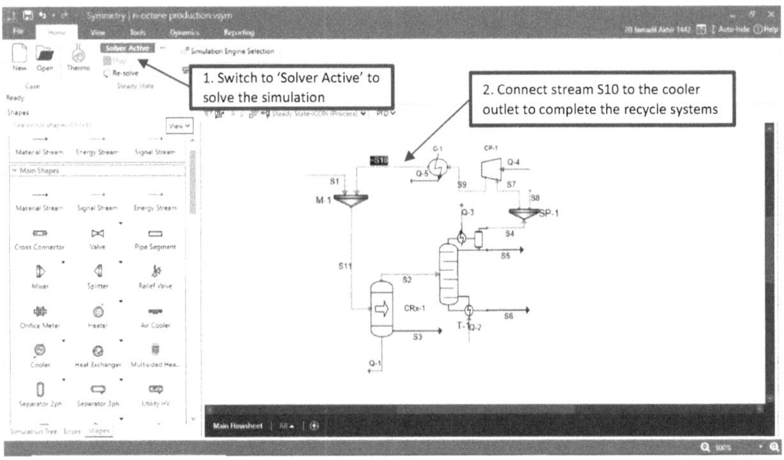

FIGURE 8.27 Steps to complete the recycle stream.

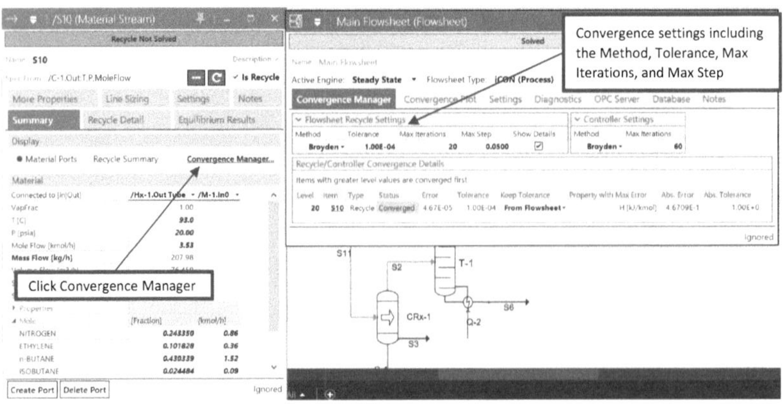

FIGURE 8.28 Error tolerance setting in convergence manager.

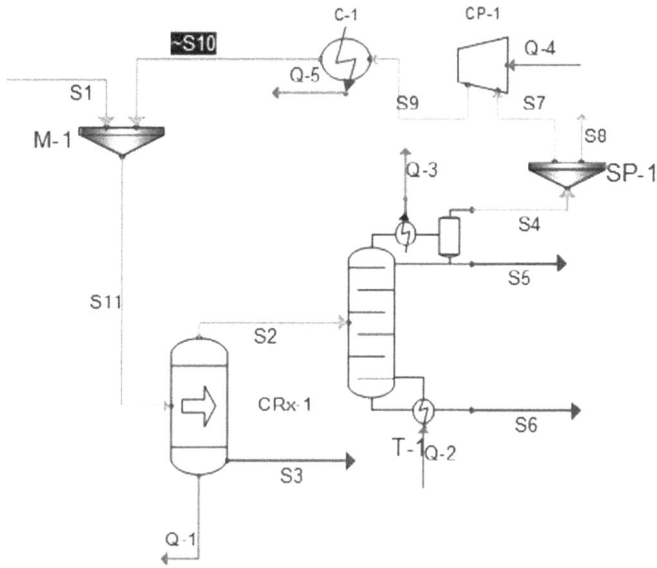

FIGURE 8.29 Process flow diagram with recycle stream.

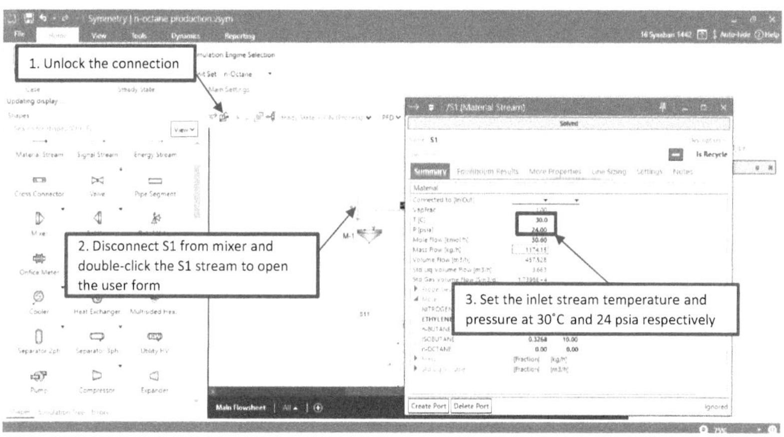

FIGURE 8.30 Steps to revise specification of the inlet stream.

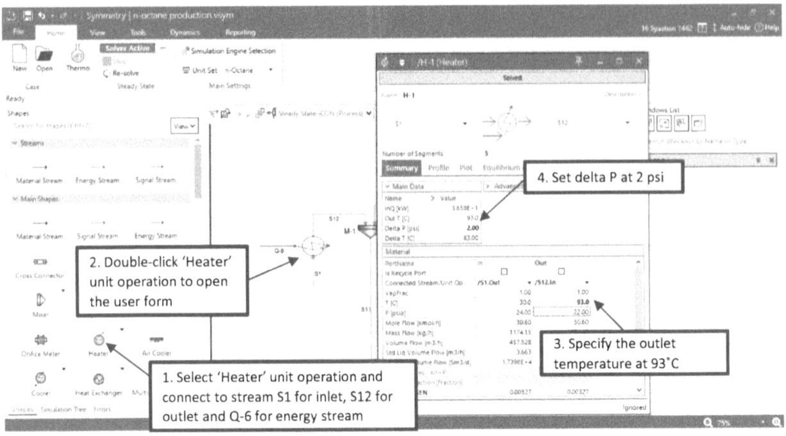

FIGURE 8.31 Steps to configure the heater.

TABLE 8.7 Specifications of the heater.

Equipment	Specifications	Details
Heater	Outlet temperature	93°C
	Delta P	2 psi

TABLE 8.8 Calculated energy streams.

	Q-1	Q-2	Q-3	Q-4	Q-5	Q-6
Calculated energy (kW)	586	1.81	13.95	3.17	0.43	36.58

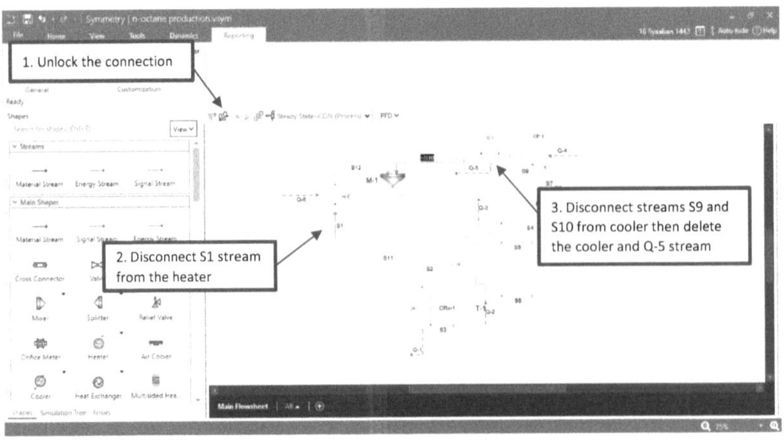

FIGURE 8.32 Steps to disconnecting the streams.

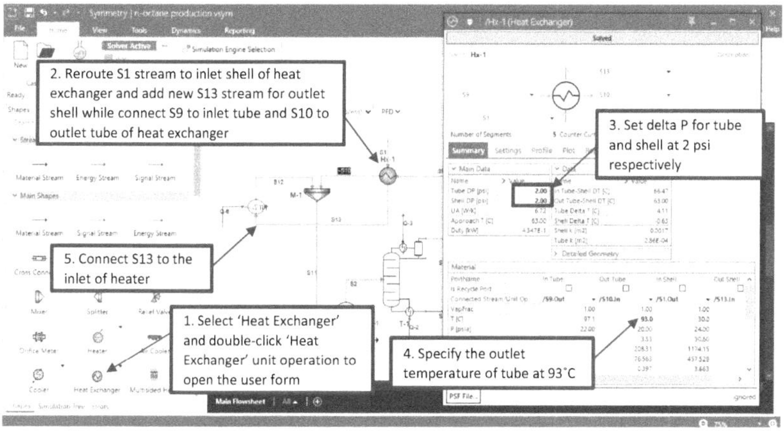

FIGURE 8.33 Selecting the heat exchanger on to the flowsheet.

TABLE 8.9 Specification of the heat exchanger.

Equipment	Specifications	Details
Heat exchanger	Delta P (tube side)	2 psia
	Delta P (shell side)	2 psia
	Outlet temperature (tube side)	93°C

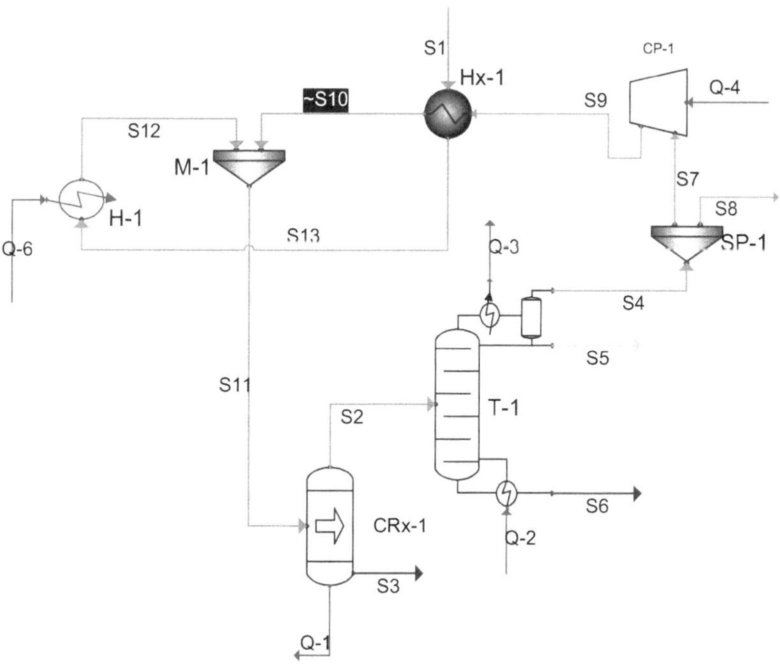

FIGURE 8.34 A converged simulation flowsheet for *n*-octane production.

TABLE 8.10 Material stream table.

Streams	S1	S2	S3	S4	S5	S6	S7
Vapor fraction	1.00	1.00	0.00	1.00	0.00	0	1.00
T [°C]	30.0	93.0	93.0	66.5	66.5	109.0	66.5
P [psia]	24.00	15.00	15.00	10.00	10.00	15.00	10.00
Mole flow [kmol/h]	30.60	4.96	9.21	3.93	0.00	1.03	3.53
Mass flow [kg/h]	1174.15	348.00	1034.13	231.46	0.00	116.54	208.31
Volume flow [m^3/h]	457.528	143.006	1.589	158.967	0.000	0.183	143.071
Nitrogen	0.0033	0.1926	0.0003	0.2434	0.0003	0	0.2434
Ethylene	0.6536	0.0806	0.0008	0.1018	0.0008	0.0001	0.1018
n-Butane	0.0163	0.3462	0.0326	0.4303	0.0448	0.0271	0.4303
Isobutane	0.3268	0.0196	0.0014	0.0245	0.0019	0.0012	0.0245
n-octane	0.00	0.3610	0.9648	0.2000	0.9523	0.9717	0.2000

Streams	S8	S9	S10	S11	S12	S13
Vapor fraction	1.00	1.00	1.00	1.00	1.00	1.00
T [°C]	66.5	97.1	93.0	93.0	93.0	30.6
P [psia]	10.00	22.00	20.00	20.00	20.00	22.00
Mole flow [kmol/h]	0.39	3.53	3.53	34.13	30.60	30.60
Mass flow [kg/h]	23.15	208.31	207.98	1382.13	1174.15	1174.15
Volume flow [m^3/h]	15.897	70.316	76.459	746.150	669.589	501.004
Nitrogen	0.2434	0.2434	0.2433	0.0281	0.0033	0.0033
Ethylene	0.1018	0.1018	0.1018	0.5966	0.6536	0.6536
n-Butane	0.4303	0.4303	0.4303	0.0591	0.0163	0.0163
Isobutane	0.0245	0.0245	0.0245	0.2955	0.3268	0.3268
n-octane	0.2000	0.2000	0.2000	0.0207	0.00	0.00

8.4 Conclusions

Basic simulation steps to model the *n*-octane production process with Symmetry has been described in detail in this chapter. The procedure for simulation is mainly guided by sequential modular approach and the Onion model (see Chapter 1 for details). Note that unit models used in this example are meant for preliminary design. Rigorous models should be used for more detailed engineering design.

Exercises

1. From the converged flowsheet in Fig. 8.34, analyze the effect of changing compressor efficiency from 70% to (a) 80% and (b) 60%.
2. Determine and analyze the effect of having a lower (0.85) or higher (0.95) recycle ratios.

Reference

iCON-Symmetry, 2021. Symmetry Special Version Created for PETRONAS Including iCON – User Guide. https://www.software.slb.com/products/symmetry.

Chapter 9

Process modeling and analysis of a natural gas dehydration process using tri-ethylene glycol (TEG) via Symmetry*

Siti Nurfaqihah Azhari, Noorhidayah Bt Hussein and Zulfan Adi Putra
PETRONAS Group Technical Solutions, Process Simulation and Optimization, Kuala Lumpur, Malaysia

Chapter outline

9.1 Introduction

When hydrocarbon is produced from the reservoir, it usually contains a large amount of water. The purpose of having water removed from natural gas is mainly to avoid downstream problems such as freezing of water, hydrates formation, and even corrosion (GPSA, 2004). Dehydration of hydrocarbon can be achieved by either liquid or solid desiccants. Typically, liquid desiccant is more economical to meet the required dehydration specification (GPSA, 2004). Glycols are usually used for such applications where dew point depression in the order of 15−49°C are required. Diethylene (DEG),

*Simulation and other electronic files of this chapter are made available on a companion website; URL: https://www.elsevier.com/books-and-journals/book-companion/9780323901680.

Chemical Engineering Process Simulation. https://doi.org/10.1016/B978-0-323-90168-0.00003-2

triethylene (TEG), and tetraethylene glycol (TREG) are commonly used as liquid desiccants, with TEG being the most favorable for natural gas dehydration.

In terms of thermodynamic models, modeling of TEG-water system showed that Peng-Robinson equation-of-state (EOS) has been found to be a good representation (Gironi et al., 2010). Others have shown that Universal Mixing Rules − Peng Robinson UNIFAC (UMR-PRU) gives a more realistic prediction of heat capacities of aqueous TEG mixtures (Petropoulou and Voutsas, 2018).

To date, there are numerous efforts in optimizing the performance of natural gas dehydration unit. Among them are modeling of TEG system for a plant in Nigeria for performance evaluation and optimization (Dagde and Akpa, 2014) or using commercial process simulation to minimize operating cost of a TEG system (Chebbi et al., 2019). Affandy et al. (2020) showed that utilizing flash gas as stripping gas lead to the reduction of TEG circulation flowrate, and hence, reduction of the reboiler duty. Process simulation on regeneration of TEG with minimum use of energy and loss of TEG has been done on a gas processing plant in Iran, combining design of experiment (DOE) and process simulation software (Kamin et al., 2017). In another study, TEG circulation mass flow and rerouting gas from TEG flash drum to TEG regenerator as stripping gas are important variables to minimize costs. Furthermore, the use of Coldfinger process has been explored to minimize the loss of TEG while still capable of reaching current water content specification in a simple and economic way (Gironi et al., 2007).

In this chapter, Symmetry is used to improve the performance of a typical natural gas dehydration process using TEG. The objective is to analyze the effect of operational parameters such as reboiler temperature and stripping gas flowrate on two performance indicators, namely water dew point and reboiler duty. This is achieved by doing sensitivity analysis via a tool called Case Study. Another objective of the model is to find optimum values for these operational parameters that minimize water dew point and/or reboiler duty. A tool called Optimizer is used to achieve this objective.

9.2 Process description

Fig. 9.1 shows an Symmetry simulation flowsheet of a typical glycol (TEG) dehydration process that is commonly found in natural gas processing facilities. As shown, the process consists of glycol contactor (C-101), glycol regenerator (C-102), and other auxiliary equipment. A stripper column (C-103) is added downstream of the regenerator to increase the purity of TEG before it is recycled to the glycol contactor.

Lean, water-free glycol (purity >99%) is fed at the top of the glycol contactor where it is contacted with the "wet" natural gas stream (stream S-3). The glycol removes water from the natural gas by physical absorption. The "dry" natural gas leaves the contactor and is then sent to a gas processing

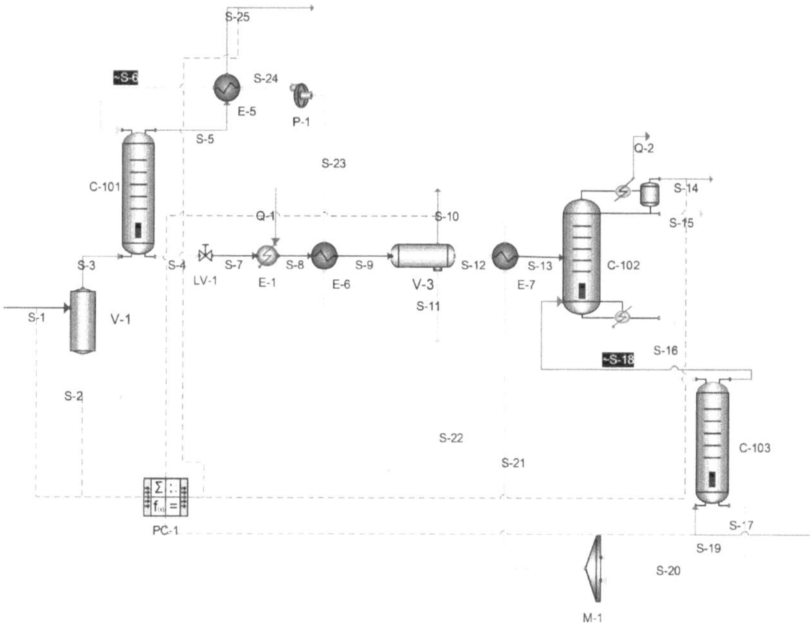

FIGURE 9.1 Simulation flowsheet for a typical TEG dehydration process in Symmetry.

plant. The internals of the glycol contactor can either be tray or packed columns.

Upon exiting the contactor, the pressure of the glycol stream (typically referred to as "rich glycol") is reduced. The glycol contactor is operated at a higher pressure than the regenerator column. This rich glycol is then heated and sent to vessel (V-3), where light hydrocarbons are removed. After the vessel, the rich glycol is heated again in a cross-exchanger (E-7), prior to entering the glycol regenerator column (C-102). This column is a typical distillation column with a partial condenser and a reboiler. The glycol is thermally regenerated to remove any excess water. To further increase its purity, the lean glycol stream is sent to the stripping column (C-103). This lean glycol stream is then cooled by the incoming rich glycol. It is then pumped back to a glycol cooler (E-5) together with a glycol make-up stream (S-20). In this flowsheet, the cooler E-5 is a heat exchanger with the dry gas (S-5), leaving the absorber as the cooling medium. Then, this cooled lean solvent enters the contactor.

9.3 Process simulation

9.3.1 Thermodynamic model and feed stream specification

In this simulation, the selected thermodynamic model is Advanced Peng Robinson (APR) for Natural Gas, which is available from the thermodynamic library of Symmetry. The model is selected due to its accuracy in predicting

TABLE 9.1 Feed gas (S-1) composition and operating conditions.

Components	Mole fraction	Conditions
Water (H_2O)	0.0010	T: 19.6°C
Nitrogen (N_2)	0.0092	
Carbon dioxide (CO_2)	0.0704	
Methane (CH_4)	0.8146	
Ethane (C_2H_6)	0.0603	
Propane (C_3H_8)	0.0201	P: 27.0 bar gas volumetric flowrate: 100 MMSCFD
Isobutane (C_4H_{10})	0.0059	
n-Butane (C_4H_{10})	0.0055	
Isopentane (C_5H_{12})	0.0030	
n-Pentane (C_5H_{12})	0.0020	
Triethylene glycol ($C_6H_{14}O_4$)	0.0000	
C6+	0.0080	

TABLE 9.2 Hypothetical component data.

Data	Value
Name	C6+
Molecular weight	86.16
NBP (°C)	68.7
Liquid density (kg/m^3)	656.0

the water content in the hydrocarbon stream. Table 9.1 shows the specification of the feed gas (stream S-1) that need to be dehydrated. Note that C6+ is Table 9.1 is hypothetical component whose properties are shown in Table 9.2.

9.3.2 Base case simulation

Specifications for some unit operations are shown in Table 9.3, while the rest are explained subsequently.

TABLE 9.3 Specifications for unit operations (refer to Fig. 9.1 for the notation).

Unit operation	Specifications	Details
Glycol contactor (C-101)	Total stages	4
	Top stage pressure (bar)	94.8
	Bottom stage pressure (bar)	95
Valve (LV-1)	Characteristics of the valve	Linear
	% Opening (%) of the valve	100
	Pressure outlet of the valve (bar)	6.01
Heater (E-1)	Outlet temperature (°C)	75
	Pressure drop (kPa)	0
Heat exchanger (E-6)	Tube pressure drop (kPa)	15
	Shell pressure drop (kPa)	15
	Outlet tube temperature (°C)	110
Heat exchanger (E-7)	Tube pressure drop (kPa)	15
	Shell pressure drop (kPa)	15
	Outlet tube temperature (°C)	150
Regeneration column (C-102)	Number of stages	3
	Liquid flowrate from condenser (kmol/h)	0
	Reboiler temperature (°C)	190
	Reflux ratio	0.15
	Pressure top stage (bar)	1.01
	Pressure bottom stage (bar)	1.04
Stripping column (C-103)	Pressure top stage (bar)	1.04
	Pressure bottom stage (bar)	1.04
	Number of stages	2
Pump (P-1)	Pump efficiency	75%
	Outlet pressure (bar)	93.35
Heat exchanger (E-5)	Tube pressure drop (kPa)	35.0
	Shell pressure drop (kPa)	35.0
	Temperature outlet tube (°C)	53.0

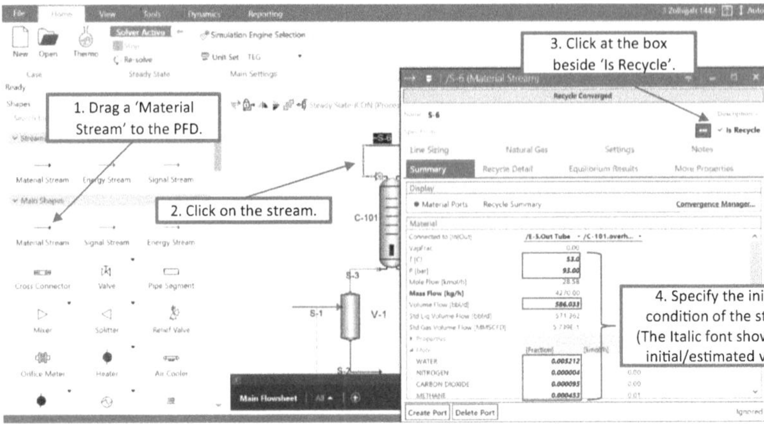

FIGURE 9.2 Steps to define recycle stream S-6.

All vessels are specified with no pressure drop (default setting) and operated adiabatically, i.e., no heat input/output.

For mixer M-1, no specification is needed. Nonetheless, the flowrate of recycled glycol stream (S-21) is specified as 4270 kg/h. The TEG make-up stream, S-20, contains 100% TEG, which is supplied at 25°C and 1.04 bar(a).

In Fig. 9.1, two recycle streams are observed, namely S-6 and S-18. A step-by-step procedure on how to define the recycle stream S-6 is shown in Fig. 9.2.[1] To simulate the flowsheet, initial conditions of these recycle streams are necessary. They are shown in Table 9.4 together with the specification for the gas stripping stream S-17. The latter is used to remove remaining water in the glycol stream that exits the regeneration column (C-102).

Once the necessary specifications are provided, the simulation flowsheet may be converged, with a calculated water dew point at 9°C (stream S-25). The amount of reboiler duty for C-102 is determined as 157 kW (readers may verify these).

9.4 Dew point evaluation with Case Study tool

Case Study tool in Symmetry allows users to create a wide range of scenarios by changing independent variables (input specifications), in order to examine their effect on those dependent variables. In other words, the Case Study tool is useful for sensitivity analysis.[2] In this example, the objective of this case study is to investigate the effect of reboiler temperature in the regeneration column

1. See Chapter 4 for more details in defining recycle stream in Symmetry.
2. Note: Case Study tool is also available in other software such as Aspen HYSYS and UniSim Design.

TABLE 9.4 Specifications of the material stream, S-6.

Components	S-6	S-18	S-17
Temperature (°C)	53	186.6	80
Pressure (bar)	93	1.04	94.8
Flowrate	586.55 bbl/day	0.066105 MMSCFD	0.028 MMSCFD
Compositions (moles)			
Water (H_2O)	0.061	0.5385	0.001
Nitrogen (N_2)	0.00	0.0037	0.0089
Carbon dioxide (CO_2)	0.00009	0.0289	0.0696
Methane (CH_4)	0.00033	0.3438	0.8167
Ethane (C_2H_6)	0.00005	0.0253	0.0604
Propane (C_3H_8)	0.000025	0.0083	0.0199
Isobutane (C_4H_{10})	0.000014	0.0024	0.0059
n-Butane (C_4H_{10})	0.000016	0.0022	0.0055
Isopentane (C_5H_{12})	0.000015	0.0014	0.0035
n-Pentane (C_5H_{12})	0.00001	0.0008	0.002
Triethylene glycol ($C_6H_{14}O_4$)	0.9384	0.0418	0.00
C6+	0.00005	0.0029	0.0075

(C-102) and the amount of stripping gas (S-17) on water dew point of the dry gas (S-25).

Once the base case model is converged, the dry gas stream (S-25) has a water mass flowrate of 6.93 kg/h, while the corresponding water dew point is 9°C (verify this yourself). The current objective is to reduce the water dew point to be lower than −35°C, so that the water does not condense when the gas is transported via long pipelines (typically hundreds of km to the onshore facility). Water condensation will lead to problems (e.g., vibration, different flow regimes) along the pipelines, which can disturb gas supplies to clients. Procedure to set up the case study is shown in Fig. 9.3.

For this case study, there are two independent variables, i.e., reboiler temperature of the regeneration column (C-102) and standard gas volume flow of stream S-17. On the other hand, the dependent variable is water dew point of stream S-25. Ranges of the independent variables are shown in Table 9.5,

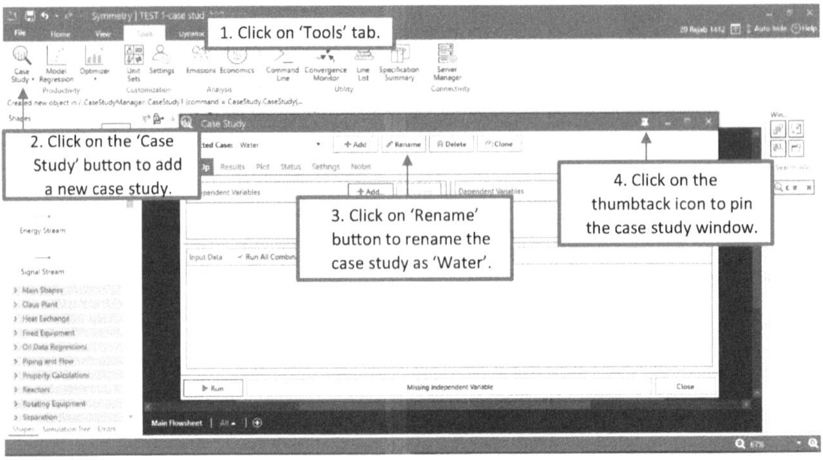

FIGURE 9.3 Steps in setting up the case study.

TABLE 9.5 Specification of reboiler temperature and stripping gas flowrate as an independent variables.

Variable	Minimum	Maximum	Steps
Reboiler temperature (°C)	185	246	11
Stripping gas flowrate (MMSCFD)	0.025	0.0485	11

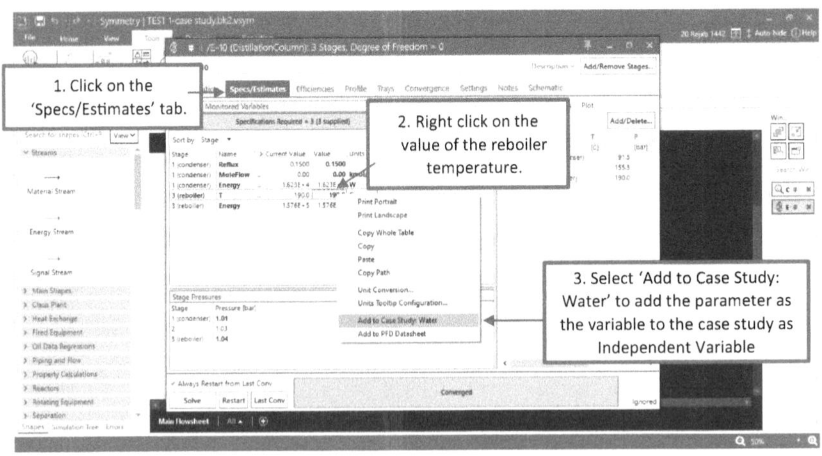

FIGURE 9.4 Steps for adding reboiler temperature as variable.

while the steps for the set up the independent variable are shown in Fig. 9.4 (declaring the variable) and Fig. 9.5 (specifying the limit).

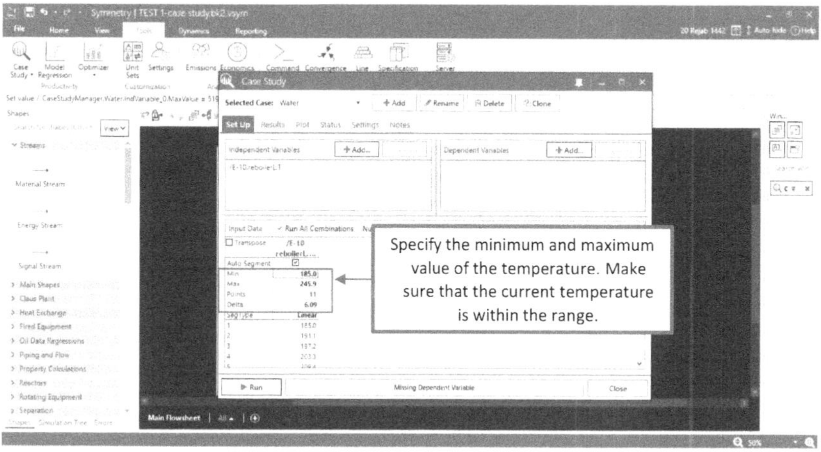

FIGURE 9.5 Steps for specifying minimum and maximum values of reboiler temperature.

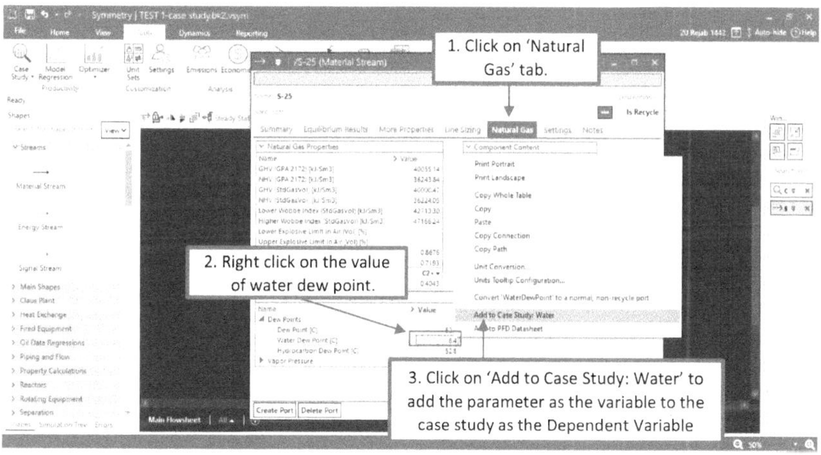

FIGURE 9.6 Steps for adding the water dew point as dependent variable.

Similar steps can be done for specifying other independent variables, i.e., stripping gas. For specifying water dew point as the dependent variable, steps in Fig. 9.6 can be followed.

Once all variables have been configured, the Case Study can be executed (see steps in Fig. 9.7), with results shown in table form (see Fig. 9.7). One may also display the results in graphical form. The table shown in Fig. 9.7 can be copied to other spreadsheet software (e.g., MS Excel) for other analysis (see details in Fig. 9.7).

The effect of reboiler temperature and stripping gas flowrate on water dew point and reboiler duty were plotted in MS Excel and are shown in Figs. 9.8

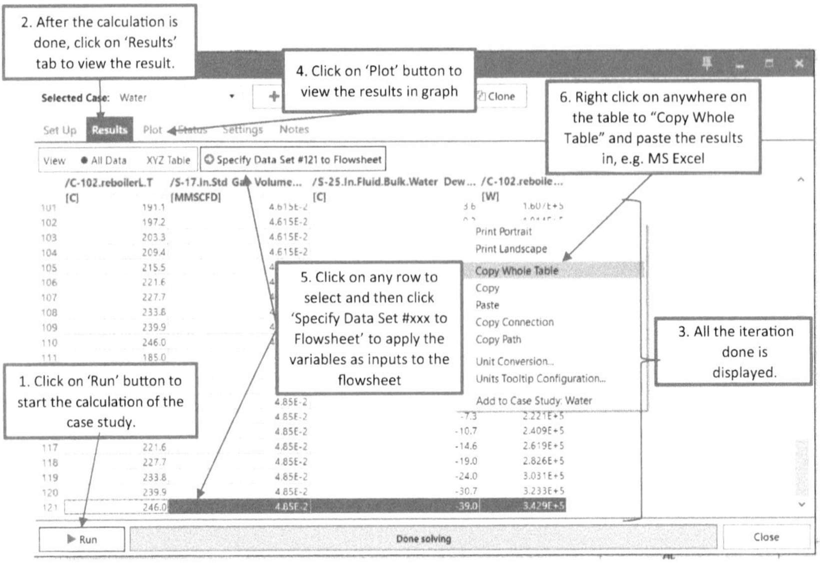

FIGURE 9.7 Results of the case study.

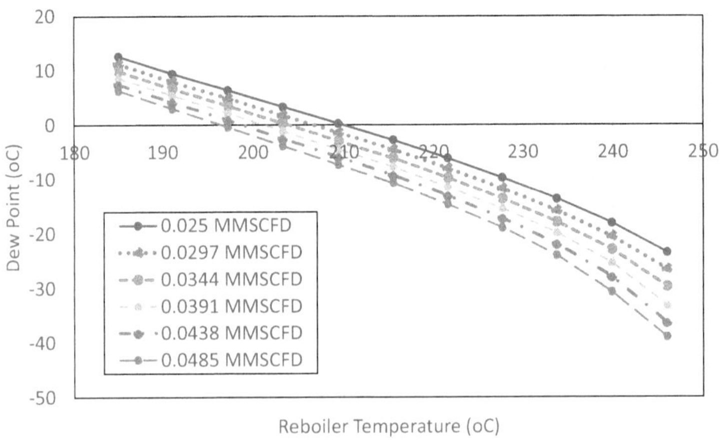

FIGURE 9.8 Water dew point (°C) of stream S-25 as a function of reboiler temperature (°C) and stripping gas flowrate (MMSCFD).

and 9.9, respectively. It can be seen that lower water dew point is observed with higher reboiler temperature and the stripping gas flowrate. Lowest water dew point of −39°C is achieved when both the reboiler temperature and the stripping gas flowrate reach their maximum limit (given in Table 9.5). It can also be observed that increasing the reboiler temperature has a higher impact in reducing water dew point, as compared to increased stripping gas flowrate.

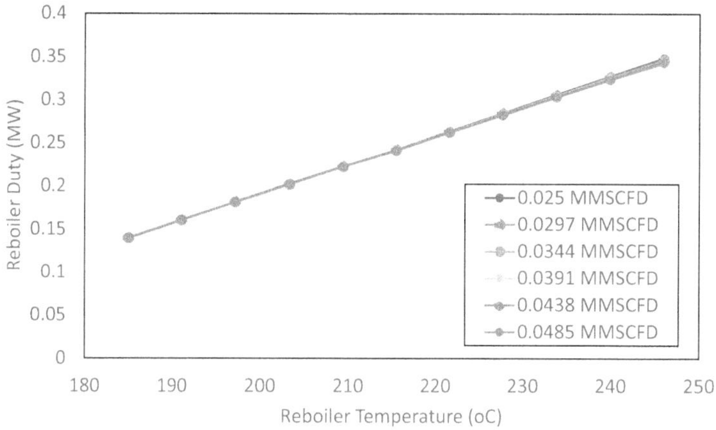

FIGURE 9.9 Reboiler duty (MW) in column C-102 as a function of reboiler temperature (°C) and stripping gas flowrate (MMSCFD).

In Fig. 9.9, changes in stripping gas flowrate do not affect the reboiler duty. It is shown that all different stripping gas flowrates are overlapping each other. From these two graphs, it is seen that reboiler temperature plays a more important role in reducing both water dew point and reboiler duty.

It is important to note that increased reboiler temperature reduces water dew point, while at the same time increases the reboiler duty. It is clear that these two objectives (i.e., water dew point and reboiler duty) are conflicting with each other. This calls for the use of optimization tool of Symmetry, which will be discussed in the following section.

9.5 Process improvement with optimizer

Optimizer is another useful tool in Symmetry for process improvement activities. It employs a general form of mathematical optimization to minimize or maximize an objective function, f.

$$\min/\max f(x, y, z)$$

subject to:

$$h_i(x, y, z) = 0, \quad \forall i \in I$$

$$g_j(x, y, z) \leq 0, \quad \forall j \in J$$

$$x \in \text{continuous}, \quad y \in \text{binary}, \quad z \in \text{integer}$$

A total of eight available local solvers are found in Symmetry v2020.3 (iCON-Symmetry User Guide, 2021). These include conventional methods such as Interior Point, Nelder-Mead, Powell, and BFGS, as well as modified methods such as LEX (simplex with a lexicographic approach), ALM (Augmented Lagrangian Multiplier), MINLP (Mixed Interior Point Non-Linear Programming), and MIGA (Mixed Integer Genetic Algorithm). Details of each solver can be found in the user guide (iCON-Symmetry User Guide, 2021) and in literature (Edgar and Himmelblau, 2001; Rhinehart, 2018).

To set up the Optimizer, three items need to be specified, i.e., objective function, manipulated variable, and constraints. Objective function is the calculated value to be maximized or minimized. Manipulated variables are to be adjusted so that the maximum/minimum objective can be achieved. Lastly, constraints are any operational or design values that need to be kept within a given limit.

For the natural gas dehydration process, the objective function is set to minimize the water dew point or the reboiler duty. Reboiler temperature and stripping gas flowrate are the manipulated variables in this case. The lower and upper limits for these variables are shown in Table 9.6. For the current step, these limits are sufficient for the optimization problem, while the constraint part of the Optimizer page can be left blank as it is.

Same procedure in the case study is used to set up the optimization problem. Detailed steps are shown in Fig. 9.10.

TABLE 9.6 Limits of independent variables.

Independent variables	Lower limit	Upper limit
Reboiler temperature (°C)	185.0	246
Stripping gas flowrate (MMSCFD)	0.025	0.0485

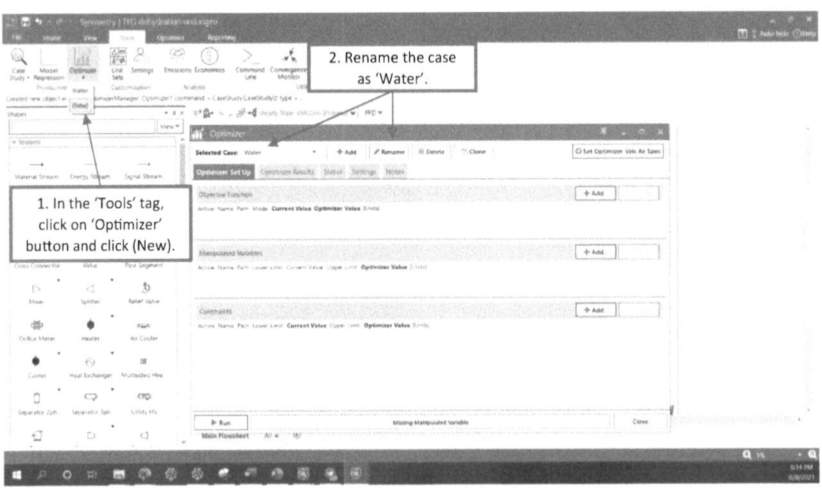

FIGURE 9.10 Steps for adding a new Optimizer case.

Figs. 9.11 and 9.12 show the detailed steps for adding reboiler temperature of glycol regenerator (C-102) as a manipulated variable. The same steps are applied for the stripping gas as shown in Fig. 9.13.

The detailed steps for adding the objective function are shown in Fig. 9.14. Once the objective function has been defined, its mode is switched to minimize as shown in Fig. 9.15.

Once the optimization formulation has been specified, users can select and run the optimization solver (see details in Fig. 9.16). Fig. 9.17 shows the optimization result by using LEX as the solver. In this particular case, the LEX

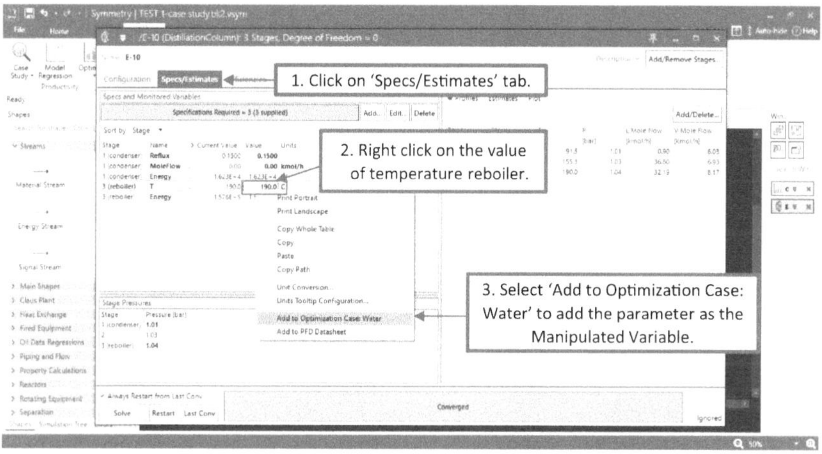

FIGURE 9.11 Steps for adding reboiler temperature as a variable.

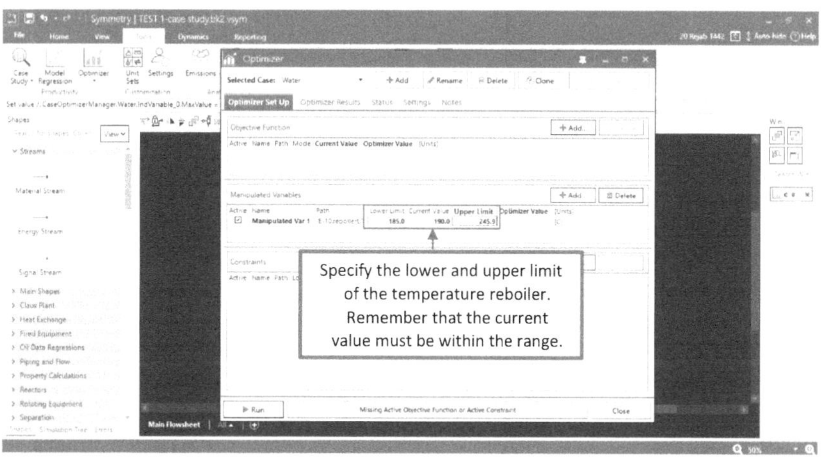

FIGURE 9.12 Steps for specifying lower and upper limits of the reboiler temperature variable.

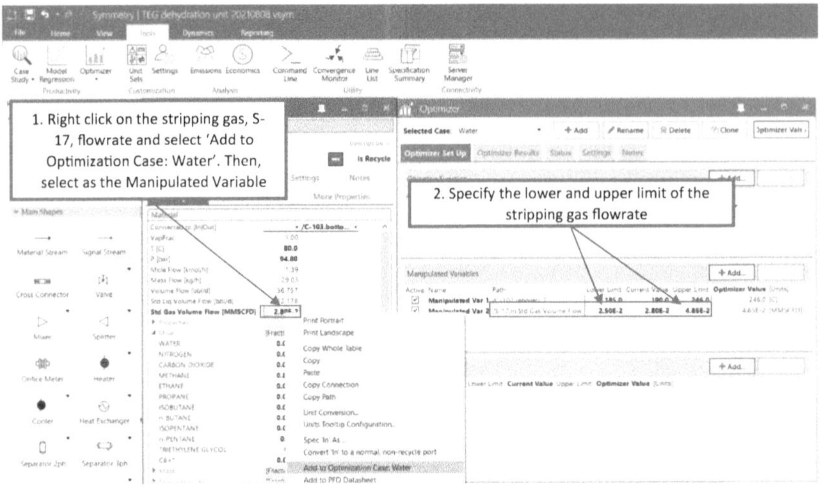

FIGURE 9.13 Specifying stripping gas flowrate as another manipulated variable.

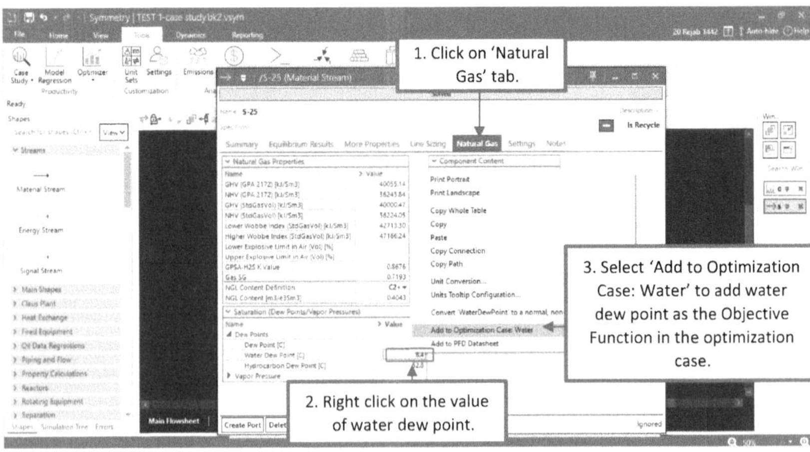

FIGURE 9.14 Steps for adding water dew point as objective function.

solver can find the solution relatively faster than others (interested readers may try with other solvers to compare the results).

After the calculation is completed, user can opt to specify the optimized value of the variables into the model (see top right of Fig. 9.17). This allows the simulator to make use of the optimized value for the TEG dehydration process model. From the results in Fig. 9.17, it is shown that the water dew point of −39°C can be achieved at regeneration reboiler temperature of 246°C and stripping gas flowrate of 0.0485 MMSCFD. These two variables are at the upper limits as defined in Table 9.6. The same exercise can be done with the

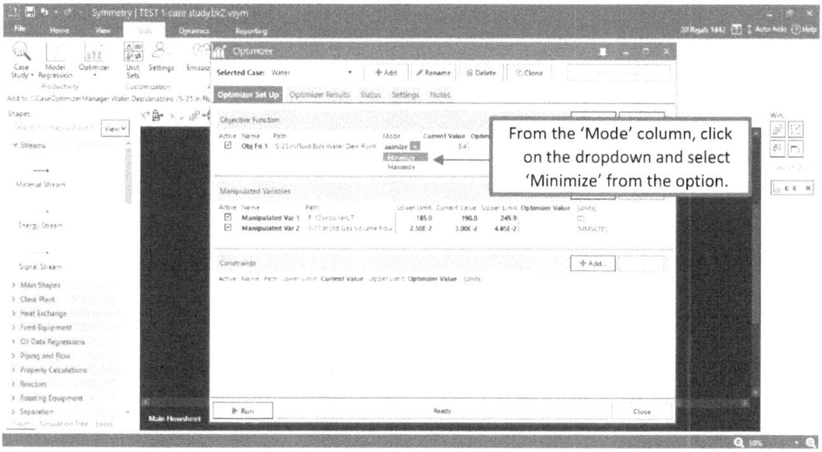

FIGURE 9.15 Steps for specifying the mode of objective function.

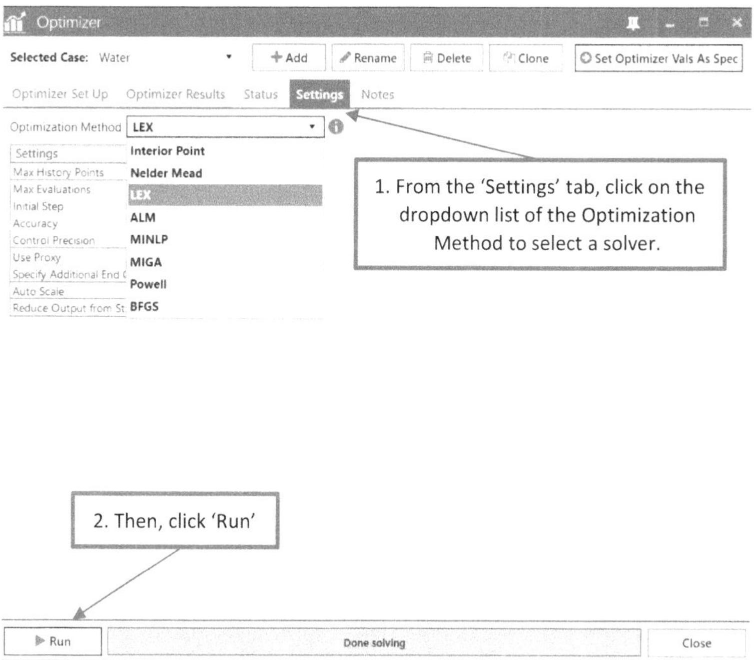

FIGURE 9.16 Selecting the optimization solver.

objective of minimizing the reboiler duty. As discussed before that minimizing water dew point will tend to maximize reboiler duty, and vice versa, it is then necessary to find a trade-off between the two objective functions.

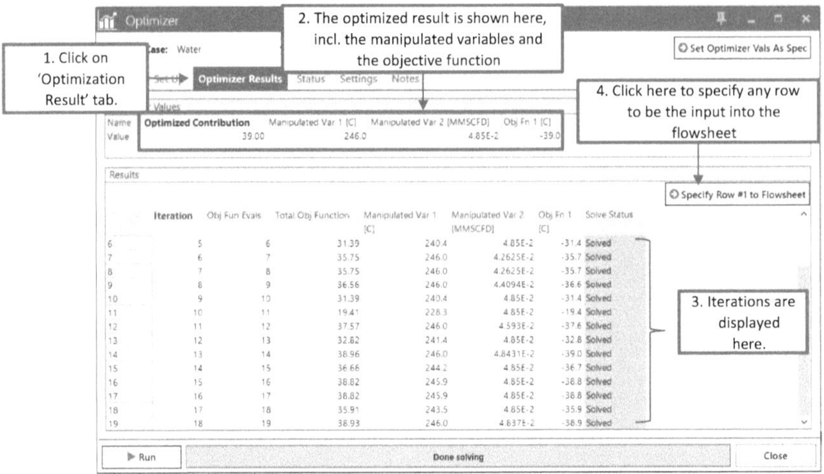

FIGURE 9.17 Optimization result using the LEX solver.

In this regard, the sensitivity analysis is revisited, with its results plotted in three-dimensional (3D) plots. The 3D plots can be done in commercial software such as MS Excel, Matlab, or Python. Fig. 9.18 shows the water dew point as a function of reboiler temperature and stripping gas flowrate. Fig. 9.19

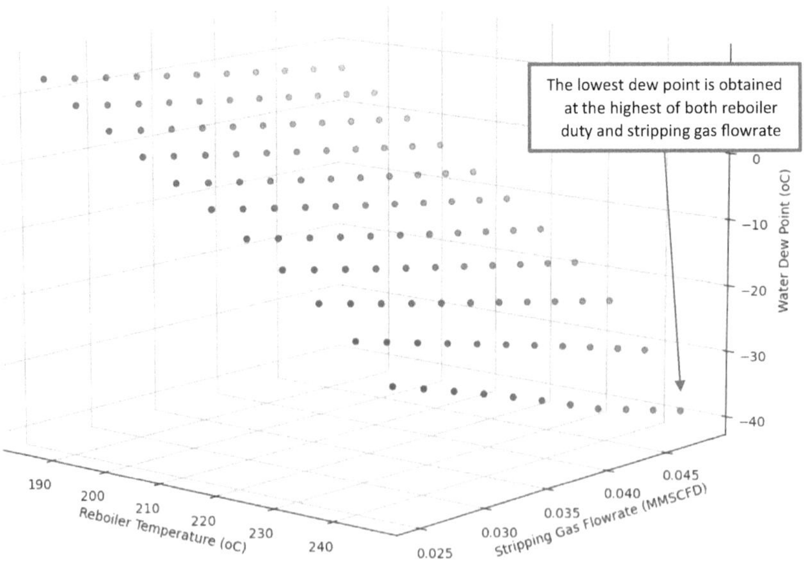

FIGURE 9.18 A 3D plot of water dew point (°C) as a function of reboiler temperature (°C) and stripping gas flowrate (MMSCFD).

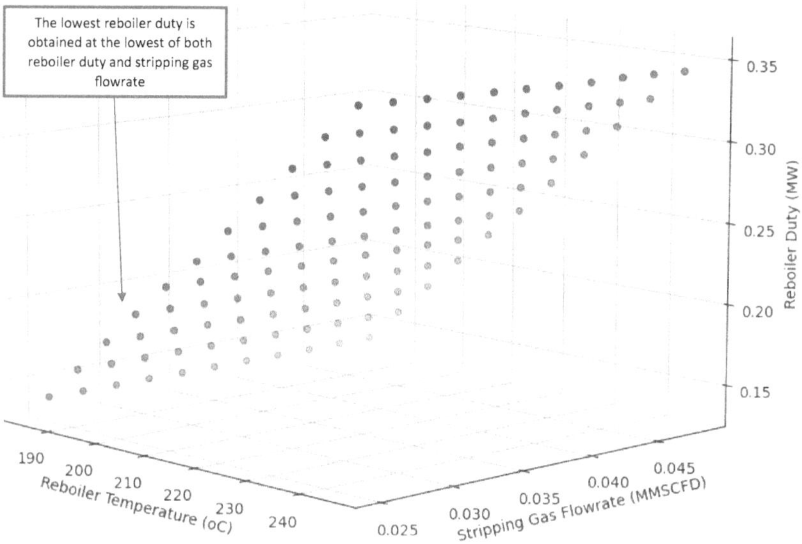

FIGURE 9.19 A 3D plot of reboiler duty (MW) as a function of reboiler temperature (°C) and stripping gas flowrate (MMSCFD).

on the other hand shows reboiler duty as the result of the same operational parameters. As shown, the two plots reiterate the conflicting objective functions. To identify the trade-off between both objectives, the concept of Pareto Front is applied here. The Pareto Front are a set of solutions where an objective function cannot be optimized without sacrificing the others. More details of Pareto Front and multiobjective optimization are available elsewhere (e.g., Rangaiah and Petriciolet, 2013).

One approach to solve an MOO problem in a process simulator is to use one objective function as a constraint and let the other objective function being optimized. In this case, water dew point is used as a constraint while reboiler duty is being minimized. The detailed setup in Optimizer is shown in Fig. 9.20. A fixed value of water dew point is chosen as a constraint. From the insights found during the sensitivity analysis previously (see Fig. 9.18), fixed values of 10, 0, −10, −20, and −30°C can be used. By selecting one value, for example using a maximum water dew point of 10°C as in Fig. 9.20, the optimizer can be run to solve for minimum reboiler duty. In this example, it is $1.394 \cdot 10^5$ W. The result in Fig. 9.20 shows that this minimum reboiler duty is obtained at water dew point of 8.5°C, which is lower than the maximum constraint of 10°C. Hence, this pair value of 8.5°C and $1.394 \cdot 10^5$ W is recorded as a result. The same exercise is repeated for other values of water dew point, which result in other corresponding values of minimum reboiler duties. It is important to note that this is a computationally demanding exercise

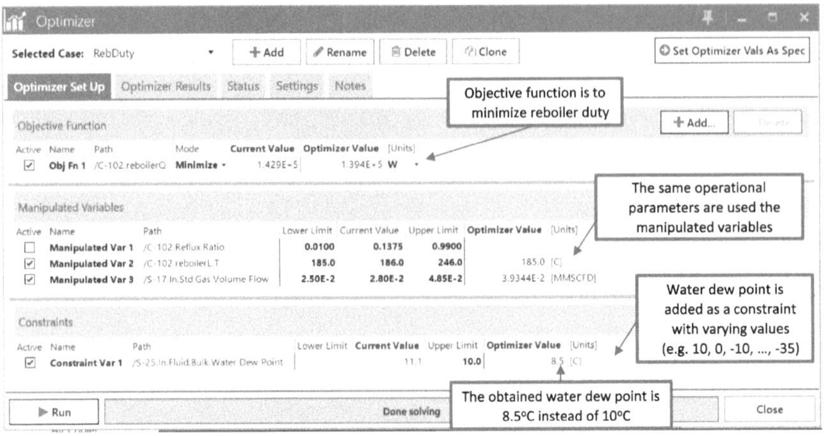

FIGURE 9.20 Optimizer setup using water dew point as the constraint.

and users' experience on optimization, selection of feasible starting points and solvers are key for successful implementation.

Once all feasible water dew points are solved, the Pareto Front is plotted in Excel and the results are shown in Fig. 9.21. In this case, water dew point cannot be reduced without increasing reboiler duty. Thus, any solution located above the Pareto Front is feasible, but not optimized. This means that for the same water dew point, the reboiler duty can be reduced until it reaches the Pareto Front. Note that any solution below the plot is infeasible. Fig. 9.21 also shows the initial condition as compared to the Pareto Front. Hence, to reduce the water dew point, reboiler duty has to be increased. In another words, this Pareto Front serves as a visual tool to show the trade-off between two objective functions.

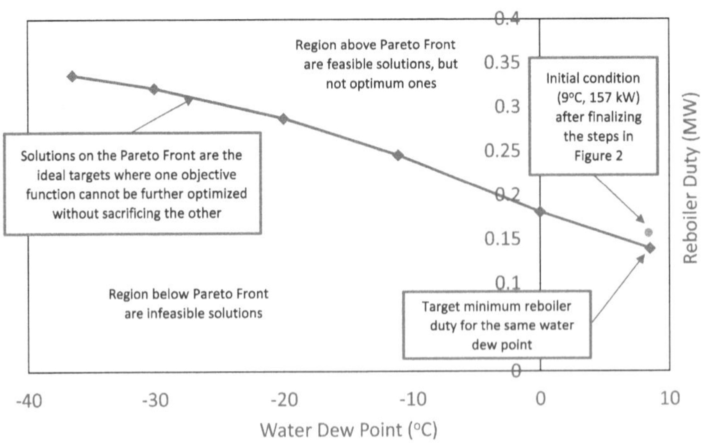

FIGURE 9.21 Pareto front for optimization problem in minimizing water dew point (°C) (*x*-axis) and reboiler duty (MW) (*y*-axis).

9.6 Conclusions

In this chapter, the effect of operational parameters (e.g., reboiler temperature and stripping gas flowrate) were evaluated on the performance of a natural gas dehydration process (e.g., water dew point and reboiler duty). The effects are evaluated using a sensitivity analysis tool called Case Study in Symmetry. Furthermore, optimum values of the parameters are obtained by the Optimizer tool to minimize either water dew point or reboiler duty. These two conflicting objectives are then solved as a multiobjective optimization problem using the concept of Pareto Front. The results can be used to show a visual trade-off between the conflicting objectives. Hence, Case Study and Optimizer are both valuable tools to evaluate process performances, which are useful in guiding engineers to improve the overall process.

Exercises

1. Analyze and discuss the consequences if the feed flowrate of the wet gas (S-1) is increased by 10%.
2. Propose, if necessary, any recommended action to minimize the makeup of TEG while meeting the dry gas specification using the 10% increase in the feed flowrate, as mentioned in Exercise 1.

References

Affandy, S.A., Kurniawan, A., Handogo, R., Sutikno, J.P., Chien, I.-L., 2020. Technical and economic evaluation of triethylene glycol regeneration process using flash gas as stripping gas in a domestic natural gas dehydration unit. Engineering Reports 2 (5), e12153. https://doi.org/10.1002/eng2.12153.

Chebbi, R., Qasim, M., Abdel Jabbar, N., 2019. Optimization of triethylene glycol dehydration of natural gas. Energy Reports 5, 723–732. https://doi.org/10.1016/j.egyr.2019.06.014.

Dagde, K.K., Akpa, J.G., 2014. Numerical Simulation of an Industrial Absorber for Dehydration of Natural Gas Using Triethylene Glycol. https://doi.org/10.1155/2014/693902. https://www.hindawi.com/journals/je/2014/693902/. (Accessed 22 December 2020).

Edgar, T.F., Himmelblau, D.M., 2001. Optimization of Chemical Processes, 2nd ed. McGraw Hill, New York, USA.

Gas Processors Suppliers Association, 2004. Chapter 20: Dehydration, GPSA Engineering Data Book, 12th Edition (Electronic), SI Volumes I & II, 12th ed, Vol. I&II. Gas Processors Suppliers Association.

Gironi, F., Maschietti, M., Piemonte, V., 2010. Natural gas dehydration: A triethylene glycol-water system Analysis. Energy Sources, Part A: Recovery, Utilization, and Environmental Effects 32 (20), 1861–1868. https://doi.org/10.1080/15567030902804756.

Gironi, F., Maschietti, M., Piemonte, V., Diba, D., Gallegati, S., Schiavo, S., 2007. Triethylene glycol regeneration in natural gas dehydration plants. In: A Study On The Coldfinger Process; Offshore Mediterranean Conference, Ravenna, Italy.

iCON-Symmetry, 2021. Symmetry Special version created for PETRONAS including iCON – User Guide. https://www.software.slb.com/products/symmetry.

Kamin, Z., Bono, A., Leong, L.Y., 2017. Simulation and optimization of the utilization of triethylene glycol in a natural gas dehydration process. Chemical Product and Process Modeling 12 (4). https://doi.org/10.1515/cppm-2017-0017.

Petropoulou, E.G., Voutsas, E.C., 2018. Thermodynamic modeling and simulation of natural gas dehydration using triethylene glycol with the UMR-PRU model. Industrial and Engineering Chemistry Research 57 (25), 8584–8604. https://doi.org/10.1021/acs.iecr.8b01627.

Rangaiah, G.P., Petriciolet, A.B. (Eds.), 2013. Multi Objective Optimization in Chemical Engineering: Developments and Applications. John Wiley & Sons, Chichester, West Sussex, United Kingdom.

Rhinehart, R.R., 2018. Engineering Optimization: Applications, Methods, and Analysis. John Wiley & Sons, New Jersey, USA.

Part IV

SuperPro designer

Chapter 10

Basics of batch process simulation with SuperPro Designer*

Dominic C.Y. Foo
University of Nottingham Malaysia, Semenyih, Selangor, Malaysia

Chapter outline

10.1 Basic steps for batch process simulation

SuperPro Designer (SPD) is commonly used for the simulation of batch processes. To construct a batch process simulation flowsheet, the following steps are necessary:

1. Basic setup: This step involves the setting up of basic information needed for simulation, which includes the process operating mode, annual operating time, registration of components, etc.
2. Operating condition: This step provides the specification needed for each *unit procedure*,[1] before the simulation may be executed.

*Simulation and other electronic files of this chapter are made available on a companion website; URL: https://www.elsevier.com/books-and-journals/book-companion/9780323901680.

1. This terminology is used to differentiate the "unit operation" that is commonly used for continuous processes. In each unit procedure, there are several operations that occur in a sequence (www.Intelligen.com).

Chemical Engineering Process Simulation. https://doi.org/10.1016/B978-0-323-90168-0.00013-5

3. Scheduling: This step ensures the individual operations in a unit procedure to start their operation at the right sequence, and to avoid conflicting operations among the process units.

Note that step 1 is only necessary at the start of the simulation exercise, while steps 2 and 3 are to be carried out for all unit procedures. The individual steps are illustrated using a simplified biochemical case study, which is described next.

10.2 Case study on biochemical production

Fig. 10.1 shows a simulation flowsheet constructed in SPD for a simplified biochemical case study taken from Athimulam et al. (2006). In the vessel procedure, water and a type of herb are first fed into the unit. Water is used as leaching agent to extract the active ingredient from the herb to produce a water-soluble extract. Water and the active ingredient are then separated from the herb solid residue and sent to the spray dryer. In the spray dryer unit, the liquid extract is contacted with hot air to produce the herb extract product. Both vessel and spray dryer procedures are operated in batch mode, with an annual operating time of 7920 h.

In the following subsections, detailed steps for constructing the simulation model in SPD are described, following the three-step procedure outlined earlier.

10.3 Basic simulation setup

In this step, basic information for the simulation model are specified in SPD. This includes the specification of the process operating mode (i.e., batch or continuous), its annual operating time, selection/registration of components for the simulation model, etc. The specification of process operating mode and its

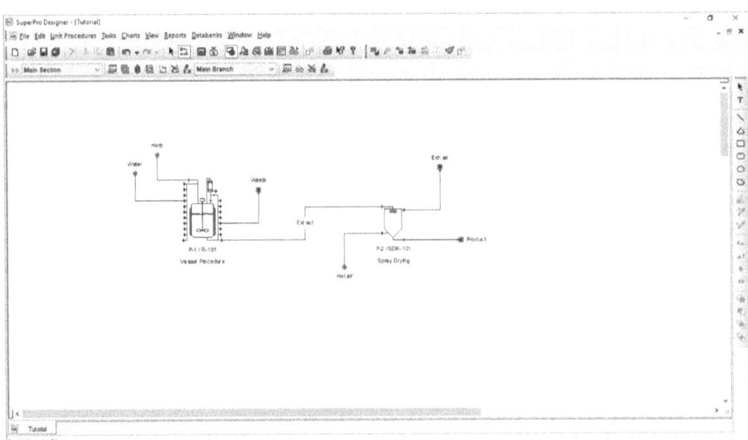

FIGURE 10.1 SPD simulation flowsheet for biochemical case study.

annual operating time are usually done when the program is first launched, as shown in Fig. 10.2.

Next, components needed for this simulation are to be registered for the flowsheet. Note that most commonly found chemicals are available in the SPD database. Fig. 10.3 shows steps to locate the SPD database. By default, water and air components (i.e., oxygen and nitrogen) are found in all SPD flowsheet.

Table 10.1 shows that four components are needed for the biochemical cases study, i.e., water, active ingredient, herb, and waste. Apart from water (found in the SPD database), user needs to register three other "user defined components" in the simulation file. For a base case model, the physical and chemical properties of these user defined components may be referred to other component found in the database; these are termed as "reference component"

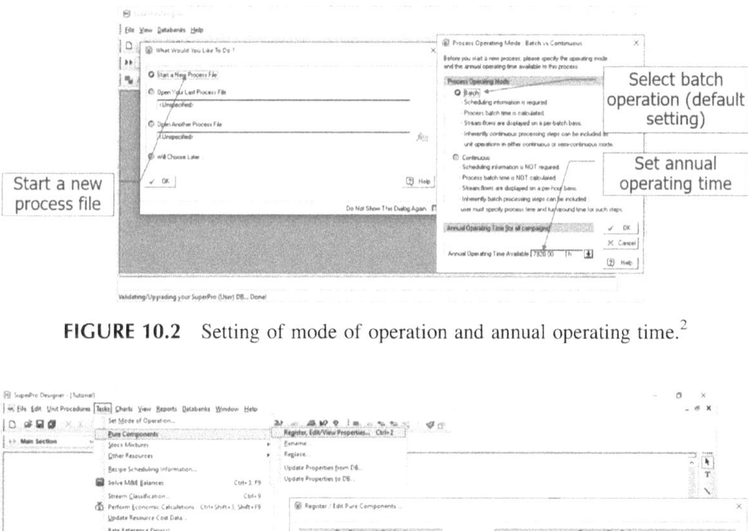

FIGURE 10.2 Setting of mode of operation and annual operating time.[2]

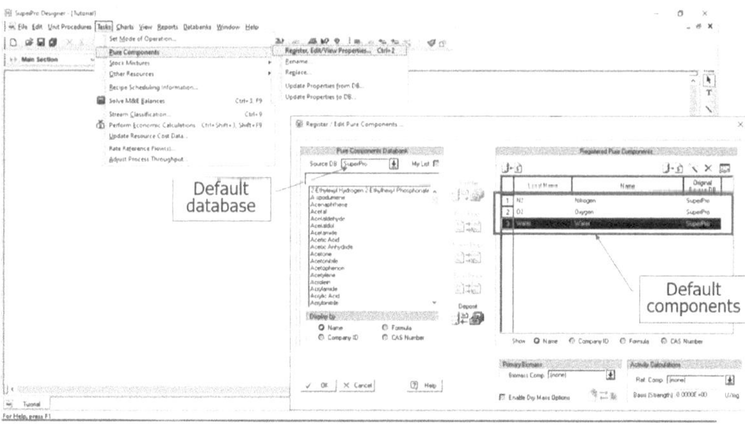

FIGURE 10.3 Steps to locate the SPD database and default components in all SPD flowsheet.

2. The mode of operation and annual operating time may be modified by visiting "Task/Set Mode of Operation" from the menu bar of SPD, once the flowsheet has been constructed.

TABLE 10.1 Component needed for biochemical production case study.

Component	Registration	Reference component
Water	SPD database (default in SPD flowsheet)	–
Active ingredient (Act Ing)	User-defined	Water
Herb	User-defined	Biomass
Waste	User-defined	Biomass

in SPD. Steps in Fig. 10.4 are to be repeated for the registration of these components. Reference component for the user-defined components are given in Table 10.1.

One may also view the property of the newly registered component. For instance, herb has been registered by referring its properties to biomass (see Table 10.1). Hence, it has the same properties as biomass, e.g., molecular weight of 24.63 g/mol, normal boiling point of 287.85°C, etc., as shown in Fig. 10.5. For user defined components with available property values, users may update these properties using the SPD interface in Fig. 10.5.

10.4 Setting for vessel procedure

To follow the *sequential modular* approach (see Chapter 1 for details), each process unit is to be converged, before the next unit is added. We first construct the flowsheet involving the first unit, i.e. vessel procedure, followed by setting

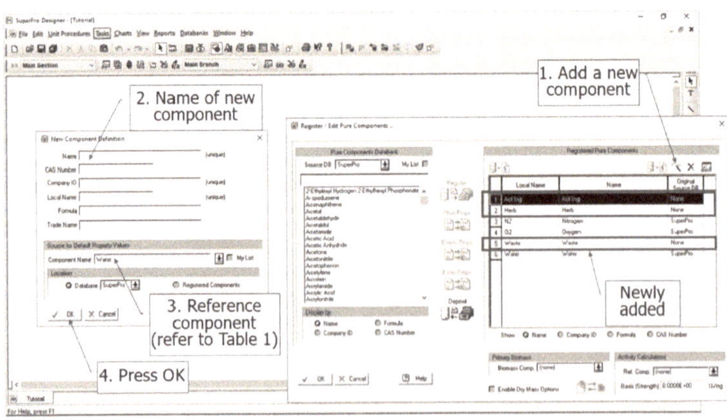

FIGURE 10.4 Detailed steps to create user-defined component (repeat these steps for herb, waste, and active ingredient).

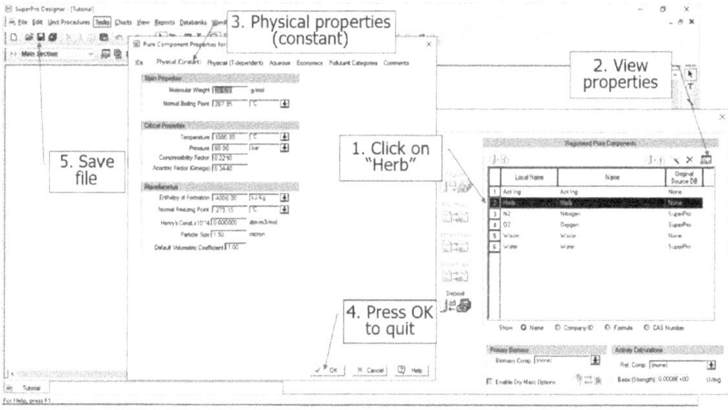

FIGURE 10.5 Detailed steps to view properties of newly registered component (herb).

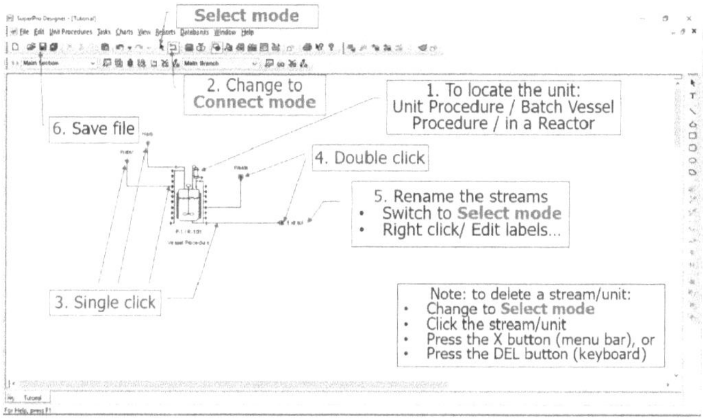

FIGURE 10.6 Detailed steps to construct flowsheet involving the vessel procedure P-1/R-101.

its specifications (step 2). Detailed steps to locate the vessel procedure and to connect its inlet and outlet streams are given in Fig. 10.6. Note that all units in SPD has a procedure tag name (P-1) and equipment ID (R-101) by default. This will help to allow the sharing of equipment unit in different unit procedures for some batch process.[3]

Note that the vessel procedure is used to approximate the leaching process, following the original work of Athimulam et al. (2006). In the vessel procedure, water is used as a leaching agent to extract the active ingredient from the herb. The two inlet streams to the vessel procedure are first specified, with detailed steps shown in Figs. 10.7 and 10.8. Note that the ratio of herb:water is to be kept at 1 kg: 60 L (Athimulam et al., 2006). Hence, for 40 kg of herb, a

3. See discussion in Chapter 11.

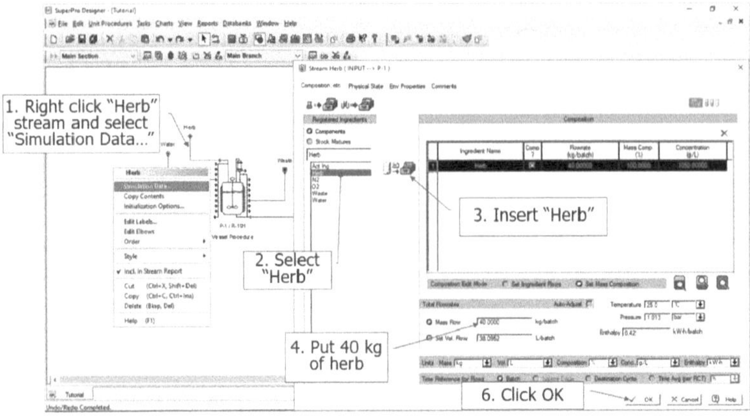

FIGURE 10.7 Detailed steps to specify the herb stream.

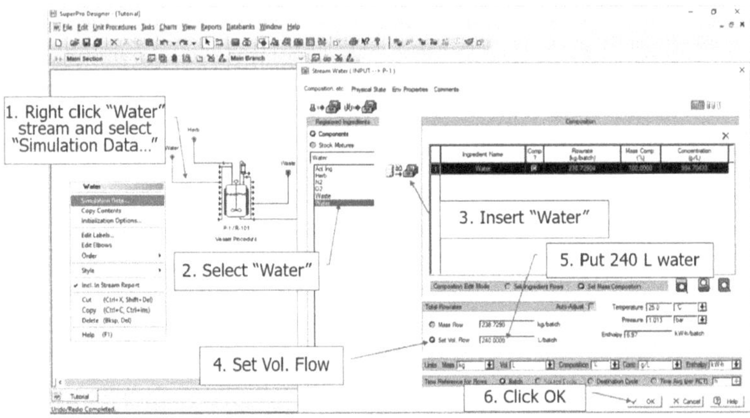

FIGURE 10.8 Detailed steps to specify the water stream.

total of 240 L water is to be used (see Figs. 10.7 and 10.8). As the process is operated at room condition, the temperature and pressure of the streams are left as default setting.

The vessel procedure is next specified. A total of six operations are necessary for this procedure, with detailed specifications given in Table 10.2. These six operations are added to the vessel procedure. Note that the sequence of operations is utmost important for batch process, as they will dictate how a batch process should operate. Hence, the users should ensure that these operations are added in the correct sequence, as shown in Fig. 10.9.

Next, the individual operations of procedure P-1 will be specified. For the two charge operations, the amount of raw material are to be loaded into the vessel using their dedicated streams (see detailed steps in Fig. 10.10). The heating operation is next specified, following steps in Fig. 10.11A.

TABLE 10.2 Specification for vessel procedure (P-1).

Operation	Operating condition	Duration
Charge-1	To charge 40 kg of herb into the vessel	Default
Charge-2	To charge 240 L of water into the vessel	Default
Heating	To raise temperature of the vessel content to 95°C	30 min
Stoichiometry reaction	Set final temperature: 95°C Set stoichiometry (mass) following Eq. (10.1), with 100% conversion	2 h
Split	Outlet pot: Waste stream Retain 100% for Act Ing, N_2, O_2, and water	10 min
Transfer out	Transfer out using: Extract stream (to spray drying procedure P-2)	Default

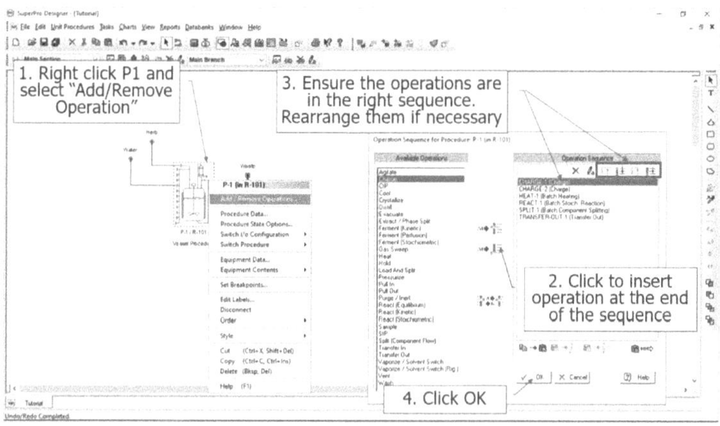

FIGURE 10.9 Detailed steps for selection of operations in vessel procedure P-1.

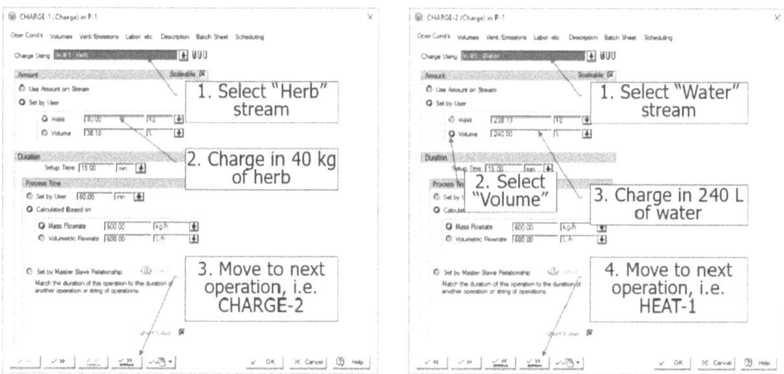

(a) (b)

FIGURE 10.10 Detailed steps to specify charging operations in P-1[4]: (A) herb and (B) water.

---

4. Access this window by doing a right click, and select "Operation Data/Charge-1."

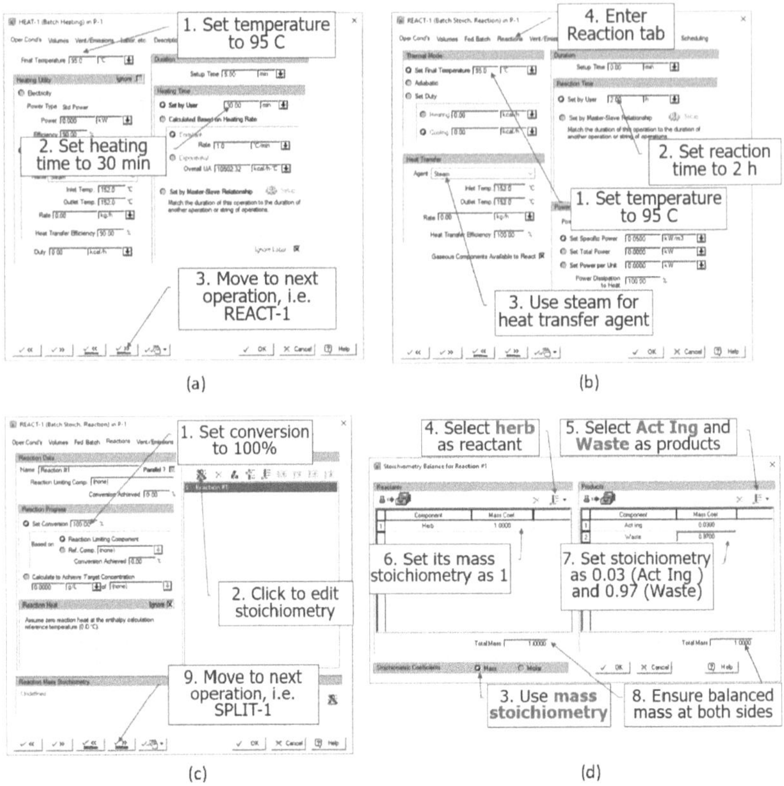

FIGURE 10.11 Detailed steps for specifying operations in P-1 (A) heating; (B) reaction (operating conditions tab); (C) reaction condition (Reaction tab); (D) setting for reaction stoichiometry.

In procedure P-1, the leaching operation is approximated using a chemical reaction, following mass stoichiometry[5] as in Eq. (10.1):

$$1 \text{ Herb} \rightarrow 0.97 \text{ Waste} + 0.03 \text{ Act Ing} \qquad (10.1)$$

where Act Ing represents active ingredient. Detailed steps for specifying this operation is shown in Fig. 10.11B–D.

Next, the splitting operation is used to approximate the separation of waste material (solid) from the active ingredient (liquid). All waste will be separated, while the extract is sent to the spray drying procedure using the transfer out

5. Mass stoichiometry is useful for specifying nonconventional chemical reactions that are difficult to be described with mole stoichiometry, particularly useful for biochemical processes.

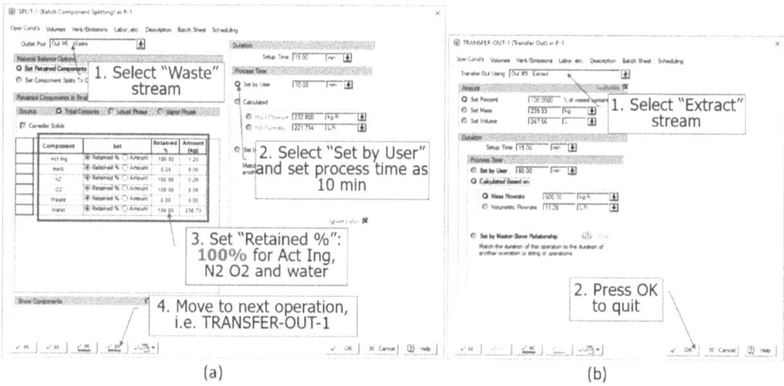

FIGURE 10.12 Detailed steps for specifying operations: (A) splitting and (B) transfer out.

operation. Detailed steps for specifying the splitting and transfer out operations are given in Fig. 10.12.

Once the transfer-out operation is specified, the simulation model may be executed.[6] The converged simulation file shows the vessel outlet stream (Extract) to have a total flow of 1.2 kg of active ingredient (see Fig. 10.13A).[7]

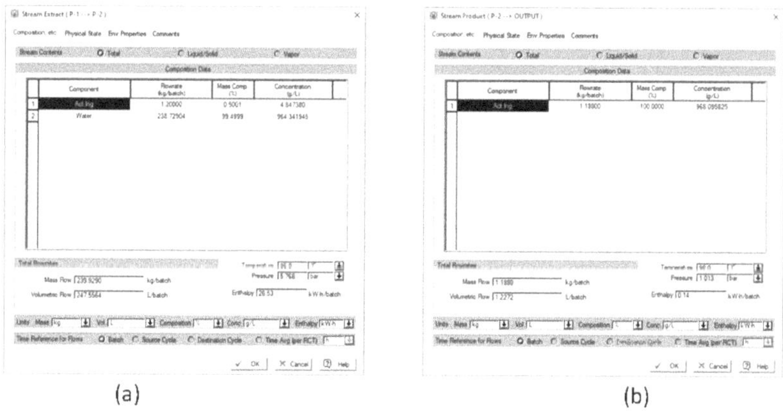

FIGURE 10.13 Simulation results for (A) extract stream; (B) product stream.

6. Simulation can be executed by going to menu bar: Tasks/Perform M&E Balances, or by pressing the button on the menu bar, or the shortcut button "F9" on keyboard.
7. Verify this from the simulation file (right click the Extract stream and select "Operational Data ...").

10.4.1 Spray drying procedure

To produce the final extract product, the liquid content from the vessel procedure is sent for spray drying in procedure P-2. The latter is first added on the SPD flowsheet, along with its inlet and outlet streams. Detailed steps for doing so are shown in Fig. 10.14. Two operations are to be configured for this procedure, i.e., drying and cleaning-in-place (CIP). Note that drying operation is found in the vessel procedure by default, while CIP operation is to be added, following the steps in Fig. 10.14.

In the spray drying procedure (P-2), a hot air stream of 170°C is used for drying. Detailed steps for specifying the hot air stream is shown in Fig. 10.15 (note that "air" is registered as "stock mixture" in SPD).

Specification of the spray drying and CIP operations in procedure P-2 are summarized in Table 10.3. As shown, the drying operation takes 22 h to

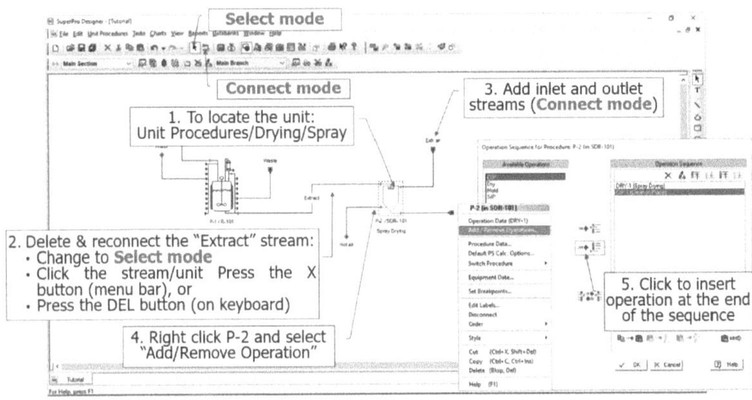

FIGURE 10.14 Detailed steps for construct flowsheet involving the spray drying procedure P-2/SDR-101, and adding of CIP operation.

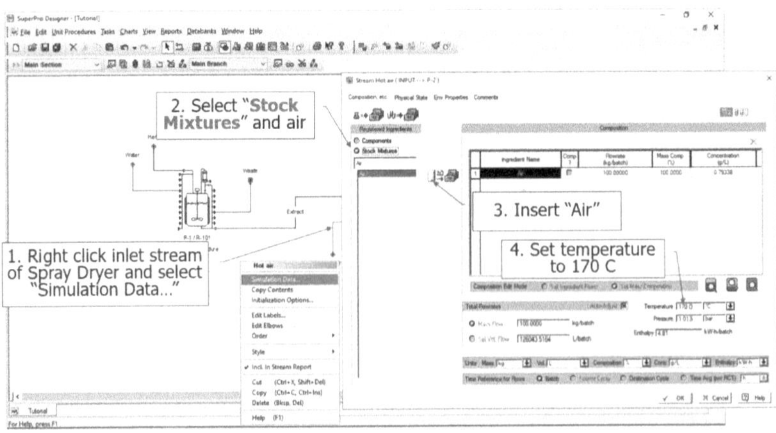

FIGURE 10.15 Detailed steps for specifying hot air stream of spray drying procedure (P-2).

TABLE 10.3 Specification for spray drying procedure P-2.

Operation	Operating condition	Duration and start time
Spray drying	Volatile liquid component evaporation Act Ing: 1% (set by user) Water: 100% (set by user) Outlet gas temperature: 150°C Dried product temperature: 98°C	Duration: 22 h Start time: Start after completion of transfer-out operation in P-1
CIP	Default setting	Duration: 15 min (default) Start time: when previous spray drying operation completed

complete, with 1% loss of the product (active ingredient), following the original work of Athimulam et al. (2006). Detailed steps for specifying the spray drying operation is shown in Fig. 10.16A. Next, CIP is to be carried out upon the completion of the spray drying operation. Default setting is to be used for this operation (see Fig. 10.16B).

Once the model has converged, the product stream would contain the active ingredient of 1.188 kg,[8] as shown in Fig. 10.13B.

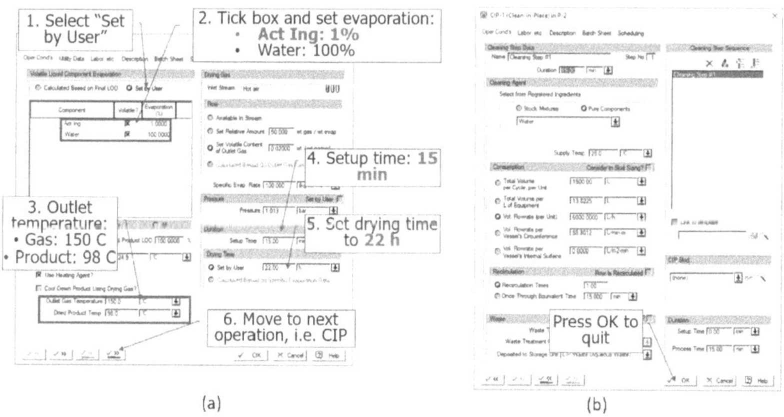

FIGURE 10.16 Detailed steps for specifying operations in P-2: (A) Spray drying[9]; (B) CIP.

8. Verify this from the simulation file (right click the Product stream and select "Operational Data …").
9. Access this window by doing a right click and select "Operation Data/Dry."

10.4.2 Process scheduling

Scheduling is an important element for any batch processes. Process scheduling avoids conflicting operations among the process units, especially for procedures who share the same equipment. The scheduling function of SPD assists user to identify such conflicts. By default, all unit procedures will start their operations from the beginning of the batch, i.e., time 0 (this includes both procedures P-1 and P-2 in the biochemical case study). Hence, user should revise the default setting if a process unit will only start its operation at a later time. For the biochemical case study, the vessel procedure (P-1) will start its operation at time 0. However, the spray drying procedure (P-2) should only start its operation when procedure P-1 completes its last operation, i.e., transfer out (see Table 10.3). Hence, the default start time of P-2 has to be altered. Detailed steps for doing so are given in Fig. 10.17A. Lastly, the CIP in P-2 will commence its operation upon the completion of drying operation, following the default setting (see Fig. 10.17B).

Next, scheduling results of the biochemical case study is examined. A convenient tool for such purpose is the Operational Gantt Chart, as shown in Fig. 10.18. Users may also display its Recipe Scheduling Information (see detailed steps in Fig. 10.18). As shown, the biochemical case study has a batch time of 26.86 h and minimum cycle time of 22.25 h. The latter is identified among units with the longest duration of its operations; such unit is known as the *time bottleneck* equipment. For the biochemical case study, the spray dryer

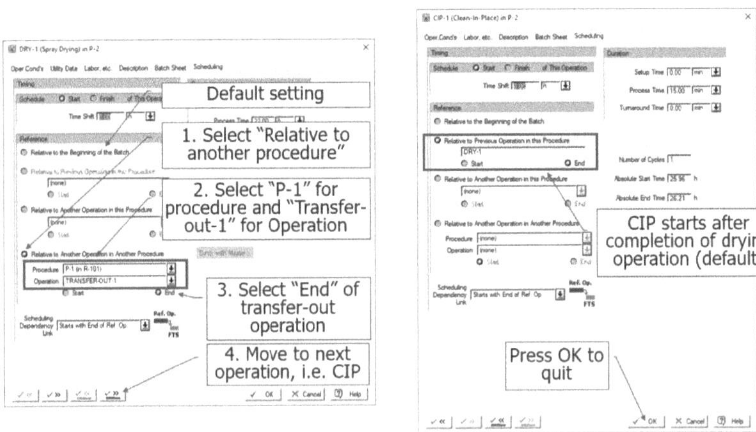

(a) (b)

FIGURE 10.17 Detailed steps to specify operation start time for procedure P-2[10]: (A) spray drying; (B) CIP.

10. Access this window by doing a right click and select "Operational Data/DRY-1," go to Scheduling tab.

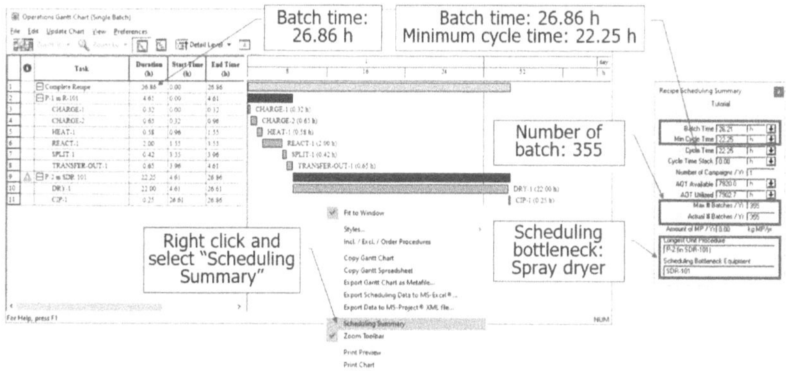

FIGURE 10.18 Operational Gantt chart for biochemical case study.[11]

(P-2/SDR-101) is identified as the time bottleneck equipment, due to its longest duration (22.25 h). In other words, the next subsequent batch process can only start once the spray dryer has completed its operation. With the annual operating time of 7920 h,[12] this translates into an annual production of 355 batches (see Fig. 10.18).

10.4.3 Strategies for batch process debottlenecking

As discussed earlier, the spray dryer (P-2/SDR-101) has been identified as the time bottleneck equipment of the biochemical case study, as it has the longest duration among the process units. In other words, the spray dryer limits the annual production of the process to be 355 batches. If one were to increase the process throughput (i.e., debottlenecking), the minimum cycle time has to be shortened. A simple strategy in doing so is to have extra spray dryer unit(s). The latter may be added by activating the "stagger mode" function of SPD, with detailed steps given in Fig. 10.19A. Doing this lead to a twofold increase of the batch production, i.e., 710 batch (cycle time reduces to 11.13 h), as shown in Fig. 10.19B. As shown, batch 2 of the process starts before the spray dryer procedure (P-2) completes its operation, since there are two units of spray dryer.

10.4.4 Economic evaluation

SPD has a build-in model to perform economic evaluation for its simulated processes (www.intelligen.com). For instance, the purchase cost of the vessel unit R-101 of the biochemical case study can be estimated using the build-in

11. Access this by going to menu bar: Charts/Gantt Charts/Operations GC ....
12. The annual operating time may be revised if necessary, by going to the menu bar: Task/Recipe Scheduling Information ....

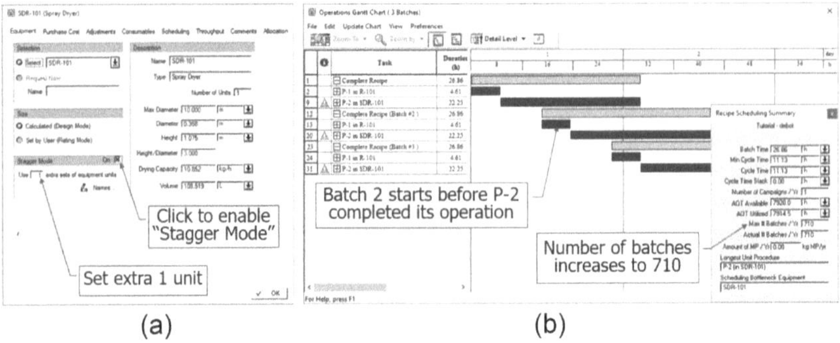

FIGURE 10.19 (A) Steps to add an additional unit for spray dryer SDR-101[13]; (B) Operational Gantt chart for biochemical case study.[14]

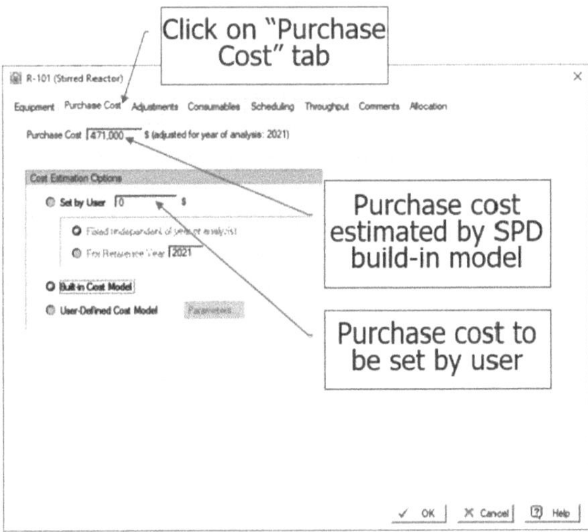

FIGURE 10.20 Purchase cost of the vessel unit R-101 estimated by SPD.

model, as shown in Fig. 10.20. Note that one may also put in the preferred purchase cost obtained externally.

To perform process economic evaluation, the purchase price of raw material, selling price of products and treatment prices of waste stream of the process are necessary. For the biochemical case study, the pricing values are given in Table 10.4. The purchase price of raw material may be provided to SPD following the detailed steps outlined in Fig. 10.21. Note that SPD will then estimate the purchase price of the inlet stream based on its stream composition.

13. Access this by doing a right click on the spray dryer unit, and choose "Equipment Data ..."
14. Access this by going to menu bar: Charts/Gantt Charts/Operations GC (Multiple Batches ...).

TABLE 10.4 Purchase, selling, and treatment prices of raw material, product, and waste streams of biochemical case study: (A) inlet streams; (B) outlet streams.

(A) Inlet streams	**Purchase price ($/kg)**	
Herb	6.50	
Water	0.001	
(B) Outlet streams	**Selling price ($/kg)**	**Stream type**
Waste	0.50	Solid waste
Product	660	Revenue
Exh air	–	Emission

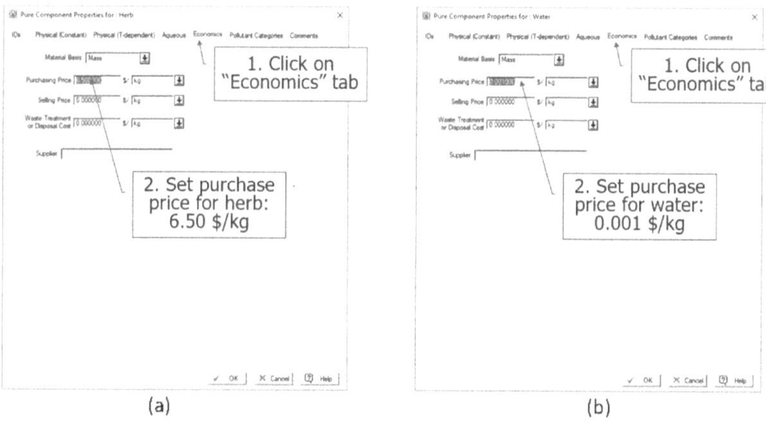

FIGURE 10.21 Setting purchase price for raw material: (A) herb; (B) water.[15]

On the other hand, all outlet streams of a flowsheet are to be classified as product or waste streams (further categorized as emission, solid waste, aqueous and organic wastes), along with their selling price (product) and disposal price (waste streams). Besides, a main product stream is to be chosen too. The setting for all outlet streams of the biochemical case study are shown in Fig. 10.22.

Upon the completion of stream classification step, economic evaluation may be performed for the process.[16] A summary of the economic evaluation

15. Access this following the steps of properties viewing of component (Fig. 10.5).
16. Economic evaluation can be executed by going to menu bar: Tasks/Perform Economic Evaluation, or by pressing the icon on the menu bar, or via shortcut key "Shift + F9" on keyboard.

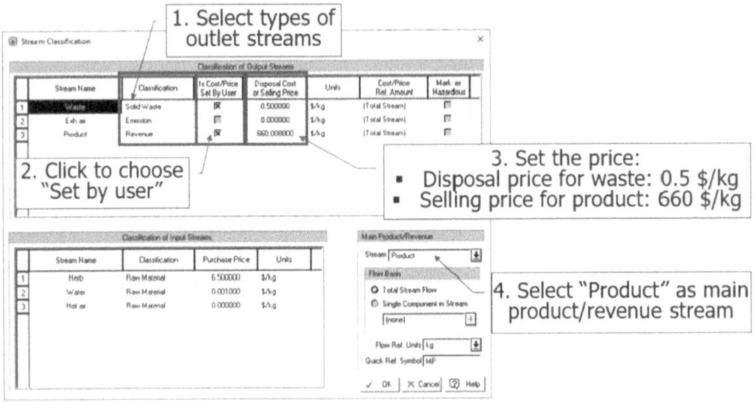

FIGURE 10.22 Setting to specify selling prices for product streams and treatment price for waste streams.[17]

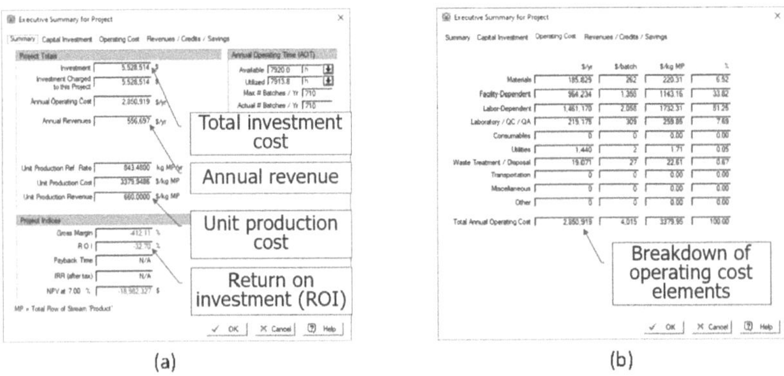

FIGURE 10.23 Economic evaluation results of the biochemical case study[18]: (A) executive summary; (B) operating cost breakdown.

results are shown in Fig. 10.23. For this case, the total capital investment of the biochemical process is estimated as $5.5 million (see Fig. 10.23A). Note that the capital investment was estimated based default parameter setting[19] based on equipment purchase cost of $851,000; the latter is given in Table 10.5. Note that the spray dryer is cost for 2 units. Besides, SPD also estimate the various elements of operating cost, with details shown in Fig. 10.23B). Due to high capital investment and low revenue ($556,697/y), this process does not have an attractive economic return. As shown in Fig. 10.23A, its return on investment (ROI) is estimated as −32.7%, with gross margin of −412%.

17. Access this window by going to menu bar: Task/Stream classification ....
18. Access this window by going to menu bar: View/Executive Summary.
19. Capital cost adjustment may be changed by pressing the icon on the menu bar.

TABLE 10.5 Breakdown of equipment cost.[a]

Equipment	Dimension	Unit cost ($)	Cost ($)
Stirred reactor (R-101)	Volume = 316.12 L	471,000	471,000
Spray dryer (SDR-101)	Volume = 108.52 L	105,000	210,000
Unlisted equipment			170,000
		Total	851,000

[a]*Access this by going to menu bar: Reports/Economic Evaluation (EER).*

10.5 Conclusion

This chapter discussed the basic steps of SuperPro Designer (SPD) in simulating batch chemical processes. Apart from mass and energy balances, process scheduling is an important aspect for batch process simulation. The chapter also describes simple strategy for process debottlenecking in order to increase process throughput. Economic evaluation can be performed using the build-in function of SPD, allowing users for preliminary assessment of the simulated process.

10.6 Further reading

Economic evaluation of the biochemical case study shows that the process has a negative return on investment (Fig. 10.23). Hence, various strategies need to be explored in order to achieve alternative production schemes with attractive economic parameters. These include strategies such as debottlecking options (Koulouris et al., 2000), and/or to produce higher value-added products (see details in Athimulam et al., 2006). Apart from biochemical process, other debottlecking studies have also been reported for pharmaceutical (Tan et al., 2006) and food production (Alshekhli et al., 2011). Besides, SPD is also useful for the analysis of process scale-up studies (Ardani et al., 2021; Lim & Foo, 2017) and uncertainty analysis (Noor et al., 2014).

Exercise

Determine the increase of batch production when two extra units of spray dryers are added to the biochemical case study. Determine also the minimum cycle time for this case.

References

Alshekhli, O., Foo, D.C.Y., Hii, C.L., Law, C.L., 2011. Process simulation and debottlenecking for an Industrial cocoa manufacturing process. Food and Bioproducts Processing 89 (4), 528–536.

Ardani, M.R., Rezan, S.A., Lee, H.L., Foo, D.C.Y., Mohamed, A.R., 2021. Synthesis of titanium powder from reduction of titanium tetrachloride with metal hydrides in the hydrogen atmosphere: thermodynamic and techno-economic analysis. Processes 9 (9), 1567.

Athimulam, A., Kumaresan, S., Foo, D.C.Y., Sarmidi, M.R., Aziz, R.A., 2006. Modelling and optimization of eurycoma longifolia water extract production. Food and Bioproducts Processing 84 (C2), 139–149.

Koulouris, A., Calandranis, J., Petrides, D., 2000. Throughput analysis and debottlenecking of integrated batch chemical processes. Computers and Chemical Engineering 24, 1387–1394.

Lim, S.S., Foo, D.C.Y., 2017. Simulation and scale-up study for a Chitosan-TiO_2 nanotubes Scaffold production. Food and Bioproducts Development 106, 108–116.

Noor, N.Z.M., Foo, D.C.Y., Rahman, B.A., Aziz, R.A., 2014. Modelling and uncertainty analysis for pilot scale monoclonal antibody production. Pharmaceutical Engineering 34 (4), 40–54.

Tan, J., Foo, D.C.Y., Kumaresan, S., Aziz, R.A., 2006. Debottlenecking of a batch pharmaceutical cream production. Pharmaceutical Engineering 26 (4), 72–84.

Chapter 11

Modeling of citric acid production using SuperPro Designer*

Alexandros Koulouris
Department of Food Science and Technology, International Hellenic University, Alexander Campus Sindos-Thessaloniki, Greece

Chapter outline

11.1 Introduction

Processing in batch or semi-continuous mode is an operational approach adopted by industries in many important sectors of economy such as pharmaceuticals, biotech, consumer goods, and food. Even though the traditional

*Simulation and other electronic files of this chapter are made available on a companion website; URL: https://www.elsevier.com/books-and-journals/book-companion/9780323901680.

Chemical Engineering Process Simulation. https://doi.org/10.1016/B978-0-323-90168-0.00001-9

chemical and petrochemical industries (mainly operating in continuous mode and producing bulk commodity products) are reaping the benefits of process simulation for many decades, batch, and semi-continuous process industries (many of them producing high-value-added products) were slower in adopting this technology. With the domination of a global competitive market, all process industries cannot afford to overlook any possible improvement in their production efficiency, including those originated from process simulation.

SuperPro Designer (www.intelligen.com) is a process simulator developed in the early 1990s to address the specific modeling needs in batch/semi-continuous processing. Even though its initial emphasis was in the biotechnological sector, its scope was expanded to cover integrated batch and continuous processes in a wide range of industries, such as pharmaceutical, specialty chemical, food processing, metallurgical, consumer goods, and environmental protection.[1] Process simulation in these industries can be used for preliminary design, retrofit of existing processes, throughput analysis, estimation of capital and operating costs, as well as to assess the environmental impact of processes.

Process simulation can play an important role in all stages of a product or process development. In biotechnology and pharmaceutical industries, by the time detailed engineering work has been initiated, a process is more than 80% fixed and the majority of important decisions for capital expenditures and product commercialization are already made. Furthermore, stringent regulatory in the biotechnology and pharmaceutical industry restricts process modifications after clinical efficacy of the drug is established. It is therefore suggested that process modeling and evaluation be initiated at the early stages of product development (Carmichael et al., 2018).

In this chapter, SuperPro Designer will be demonstrated for a citric acid production case study. Citric acid is a commodity organic acid used in the food and beverage industries to preserve and enhance flavor (Crueger & Crueger, 1989). Among all fruit acids used in the process industries, citric acid plays an exceptional role with a global annual production that surpassed 2 million tons in year 2018, and is projected to reach a volume of nearly 3 million tons by year 2024 (Cision, 2021). Around half of this production is consumed by the beverages industry. Citric acid is the main organic acid produced today by fermentation (Berovic & Legisa, 2007). In the food industry, citric acid is used extensively in carbonated beverages to provide taste and complement fruit flavors, in jams and jellies for taste and for pH adjustment, in frozen fish and shellfish to prolong their shelf-life. It is also used in various foods to increase the effectiveness of antimicrobial preservatives, as well as to enhance the action of antioxidants and inactive enzymes. Since it is biodegradable, it is readily metabolized and safe for both consumers and industry. It is also used in the pharmaceutical industry for various purposes, e.g., to maintain the stability

1. See Chapter X for the modeling of wastewater treatment plant.

of active ingredients, to enhance the activity of preservatives, pH adjustment in the cosmetics industry, and as metallic-ion chelator in antioxidant systems (Apelblat, 2014; Berovic & Legisa, 2007).

Fermentation is the most commonly used method for bulk production of citric acid. Low-cost and readily available carbohydrates are used as raw materials for its production. Beet and cane molasses, maltose syrups, hydrolyzed starch (corn, wheat, tapioca, and potatoes), cellulose hydrolysates, and waste products of the sugar industry are utilized in different countries depending on their local availability (Apelblat, 2014). High quality molasses is preferred because it reduces the pretreatment costs before it is used in fermentation (Berovic & Legisa, 2007).

Even though different microorganisms have been studied, strains of filamentous fungus *Aspergillus niger* are by far the most commonly used microorganisms. The most important advantages that led to its extensive use were their ability to grow at low pH and the high product yields. *Aspergillus niger* is easy to handle, tolerant to high concentrations of substrates, and can ferment various cheap raw materials. Mutagenesis and strain selection have been carried out to create industrial strains with even better performance (Show et al., 2015).

Citric acid is commonly precipitated as calcium citrate by the addition of lime (calcium hydroxide) and recovered after treatment with sulfuric acid. The disadvantage of this method is that gypsum is produced as an undesired by-product. Alternative technologies have been proposed to eliminate the formation of gypsum such as solvent extraction, ion-exchange adsorption, membrane filtration, and electrodialysis (Apelblat, 2014). Butanol has been used as an extractant, as has tributyl phosphate (Roberts, 1979). Ion pair extraction by means of secondary or tertiary amines dissolved in a water-immiscible solvent (e.g., octyl alcohol) provides an alternative route. The use of electrodialysis membranes which allows the recovery of citric acid directly from the fermentation broth is also an attractive alternative (Blanch & Clark, 1997).

In this case study, citric acid is produced via fermentation using *Aspergillus niger*. The data used in the development of the model were obtained from the open research literature. The case study is a modified version of an example found in SuperPro Designer software (Intelligen Inc., 2021) with a more detailed depiction of the steps involved in the development of the model and extended analysis of the results.

11.2 Process description

Fig. 11.1 shows the complete process flowsheet developed in SuperPro Designer for the production of citric acid with the use of molasses as the carbon source for fermentation. It is a large-scale batch process utilizing approximately 139.8 metric tons (MT) of molasses per batch and producing 54.8 MT of dehydrated citric acid crystals of high purity (99.5%). The

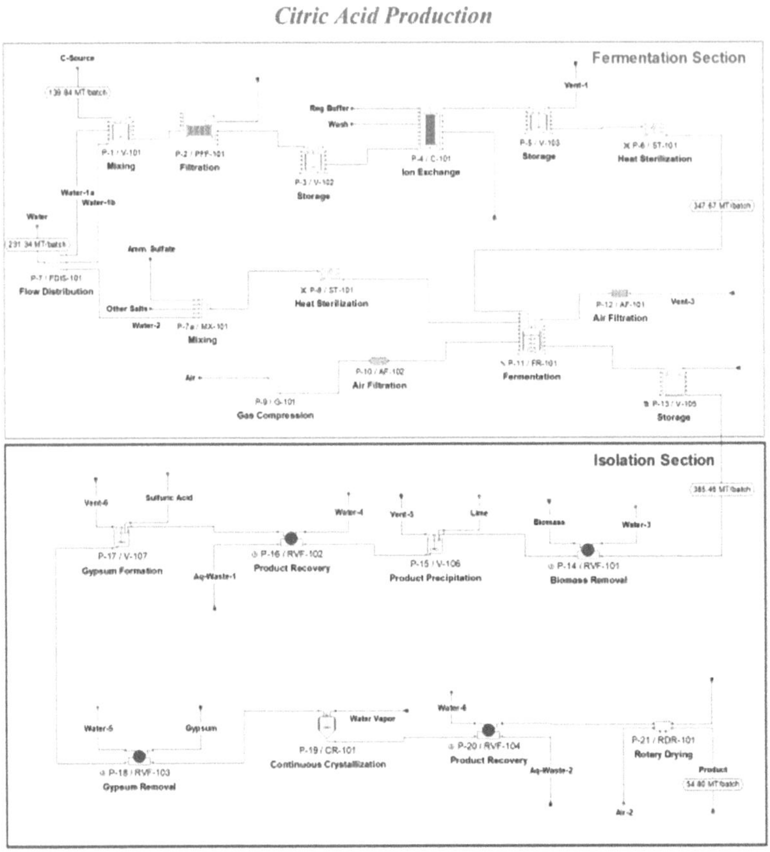

FIGURE 11.1 The citric acid production flowsheet.

corresponding annual amounts are 45,868 MT of molasses and 17,973 MT of citric acid. Table 11.1 shows the assumed (simplified) composition of the stock mixture Molasses. As shown, "Glucose," $C_6H_{12}O_6$, is used to represent all fermentable sugars and "NFS" represents the nonfermentable substances. The component "Impurities" collectively represents all heavy metals and other impurities that must be removed because they are detrimental to the growth of the microorganism. Apart from molasses, three additional stock mixtures are registered in the model representing buffers used in the process, e.g., "H_2SO_4 (10% w/w)," "Lime (33%)," and "NaOH (1 M)"; their compositions are also given in Table 11.1.

The declaration of components in the SuperPro Designer model is completed with the inclusion of ammonium sulfate, $(NH_4)_2SO_4$, used as a nitrogen source in the fermentation, "Nutrients" representing all other salts fed into the fermenter, citric acid ($C_6H_8O_7$) in diluted from, citric acid in crystal form ("CA Crystal") and various reagents used (sodium hydroxide, NaOH;

TABLE 11.1 Composition of stock mixture used in simulation.

Streams	Component	(%w/w)
Molasses	Glucose	50
	NFS	20
	Impurities	1
	Water	29
H_2SO_4 (10% w/w)	H_2SO_4	10
	Water	90
Lime (33%)	$Ca(OH)_2$	33
	Water	67
NaOH (1 M)	NaOH	3.84
	Water	96.16

calcium hydroxide, $Ca(OH)_2$; sulfuric acid, H_2SO_4) or produced (calcium citrate, $C_6H_5O_7Ca_3$; gypsum, $CaSO_4$) in the downstream process. The components CA Crystal, Calcium Citrate, Impurities, NFS, and Nutrients are user-defined components. The component Water was used as a reference component for the initialization of the properties of all user-defined components with the exception of Ca Crystal for which citric acid was used. All other components used in the model and their properties are originated from the SuperPro Designer database.

The process is divided into two sections: "Fermentation" and "Isolation." In SuperPro Designer, a section represents a group of procedures that serve a common processing objective (e.g., the section "Isolation" in this case study encompasses all product purification procedures). Sections can also have different economic parameters. It follows that reporting of all model outputs (yields, resources consumption, etc.) and economic results can be done on a per-section level. In this example, the Fermentation section includes the media preparation, treatment and fermentation procedures. The inoculum preparation steps were ignored for brevity. The Isolation section processes the fermentation broth and includes all purification and isolation procedures that lead to the production of high-purity crystal citric acid. Details for all processes in the "Fermentation" and "Isolation" sections are described below.

Note that all procedures in SuperPro Designer are tagged by their ID, the hosting equipment ID, and a textual description of the procedure. All these three tags are shown below every procedure icon (see Fig. 11.1). In the model formulation of batch processes, it is important to differentiate the processing steps (comprising the "recipe") from the equipment used; this is because the same equipment could be used to carry out multiple process steps at different times. This is why SuperPro Designer adopts a dual naming approach in each process icon with separate IDs for the procedure and the equipment (SuperPro Designer, User Guide, 2021).

The important input parameters for all operations in the process are shown in Table 11.2. It should be noted that in the context of batch processing, scheduling information (i.e., the duration of operations and how their start/end is defined) is equally important with operating condition parameters. The set-up time for all operations has been set to 0. Also, all input streams are fed in ambient conditions (25°C and 1 atm).

11.2.1 Fermentation section

Molasses is used as carbon source for the fermentation process. It is diluted with water from 50% fermentable sugar content to 20% which is favorable for the growth of *Aspergillus niger*. The process is carried out in mixing tank P-1/V-101, and includes three operations: charging of 100 MT of water into the tank, charging of molasses, and charging of extra water so that the concentration of sugars in the tank drops to 20%. Suspended particulate material is then removed by filtration (P-2/PFF-101) and the media solution is stored in tank P-3/V-102.

Metal ions are subsequently removed by an ion exchange chromatography column (P-4/C-101). The column operates in a cycle consisting of the load, elution, and washing steps. NaOH (1M) is used as the elution buffer and water is used for column washing. The purified solution is collected in a tank (P-5/V-103), heat-sterilized (P-6/ST-101), and fed into the fermenter (P-11/FR-101). In parallel, nutrients (ammonium sulfate and nutrients) are dissolved in water (P-7a/MX-101), heat-sterilized (P-8/ST-101), and fed to the fermenter (P-11/FR-101) after molasses are fed. The same sterilizer (ST-101) is used for the sterilization of both the molasses (P-6) and the nutrients (P-8), but at different times.

Fermentation is carried out according to the following mass stoichiometry:

$$\begin{aligned}&2(\text{Ammonium sulfate}) + 180(\text{Glucose}) + 12(\text{Nutrients}) + 56\ O_2 \rightarrow \\ &16(\text{Biomass}) + 30\ CO_2 + 154(\text{Citric Acid}) + 50\ H_2O\end{aligned} \tag{11.1}$$

The extent of the fermentation reaction is set to 99% and the fermentation is carried out over a period of 5 days at a temperature of 35°C. Air is supplied by a compressor (P-9/G-101) and purified by filtration (P-10/AF-102) at a rate of 0.3 VVM (volume of air per volume of liquid per minute). Cooling water removes the heat produced by the exothermic process and maintains a constant temperature. Once fermentation is completed, the broth is discharged into the holding tank P-13/V-105, which acts as a buffer tank between the upstream

TABLE 11.2 Operating and scheduling data used in the model.

Operation	Operating conditions	Scheduling data
P-1/V-101		
Charge water	100 MT of water	30 min, start with batch start
Charge carbon source	139.8 MT of molasses	60 min, start at end of Charge water
Dilute to 20% (pull-in)	109.4 MT of water	30 min, start at end of Charge carbon source
Transfer to filter	349.2 MT of contents	Duration set by filtration, start at end of Dilute to 20%
P-2/PFF-101		
Filter	Retained: 90% of impurities Cake porosity: 0.3 v/v Filtration flux: 200 L/m^{2}h	4 h, start with Transfer to filter in P-1
P-3/V-102		
Transfer-in		Duration and start set by Filter in P-2
Transfer-out		Duration set by load in P-4, start at end of Transfer-in
P-4/C-101		
Load	Resin binding capacity: 30 g/L Linear velocity: 1000 cm/h Retained: 100% of impurities	11.5 h, start with Transfer-out in P-3
Elute	2 BV of NaOH (1M) buffer	24 min, starts with end of Load
Wash	2 BV of water	24 min, starts with end of Elute
P-5/V-103		
Transfer-in		Duration and start set by Load in P-4
Transfer-out		Duration set by Sterilize molasses in P-6, start at end of Transfer-in
P-6/ST-101		
Sterilize molasses	Sterilization temp: 140°C Regenerator cold stream out: 60°C Exit temp: 35°C Heat transfer coefficient: 2700 W/m^{2}K	8 h, start with Transfer-out in P-5

Continued

TABLE 11.2 Operating and scheduling data used in the model.—cont'd

Operation	Operating conditions	Scheduling data
P-8/ST-101		
Sterilize salts	Sterilization temp: 140°C Regenerator cold stream out: 60°C Exit temp: 35°C Heat transfer coefficient: 2700 W/m^2K	4 h, start with Pull-in nutrients in P-11
P-11/FR-101		
Transfer-in molasses solution		Duration and start set by Transfer-out in P-6
Pull-in nutrients	Nutrients equal to 8% of molasses	Duration set by Sterilize salts in P-8, start at end of Transfer-in molasses solution
Inoculate		30 min, start at end of Pull-in nutrients
Ferment	Fermentation temperature: 35°C Air flow: 0.3 VVM Conversion: 99% of limiting component Reaction Enthalpy: 2990 kcal/kg of citric acid at 35°C Venting pressure: 1 atm	5 days, start at end of Inoculate
Transfer-out	Rate: 140,000 L/h	2.24 h, start at end of Ferment
CIP	Agent flowrate: 30 L/min·m	Start at end of Transfer-out
	Water flush 1	10 min
	NaOH (1M) flush	30 min
	Water flush 2	10 min
P-13/V-105		
Store		Duration and start set by Transfer-out in P-11

TABLE 11.2 Operating and scheduling data used in the model.—cont'd

Operation	Operating conditions	Scheduling data
P-14/RVF-101 (operating in batches with cycle time of 31 h)		
Apply pre-coat		1 h
Filter	Retained: 100% of biomass LOD: 30% Filtrate flux: 250 L/m^{2}h Water wash: 0.6 vol/cake vol	30 h
P-15/V-106 (operating continuously)		
Neutralize	Neutralization temperature: 50°C Neutralizing agent: $Ca(OH)_2$ 33% w/w Agent excess: 5%	
P-16/RVF-102 (operating in batches with cycle time of 31 h)		
Apply pre-coat		1 h
Filter	Retained: 98% of calcium citrate LOD: 30% Filtrate flux: 175 L/m^{2}h Water wash: 0.6 vol/cake vol	30 h
P-17/V-107 (operating continuously)		
Neutralize	Neutralization temperature: 35°C Neutralizing agent: H_2SO_4 10% w/w Agent excess: 4%	
P-18/RVF-103 (operating in batches with cycle time of 31 h)		
Apply pre-coat		1 h
Filter	Retained: 95% of $CaSO_4$, 1% citric acid LOD: 30% Filtrate flux: 250 L/m^{2}h Water wash: 0.6 vol/cake vol	30 h

Continued

TABLE 11.2 Operating and scheduling data used in the model.—cont'd

Operation	Operating conditions	Scheduling data
P-19/CR-101 (operating continuously)		
Crystallize	Evaporation: 50% of H_2O at 100°C Crystallizing component: Citric acid Crystallization yield: 98%	
P-20/RVF-104 (operating in batches with cycle time of 31 h)		
Apply pre-coat		1 h
Filter	Retained: 98% of crystal citric acid LOD: 30% Filtrate flux: 250 L/m^2h Water wash: 0.6 vol/cake vol	30 h
P-21/RDR-101 (operating continuously)		
Dry	Final LOD: 0.5% Specific evaporation rate: 20 $(kg/h)/m^3$ Product temperature: 110°C	

"Fermentation" section and the downstream "Isolation" section operated in continuous mode. This means that, even though the overall process is still batch, these procedures receive and process material in a continuous way with no breaks in-between batches (see Gantt chart in Fig. 11.16 later in this chapter).

11.2.2 Isolation section

Purification starts with the removal of biomass by a rotary vacuum filter (P-14/RVF-101). The clarified fermentation liquor subsequently flows into an agitated reactor vessel (P-15/V-106) where approximately one part of hydrated lime ("Lime (33%)"), for every three parts of liquor, is gradually added in order to precipitate calcium citrate according to the reaction (in molar stoichiometry) in Eq. (11.2):

$$2(\text{Citric Acid}) + 3\ Ca(OH)_2 \rightarrow 1(\text{Calcium Citrate}) + 6\ H_2O \quad (11.2)$$

The lime solution must be very low in magnesium content in order to avoid the formation of magnesium citrate which, unlike calcium citrate, is relatively soluble to water. Calcium citrate is then separated by a second rotary vacuum filter (P-16/RVF-102) and the citrate-free filtrate is sent to a wastewater collection tank. The calcium citrate cake is sent to another agitated reactor vessel (P-17/V-107), where it is acidified with dilute sulfuric acid ("H_2SO_4 (10% w/w)") to form a precipitate of calcium sulfate (gypsum):

$$1(\text{Calcium Citrate}) + 3\ H_2SO_4 \rightarrow 2(\text{Citric Acid}) + 3\ CaSO_4 \qquad (11.3)$$

A third filter (P-18/RVF-103) is used to remove the precipitated gypsum and yields an impure citric acid solution. Careful control of pH and temperature of the precipitation steps is important for maximizing the yield of citric acid. The resulting solution is concentrated and crystallized using a continuous evaporator/crystallizer (P-19/CR-101). The crystals are recovered by filtration (P-20/RVF-104) and dried in a rotary dryer (P-21/RDR-101). If the final product is required in high purity (e.g., for pharmaceutical applications), treatment with activated carbon may precede crystallization in order to remove colorants. In addition, ion exchange is sometimes used to remove ionic species (Berovic & Legisa, 2007). These operations are not included in this model.

11.3 Model setup highlights

The basics of setting up a batch process model in SuperPro Designer have been explained in Chapter 10. In this section, some advanced process modeling features are discussed.

11.3.1 Material charges

Modeling of material charge operations is done with the help of a Charge (for raw materials) or Transfer-In (for materials fed from another procedure) operations. However, if the required amount of fed material is not known a priori, a "Pull-In" operation can be used instead. Pull-In operations calculate the material amount based on some user-defined specification. For instance, Fig. 11.2 shows how the second water charge operation is set up in P-1/V-101 in order to dilute the glucose contents to 20%. Other specification objectives that could be used in a Pull-In operation include the specification of target feed/contents ratio, target final amount, target temperature, etc. (see Fig. 11.2).

Similar setup is necessary for the fermenter P-11/FR-101 where the amount of nutrients needed is proportional to the amount of molasses fed. Therefore, nutrients need to be *pulled* in the fermenter at a required amount which must be related to the flow of molasses. This is achieved by modeling the charging of nutrients in the fermentation procedure (P-11/FR-101) as a Pull-In operation where a 0.08 nutrients/molasses ratio must be achieved. The required amount for nutrients, as calculated at solution time, is back-propagated to the mixture preparation procedure P-7a/MX-101. This is a mixing step with regulated inputs so that a certain mixture amount and composition are achieved in the

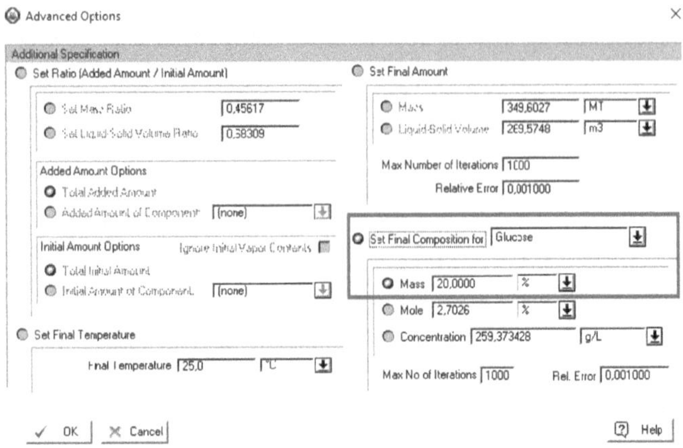

FIGURE 11.2 The advanced options dialog of a Pull-In operation for mixing unit P-1/V-101.

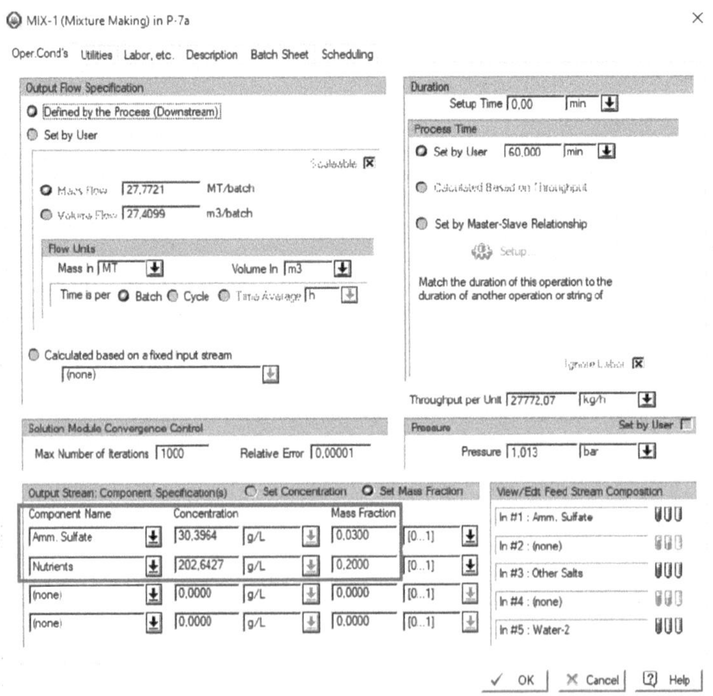

FIGURE 11.3 The operating conditions dialog of a Pull-In operation in mixing unit P-7a/MX-101.

output. As shown in Fig. 11.3, Ammonium Sulfate and Nutrients should be fed in P-7a/MX-101 so as to achieve at the output mass fractions of 0.03 and 0.2, respectively. Input streams whose amount is to not be determined at solution

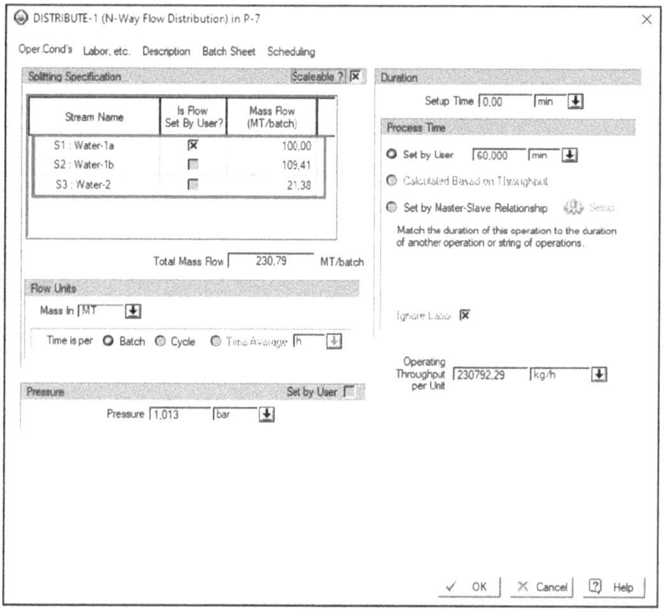

FIGURE 11.4 The operating conditions dialog of Flow Distribution operation P-7/FDIS-101.

time by the process (e.g., "Amm. Sulfate" and "Other Salts" streams in P-7a/MX-101) are labeled as "Auto-Adjust."

Water used in the plant originates from a central distribution system modeled by a flow distribution procedure (P-7/FDIS-101) which is used to feed water to the molasses tank (P-1/V-101) and the nutrients mixing process (P-7a/MX-101). In Fig. 11.4, it is shown that in the Flow Distribution operation, the amount of one output stream (Water-1a) is fixed at 100 MT/batch (the fixed amount of water fed initially into the molasses tanks), but the amounts of the other two output streams ("Water-1b" and "Water-2") are determined by the downstream process. With this set-up of pulling operations, the amounts of water, nutrients, and ammonium sulfate needed per batch are determined by the amount of molasses fed.

11.3.2 Modeling the fermentation step

The batch fermentation of molasses to citric acid is executed through a series of consecutively executed operations (Fig. 11.5): transfer-in of the molasses, pull-in of nutrients, inoculation, fermentation, transfer-out of broth, and CIP (clean-in-place) of the fermenter.

The fermentation operation is executed while maintaining a constant temperature of 35°C; since the fermentation is exothermic, cooling with Cooling Water is required to maintain the above temperature. The required

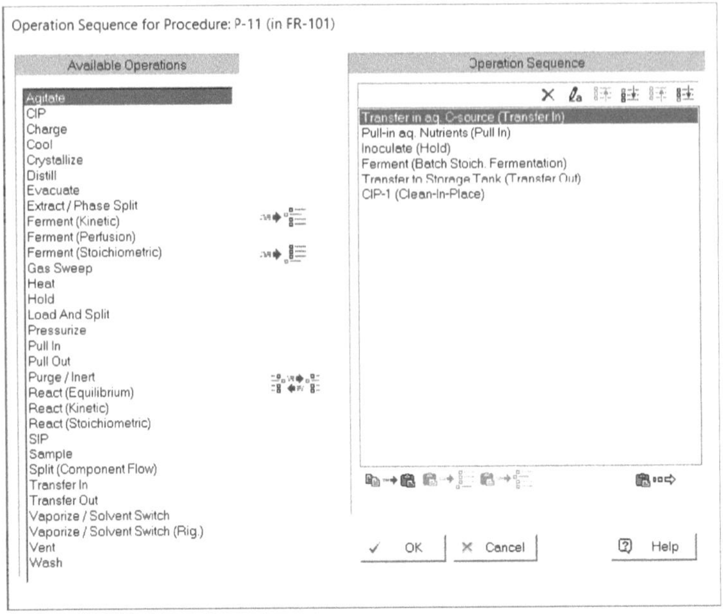

FIGURE 11.5 The declaration of operations in the Fermentation procedure P-11.

cooling duty and agent rate are calculated by the energy balance. Fermentation lasts for 5 days.

The oxygen required for fermentation is supplied by an air stream (S-114). Air is first compressed and filtered before fed into the fermenter. The rate of air needed is calculated to achieve 0.3 VVM (volume of air per volume of broth per minute). All these specifications are entered through the fermentation operating conditions dialog shown in Fig. 11.6.

Fig. 11.7 shows how the reaction stoichiometry is defined with the reactants in the left table and products in the right. The stoichiometry can be declared either on a mass or molar basis; in the latter case, proper molecular weights are required for all components participating in the reaction to check that the mass balance between reactants and products is satisfied.

In SuperPro Designer, there exist multiple options for entering a reaction scheme depending on the type of information available. In the simplest case, a "stoichiometric" operation allows the definition of (parallel or sequential) reactions using only their stoichiometry and final extent. If the kinetics of the reactions is known, then a "kinetic" operation can be used. In this case, the kinetic expression type and parameters must be set (Fig. 11.8). For fermentation kinetics, the generic reaction rate expression contains different terms (Monod, Haldane, Inhibition, etc.) which are activated and their parameters supplied as needed. In the kinetic case, the reaction extent is predicted by the model by solving the differential conservation equations over time.

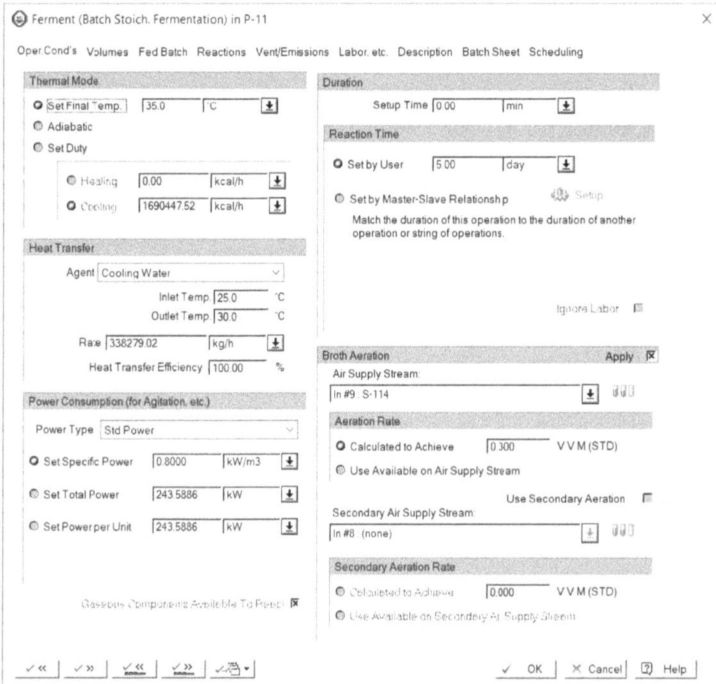

FIGURE 11.6 The operating conditions dialog of the Fermentation operation in P-11.

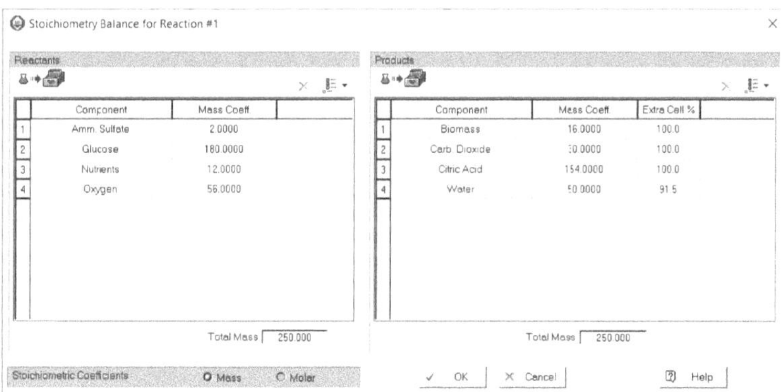

FIGURE 11.7 The fermentation reaction stoichiometry dialog for the Fermentation operation in P-11.

In this case study, the "stoichiometric" approach was used; as shown in Fig. 11.9, a conversion rate of 99% with respect to the reaction limiting component (determined at solution time) was assumed. An exothermic heat of reaction of −2990 kcal/kg of citric acid at 35°C is declared.

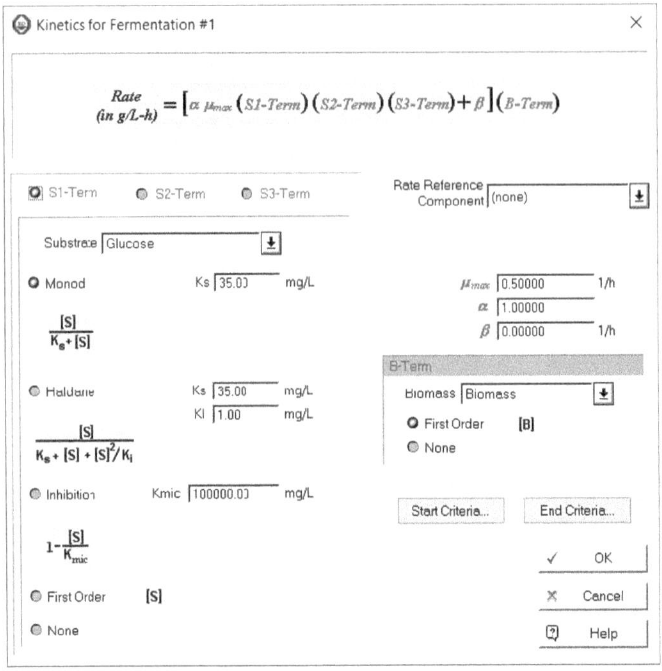

FIGURE 11.8 The kinetics dialog for a Fermentation operation.

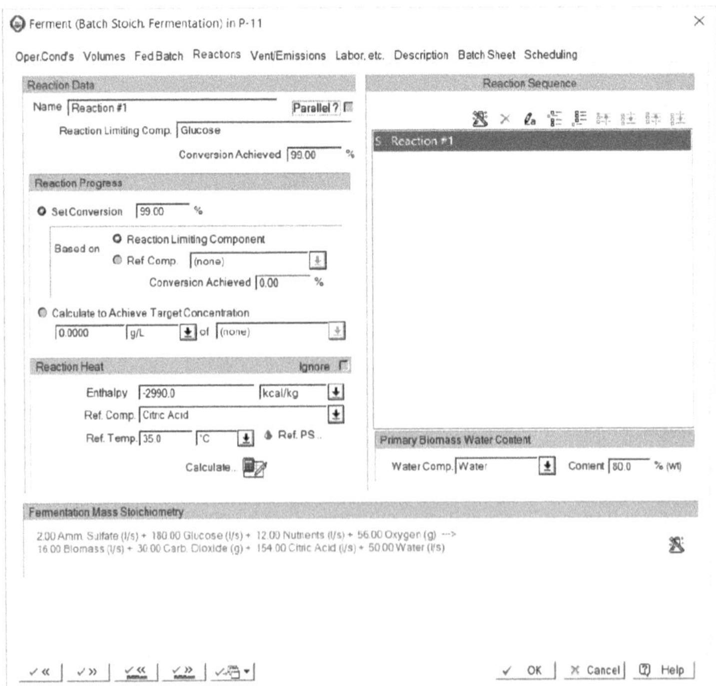

FIGURE 11.9 The stoichiometric fermentation reaction dialog for the Fermentation operation in P-11.

11.3.3 Modeling the cleaning operations

The fermentation and holding tanks need to be cleaned at the end of their use in every batch. Even though cleaning operations are not processing steps, they have to be accounted because they utilize material resources (cleaning agents) and they occupy time of the cleaned equipment and any auxiliary equipment (e.g., CIP-skids) used.

A cleaning operation may involve multiple steps. As shown in Fig. 11.10, the cleaning of fermenter includes a water flush, a caustic solution wash and a second water flush. The type and amount of cleaning agent used in each step along with its duration are declared through the dialogue in Fig. 11.10. A CIP-Skid can be also defined as auxiliary equipment used.

The waste generated by the CIP process is collected in the "CIP Waste" storage unit. Storage units are material collection points for the supply of raw material, or the deposit of product/waste streams. By specifying a replenishment/discharge strategy, it is possible to create the inventory profile of these units and use this information to properly design their capacity.

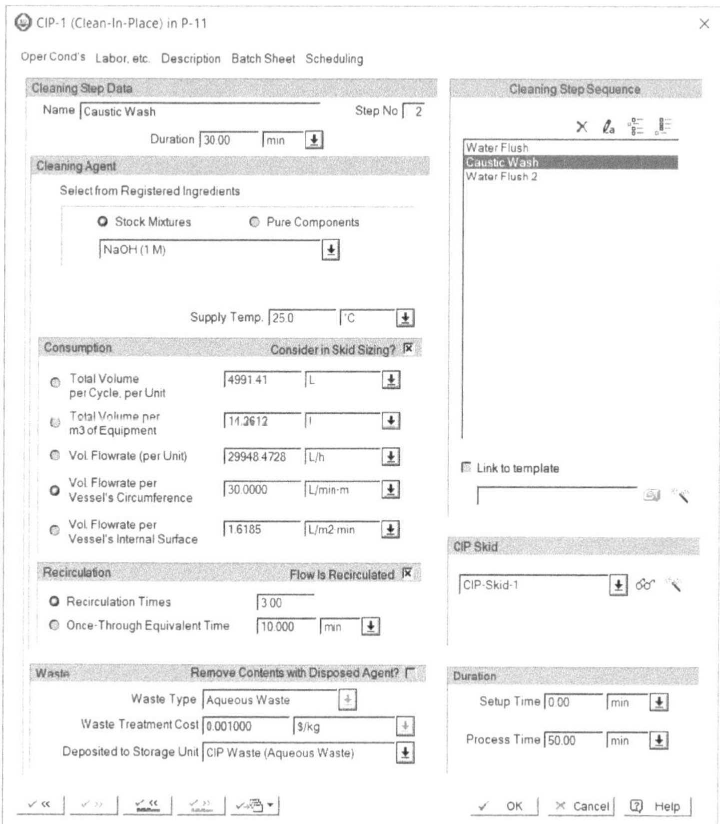

FIGURE 11.10 The CIP operation definition dialog in P-11.

11.4 Scheduling setup

As explained in Chapter 10, in a cyclically repeating batch process, the *recipe cycle time* is defined as the time period between the start of two consecutive batches. The *minimum recipe cycle time* is the cycle time threshold value below which overlaps in the utilization of equipment are unavoidable. If there is no equipment sharing among procedures of the same batch or batch integrity is maintained (i.e., equipment cannot be used by subsequent batch unless it has executed all procedures belonging to the current batch), the minimum cycle time can be calculated analytically. In all other cases, the minimum cycle time cannot be expressed by a simple equation. The subsequent sections describe how SuperPro Designer handles such complex scheduling scenarios.

11.4.1 Operating in staggered mode

Fermentations are typical examples of processes that operate in *staggered mode* (i.e., using equipment alternating among batches). Because of their long process times, they are usually the process bottlenecks. The use of multiple fermenters in staggered mode reduces the process cycle time and makes more efficient utilization of the downstream equipment.

In the citric acid case, the fermentation procedure lasts for 6.65 days including the actual fermentation and all other operations (e.g., material transfers). Fig. 11.11 shows the Equipment Data dialog for the fermentation procedure where six extra fermenters are assumed to be used in staggered mode. With a total of seven fermenters available, the effective cycle time of the fermentation procedure drops below 1 day, which makes possible the execution of a new batch every day (recipe cycle time = 1 day).

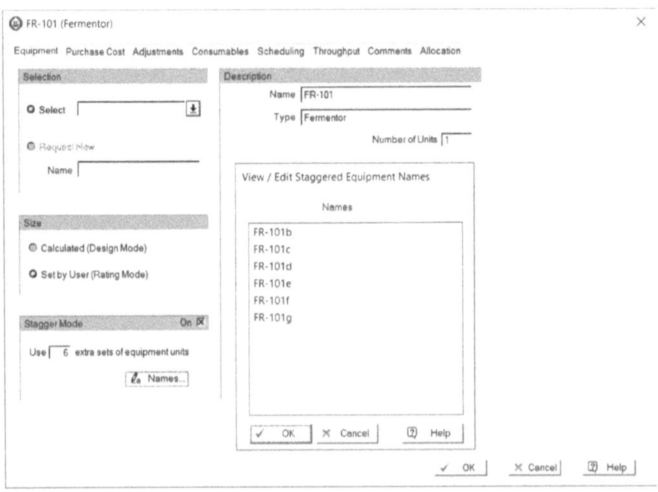

FIGURE 11.11 The equipment properties dialog of the Fermentation procedure.

It should be noted that staggering of extra equipment units is possible not only among batches but also among multiple cycles of the same procedure in the same batch (e.g., when the material to be processed cannot be handled by a single equipment unit and must be split into different "cuts"). A "carousel" of chromatography columns is a typical example of such a scenario: at every instant, one column is loaded, another is eluted, a third is regenerated etc. and the process is cyclically repeated by alternating columns. In SuperPro Designer, staggered equipment can be used to handle such cases as well.

11.4.2 Operating with independent cycling

Independently cycling procedures repeat themselves with a frequency which is different than that of the main process. This means that an independently cycling procedure processes a material amount which is different than the batch amount. If, for example, the overall process has a cycle time of 24 h and there is an independently-cycling procedure with a cycle time of 48 h, then the amount of material processed in one batch of this procedure must correspond to two batches of the main process. In this case, time is traded for capacity since the cycle time is longer but the batch size (and, consequently, the capacity of the used equipment) is larger than that of the main process.

In this case study, the vacuum filtration procedures (P-14, P-16, P-18, and P-20) are assumed to operate in independent cycling mode. In the Procedure Data dialog (Fig. 11.12), users can define a procedure's cycle time either directly or set it equal to the duration of some operation or other independently cycling procedure. Using the latter option, one can declare a set of procedures that are synchronized with the same "clock." Instead of specifying the cycle time, it is possible to define independent cycling by setting the ratio that represents how many cycles of this procedure correspond to one cycle (batch) of the main process.

It should be noted that through the same dialog, it is possible to set the number of cycles of a "normal" procedure or declare a procedure as "continuous"; in the latter case, the procedure is assumed to be operating all the time and is ignored in cycle time calculations. In this example, the downstream reactors (P-15/V-106, P-17/V-107), crystallization (P-19/CR-101), and drying procedures (P-21/RDR-101) have been defined as continuous.

11.4.3 Calculating the minimum cycle time

Using the operation scheduling data, SuperPro Designer calculates the duration of every procedure and estimates the minimum cycle time and the process bottleneck (see Fig. 11.13). As indicated earlier, if there exist procedures that share the same equipment across batches and/or sharable auxiliary equipment, the calculation of the minimum cycle time is a more involved process which is done through the Cycle Time Calculator (Fig. 11.14). Using this calculator, all

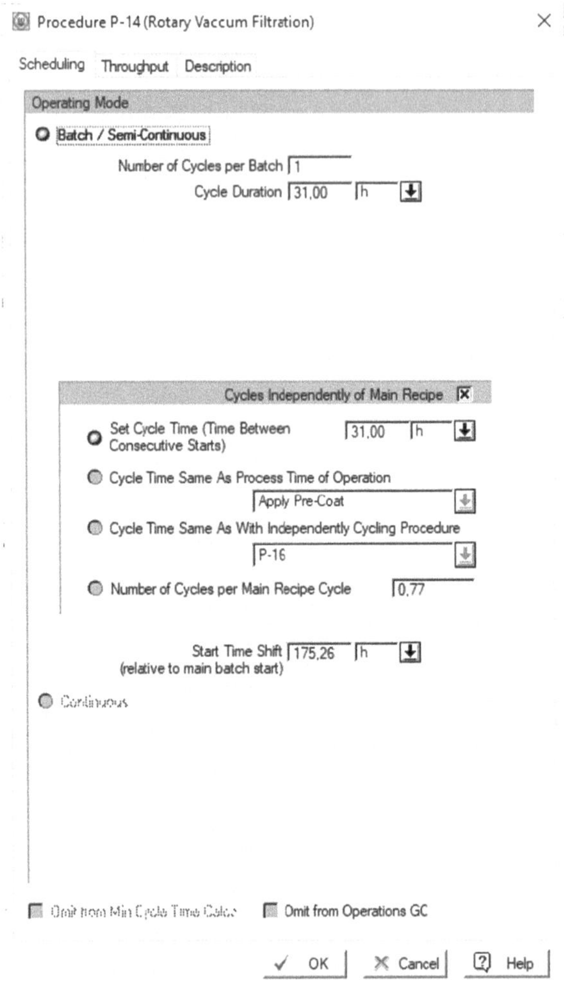

FIGURE 11.12 The procedure data dialog.

ranges of feasible cycle time values can be calculated and displayed. In this example, equipment sharing exists for the skid used for cleaning the fermenters and mixing vessels, so Fig. 11.14 shows all feasible cycle times that do not cause any skid overlap.

Users can still select a different cycle time from the calculated minimum. In this example, the minimum cycle time was calculated at 22.8 h; however, a cycle time of 24 h (a new batch every day) has been set (Fig. 11.13). This cycle time is within the feasible ranges shown in Fig. 11.14, so, no equipment utilization conflicts are expected when scheduling multiple batches of this process.

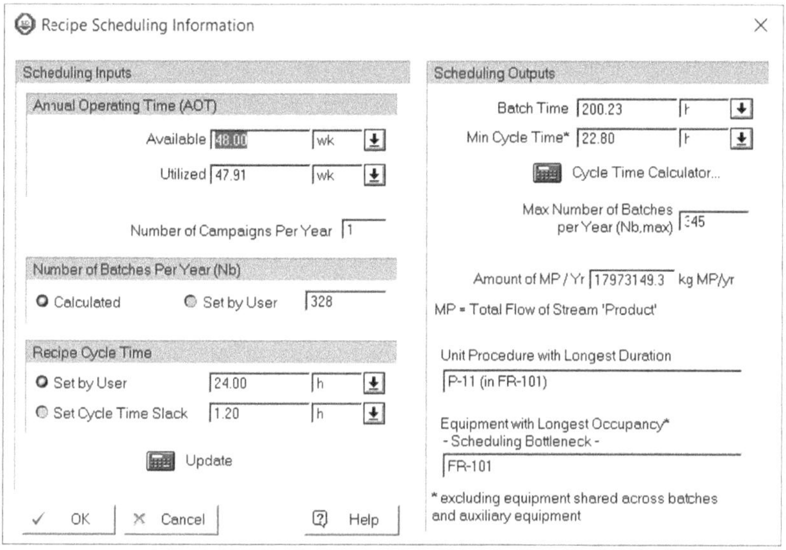

FIGURE 11.13 The scheduling information dialog and the definition of the process cycle time.

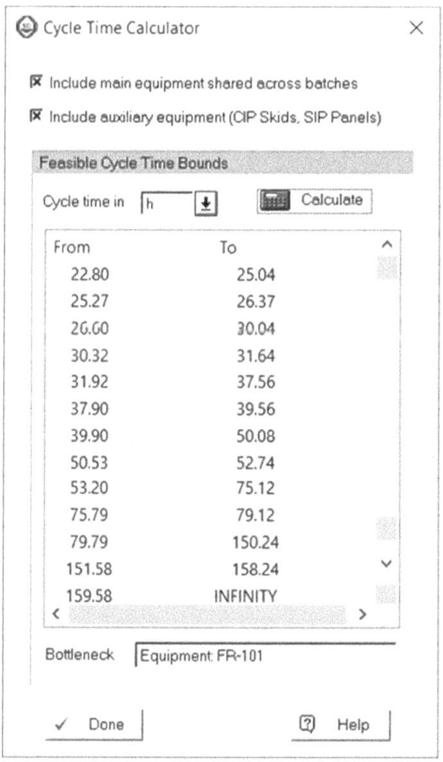

FIGURE 11.14 The process cycle time calculator.

11.5 Process simulation results

SuperPro Designer implements a sequential-modular strategy for solving the model, i.e., procedures are solved sequentially following the material flow (see Chapter 1 for details). If recycle loops exist, the tear-stream algorithm is implemented with the use of an iterative numerical solver until some convergence criteria are satisfied (see Chapter 4 for details).

The solution of balance equations for all procedures results in the calculation of the state (composition, amount, and conditions) of all intermediate/output streams and of all internal procedure states, the required amounts of utilities and the required size of all equipment.

In the citric acid case under study, for an annual production of 17,973 MT of citric acid, the calculated amounts of raw materials needed are shown in Table 11.3.

The size of all process equipment is shown in Table 11.4. Most equipment were set in design mode and their size was calculated; a few (e.g., holding tanks) were set in rating mode and their size was provided by the user. Equipment in design mode are sized to fit the requirements of the hosted operations through a "bidding" process (i.e., the most demanding operation sets the equipment size). For equipment in rating mode, SuperPro Designer checks if the provided size can satisfy the process requirements.

11.6 Process scheduling and debottlenecking

The scheduling data of every operation is the basis for determining the recipe cycle time, identifying bottleneck(s), and any potential conflict in the use of main and auxiliary equipment. The Operations Gantt chart in Fig. 11.15 shows how all operations are lined up in time for the execution of a single batch for the citric acid case.

TABLE 11.3 Raw material requirements for the entire process (MP: main product = citric acid).

Material	MT/year	MT/batch	MT/MT of MP
Water	143,024	436.049	7.96
Molasses	45,868	139.841	2.55
Amm. Sulfate	273	0.833	0.02
Nutrients	1822	5.554	0.10
NaOH (1 M)	4897	14.930	0.27
Air	273,105	832.636	15.20
Lime (33%)	35,505	108.248	1.98
H_2SO_4 (10% w/w)	149,793	456.687	8.33

TABLE 11.4 The process equipment size and cost.

Name	Type/size	Unit cost ($)
AF-101	Air filter	94,000
	Rated throughput = 27,735.4 m^3/h	
AF-102	Air filter	34,000
	Rated throughput = 10,104.8 m^3/h	
C-101	PBA column	150,000
	Column volume = 4.66 m^3	
CR-101	Crystallizer	507,000
	Vessel volume = 127.01 m^3	
FR-101	Fermenter	600,000
	Vessel volume = 350.00 m^3	
G-101	Centrifugal compressor	496,000
	Compressor power = 1468.18 kW	
PFF-101	Plate and frame filter	145,000
	Filter area = 334.72 m^2	
RDR-101	Rotary dryer	441,000
	Drying area = 82.81 m^2	
RVF-101	Rotary vacuum filter	150,000
	Filter area = 51.24 m^2	
RVF-102	Rotary vacuum filter	225,000
	Filter area = 101.53 m^2	
RVF-103	Rotary vacuum filter	194,000
	Filter area = 78.74 m^2	
RVF-104	Rotary vacuum filter	138,000
	Filter area = 41.25 m^2	
ST-101	Pasteurizer	101,000
	Rated throughput = 33,454.63 L/h	
V-101	Blending tank	340,000
	Vessel volume = 300.00 m^3	
V-102	Blending tank	340,000
	Vessel volume = 300.00 m^3	

Continued

TABLE 11.4 The process equipment size and cost.—cont'd

Name	Type/size	Unit cost ($)
V-103	Blending tank	340,000
	Vessel volume = 300.00 m^3	
V-105	Flat bottom tank	209,000
	Vessel volume = 174.39 m^3	
V-106	Neutralizer	118,000
	Vessel volume = 41.32 m^3	
V-107	Neutralizer	87,000
	Vessel volume = 14.61 m^3	

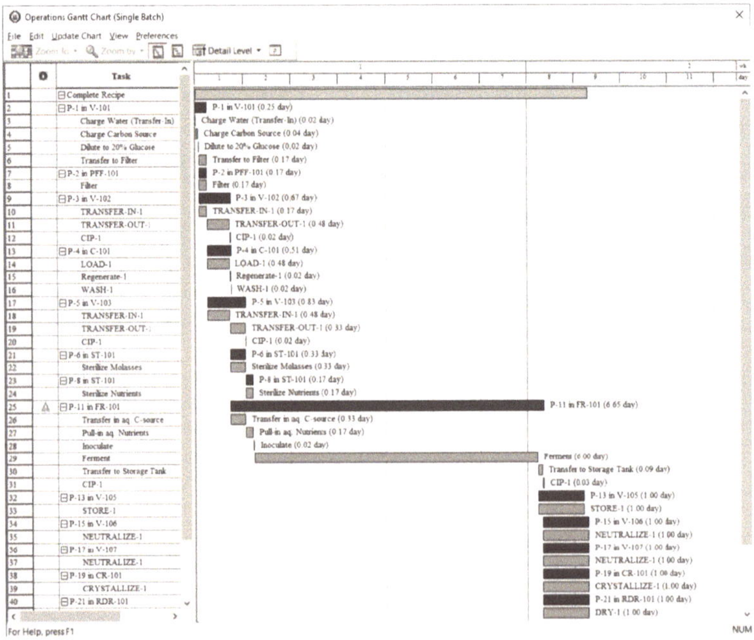

FIGURE 11.15 The operations Gantt chart.

Fig. 11.16 displays the Equipment Occupancy Gantt chart for 14 consecutive batches assuming a cycle time of 24 h; despite the long fermentation time (6.65 days), this cycle time is possible because of the use of multiple fermenters operating in staggered mode. The batch time (i.e., completion time of a batch) is approximately 8.3 days.

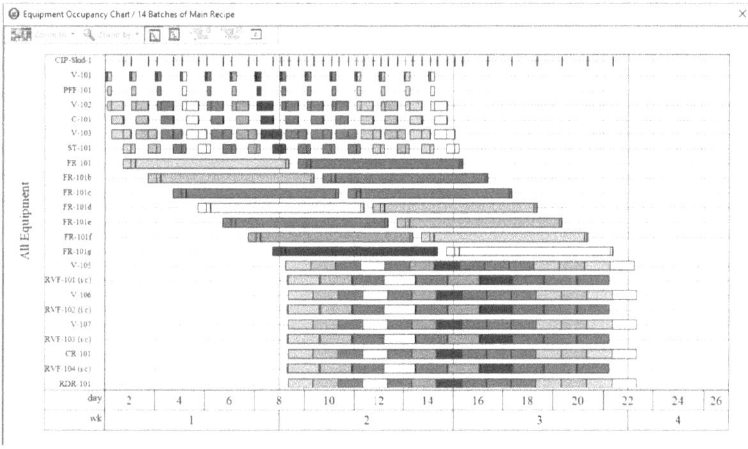

FIGURE 11.16 The equipment occupancy Gantt chart—14 batches.

Schedule-wise, the overall process is divided into the upstream which is purely in batch mode and the downstream which operates continuously. The two parts are connected through the holding tank P-13/V-105, which receives the broth from all fermenters and feeds the purification train. The absence of any gap in the utilization of the downstream equipment signifies that they operate continuously in a "back-to-back" mode. In the upstream section, fermenters are the process bottlenecks; further reduction in the cycle time is possible if more fermenters become available. In this case, extra holding tanks (e.g., V-103, V-102) and chromatography columns may also be needed (see Exercise 1). If acquiring new equipment is not possible, reduction in the cycle time may be possible by decreasing the duration of operations in bottleneck equipment by proper modifications. For example, technological advancements in increasing the fermentation efficiency and reducing the fermentation time could have significant impact on the cycle time and plant throughput.

11.7 Process economics

SuperPro Designer performs thorough economic analysis which includes the estimation of capital expenditure (CAPEX), operational expenditures (OPEX), cash flows over the project lifetime, and the values of economic evaluation indices.

11.7.1 Capital investment costs

The estimation of CAPEX is done on the basis of the total equipment cost using user-provided multiplication factors for all other costs such as piping, instrumentation, and engineering. SuperPro Designer provides meaningful default values for all these multipliers; however, the actual values may vary considerably with the type of the plant (please refer to the SuperPro User Guide for more details).

The estimated equipment cost is shown in Table 11.4; the total equipment cost is calculated at around \$9.5 million which translates to a direct fixed capital (DFC) of around \$39.2 million. The total capital investment (including working capital and start-up costs) is estimated as \$42.8 million.

11.7.2 Operating costs

OPEX is divided into variable and fixed costs. Variable costs are directly dependent upon the level of production and include the cost of raw materials, utilities, consumables, labor, and waste treatment or disposal. Fixed costs are facility-dependent invariable costs such as depreciation, maintenance, and insurance.

To calculate revenues, material, and waste treatment costs, SuperPro Designer needs to know the economic role all input and output streams. Through the dialog of Fig. 11.17, users can classify input streams as "raw material" or "revenue" (in case plant revenues originate from the treatment of input streams) and output streams as "revenue" or "waste" (organic, aqueous, sold, or emissions). The selling price of revenue streams or the disposal cost of waste streams can also be set in this dialogue. In this study, a selling price of \$2/kg for crystal citric acid was assumed along with a disposal cost of \$1/m^3 and \$50/MT for liquid and solid (gypsum) waste streams, respectively.

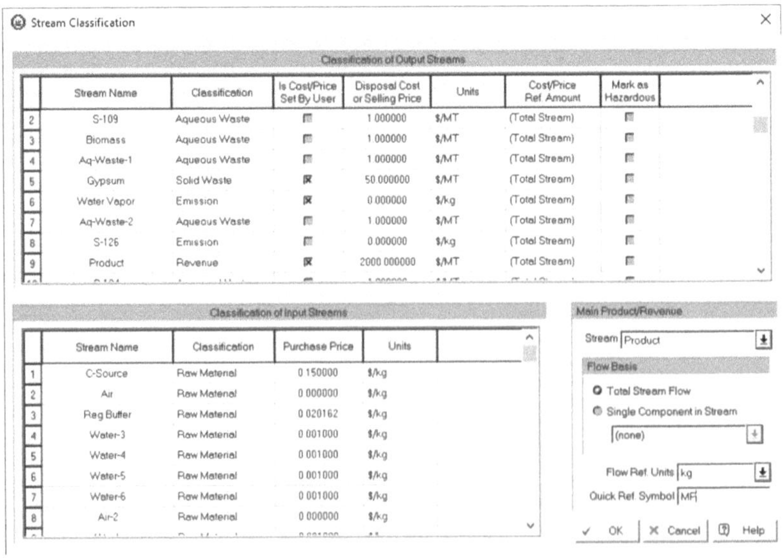

FIGURE 11.17 The stream classification dialogue for citric acid case study.

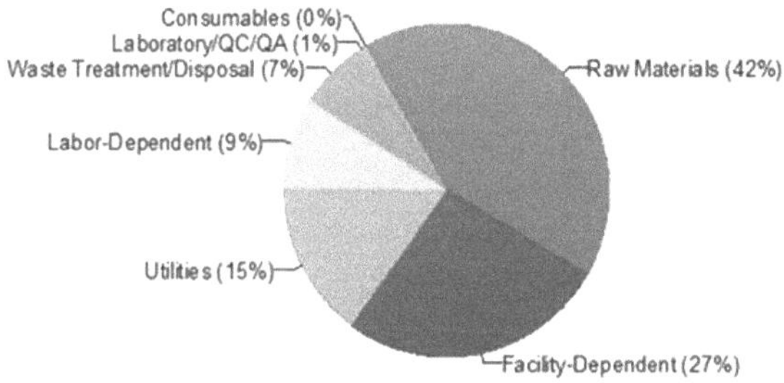

FIGURE 11.18 The annual operating cost breakdown.

Utility (heating, cooling, power) and consumable requirements are calculated from the process balances; labor requirements can be set on a per-operation or lumped-process basis. The annual costs of all these components of OPEX are calculated based on user-provided values of unit costs (e.g., per MT of steam, per labor-hour, etc.). Laboratory/QC/QA, transportation, R&D, and other costs can optionally be included.

In this case study, the total annual operating cost is estimated as $24.7 million. Fig. 11.18 shows the breakdown on all cost components in the form of a pie chart (generated within the SuperPro Designer reports). The cost of raw materials is the greatest component (42%) followed by the facility-dependent fixed costs (27%). The cost of molasses ($0.15/kg) accounts for approximately 67% of the total materials cost. The utility cost is also significant (15%). The disposal of gypsum accounts for 85% of the total environmental cost.

11.7.3 Economic evaluation

With the use of the cost and revenue values, SuperPro Designer calculates and reports various economic evaluation indices, i.e., gross margin, return on investment (ROI), payback time (PBT), internal rate of return (IRR), and net present value (NPV).

Fig. 11.19 shows all the key economic figures for the citric acid case study. The unit production cost is approximately $1.4/kg. For a selling price of $2/kg, the economic potential of the investment seems promising, with ROI at 21.6%, PBT at 4.6 years, IRR at 18.2%, and NPV 7% at approximately $41.3 million.

The effort to improve the process in order to safeguard against economic uncertainties could take different directions. One approach would be to investigate different technologies that may be more efficient and economical. As indicated before, the formation of gypsum is both environmentally and economically undesirable. Alternative flowsheets could be generated encompassing technologies such as extraction or electrodialysis membranes. In the context of process design, such alternative flowsheets could be evaluated and compared from a technological economic and environmental point of view.

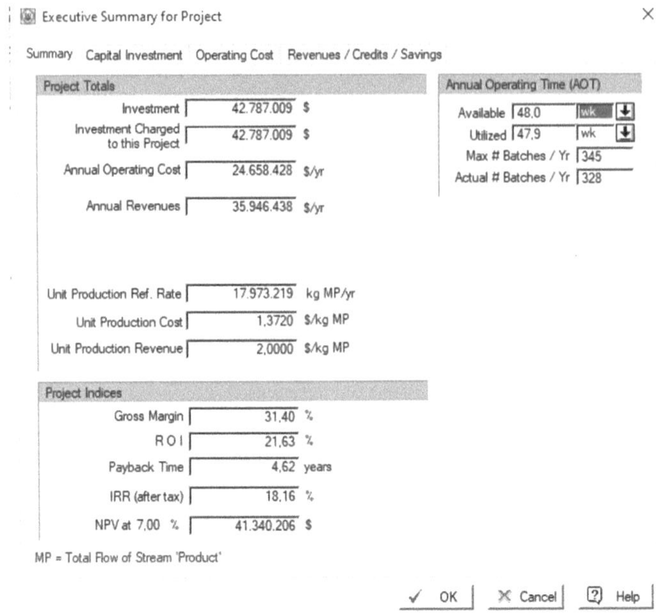

FIGURE 11.19 The executive summary dialogue with the key economic figures.

Another approach would be to seek ways to reduce operating costs, which may be done through acquiring used equipment and, thus, reducing the depreciation cost. Finding a source of cheap molasses and locating the plant near that source can reduce the raw materials cost. These types of what-if analyses are easy to perform when a simulation model of the integrated process is available.

11.8 Variability analysis

It is important to assess the effect of process variabilities on key operating, economic, and environmental variables. In SuperPro Designer, it is possible to run variability analyses with external tools with Application Programming Interface (API) functionality. Using this functionality, it is possible to change the values of input parameters, solve the balance equations and retrieve the values of output parameters through an external application while using SuperPro Designer in the background. This set-up can be used to perform sensitivity or variability analyses.

A typical way to exploit this functionality in variability analysis is to use Microsoft Excel and some add-in utilities that perform Monte-Carlo simulations such as Crystal Ball (Oracle Crystal Ball, 2021). In this case, the probability distributions of uncertain variables are provided and the Monte-Carlo simulator produces sets of random values following the provided distribution. For each set of values, the SuperPro Designer model is solved, the results are read back and the distribution of the uncertain output variables is estimated. These analyses can be used to assess the degree to which a production system is capable of achieving specific production goals in the face of uncertainty.

To illustrate this approach, a variability analysis is executed on the citric acid case study with the objective being to identify the probability of achieving favorable economic results under uncertainty in the cost of molasses, the selling price of citric acid and the batch size. These three input variables were selected because of their significant impact on the economic bottom line and because their values are expected to show fluctuations (e.g., because of variability in the market demand and in the availability of molasses). For demonstration purpose, it is assumed that all uncertain input variables follow a triangular distribution. As shown in Fig. 11.20, the probability distribution is assumed symmetrical for batch size (within the range of 100−180 MT/batch), and skewed for the other two variables ($0.10−$0.30/kg for the molasses cost and $1.2−$2.2/kg for the product selling price). The charts in Fig. 11.20 show the assumed theoretical distribution of each variable (continuous line) and the actual distribution over 2000 samples created by the Monte-Carlo simulation.

The estimated output variables are the annual throughput (MT/yr of citric acid produced), unit production cost ($/kg of product), ROI and NPV at 7% interest rate. In the charts of Fig. 11.21, the range of expected values and the most probable values of the output variables can be extracted. Cumulative

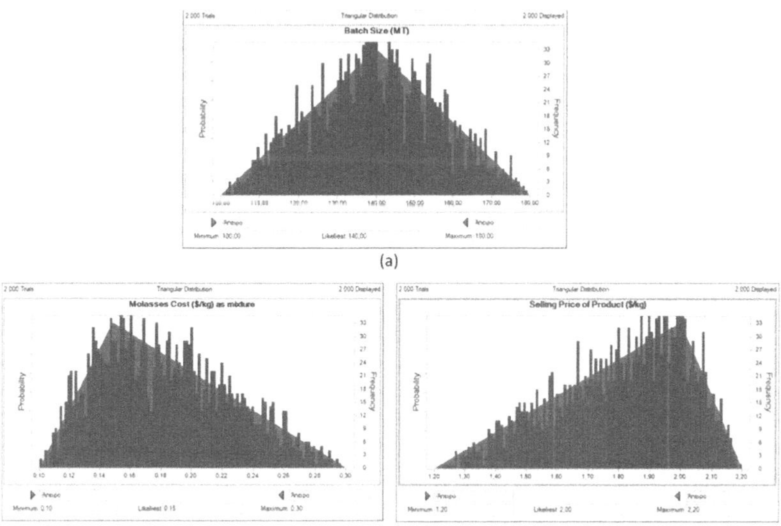

FIGURE 11.20 Probability distributions of input variables: (A) batch size (MT), (B) cost of molasses ($/kg), and (C) citric acid selling price ($/kg).

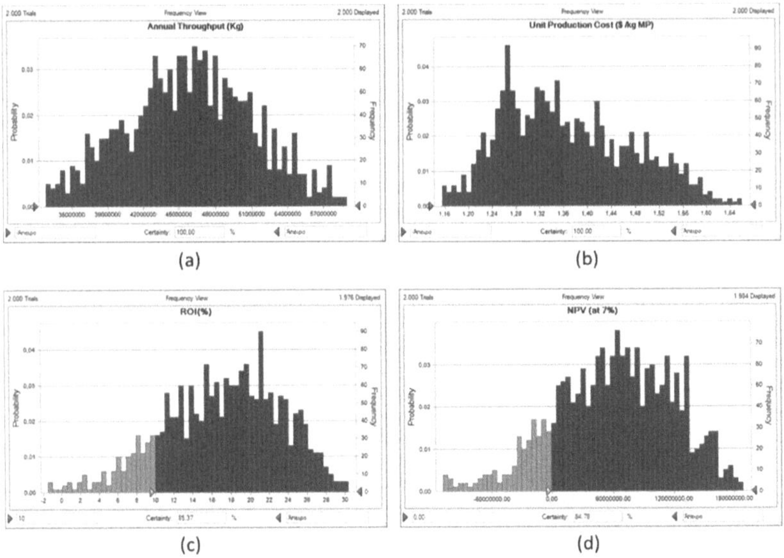

FIGURE 11.21 Predicted probability distributions of output variables: (A) annual throughput (kg/year), (B) unit production cost ($/kg of citric acid), (C) ROI, and (D) NPV at 7% interest rate ($).

probability values represent the certainty by which some target value can be achieved. For example, Fig. 11.21D shows that a positive NPV value (at interest rate of 7%) can be achieved with a certainty of about 85%. The same certainty is reported for ROI for a cut-off value of 10% (Fig. 11.21C). This threshold value, however, is very conservative in an economic viability assessment. If a higher value of ROI at 20% is sought, the certainty of achieving it in the face of uncertainty drops to 35%, indicating that high returns are not likely to be achieved with the current design. Therefore, process modifications may be needed in order to optimize the process both technologically and economically.

11.9 Conclusions

Setting up a process model in SuperPro Designer requires the declaration of all unit procedures and operations in the model, specification of their operating conditions and scheduling-related data. This chapter demonstrated how to model special operations (such are reactions, CIP, material pull-ins, flow distribution etc.) and how to set-up exceptional scheduling scenarios (such as procedures with independent cycling). It was shown how the process scheduling information can be used to estimate the process cycle time. With the latter, the annual throughput can be calculated. Process bottleneck can then be identified, which allows the reduction of cycle time to increase throughput.

Finally, it was demonstrated how modeling and scheduling results can be used as the basis for a thorough economic analysis within SuperPro Designer. With the help of an external Monte-Carlo simulation tool, variability analysis can be used to assess the effect of process uncertainties on economic and other results.

Exercises

Exercise 1: Decreasing the cycle time

As indicated in this chapter, reducing the recipe cycle time (RCT) is one way to increase the annual throughput since more batches can be executed within a year. In order to reduce the RCT, the bottleneck must be identified and eliminated with proper process modifications. In the citric acid case study, the bottleneck is the fermentation procedure; its effect on the RCT was reduced by incorporating six additional fermenters operating in staggered mode. The reduction in the minimum RCT is proportional to the number of fermenters. For example, the minimum RCT is 159.6 h (=6.65 days) when one fermenters used, and is reduced to 22.8 with 7 fermenters. It may be expected that further decrease in the RCT could be possible if the number of fermenters was increased further.

In this context, use the SuperPro Designer citric acid model, set the number of staggered fermenters to eight and re-solve the model. [NOTE: before resolving, you will need to reset the Recipe Cycle Time option from "Set by user" (24 h) to "Set Cycle Time Slack" and set this slack to 0.0. In other words, force the model to use the minimum cycle time as calculated in each case.]

a. Did the minimum RCT decrease further? If yes, was the decrease again proportional to the number of fermenters? If not, can you explain why?
b. What process/equipment modification would you propose to further decrease the RCT?
c. The above changes in the RCT are not expected to affect the equipment sizes. Yet, if you look at the estimated volume of P-15/V-106 you will notice that it increases with every reduction in the RCT. Why is that the case? (Hint: P-15 is declared as "Continuous")

Exercise 2: Increasing the batch size

Increasing the batch size may be another way to achieve higher throughput. In most cases, larger batch size means longer process times and, consequently, increased recipe cycle time (RCT). It is expected, however, that the increase in the RCT is not proportional to the increase in the batch size and the net effect on throughput will be positive. Larger batch sizes also mean increased equipment size and capital expenditures.

Use the scaling functionality of SuperPro Designer (option to "Adjust Process Throughput") and set different target batch sizes with respect to the nominal value (54.8 MT/batch of product).

a. Does the minimum RCT change when the batch size is changed? If yes, is this change proportional to the batch size change (i.e., if the batch size is doubled, is the cycle time doubling as well)? If not, why?
b. What happens to capital investment cost, operating cost per kg of final product, and the economic evaluation indices if you increase the batch size?

Acknowledgments

The author wishes to acknowledge the contribution of Jim Prentzas (Intelligen, Inc.) on the variability analysis section and of Demetri Petrides (Intelligen, Inc.) on helpful discussions.

References

Apelblat, A., 2014. Citric acid. Springer, Cham.

Berovic, M., Legisa, M., 2007. Citric acid production. In: Biotechnology Annual Review, Vol 13, pp. 303–343.

Blanch, H.W., Clark, D.S., 1997. Biochemical Engineering. Dekker, New York.

Carmichael, D., Siletti, C.A., Koulouris, A., Petrides, D., 2018. Bioprocess simulation and scheduling. In: Chang, H.N. (Ed.), Emerging Areas in Bioengineering. Wiley-VCH Verlag GmbH & Co.

Cision, 2021. http://www.prnewswire.com/news-releases/global-citric-acid-markets-report-2011-2018-2019-2024-300814817.html. (Accessed September 2021).

Crueger, W., Crueger, A., 1989. Biotechnology—A Textbook of Industrial Microbiology, 2nd ed. Sinauer, Sunderland, MA.

Intelligen Inc, 2021. https://www.intelligen.com/products/superpro-examples/. (Accessed September 2021).

Oracle Crystal Ball, 2021. https://www.oracle.com/applications/crystalball/. (Accessed September 2021).

Roberts, L.R., 1979. Citric acid. In: McKetta, J.J., Cunningham, W.A. (Eds.), Encyclopedia of Chemical Processing and Design, vol. 8. Dekker, New York, p. 324.

Show, P.L., Oladele, K.O., Siew, Q.Y., Zakry, F.A.A., Lan, J.C.-W., Ling, T.C., 2015. Overview of citric acid production from Aspergillus Niger. Frontiers in Life Science 8 (3), 271–283.

SuperPro Designer v12, 2021. User Guide. Intelligen Inc., USA.

Further reading

Roehr, M., 1998. A century of citric acid fermentation and research. Food Biotechnology 36, 163–171.

Chapter 12

Design and optimization of wastewater treatment plant (WWTP) for the poultry industry*

Chien Hwa Chong[1], Rui Ma[1] and Dominic C.Y. Foo[2]

[1]*Department of Chemical and Environmental Engineering, University of Nottingham Malaysia, Semenyih, Selangor, Malaysia;* [2]*Department of Chemical and Environmental Engineering/Centre of Excellence for Green Technologies, University of Nottingham Malaysia, Semenyih, Selangor, Malaysia*

Chapter outline

12.1 Introduction

Poultry industry effluent is regarded as nontoxic waste. However, the untreated effluent may still lead to serious environmental pollution due to its large quantity of organic content. The latter includes undigested food, blood, fat, lard, loose meat, paunch, colloidal particles, excrement, soluble proteins, and suspended materials, which can lead to river deoxygenation and groundwater pollution. Therefore, the poultry industry effluent is often monitored for its water quality parameters, such as the concentrations of biochemical oxygen demand (BOD), total suspended solids concentration (TSS), chemical oxygen demand (COD5), fats, oil, and grease. A typical wastewater treatment plant (WWTP) for the poultry industry is divided into four steps, i.e., pretreatment,

*Simulation and other electronic files of this chapter are made available on a companion website; URL: https://www.elsevier.com/books-and-journals/book-companion/9780323901680.

Chemical Engineering Process Simulation. https://doi.org/10.1016/B978-0-323-90168-0.00012-3

primary, secondary, and tertiary treatment (Bingo et al., 2019). The pretreatment systems include screening, settling, catch basins, and floatation processes, where offal materials are removed. The primary step consists of physiochemical treatment to remove fats, oil, and grease, TSS, COD, and BOD. In secondary step, biological treatment is used to breakdown remaining organic material. Lastly, tertiary step usually consists of a recycle system, where filtration, adsorption, membrane processes, etc. are used to remove the remaining suspended or dissolved substances. To improve water quality parameters during the treatment, Kobya et al. (2006) recommended the use of iron and aluminum as anode materials in the electrocoagulation unit at low pH of 2–3. However, this operating condition is not suitable for the growth of microbes. Similar findings were reported by Tezcan et al. (2009), but with higher pH value of 7.8. More recently, it was found that sequential batch reactor with fibers was recommended by Aziz et al. (2018) to enable the attachment of suspended solids to the fibers; doing this increases the biomass concentration in the reactor and allows higher removal rate of COD and BOD5 (in the range of 96%–98%). Most of the above research focuses on new technology developments rather than optimizing the exiting WWTP. Note that the introduction of new technology will lead to high capital cost. It is suitable for new WWTP rather than for upgrading of existing system.

In this chapter, design and optimization for an existing WWTP for the poultry industry is demonstrated using SuperPro Designer v9 (Intelligen, 2014). The effects of recycle ratio on total annualized cost (TAC) was examined. Trade-off between TAC and final discharge quality was also explored.

12.2 Case study: poultry WWTP

A case study involving a poultry industry in Malaysia is demonstrated. All poultry industry effluent discharge must comply with the Environmental Quality Act (Effluent, Revision 2009) set by Malaysian Department of Environment (DOE, 2009) to prevent, abate, and control pollution to the environment. It is restricted by the discharge of waste in contravention of the acceptable conditions and highly dependent on the location of the plant where the acceptable values are shown in Table 12.1.

TABLE 12.1 Water discharge quality based on standard A and standard B (DOE, 2009).

Parameters	Standard A (mg/L)	Standard B (mg/L)
TSS	100	50
COD	200	120
BOD5	50	20
Nitrate nitrogen	50	20
Oil and grease	10	5

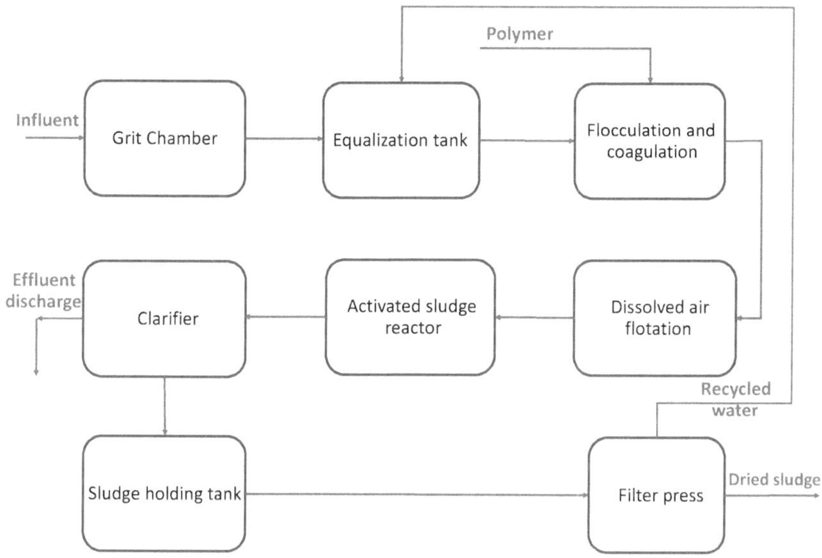

FIGURE 12.1 Process flow diagram (PFD) of existing poultry WWTP.

Fig. 12.1 shows an existing WWTP that was designed to treat 72 m^3/day of effluent prior to final discharge. As shown, the WWTP consists of oil and grease trap, influent tank, equalization tank, dissolved flotation unit (DAF), activated sludge reactor, clarifier and filter press.

A grit chamber was designed to remove part of the fixed suspended solids including fats, oil, and grease from the influent. The wastewater is then pumped into the equalization tank which functions as a buffer unit. Even where there is high fluctuation of the influent, the equalization tank provides consistent flow to the downstream treatment processes. An assumption is made that no reaction occurs in the equalization tank. Effluents then enters the coagulation and flocculation unit where polymer are added. Wastewater then enters the DAF unit followed by an activated sludge reactor. The latter was designed for removal of COD and BOD where the wastewater was treated with bacterial aerobically. The treated water was then discharged from the WWTP after it is separated from its solid sludge in the clarifier. The sludge was then sent to the filter press for dewatering process. Due to aging factor, the treated wastewater does not meet the required specification (Standard B). Hence, the replacement of the activated sludge reactor is being considered by the plant authority.

FIGURE 12.2 Reaction steps in the activated sludge reactor.

Biomass-n Decay Reaction

$$\text{Biomass-n} + O_2 \rightarrow NH_3 + CO_2 + H_2O + \text{Biomass-i}$$

Biomass-h Decay Reaction

$$\text{Biomass-h} + O_2 \rightarrow NH_3 + CO_2 + H_2O + \text{Biomass-i}$$

Denitrification

$$\text{BioWaste} + NO_3 \rightarrow CO_2 + N_2 + H_2O + \text{Biomass-h}$$

BioWaste Degradation

$$\text{BioWaste} + NH_3 + O_2 \rightarrow \text{Biomass-h} + H_2O + CO_2$$

Nitrification

$$NH_3 + O_2 + CO_2 \rightarrow \text{Biomass-n} + H_2O + NO_3$$

12.3 Base case simulation model

Fig. 12.3 shows the simulation flowsheet for the WWTP in Fig. 12.1, developed using SuperPro Designer V9.0 (Intelligen, 2014). Table 12.2 shows the components used in the simulation flowsheet. As shown, some components are found in the software database, while some are user-defined by referring to other component with similar properties in SuperPro Designer database. Table 12.3 shows the influent stream compositions of the poultry plant. The component in discharge stream 1 such as carbon dioxide (CO_2) and ammonia (NH_3) are not traceable. In Fig. 12.3, the WWTP consists of equipment as described in the previous subsection. In the new design, the effluent discharged from the DAF unit is channeled to a newly designed activated sludge reactor (Aerobic Bio-oxidation). The latter is used to degrade the biomass. Fig. 12.2 shows the reactions steps that take place in the activated sludge reactor. Effluents from the activated sludge reactor is then transferred to the newly replaced clarifier tank. The latter separates the solid sludge from the liquid phase. Table 12.4 shows the specifications of all unit operations in SuperPro Designer. Figs. 12.4 and 12.5 show detailed steps for setting reactions schemes and their kinetics data for the activated sludge reactor (AB-101).

In this study, the grit chamber, equalization tank and DAF units are existing units and hence they are excluded from the calculation of the capital cost. The latter is mainly contributed by the newly added units, i.e., active sludge reactor (AB-101), clarifier (CL-102), storage tank (V-101), and belt filter (BF-101), are therefore predominantly analyzed. These units are added as the quality of the discharge has exceeded the limit (Standard B; see Table 12.1) where the BOD_5 value was reported as 76 mg/L. Note that the storage tank (V-101) was

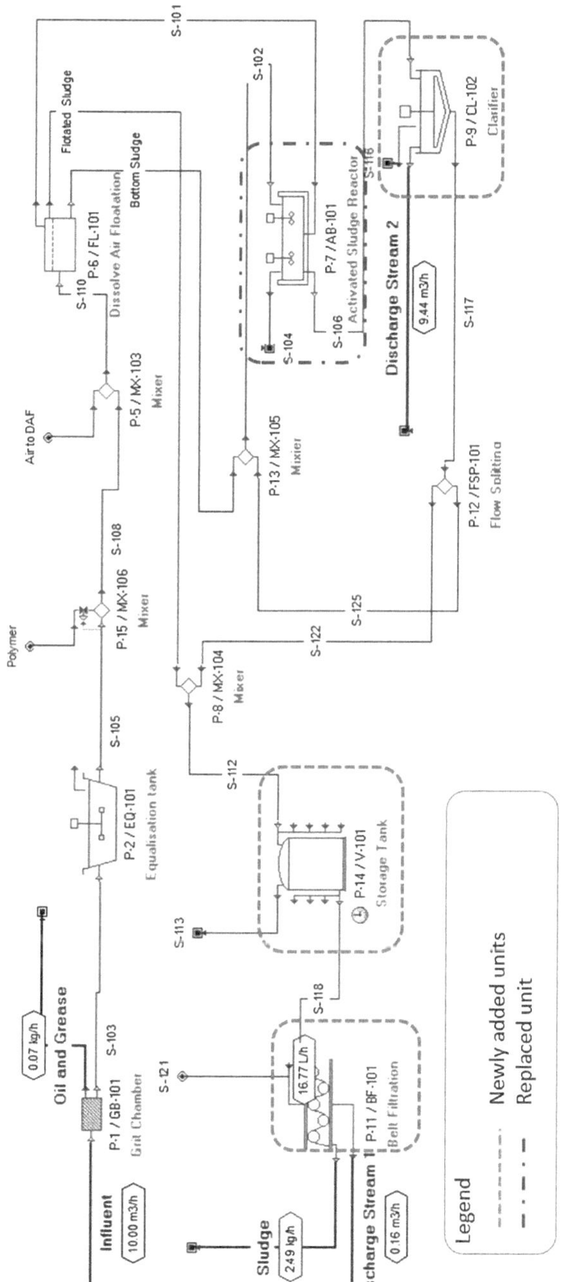

FIGURE 12.3 Simulation flowsheet for poultry WWTP. The newly added and replaced units are active sludge reactor (AB-101), clarifier (CL-102), storage tank (V-101), and belt filter (BF-101).

TABLE 12.2 Components used in process simulation model (Ma et al., 2021).

Components	User defined	Definition
Carbon dioxide	No	CO_2 or dissolved in the form of HCO_3 or/and H_2CO_3
Ammonia	No	NH_3 or dissolved NH_4
Biowaste	Yes[a]	Bio-degradable waste
Water	No	Pure water
Oxygen	No	Oxygen diffused from aeration/air supplied
Nitrogen	No	Product from denitrification and air supplied
SS	Yes[a]	Fixed suspended solid (non-degradable)
NO_3	Yes[a]	Nitrate in whole nitrification
NO_2	Yes[a]	Nitrite in partial nitrification
TDS	Yes[a]	Non-biodegradable dissolved solids
Biomass-*h*	Yes[a]	Active heterotrophic biomass used in denitrification
Biomass-*n*	Yes[a]	Active nitrifiers autotrophic biomass
Biomass-*i*	Yes[a]	Inert biomass represents biomass decay

[a]*Reference properties are based on water for user-defined components.*

TABLE 12.3 Details of the influent stream.

Components	Flowrate (kg/h)	Concentration (mg/L)
Carbon dioxide Nitrogen Oxygen	0.48 0.91 0.24	48.14 913.5 242.8
Ammonia	0.20	19.82
Biowaste	0.49	49.08
Water	9434	943,464
SS	0.09	9.44
TDS	2.89	288
Biomass-*h*	0.094	9.44
Biomass-*n*	0.094	9.44
Biomass-*i*	0.057	5.66

TABLE 12.4 Specification of all unit operations.

Equipment	Unit No.	Specification: Operating conditions	Specification: Others (splitting/sampling/vent/emission)
Grit chamber	GB-101	Set adiabatic Pressure: 101.325 kPa	Top stream Component split SS: 70%
Equalization tank	EQ-101	Select constant outlet flow Liquid/Vessel volume: 75% Liquid viscosity: 1.0 cP	Sampling interval: 22 h Average volume in (m^3/h): 150
Mixer	MX-106	Select set output composition for: Polymer Concentration: 5 mg/L	Nil.
Dissolved air floatation	FL-101	Separation efficiency (flotation %): SS: 90 Biomass-*h*: 70 Biomass-*i*: 70 Biomass-*n*: 70	Vent/Emissions Check box to perform emission calculations (%): Ammonia: 100 Carbon dioxide: 100 Nitrogen: 100 Oxygen: 100
Activated sludge reactor (kinetic aerobic bio-oxidation)	AB-101	Set adiabatic Residence time: 1.2 day Working to vessel volume ratio (%): 70 Refer to Fig. 12.4 to set the reaction scheme Biomass-*n* decay reaction	Check box to perform emission calculations (%): Carbon dioxide: 95 Nitrogen: 100 Oxygen: 100

Continued

TABLE 12.4 Specification of all unit operations.—cont'd

Equipment	Unit No.	Specification	
		Operating conditions	Others (splitting/sampling/vent/emission)
		Stoichiometry (mass coefficient): 1.05Biomass-*n* + 1.15O_2 → 0.1NH_3 + 1.45CO_2 + 0.45H_2O + 0.2Biomass-*i* Rate reference component: Biomass-*n* Substrate: Biomass-*n* *S*-term: first order *O*-term: none *B*-term: none Refer to Fig. 12.5 to view/edit kinetic rate Biomass-*h* decay reaction 1.05Biomass-*h*+ 1.15O_2 → 0.1NH_3 + 1.45CO_2 + 0.45H_2O + 0.2Biomass-*i* Denitrification 34.6Biowaste + 62NO_3 → 33.4CO_2 + 13.2N_2 + 42.6H_2O + 7.4Biomass-*h* Biowaste degradation Biowaste degradation Biowaste + 0.1NH_3 + 1.4O_2 → 0.8Biomass-h + 0.8H_2O + 0.9CO_2 Nitrification 7.5NH_3 + 29.46O_2 + 2.19CO_2 → 1.12Biomass-*n* + 11.29H_2O + 26.74NO_3	

Clarifier	CL-102	Particle removal (%): Ammonia: 27.97 Carbon dioxide: 27.97 Biowaste: 99.9 SS: 99.9 NO_3: 90 Polymer: 90 TDS: 27.96 Water: 3 Biomass-*h*: 99.9 Biomass-*i*: 99.9 Biomass-*n*: 99.9.	Check box to perform emission calculations (%): Carbon dioxide: 98 Nitrogen: 99 Oxygen: 99
Flow splitting	FSP-101	Splitting for bottom stream (%): 50	Nil.
Storage tank (operation sequence for procedure)	V-101	Select TRANSFER-IN-1(Transfer in), STORE-1 (batch storage), to store and then TRANSFER-OUT-1(Transfer out)	Nil.
Belt filtration	BF-101	Particle removal: 99% for Biowaste, SS, polymer, TDS, Biomass-*h*, Biomass-*i*, Biomass-*n*. Solids in cake (wt%): 40	Nil.

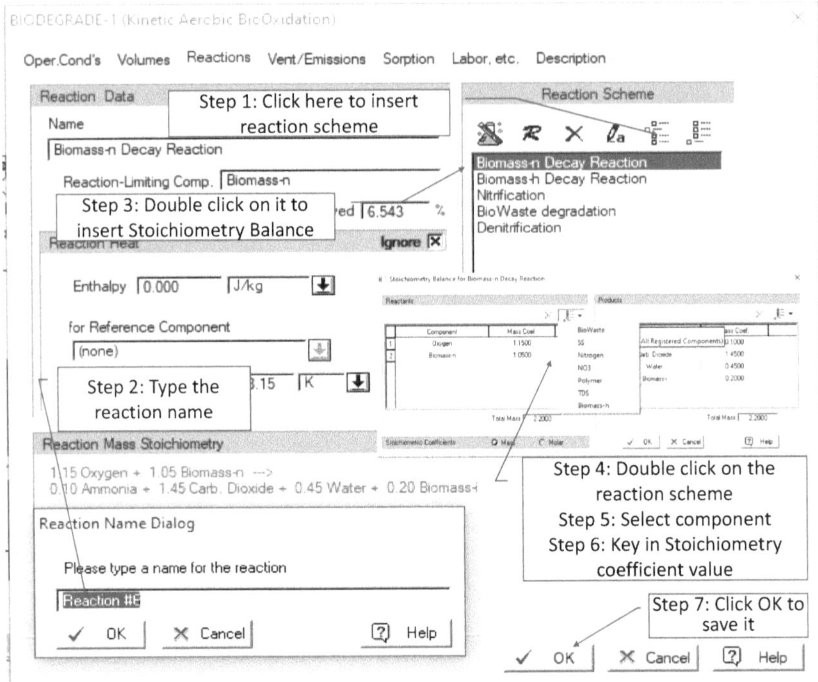

FIGURE 12.4 Procedure of setting reactions schemes.

set to operate in batch mode, as the input (sludge) is too little in order for the belt filter to operate continuously.

The simulated composition at the final discharge streams are shown in Table 12.5. At discharge stream 1, only Biowaste, SS, TDS, Biomass-*h*, Biomass-*n*, Biomass-*i* presents ranged from 2 to 28 mg/L with flowrate ranged from 0.00027 to 0.00378 kg/h. For discharge stream 2, only water and small amount of TDS are traceable. In terms of BOD, COD, and TDS, these streams fulfill the discharge limit (Standard B).

12.4 Process optimization

In this section, the effect of recycle ratio of sludge on TAC and wastewater quality is analyzed. The recycle ratio affected the TAC due to the addition of a belt filter, a storage tank, reactor and a clarifier tank. In addition, there was a major impact on TOC due to the production of sludge from the belt filter. The TAC is given as the summation of annualized capital cost (ACC) and annual operating cost (AOC), calculated using Eq. (12.1) (Smith, 2016). The AOC is contributed by materials, facility-dependent, labor-dependent, laboratory/QC/QA, utilities, waste-treatment/disposal cost (Intelligen, 2014). The ACC for

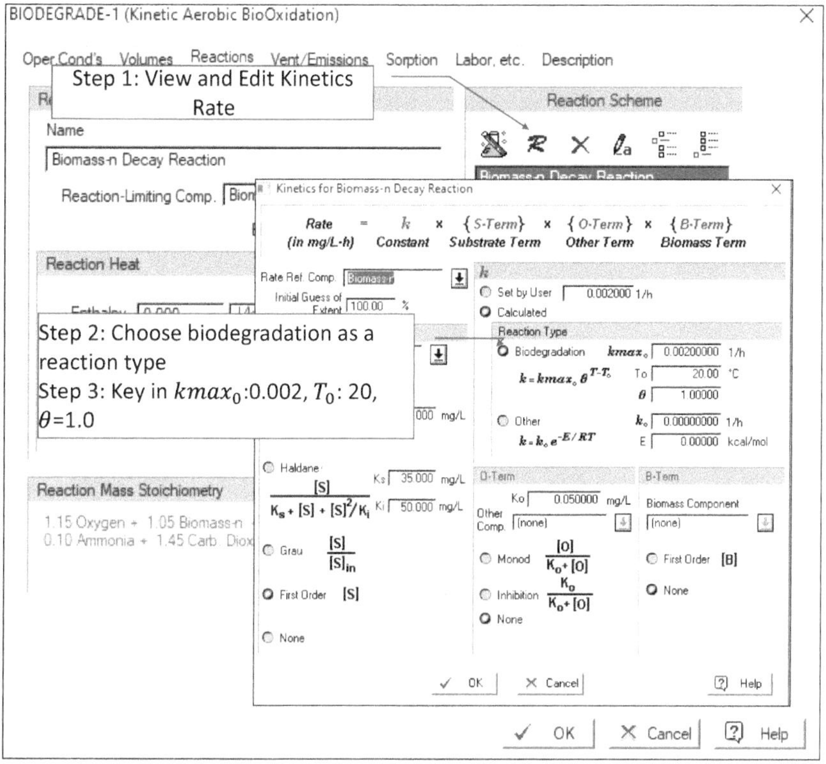

FIGURE 12.5 Procedure of editing kinetics data setting for reactions schemes.

the active sludge reactor, clarifier, storage tank and belt filter were calculated based on their capital cost (CC) using Eq. (12.2). Note that the CC is determined using the built-in model of SuperPro Designer (see Table 12.6).

$$\text{TAC} = \text{ACC} + \text{AOC} \tag{12.1}$$

$$\text{ACC} = \text{CC} \times \frac{i(1+i)^n}{(1+i)^n - 1} \tag{12.2}$$

where i is the fractional interest rate per year, and n is the number of years. For this case, n was set for 30 years and i was taken as 0.07 (Lok et al., 2020). This leads to the annualized factor (last term in Eq. 12.2) of 0.08.

As shown in Fig. 12.6, the lowest TAC was determined as \$4.52 million, at the optimum recycle ratio of 50%. Detailed costing data (TOC and ACC) for the analysis is shown in Table 12.7 in Appendix A. It is worth nothing that clarifier and belt filtration units do not experience any changes, as their dimension stay identical despite of the varying recycle ratio. Fig. 12.6 also shows the discharge qualities of Stream 1, i.e., COD, BOD and TSS. Fig. 12.6 shows that the COD, BOD_5 and TSS values of Stream 1 are reported at 42.5,

TABLE 12.5 Simulated composition at final discharge streams 1 and 2 for recycle ratio set at 50%.

	Discharge stream 1		Discharge stream 2	
Components	**Flowrate (kg/h)**	**Concentration (mg/L)**	**Flowrate (kg/h)**	**Concentration (mg/L)**
Carbon dioxide			N/T	N/T
Ammonia			N/T	N/T
Biowaste	0.001	5.8	N/T	N/T
Water	163.6	999,933	9284	983,818
SS	0.0003	1.7	N/T	N/T
TDS	0.005	28.2	2.4	251
Biomass-*h*	0.003	15.5	N/T	N/T
Biomass-*n*	0.0008	5.1	N/T	N/T
Biomass-*i*	0.0006	3.3	N/T	N/T

N/T, the value is too small to consider.

TABLE 12.6 Dimension of important unit operations at sludge recycle ratio of 50%.

Equipment	Unit No.	Dimension	Capital cost ($)
Equalization tank	EQ-101	Rectangle Tank depth: 5 m Length/Width: 4 m	0
Dissolved air floatation	FL-101	Surface area: 23.18 m^2	0
Activated sludge reactor (kinetic aerobic bio-oxidation)	AB-101	Volume: 394 m^3 Length: 18 m Width:7.3 m	1,260,000
Clarifier	CL-102	Circular Surface area: 7.32 m^2 Tank diameter: 3.05 m Tank volume: 21.95 m^3	38,000
Storage tank	V-101	21.1 m^3	39,000
Belt filtration	BF-101	Belt width: 0.003 m Max belt width: 3.5 m	294,000

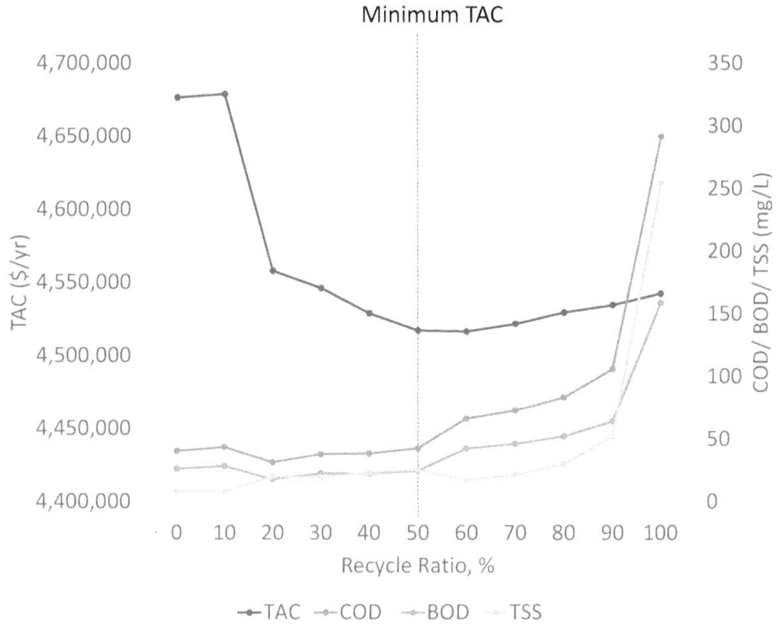

FIGURE 12.6 Trade-off of TAC, TSS, COD, and BOD of Stream 1 as an effect of recycle ratio of sludge.

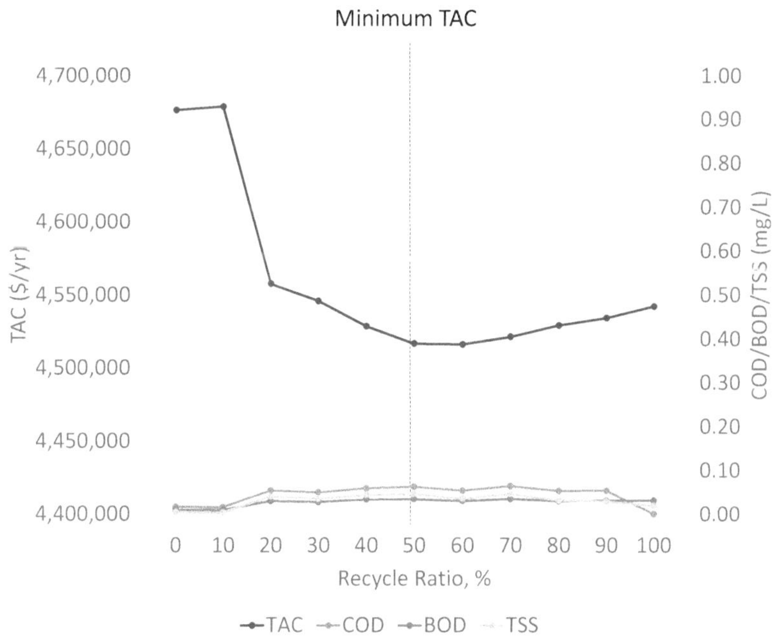

FIGURE 12.7 Trade-off of TAC, TSS, COD, and BOD at final discharge qualities Stream 2 as an effect of recycle ratio of sludge.

24.7 mg O/L and 25.5 mg solids/L, respectively for a recycle ratio of 50%. These parameters are well below the discharge limits (Standard B; see Table 12.1) of 20 (BOD), 120 (COD) and 50 mg/L (TSS), respectively.

For discharge stream 2, Fig. 12.7 shows that any changes of recycle ratio does not have much effect on its water discharge quality. The concentrations of BOD, COD and TSS of this stream are well below the limits in DOE Standards A and B (Table 12.1).

12.5 Conclusion

In this chapter, modeling and optimization of a WWTP for poultry industry has been demonstrated. New units are added to troubleshoot the WWTP in order for its effluent to comply with the discharge limit. By manipulating the recycle ratio from the bottom discharge stream of the clarifier, the performance and cost of the WWTP could be optimized while fulfilling the effluent discharge limits. The optimum sludge recycle ratio was identified at 50% as it can minimize the total annualized cost.

12.6 Appendix A

TABLE 12.7 Cost trade-off for AB-101, CL-102, BF-101, and V-101 as an effect of the recycle ratio.

			Capital cost			
Recycle ratio	ACC ($/year)	TOC ($/year)	Activated sludge reactor ($)	Clarifier ($)	Belt filtration ($)	Storage tank ($)
0	2,266,906	2,409,563	1,248,000	37,000	294,000	294,000
10	2,268,250	2,410,578	1,250,000	38,000	294,000	106,000
20	2,210,468	2,347,287	1,252,000	37,000	294,000	61,000
30	2,205,093	2,340,882	1,255,000	38,000	294,000	54,000
40	2,197,031	2,331,768	1,257,000	38,000	294,000	46,000
50	2,191,656	2,325,327	1,260,000	38,000	294,000	39,000
60	2,191,656	2,324,806	1,262,000	38,000	294,000	37,000
70	2,194,343	2,327,245	1,262,000	38,000	294,000	37,000
80	2,198,375	2,331,012	1,267,000	38,000	294,000	37,000
90	2,201,062	2,333,433	1,269,000	38,000	294,000	37,000
100	2,205,093	2,337,140	1,272,000	38,000	294,000	37,000

12.7 Exercise

1. Revisit the simulation flowsheet of the poultry WWTP in Figure 12.3. Determine the increase of capital cost when the recycle ratio is changed to 25%, 45% and 75%.

References

Aziz, H.A., Puat, N.N.A., Alazaiza, M.Y.D., Hung, Y.T., 2018. Poultry slaughterhouse wastewater treatment using submerged fibers in an attached growth sequential batch reactor. International Journal of Environmental Research and Public Health 1734.

Bingo, M.N., Basitere, M., Ntwampe, S.K.O., 2019. Poultry slaughterhouse wastewater treatment plant design advancements. In: 16th South Africa International Conference on Agricultural, Chemical, Biological and Environmental Sciences (ACBES-19) November 18–19. Johannesburg (S.A.).

Department of Environment (DOE), 2009. Environmental Quality Act 1974—Environmental Quality (Sewage) Regulations 2009. Government of Malaysia, Ministry of Natural Resources and Energy.

Intelligen, Inc, 2014. SuperPro Designer User Guide.

Kobya, M., Senturk, E., Bayramoglu, M., 2006. Treatment of poultry slaughterhouse wastewaters by electrocoagulation. Journal of Hazardous Materials 172–176.

Lok, X., Chan, Y.J., Foo, D.C.Y., 2020. Simulation and optimisation of full-scale palm oil mill effluent (POME) simulation and optimisation of full-scale palm oil mill effluent (POME). Journal of Water Process Engineering 38 (1), 101558.

Ma, R., Chong, C.H., Foo, D.C.Y., 2021. Industry, design and optimisation of wastewater treatment plant for the poultry. In: MATEC Web of Conferences. EDP Sciences, p. 333.

Smith, R., 2016. Chapter 2 Process economics. In: Chemical Process Design and Integration, 2nd Edn. Wiley, pp. 26–27.

Tezcan Un, U., Koparal, A.S., Bakir Oğütveren, U., 2009. Hybrid processes for the treatment of cattle-slaughterhouse wastewater using aluminum and iron electrodes. Journal of Hazardous Materials 580–586.

Part V

aspenONE engineering

Chapter 13

Basics of process simulation with Aspen HYSYS*

Nishanth Chemmangattuvalappil[1], Siewhui Chong[1,2] and Dominic C.Y. Foo[1]
[1]*University of Nottingham Malaysia, Semenyih, Selangor, Malaysia;* [2]*Current affiliation: Xodus group, Perth, Australia*

Chapter outline

In this chapter, a step-by-step guide is provided for the simulation of an integrated process flowsheet using Aspen HYSYS. The concept of simulation is based on *sequential modular* approach and follows the onion model for flowsheet synthesis (see Chapter 1 for details). The case study on *n*-octane production is used for illustration throughout the chapter.

13.1 Example on *n*-octane production

A simple example that involves the production of *n*-octane (C_8H_{18}) (Foo et al., 2005) is demonstrated, with detailed descriptions given in Example 1.1. The basic simulation setup involving registration of components, thermodynamic models, and reaction stoichiometry is to be carried out in the *Basis Environments* of Aspen HYSYS, while the modeling of reactor, separation, and recycle system are to be carried out in the *Simulation Environments*. The individual steps are discussed in the following subsections.

Step 1. Basic Simulation Setup

The first step in simulation using Aspen HYSYS is the definition of components. All components involved in the process are entered into the flowsheet by selecting from the component database, as shown in Fig. 13.1.

*Simulation and other electronic files of this chapter are made available on a companion website; URL: https://www.elsevier.com/books-and-journals/book-companion/9780323901680

Chemical Engineering Process Simulation. https://doi.org/10.1016/B978-0-323-90168-0.00009-3

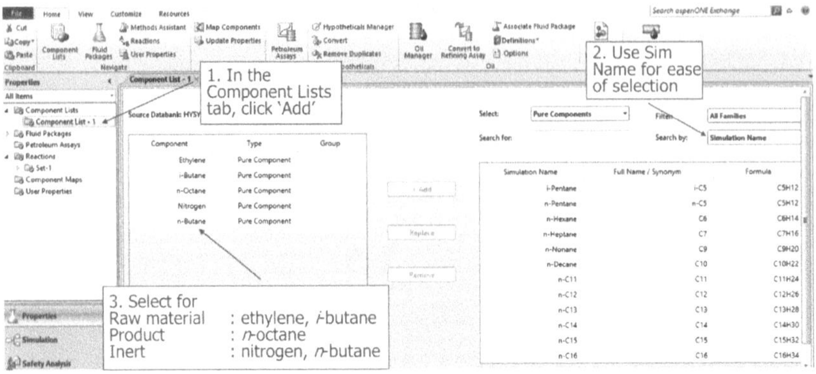

FIGURE 13.1 Defining components.

Once the components are chosen, the appropriate thermodynamic model (this is known as "fluid package" in Aspen HYSYS terminology) for the system is chosen.[1] Because this process involves the hydrocarbons at high pressure, "Peng–Robinson" equation of state has been chosen as the fluid package, with steps shown in Fig. 13.2.

In the next step, the reactions involved in the process must be defined at the "Reaction" tab (Fig. 13.3). The first step is the selection of reactor model. In this example, the reaction is modeled as a "conversion reactor." The conversion reactor model will treat the reactor as a stoichiometric problem and solve the mass balance based on the specified conversion (see steps in Fig. 13.3A).

Once the reactor model is chosen, all components taking part in the reaction are selected accordingly and their stoichiometric coefficients are entered. The limiting component must be chosen along with the conversion.

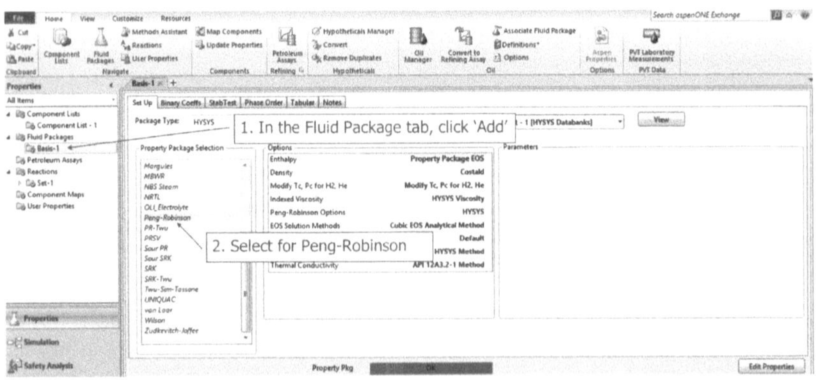

FIGURE 13.2 Selecting thermodynamic model (fluid package).

1. See Chapter 3 for guidelines on thermodynamic package section.

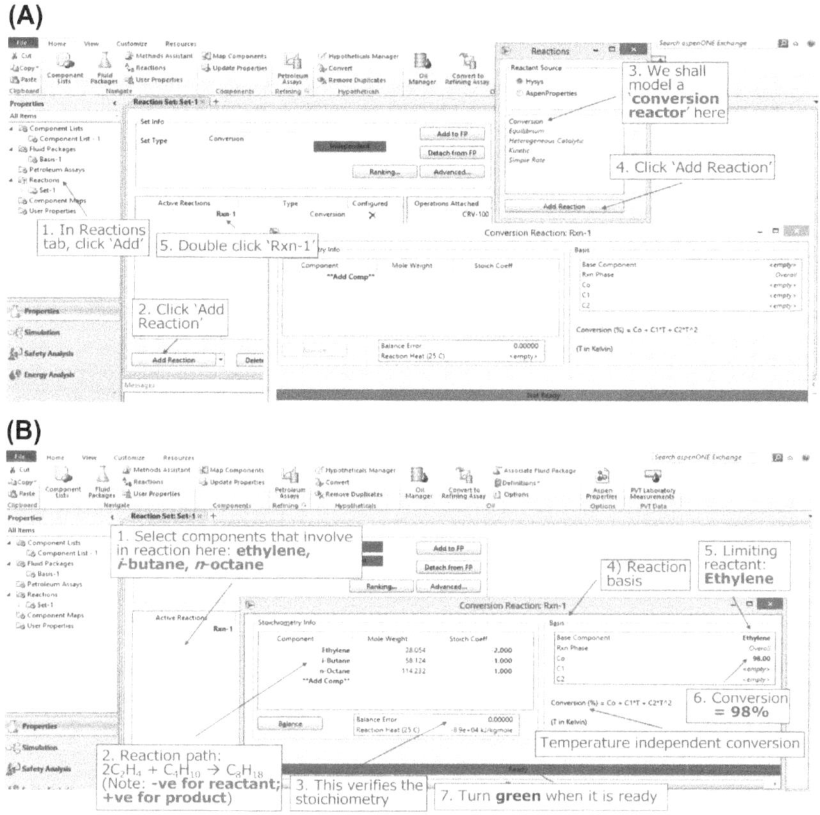

FIGURE 13.3 Steps in specifying reaction sets. (A) select reactor type, (B) enter reaction details, (C) assign a fluid package, (D) enter the Simulation Environments.

It is to be noted that, in Aspen HYSYS, the conversion must be specified in percentage (see Fig. 13.3B). Once all information is entered, a thermodynamic package must be assigned to this reaction to estimate the conditions after the reaction. In this example, the reaction is linked to the Peng–Robinson equation of state model (see steps in Fig. 13.3C).

These are the basic information required for creating the simulation flowsheet. Now we may proceed to the *Simulation Environments* to perform modeling of the unit operations. To enter the *Simulation Environments*, we shall press the "Enter Simulation Environment" button on the *Simulation Basis Manager* (Fig. 13.3D).

Step 2. Modeling of Reactor

The *Simulation Environments* of Aspen HYSYS consist of main flowsheet, subflowsheet, and column subflowsheet environments. For the *n*-octane production example, only the main flowsheet is used. In this stage, the topology of

(C)

(D)

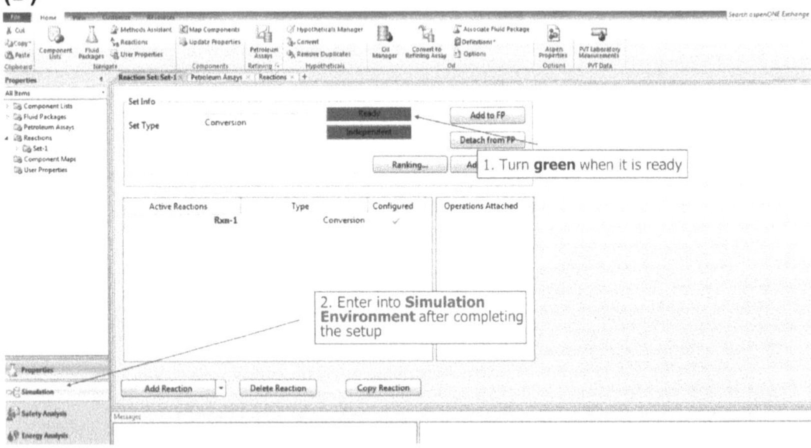

FIGURE 13.3 cont'd

the flowsheet must first be defined to identify the sequence of unit operations. Now, according to the onion model,[2] the reactor system is simulated in the first step. It is necessary to define the reactions in a flowsheet before entering the Simulation Environment. At this stage, the type of reactor, information on conversion, kinetics, equilibrium, etc., (depend on the available information) have to be specified as well. Although there are different types of reactor models available in Aspen HYSYS such as continuous stirred tank reactor and plug flow reactor, a conversion model is chosen for this example. The kinetic reaction models require the information on the reaction kinetics. For the *n*-octane production example, we shall utilize the "conversion reactor" model.

2. See Example 1.1 for details.

It is to be noted that the conversion reactor model can only perform mass and energy calculations based on the stoichiometry.

Detailed steps to draw the flowsheet on the process flow diagram (PFD) are shown in Fig. 13.4, where the conversion reactor consists of two inlet (Streams 1 and Q-101) and two outlet streams (Streams 2 and 3). Note that Streams 1, 2, and 3 are the actual process streams that consist of material (termed as Material Stream in HYSYS), while Stream Q-101 is actually virtual stream that is used for performing heat balances.

After the connections are made, the conditions need to be specified in the incoming stream according to Table 13.1. It is advisable to conduct a degree of

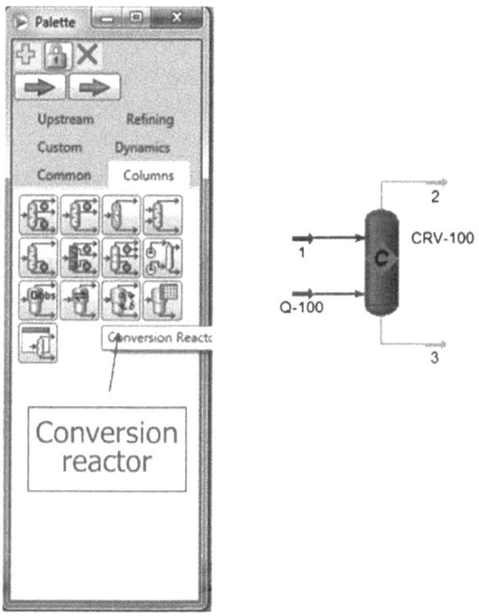

FIGURE 13.4 Setting up feed.

TABLE 13.1 Feed condition.

Component	Flowrate (kmol/h)	Condition
Nitrogen, N_2	0.1	T = 93°C
Ethylene, C_2H_4	20	P = 20 psia
n-Butane, C_4H_{10}	0.5	
i-Butane, C_4H_{10}	10	
n-Octane, C_8H_{18}	0	

freedom analysis before designing the equipment to make sure that there are enough process parameters for the design of equipment and the variables are not overspecified. To do a degree of freedom analysis, we list all the variables involved in the process units. These variables can be operating conditions, such as temperature and pressure, flowrates, and compositions. Once sufficient conditions, which are molar flowrates, temperature, and pressure in this case, are entered, the incoming stream is completely defined.

In the next step, the reactor model must be completely specified. We may enter the pressure drop and also define the reaction set to completely define the reaction (see detailed steps in Fig. 13.5). In this example, there is only one reaction set that is chosen. Because the reaction is conducted at isothermal mode, the outlet stream temperature should be selected and set to the same temperature as the incoming stream. We may notice that the energy stream now indicates a heat flow to maintain the reactor at isothermal conditions. In case of adiabatic reactors, no energy stream must be connected to the reactor. At this stage, we may notice that the reactor model is converged because all variables have been specified.

Aspen HYSYS has been set to solve once the necessary data are sufficient. A convenient way of displaying the simulation results is via the *Workbook*. Fig. 13.6 shows the detailed steps in displaying molar flowrates of all components on the Workbook. One may also insert the *Workbook Table* within the PFD. This is illustrated with Fig. 13.7, where material and energy streams are displayed (one may also choose to display the stream compositions).

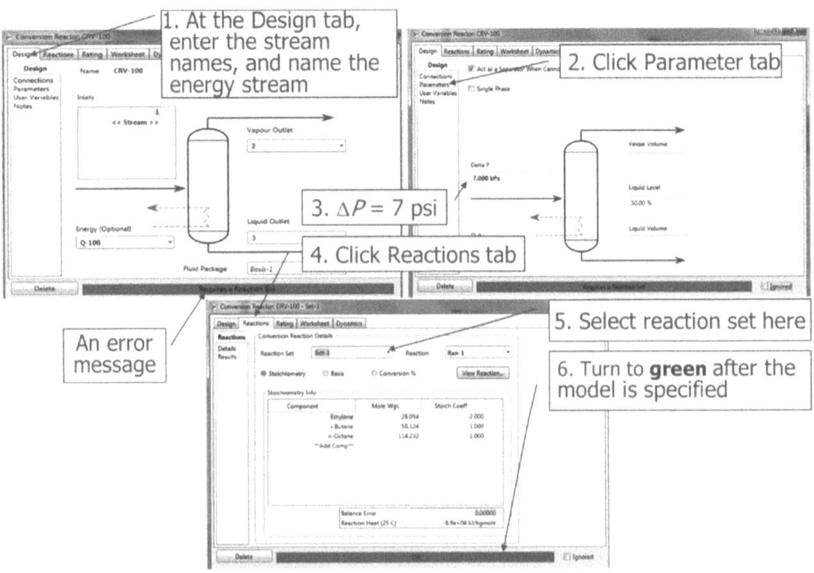

FIGURE 13.5 Setting up reactor connection.

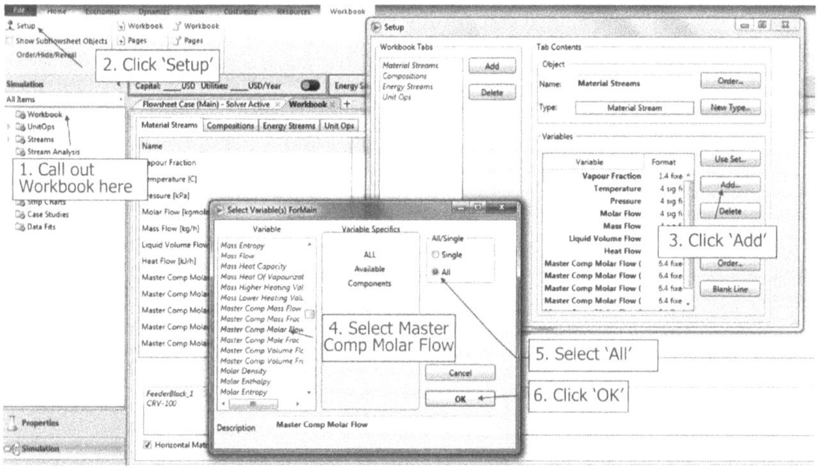

FIGURE 13.6 Setting up Workbook in Aspen HYSYS.

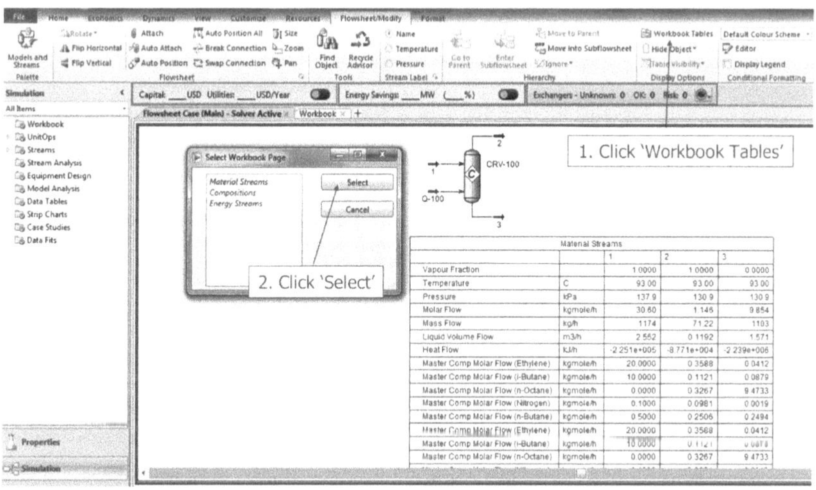

FIGURE 13.7 Adding Workbook Table to process flow diagram in Aspen HYSYS.

Note that the energy stream in the converged reactor model has a negative value, which indicates that energy must be removed to maintain the isothermal conditions. This indicates an exothermic reaction in the reactor. In the final step, the basic mass balance calculations can be performed to see that the outlet composition from the reactor is in agreement with the stoichiometry. However, in more complex flowsheets, it may not be easily verified through hand calculations.

Step 3. Separation Units

In this process, the only separation system is a distillation column. The preliminary design of the distillation column can be done using the "shortcut distillation" model in HYSYS. This model is based on the Fenske–Underwood–Gilliland model, which is useful for conducting the preliminary design of a distillation column. The parameters obtained from shortcut distillation model can be used as initial estimates in the rigorous distillation model, which performs stage-by-stage calculations. To continue in building the topology of the flowsheet, one may refer to Fig. 13.8 for the alternative steps in connecting the material and energy streams. Fig. 13.8 also shows that the column is set to operate with partial condenser, where the column top stream exists in vapor form.

The following design parameters are defined for this separation unit (Fig. 13.9):

1. Condenser type: partial condenser
2. Light key and mole fraction: ethylene (0.0015 in bottom stream)
3. Heavy key and mole fraction: *n*-octane (0.28 in distillate stream)
4. Column pressure: 25 psia in condenser
5. Pressure drop: 10 psia

In the next step, the specifications of final product stream are added (mole fractions of the key component), and the pressure of condenser and reboiler is defined as shown in Fig. 13.10. After entering the given specifications, it can be noticed that the minimum reflux ratio is automatically calculated. In this case it is 0.007. So, an actual reflux ratio must be chosen to estimate the number of trays required to obtain the desired separation. In this case, we use an actual reflux ratio of 1. Once this information is entered, the column

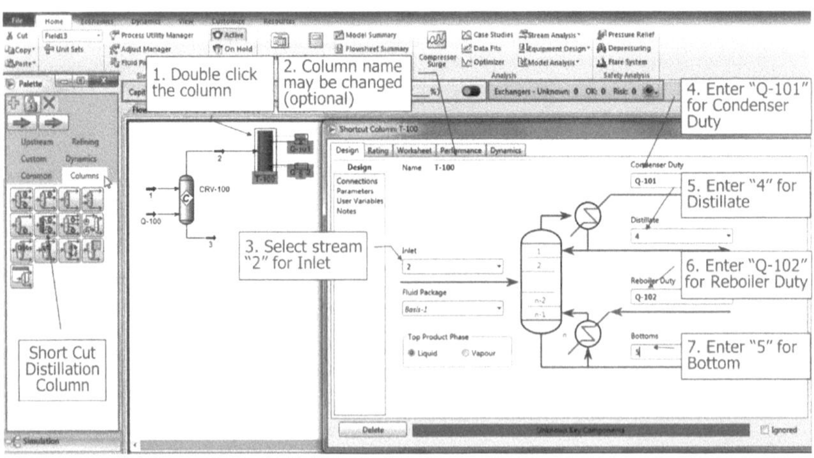

FIGURE 13.8 Adding shortcut distillation model to the process flow diagram.

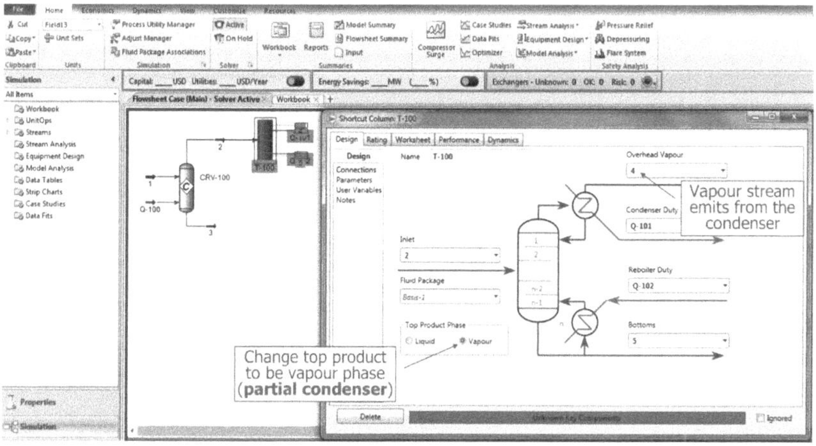

FIGURE 13.9 Changing the type of condenser.

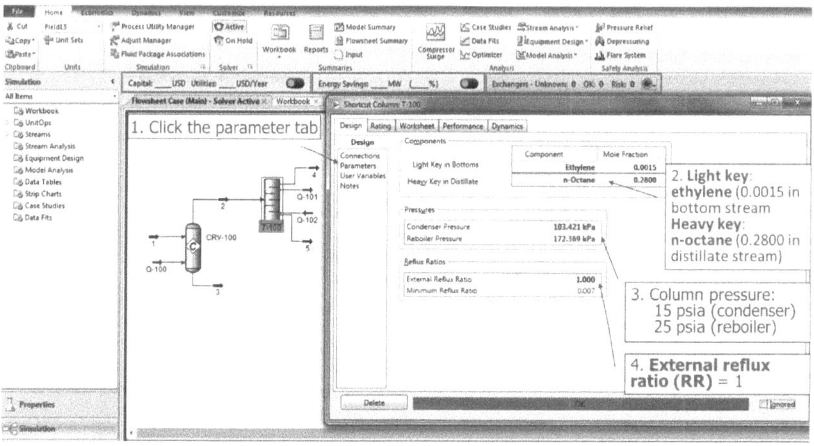

FIGURE 13.10 Specifications for the shortcut distillation model.

simulation converges. The calculated number of trays and column conditions can be seen on the tab "Performance" as shown in Fig. 13.11.

We next move on to display the simulation results using the *Workbook Table*, as shown in Fig. 13.12.

Step 4. Recycle System (Materials)

Example 1.1 shows that the n-octane case contains both material and heat recycle streams. In this step, the convergence of these recycle streams is illustrated. Fig. 13.12 indicates that the distillate stream (Stream 4) contains some unconverted raw material that can be recycled to the reactor. Hence, we need to insert a recycle loop into the simulation to model the material recycle

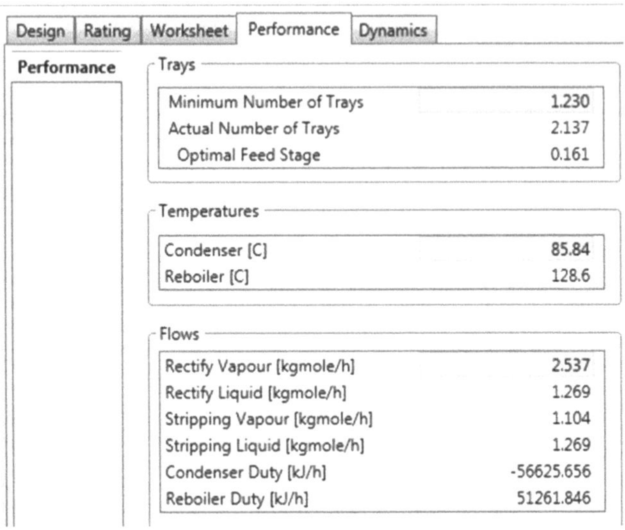

FIGURE 13.11 Results of shortcut distillation column design.

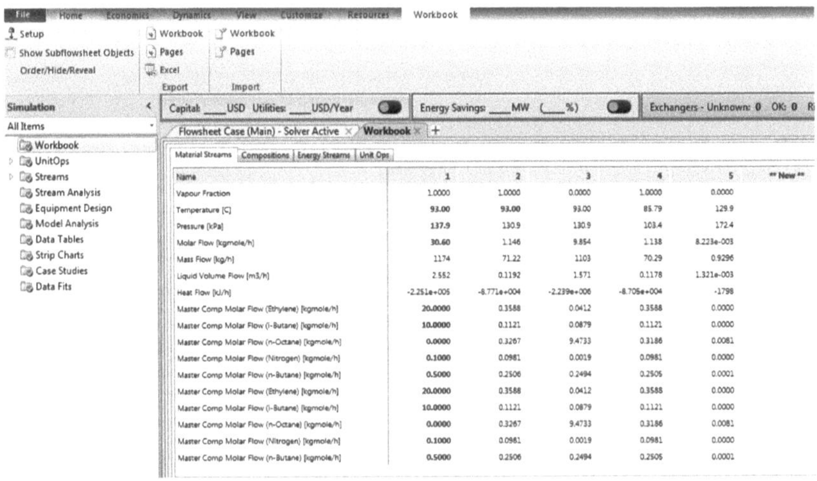

FIGURE 13.12 Simulation results.

stream, which involves a purge stream unit, compressor, and cooler. Purging unit is necessary to avoid the trapping of materials inside the system. In this case, it is assumed that 10% of the distillate is purged. The compressor and a cooler are then used to adjust the pressure and temperature of the recycle stream to match those of the reactor. Specifications for these units are given in Table 13.2.

TABLE 13.2 Specifications for units in the material recycle system.

Equipment	Specifications
Purge unit	Flow ratio for recycle stream: 0.9
Compressor	Outlet P: 22 psia
Cooler	Outlet T: 93°C
	Delta P: 2 psi

To model the purge unit, a stream splitting model (called the "Tee" in Aspen HYSYS) is introduced in PFD. Detailed steps to connect the Tee model (to distillation top stream) and to provide its model specifications are shown in Fig. 13.13. Because 10% of the distillate is purged, the flow ratio for recycle stream is set to 0.9.

The distillate is collected at a pressure of 15 psia, whereas the reactor operates at 20 psia. To match the pressure of the distillate steam to that of reactor, a compressor is added to raise its pressure to 20 psia. Detailed steps to connect the compressor model and to provide its specifications are shown in Fig. 13.14.

Compression will increase the temperature of streams and in this case, the temperature of recycle stream after compression is raised to 99°C, which is higher than the reactor operating temperature. A cooler unit is added to reduce the temperature to the reactor temperature of 93°C. Fig. 13.15 shows the detailed steps to connect the cooler model and to provide its specifications.

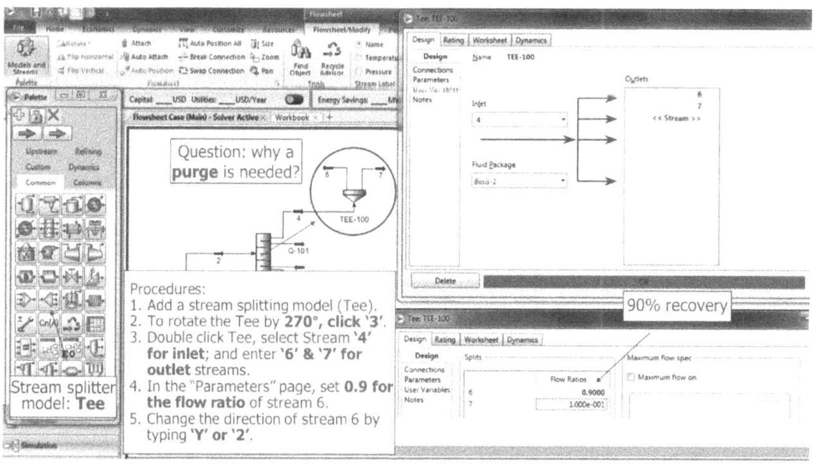

FIGURE 13.13 Adding a Tee for purge stream.

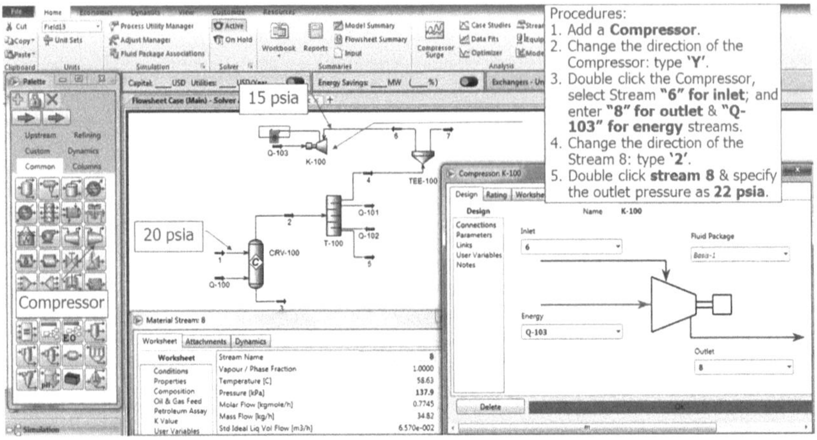

FIGURE 13.14 Adjusting the stream pressure.

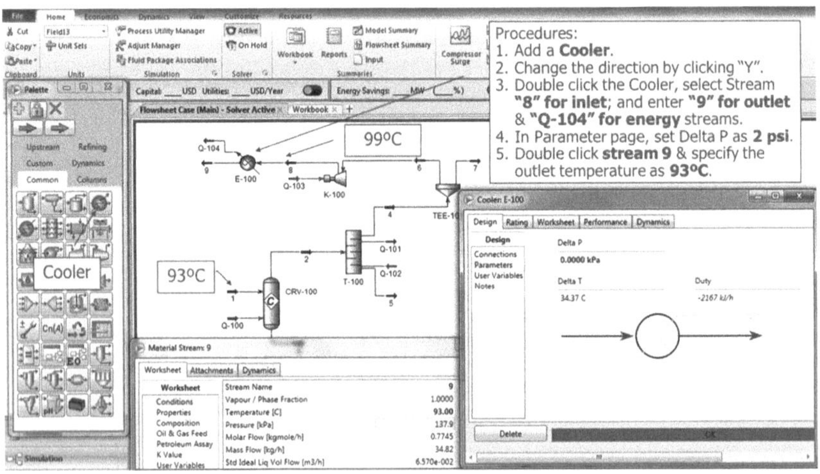

FIGURE 13.15 Adjusting the stream temperature.

Because both pressure and temperature of the recycle stream are now adjusted to match those of the reactor, we can connect the recycle stream to the reactor. To converge this material recycle stream, we can make use of the "Recycle" unit. The latter facilitates the convergence of a recycle loop following the "tear stream" concept.[3] Detailed steps to converge the recycle stream to the reactor with the Recycle unit are shown in Figs. 13.16, 13.17,

3. The Recycle unit in Aspen HYSYS performs the tear-stream calculation (see Chapter 4 for details) to converge the recycle stream automatically.

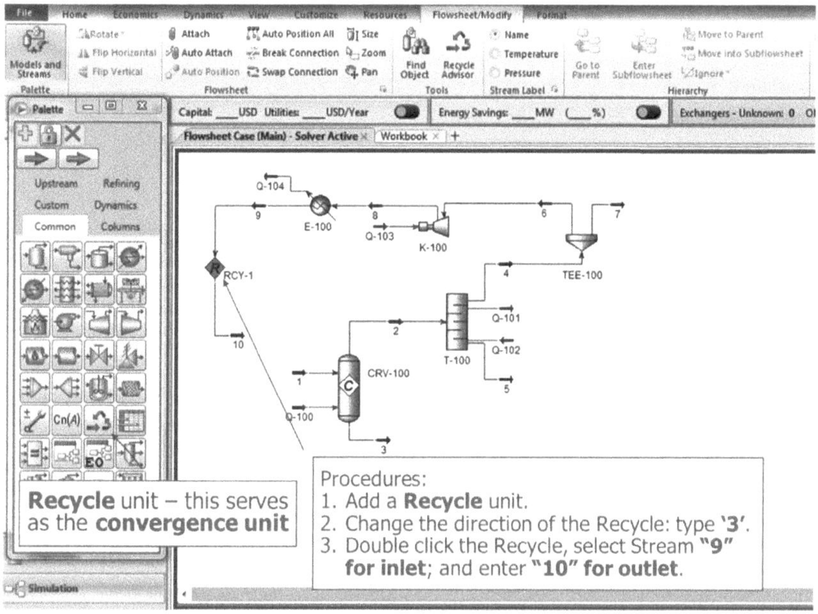

FIGURE 13.16 Adding the Recycle unit.

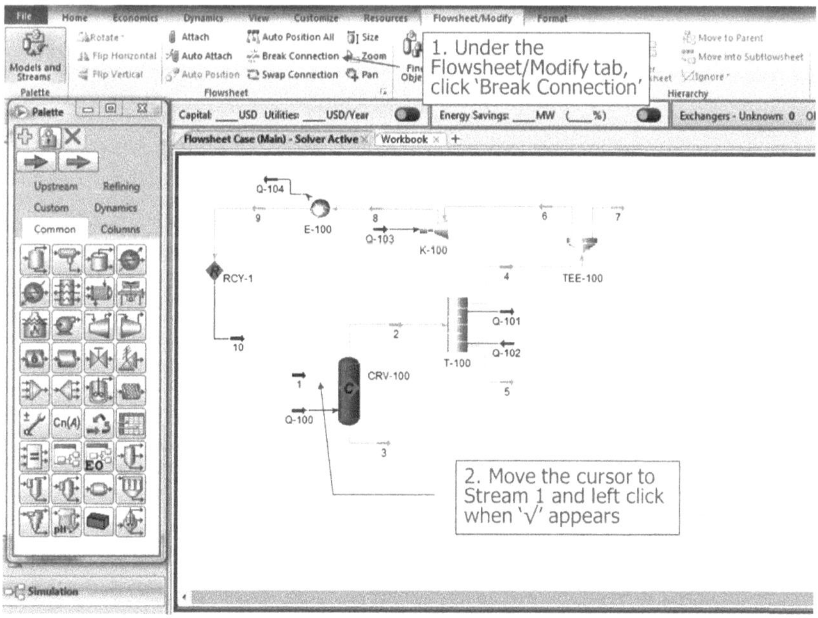

FIGURE 13.17 Break the connection of fresh feed to the reactor.

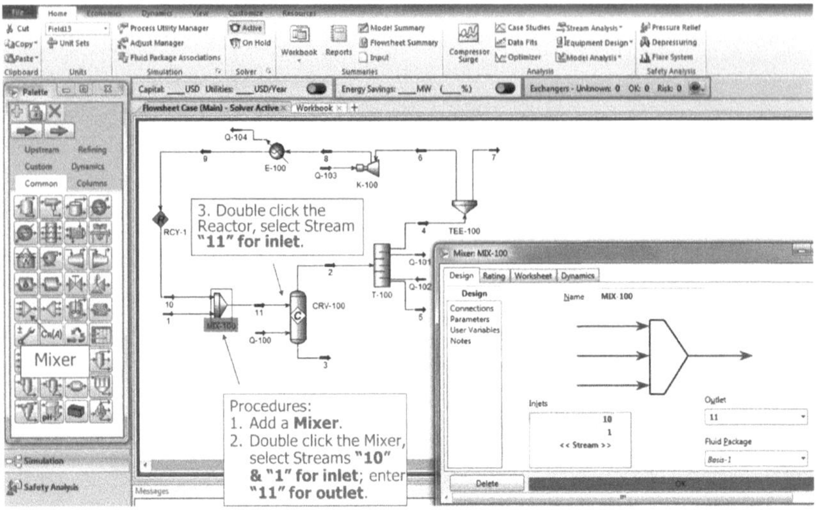

FIGURE 13.18 Add the fresh feed and recycle stream to the reactor.

13.18 and 13.19. Note that the Recycle unit shows a yellow outline when the recycle stream is first connected to the reactor. This means that some parameters are not converged after 20 rounds of iteration (default setting in Aspen HYSYS). Hence, more iteration is needed to ensure all parameters are converged completely (by pressing the "Continue" button in its Connections page). The simulation results are also displayed in the *Workbook Table* in Fig. 13.20.

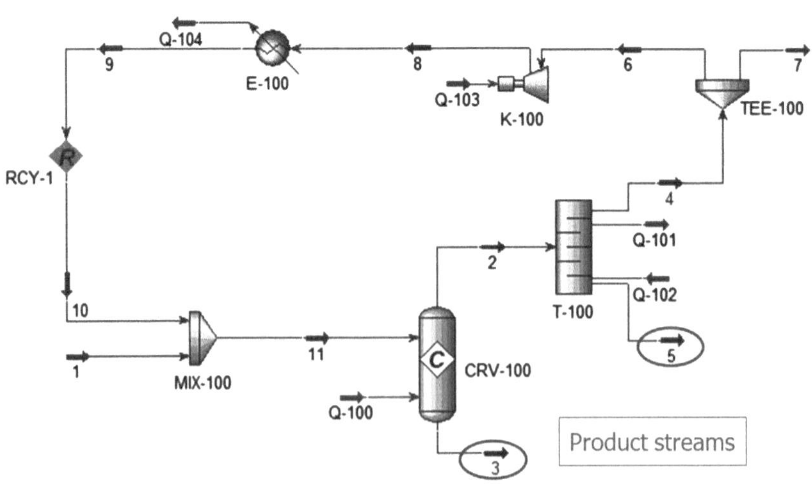

FIGURE 13.19 Final process flow diagram with recycle stream.

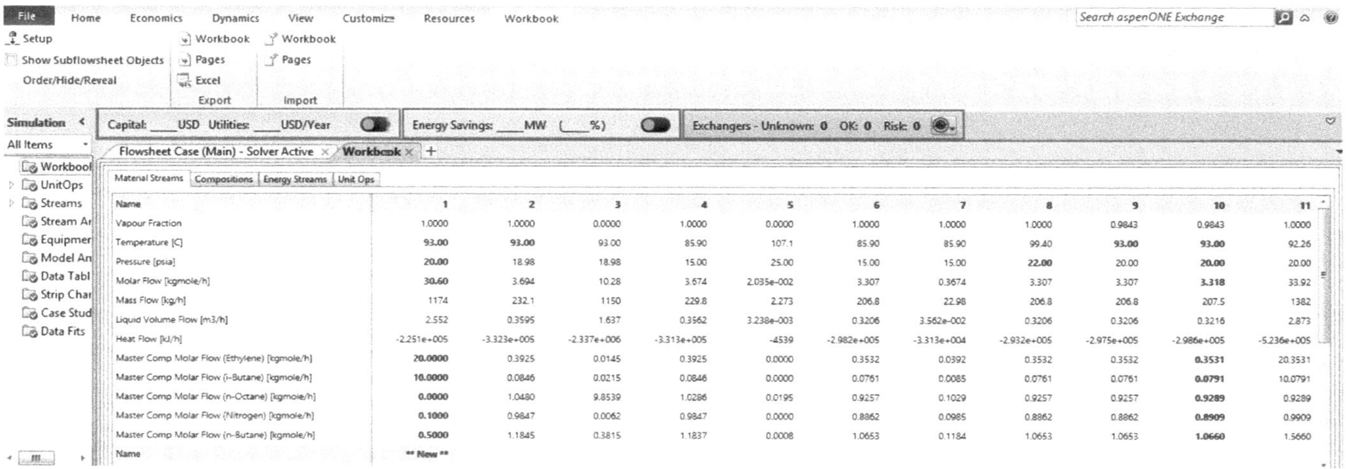

Name	1	2	3	4	5	6	7	8	9	10	11
Vapour Fraction	1.0000	1.0000	0.0000	1.0000	0.0000	1.0000	1.0000	1.0000	0.9843	0.9843	1.0000
Temperature [C]	**93.00**	**93.00**	93.00	85.90	107.1	85.90	85.90	99.40	**93.00**	**93.00**	92.26
Pressure [psia]	**20.00**	18.98	18.98	15.00	25.00	15.00	15.00	**22.00**	20.00	**20.00**	20.00
Molar Flow [kgmole/h]	**30.60**	3.694	10.28	3.674	2.035e-002	3.307	0.3674	3.307	3.307	**3.318**	33.92
Mass Flow [kg/h]	1174	232.1	1150	229.8	2.273	206.8	22.98	206.8	206.8	207.5	1382
Liquid Volume Flow [m3/h]	2.552	0.3595	1.637	0.3562	3.238e-003	0.3206	3.562e-002	0.3206	0.3206	0.3216	2.873
Heat Flow [kJ/h]	-2.251e+005	-3.323e+005	-2.337e+006	-3.313e+005	-4539	-2.982e+005	-3.313e+004	-2.932e+005	-2.975e+005	-2.986e+005	-5.236e+005
Master Comp Molar Flow (Ethylene) [kgmole/h]	**20.0000**	0.3925	0.0145	0.3925	0.0000	0.3532	0.0392	0.3532	0.3532	**0.3531**	20.3531
Master Comp Molar Flow (i-Butane) [kgmole/h]	**10.0000**	0.0846	0.0215	0.0846	0.0000	0.0761	0.0085	0.0761	0.0761	**0.0791**	10.0791
Master Comp Molar Flow (n-Octane) [kgmole/h]	**0.0000**	1.0480	9.8539	1.0286	0.0195	0.9257	0.1029	0.9257	0.9257	**0.9289**	0.9289
Master Comp Molar Flow (Nitrogen) [kgmole/h]	**0.1000**	0.9847	0.0062	0.9847	0.0000	0.8862	0.0985	0.8862	0.8862	**0.8909**	0.9909
Master Comp Molar Flow (n-Butane) [kgmole/h]	**0.5000**	1.1845	0.3815	1.1837	0.0008	1.0653	0.1184	1.0653	1.0653	**1.0660**	1.5660
Name	** New **										

FIGURE 13.20 Simulation results after material recycle.

After the material recycle system is converged, we next proceed to converge the energy recycle stream. This will be done using the "tear stream" concept, i.e., without the use of the Recycle unit. Specifications for heat exchanger and heater in the energy recycle system are given in Table 13.3.

In earlier steps, it has been assumed that the fresh feed stream is available at 93°C (see Table 13.2). This assumption is now relaxed. A heater is added to raise the temperature of the fresh feed stream from 30°C. Detailed steps to do so are given in Fig. 13.21. The simulated results indicate that the heater requires a total heating duty of 131 MJ/h (indicated by energy stream Q-105), while 5.4 MJ/h of energy needs to be removed by the cooler (indicated by energy stream Q-104).

TABLE 13.3 Specifications for units in the energy recycle system.

Equipment	Specifications
Heat exchanger	Delta P: 2 psi (tube side)
	Delta P: 2 psi (shell side)
Heater	Outlet T: 93°C
	Delta P: 2 psi

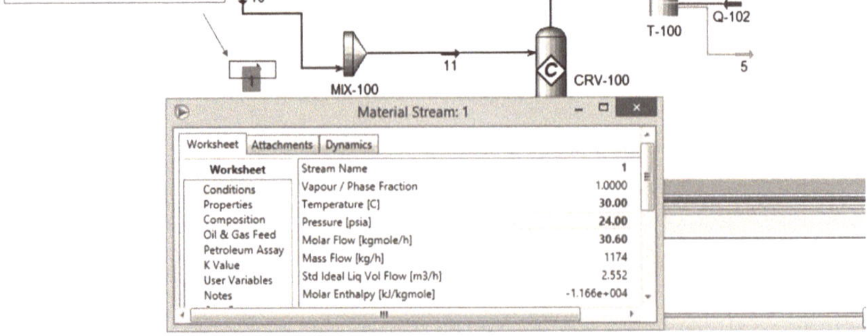

FIGURE 13.21 Set the inlet condition of feed.

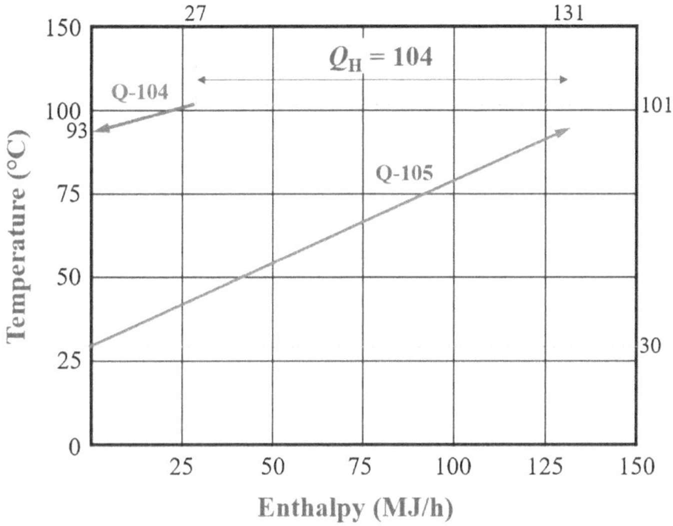

FIGURE 13.22 Temperature–enthalpy plot for heat recovery system.

Fig. 13.22 shows a *temperature–enthalpy plot*[4] for the streams undergoing heating and cooling in the heaters and cooler. As shown, the temperature profiles of the cooler (Q-104—the material recycle stream) are higher than those of the heater (Q-105—fresh feed). Hence, energy released from the heater can be completely recovered to the cold stream. In other words, part of the heating requirement of the heater (27 MJ/h) is to be fulfilled by the cooling duty of the cooler, through a process-to-process heat exchanger. The remaining heating duty of the cold stream ($Q_H = 104$ MJ/h) is to be supplied by the heater, as shown in Fig. 13.22.

With the heating and cooling requirements identified, we can now move on to simulate the process-to-process heat exchanger in the original PFD (Fig. 13.23). The "heat exchanger" model is utilized and added to replace the cooler model. Because the recycle model is not utilized in this case, we shall create a *tear stream*[5] for the energy recycle system, for the stream connecting the heat exchanger and heater. Detailed steps for doing so are shown in Fig. 13.24. Note that the flowsheet is unconverged at this stage.

We next proceed to provide the missing parameters to converge the flowsheet. These include the estimation of values for the tear stream. Because this stream is essentially the same fresh feed stream that enters the heat exchanger at the shell side, its condition should be very similar (except that

4. This is the most basic form of *heat transfer composite curves* in *process integration*; see Linnhoff et al. (1982) or Smith (2016) for more details.
5. See Chapters 1 and 4 for detailed discussion on the use of tear stream for recycle simulation.

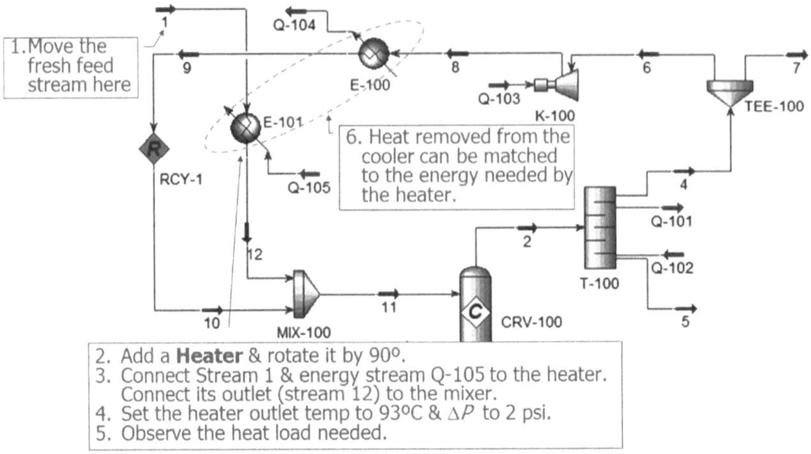

FIGURE 13.23 Identification of heat load.

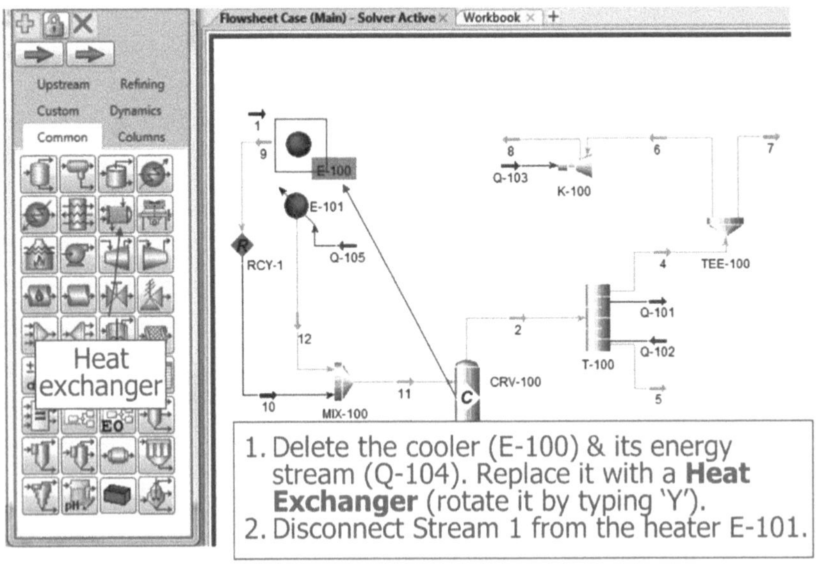

FIGURE 13.24 Delete cooler to bring heat exchanger.

with different temperature). Once the tear stream is specified, the "open loop" flowsheet is converged (Fig. 13.25). In the final step, the tear stream is removed and the outlet stream from the heat exchanger is connected to the heater. A converged "close loop" flowsheet is resulted and is shown in Fig. 13.26. The material and energy stream conditions are shown in Table 13.4,

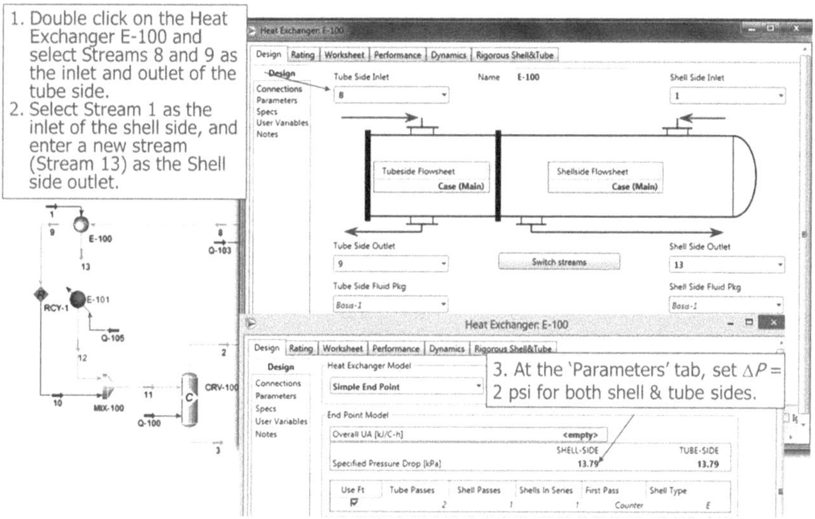

FIGURE 13.25 Bring heat exchanger to make use of the process stream temperature.

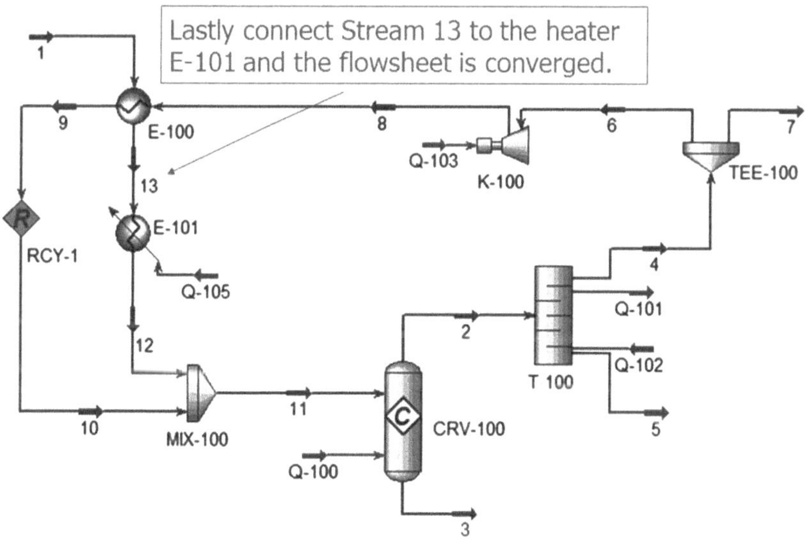

FIGURE 13.26 Final process flow diagram.

exported from the *Workbook Table*. As shown, the energy stream of heater E-101 (Q-105) has a heat flow of 104 MJ/h.[6]

6. Readers should verify the heat exchanger E-100 has a heat duty of 27 MJ/h.

TABLE 13.4 Stream conditions of final simulation flowsheet exported from Workbook Table: (a) mass streams; (b) energy streams.

(a) Mass streams	1	2	3	4	5	6	7	8	9	10	11	12	13
Vapor fraction	1	1	0	1	0	1	1	1	0.88	0.88	1	1	1
Temp (°C)	30.0	93.0	93.0	92.4	98.5	92.4	92.4	101.3	93.0	93.0	85.5	93.0	43.7
Pressure (psia)	20	13	13	15	25	15	15	20	20	20	20	20	20
Molar flow (kgmol/h)	30.6	6.141	9.443	5.486	0.655	4.938	0.549	4.938	4.938	4.938	35.538	30.6	30.6
Mass flow (kg/h)	1174.18	457.68	1062.86	384.98	72.70	346.49	38.50	346.49	346.49	346.35	1520.53	1174.18	1174.18
Component flowrate (kgmol/h)													
Ethylene, C_2H_4	20	0.402	0.006	0.401	0.001	0.361	0.040	0.361	0.361	0.362	20.362	20	20
i-Butane, C_4H_{10}	10	0.113	0.011	0.112	0.001	0.101	0.011	0.101	0.101	0.101	10.101	10	10
Nitrogen, N_2	0.1	0.982	0.002	0.982	0.001	0.883	0.098	0.883	0.883	0.884	0.984	0.1	0.1
n-Butane, C_4H_{10}	0.5	2.107	0.259	2.072	0.034	1.865	0.207	1.865	1.865	1.865	2.365	0.5	0.5
n-Octane, C_8H_{18}	0	2.538	9.165	1.920	0.618	1.728	0.192	1.728	1.728	1.726	1.726	0	0

(b) Energy Streams	Q-100	Q-101	Q-102	Q-103	Q-105
Heat flow (MJ/h)	−2098.3	−236.0	213.1	5.7	104.0

Exercise

In a styrene (MW = 104.15) production process, toluene (MW = 92.14) and methanol (MW = 32.04) are used as feedstock. Apart from the main product of styrene, by-product of ethylbenzene (MW = 106.17) is also product from the side reaction, following the reaction stoichiometry in Eqs. (13.1) and (13.2) (Boehm, 1997).

$$\text{Toluene} + \text{methanol} \rightarrow \text{styrene} + \text{water} + \text{hydrogen (main reaction)} \quad (13.1)$$

$$\text{Toluene} + \text{methanol} \rightarrow \text{ethylbenzene} + \text{water (side reaction)} \quad (13.2)$$

The styrene reactor is operated at 400 kPa and 525°C, with toluene conversion reported as 0.82 mol/mol of toluene feed, while styrene yield is given as 0.72 mol/mol of toluene reacted (Boehm, 1997). The styrene product stream should have high purity, preferable with ethylbenzene composition of less than 300 ppm (Boehm, 1997).

Fig. 13.27 shows the converged simulation flowsheet for the base case process, with Table 13.5 summarizing condition of the two feed streams. The basic specification for all process units (e.g., outlet temperature/pressure, pressure drop (ΔP), etc.) is summarized in Table 13.6. For the basic simulation setup, the following information is necessary:

- Component from database: hydrogen, methanol, water, toluene, ethylbenzene, styrene.
- Thermodynamic model: Peng–Robinson

Develop the base case process by solving the following tasks:

1. Following the concept of the onion model*, the reactor unit is first simulated using the conversion reactor model in Aspen HYSYS, with feed stream data in Table 13.5 (i.e., only a single reactor model with two heated

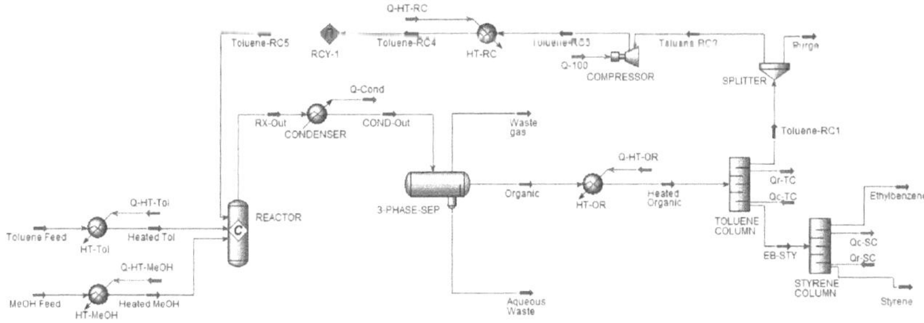

FIGURE 13.27 Converged flowsheet of styrene production process.

*. Refer to Chapter 1 for the details of onion model.

TABLE 13.5 Condition of feed streams.

Feed streams	Component flowrate[a]	Stream condition
Toluene feed	Toluene—490 kmol/h	450 kPa, 20°C, liquid
Methanol (MeOH) feed	Methanol—495 kmol/h	450 kPa, 20°C, liquid

[a]*Note: calculated based on stoichiometry balance, without considering product losses.*

TABLE 13.6 Specification of equipment.

Equipment model	Setting
Toluene heater (HT-Tol)	$\Delta P = 50$ kPa Outlet temperature = 525°C
Methanol heater (HT-MeOH)	$\Delta P = 50$ kPa Outlet temperature = 525°C
Reactor	$\Delta P = 70$ kPa Reaction: • Styrene formation (conversion: 59%) • Ethylbenzene formation (conversion: 23%)
Condenser	$\Delta P = 50$ kPa Outlet temperature = 38°C
Three-phase separator	$\Delta P = 50$ kPa (all vapor and liquid outlets)
Organic stream heater (HT-OR)	$\Delta P = 50$ kPa Outlet temperature = 100°C
Toluene column (shortcut model)	Component • Light key (toluene) at bottom: 0.0010 • Heavy key (ethylbenzene) at bottom: 0.0012 Pressure: • Condenser: 110 kPa • Reboiler: estimate its pressure if ΔP is set to 0.7 kPa/tray • External reflux ratio: 5
Styrene column (shortcut model)	Component • Light key (ethylbenzene) at bottom: 0.0003 (note: equivalent to 300 ppm) • Heavy key (styrene) at bottom: 0.0010 Pressure: • Condenser: 55 kPa (vacuum) • Reboiler: estimate its pressure with ΔP of 0.6 kPa/tray • External reflux ratio: 20
Splitter	5% purge
Compressor	Outlet pressure: 450 kPa
Recycle stream heater (HT-RC)	$\Delta P = 50$ kPa Outlet temperature = 525°C

Adapted from Boehm, R.F. 1997. Developments in the Design of Thermal Systems. Cambridge University Press, Cambridge.

feed streams is simulated, without other units). Note that since there are two reaction stoichiometries for this case, the exact conversion for the main product (styrene) and by-product (ethylbenzene) are to be calculated for the conversion reactor model, based on the specified conversion and yield. Perform a manual mass balances to verify the conversion setting given in Table 13.6 for the main product and by-product.

2. Separation units are next added at the downstream of the conversion reactor model (layer two of the onion model*), i.e., condenser, three-phase separator, heater, and distillation columns (shortcut distillation model may be used). The specifications of these units are found in Table 13.6. For both distillation columns, their bottom pressure settings are to be determined using the assumed pressure drop (see Table 13.6).
3. Complete base case simulation model by adding the remaining units in the material recycle stream (layer three of the onion model*). The complete flowsheet should look like that in Fig. 13.27. Modeling of heat recovery system may be omitted for this case. Bottom pressure of the distillation columns needs to be readjusted with the new operating condition. Determine the styrene production rate for the converged flowsheet.
4. For an annual styrene production of 300,000 MT, a production rate of 346 kmol/h is desired, assuming an annual operating time of 8320 h (Boehm, 1997). Suggest a strategy that will lead to such production rate.
5. It is desired to avoid temperature beyond 145°C for streams with styrene composition higher than 50 mol%, as styrene may polymerize beyond its boiling temperature (Boehm, 1997). Determine the operating pressure of the both distillation columns so to avoid their bottom product stream (with styrene content ≥50 mol%) from reaching 145°C, while their pressure drop (ΔP) is to be kept at 0.6 kPa/tray for vacuum operation.

References

Boehm, R.F., 1997. Developments in the Design of Thermal Systems. Cambridge University Press, Cambridge.

Foo, D.C.Y., Manan, Z.A., Selvan, M., McGuire, M.L., 2005. Integrate process simulation and process synthesis. Chemical Engineering Progress 101 (10), 25–29.

Linnhoff, B., Townsend, D.W., Boland, D., Hewitt, G.F., Thomas, B.E.A., Guy, A.R., Marshall, R.H., 1982. A User Guide on Process Integration for the Efficient Use of Energy. IChemE, Rugby, UK.

Smith, R., 2016. Chemical Process: Design and Integration, second ed. John Wiley and Sons, New York.

Chapter 14

Process simulation and design for acetaldehyde production*

Lik Yin Ng[1], Jie Yi Goo[2], Rebecca Lim[3] and Mijndert Van der Spek[4]
[1]*Department of Chemical and Petroleum Engineering, Faculty of Engineering, Technology and Built Environment, UCSI University, UCSI Heights, Cheras, Kuala Lumpur, Malaysia;* [2]*Heriot-Watt University Malaysia, Putrajaya, Wilayah Persekutuan Putrajaya, Malaysia;* [3]*Heriot-Watt University Dubai, Dubai Knowledge Park, Dubai, United Arab Emirates;* [4]*Heiort-Watt University, Edinburgh, United Kingdom*

Chapter outline

14.1 Introduction

Acetaldehyde (CH_3CHO) is a petrochemical intermediate that is widely used in the production of acetic acid, pentaerythritol, acetic anhydride, and *n*-butanol (Speight, 2019). Under normal room condition, acetaldehyde appears as a pungent and colorless liquid.

The major acetaldehyde production processes developed on industrial scale include the oxidation of ethylene, hydration of acetylene, and dehydrogenation of ethanol. In this chapter, production of acetaldehyde through dehydrogenation of ethanol will be demonstrated as it is a well-established and mature

*Simulation and other electronic files of this chapter are made available on a companion website; URL: https://www.elsevier.com/books-and-journals/book-companion/9780323901680.

Chemical Engineering Process Simulation. https://doi.org/10.1016/B978-0-323-90168-0.00010-X

process. Compared to other processes, its high selectivity of main reaction results in the production of lesser by-products (see discussion later). Hydrogenation of ethanol is also advantageous over other processes for its ease of operation, since it is relatively stable and operates at moderate temperature and pressure. In addition, product purification can be carried out through conventional unit operations which are simpler and less costly.

14.2 Process simulation

The main goal of the process is to produce 50,000 MTA of acetaldehyde with a purity of at least 99.5 wt%. The approximated flow rate of acetaldehyde product is 6250 kg/h, assuming 8000 h of plant's annual operation. In addition, a by-product of gaseous hydrogen with at least 95 wt% purity should be produced, while its production rate will be determined as part of the simulation exercise.

14.2.1 Simulation setup

While performing process simulation using Aspen HYSYS, some basic simulation setup are necessary, such as definition of components, thermodynamic selection, and reaction stoichiometry definition.[1] For this case, all pure components needed for the process are taken from the software database, while the thermodynamic model (termed as Fluid package in Aspen HYSYS) selected is NRTL (see Table 14.1). The latter is suitable for this process as it involves polar compounds and multiple phases.

Dehydrogenation of ethanol is accompanied by three other side reactions (Chauvel & Lefebvre, 1989). Basic information for the reactions may be

TABLE 14.1 Basic setup for process simulation.

Component	Fluid package
Acetaldehyde	NRTL
Acetic acid	
2-Butanol	
Water	
Ethanol	
Ethyl acetate	
Hydrogen	

1. Refer to Chapter 13 for the basic steps of process simulation with Aspen HYSYS.

obtained from Church and Joshi (1951), with reaction stoichiometry shown in Eqs. (14.1)–(14.4).

$$C_2H_5OH \rightarrow CH_3CHO + H_2 \quad \text{conversion} = 84.00\% \tag{14.1}$$

$$2C_2H_5OH \rightarrow CH_3COOC_2H_5 + 2H_2 \quad \text{conversion} = 6.58\% \tag{14.2}$$

$$2C_2H_5OH \rightarrow C_4H_9OH + H_2O \quad \text{conversion} = 3.20\% \tag{14.3}$$

$$C_2H_5OH + H_2O \rightarrow CH_3COOH + 2H_2 \quad \text{conversion} = 0.56\% \tag{14.4}$$

where ethanol (C_2H_5OH) is the raw material, acetaldehyde is the main product while hydrogen (H_2), ethyl acetate ($CH_3COOC_2H_5$), butanol (C_4H_9OH), water (H_2O), and acetic acids (CH_3COOH) are the by-products of the process.

14.2.2 Process flowsheeting

The general processes involved in the production of acetaldehyde include dehydrogenation of ethanol, phase separation, hydrogen recovery, and acetaldehyde purification (Chauvel & Lefebvre, 1989). This is shown in the block flow diagram in Fig. 14.1.

Fig. 14.1 shows that the process does not involve recycle stream. Since the process to recover unreacted raw material (ethanol) is costly, the economic potential is not appealing. Hence, recycle stream is not considered in this case study. As raw material requirement is still unknown at this stage, a flow rate of 1000 kg/h of 85 wt.% ethanol feed is assumed as the basis. Condition of the ethanol feed stream (S1) is shown in Table 14.2.

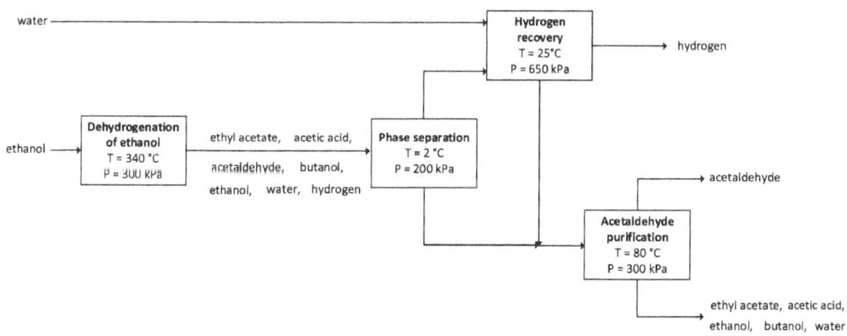

FIGURE 14.1 Block flow diagram of dehydrogenation of ethanol.

TABLE 14.2 Ethanol feed condition.

Component	Flow rate (kg/h)	Condition
Ethanol	850	$T = 25°C$ $P = 101.3$ kPa
Water	150	

14.2.2.1 Dehydrogenation of ethanol and phase separation

As shown in Fig. 14.1, the dehydrogenation reactor is to operate at 340°C and 300 kPa. Hence, the ethanol feed will undergo pumping and heating operations before entering the reactor. The latter is modeled as an isothermal reactor at 340°C. It should be noted that as all products are in gaseous phase, and there is no liquid outlet from the reactor, as shown in Fig. 14.2.

Hydrogen by-product is to be separated from the reactor effluent. This is done through phase separation that is operated at 2°C and 200 kPa, which is placed after a condenser. For illustration purpose, pressure drop across all equipment in the flowsheet is assumed as 50 kPa.

14.2.2.2 Hydrogen recovery

The top stream of the flash separator is rich in hydrogen (stream S7); it is fed to the hydrogen recovery process (Fig. 14.3). This stream is sent to compressor, followed by a cooler for pressure and temperature adjustment, before it is fed to a scrubber. In the latter, water is used as scrubbing solution (stream S11) to remove compounds other than hydrogen (Chauvel & Lefebvre, 1989). Since the amount of water required for the scrubbing process is unknown at this stage, a flow rate of 100 kg/h is assumed as the basis, where a pump is used for its pressure adjustment. Table 14.3 shows the specifications of the units in hydrogen recovery process.

A pressure drop of 0.1 psi/tray is assumed for the distillation and gas absorption columns (Couper et al., 2012), which gives a total of 17 kPa across

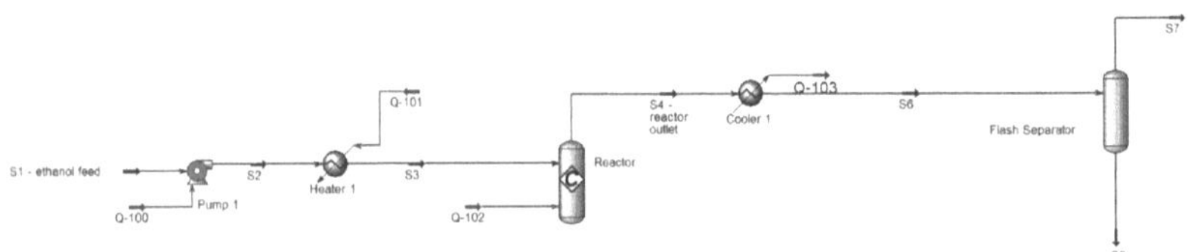

FIGURE 14.2 Dehydrogenation of ethanol and phase separation.

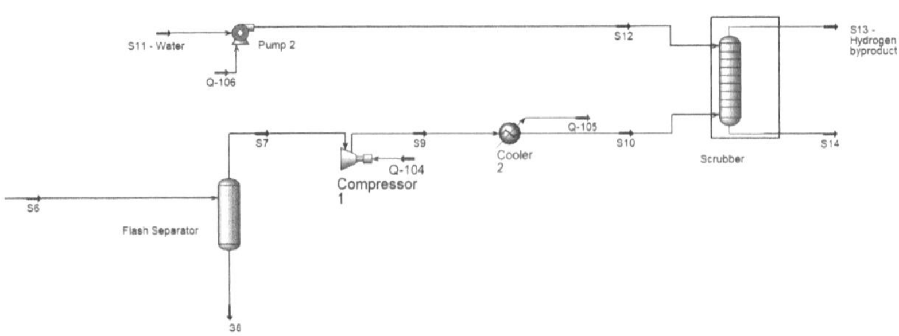

FIGURE 14.3 Hydrogen recovery process.

TABLE 14.3 Specification of units in the hydrogen recovery process.

Unit operation/stream	Condition
Stream (S11)	$T=25°C$ $P=101.3$ kPa
Compressor 1 outlet	$P=680$ kPa
Cooler 2 outlet	$T=25°C$
Pump 2 outlet	$P=650$ kPa
Scrubber	No. of stages = 25 Pressure drop = 0.1 psi per tray/stage Top stage pressure = 653 kPa Bottom stage pressure = 670 kPa

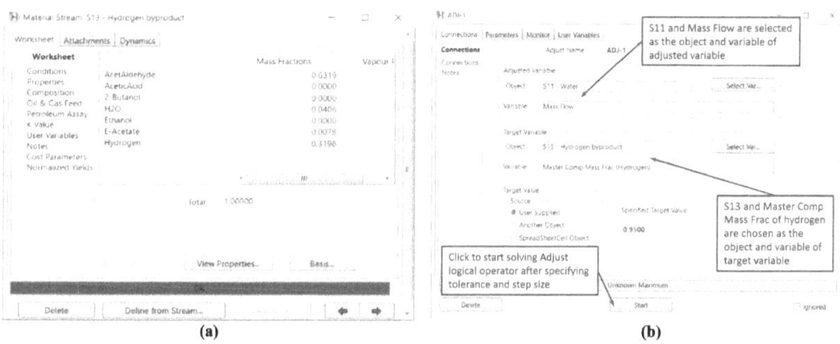

FIGURE 14.4 (A) Composition of Stream (S13); (B) Specification of Adjust logical operation for hydrogen purity.

the column. Once these information for scrubber is specified, calculation for scrubber can be initiated.

The mass fraction of hydrogen in the absorption column top stream (S13) is determined as 0.3196, as shown in Fig. 14.4A, which does not satisfy the required specification (a minimum purity of 95 wt% is needed). This indicates that the amount of water used for the absorption column is insufficient. Instead of estimating through trial and error, the amount of water required to produce hydrogen by-product of 95 wt.% purity is determined through Adjust logical operation.

It is intended to adjust the flow rate of water in stream S11 to ensure that the hydrogen stream (S13) will meet the mass fraction of 0.95. Hence, adjusted and target variables are specified in the Adjust logical operation (see detailed steps and specifications in Fig. 14.4B). In specifying the Adjust logical operation ADJ-1, it is important to provide reasonable tolerance and practical step size to ensure that the balance between result accuracy and computation effort is acceptable. For this case study, tolerance of 1.0000×10^{-4} is specified, while step size of 2 kg/h is used. It is optional to provide the range bounded by minimum and maximum value of adjusted variable. Once the required specification is provided, the calculation can be initiated. With the Adjust logical operation, flow rate for stream S11 is increased by sevenfold to 707 kg/h (from the original value of 100 kg/h); doing this ensures the hydrogen stream (S13) to reach the mass fraction of 0.95.

14.2.2.3 Acetaldehyde purification

The simulation determines that stream S14 has a significant amount of acetaldehyde, which can be recovered. Hence, stream S14 is combined with the acetaldehyde-rich liquid outlet of the flash separator bottom, i.e., stream (S8) to be sent for acetaldehyde purification. A valve is used to reduce the pressure of Stream (S14) to 350 kPa while a pump is utilized to increase the pressure of flash separator bottom (S8) to 350 kPa. The effluent streams from valve (S15) and pump (S16) are then mixed prior to be heated to the operating temperature of distillation column, i.e., 80°C. Simulation results of streams entering the distillation column are shown in Fig. 14.5.

Acetaldehyde purification will be carried out in a distillation column. As basic information are currently unknown, Shortcut Distillation model is used, with its important parameters given in Table 14.4.

Couper et al. (2012) reported that optimum reflux ratio can be estimated by multiplying the value of minimum reflux ratio (R_m) with a factor of 1.2. Hence, with the R_m of 2.343 given by Aspen HYSYS, an external reflux ratio of 2.812 is specified. From the simulation result of shortcut distillation column, the

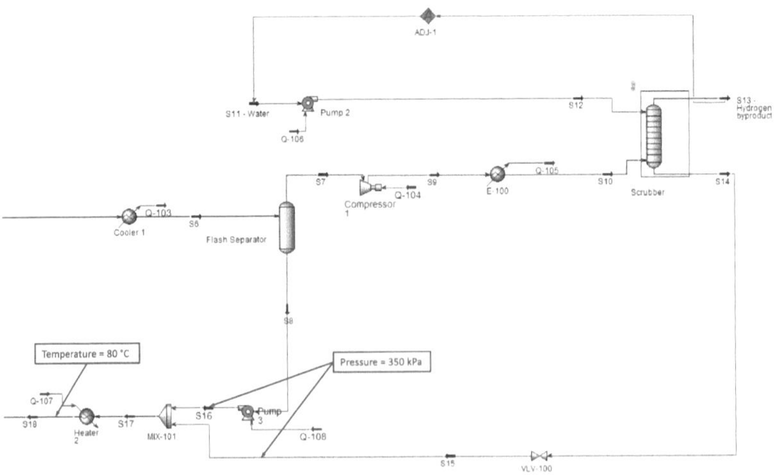

FIGURE 14.5 Temperature and pressure of streams entering the distillation column.

TABLE 14.4 Specification for shortcut distillation column.

Parameter	Condition/value
Distillate	Vapor phase
Light key in bottoms	Acetaldehyde, mole fraction 0.0150
Heavy key in distillate	Ethyl acetate, mole fraction 0.0020
Condenser pressure	290.000 kPa
Reboiler pressure	310.000 kPa

TABLE 14.5 Specification of adjust logical operation for acetaldehyde production rate.

Parameter	Condition/value
Adjusted variable	Mass flow rate of stream (S1)
Target variable	Mass flow rate of stream (S19)
Specified target value	6250 kg/h
Tolerance	1.0000 kg/h
Step size	2.0000 kg/h

mass fraction of acetaldehyde is determined as 0.9960, which satisfies the required purity of 99.5 wt.%. The simulation results also determine that a total of 38 trays are needed, with the optimal feed stage at tray 23. Note that one may also utilize two distillation columns for this purification task, which will incur higher capital and operating expenditures. The simulation results indicate that the distillate product (S19) has a mass flow rate of 649.6 kg/h, which is below the required production rate of 6250 kg/h. Thus, process scale-up is required, which can be done using Adjust logical operation. Specification of the latter is provided in Table 14.5.

With the Adjust logical operation, the amount of raw material (ethanol) is adjusted to 9623 kg/h; doing this leads to the production of 6250 kg/h acetaldehyde (with 99.5 wt.% purity). The simulation model also determines that 6798 kg/h of water is required to produce the hydrogen by-product of 95 wt.% purity. Note also that all process units (e.g., scrubber and distillation columns) are recalculated.

At this stage, the overall process flowsheet of acetaldehyde production through dehydrogenation of ethanol is completed. However, accuracy of the process can be enhanced through modeling distillation using rigorous distillation column. A virtual material stream that has identical specification as the inlet of shortcut distillation column is created, and used as the inlet to rigorous distillation column, as shown in Fig. 14.6.

Specification of rigorous distillation column is performed using the information provided in Table 14.6. Basic information such as number of stages, feed stage, etc. are obtained from shortcut distillation column. Pressure drop of the column is specified based on the assumption of 0.1 psi pressure drop per tray/stage, while pressure drop across condenser and reboiler is also assumed as 0.1 psi.

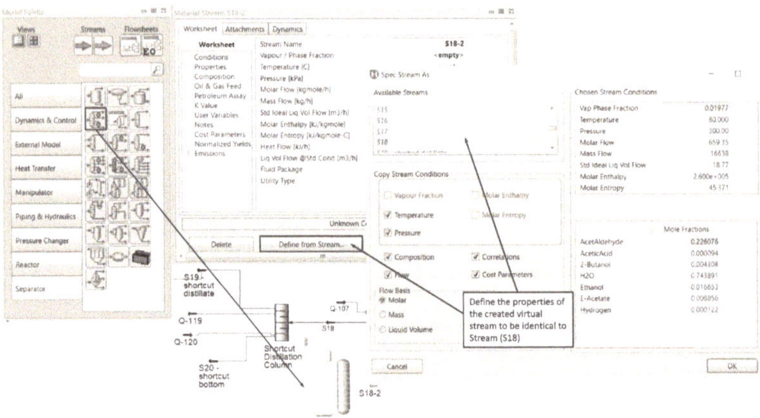

FIGURE 14.6 Specification of virtual stream (S18-2).

TABLE 14.6 Specification of rigorous distillation column.

Parameter	Condition/value
Stage numbering	Top down
Number of stages	35
Feed stage	22
Condenser	Full Rflx (distillate in vapor phase)
Condenser pressure	285 kPa
Reboiler pressure	309 kPa
Vapor rate	6250 kg/h
Reflux ratio	2.675

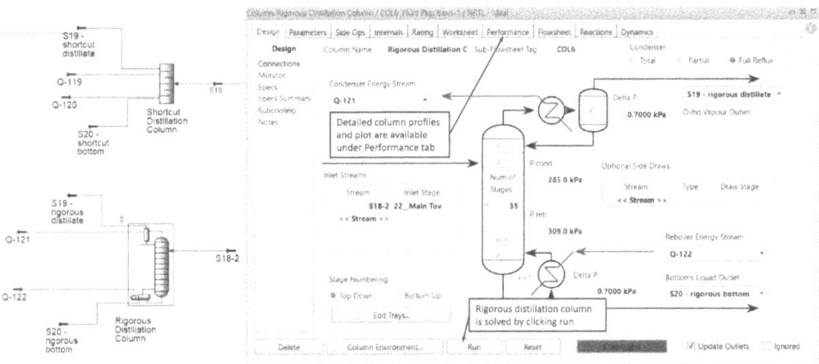

FIGURE 14.7 Converged rigorous distillation column 1T-100.

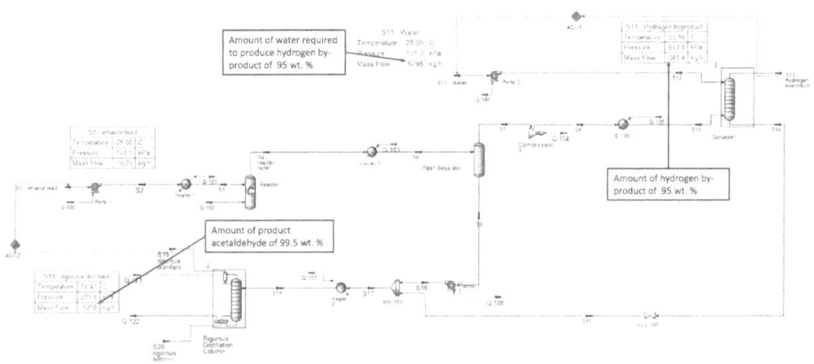

FIGURE 14.8 Complete flowsheet of acetaldehyde production through ethanol dehydrogenation.

Fig. 14.7 shows the column window of the converged Rigorous Distillation Column. As rigorous distillation column gives greater detail compared to shortcut distillation column, different profiles and plots can be generated under the Performance tab (omitted here for brevity).[2]

With the rigorous distillation column converged, the flowsheet is completed. Fig. 14.8 shows the complete flowsheet of acetaldehyde production through dehydrogenation of ethanol, with specification of raw material ethanol stream (S1), water stream (S11) as scrubbing solution, hydrogen by-product stream (S13), and acetaldehyde final product stream (S19).

14.3 Process analysis/potential process enhancement

The completed process flowsheet satisfies the process requirement of producing 50,000 MTA of acetaldehyde with 99.5 wt.% purity, and hydrogen

2. Readers may refer to Chapter 6 for detailed discussion on shortcut and rigorous distillation models.

by-product of 95 wt.% purity. With the aid of different built-in functions available in Aspen HYSYS, analyses are carried out to explore potential process enhancements. These are discussed in detail in the following subsections.

14.3.1 Energy recovery

While operating a chemical production plant, utilities can be a major contributor in terms of operating cost. In order to cut down the operating cost, it is a common practice to explore opportunities for energy recovery among plant utilities. The heating load of Heater 1 and cooling load of Cooler 1 of acetaldehyde production process are shown in Fig. 14.9. It is possible to recover energy from reactor outlet stream (S4) to provide heating to reactor inlet stream (S2) with the use of heat exchanger.

For illustration purpose, a 1-2 shell and tube heat exchanger is used. As the temperature of reactor outlet stream (S4) is higher, it is connected to the tube side while reactor inlet through Heater 1 stream (S2) is connected to the shell side of heat exchanger. Pressure drops for both tube side and shell sides of heat exchanger are assumed as 50 kPa, while compressor is utilized to increase the pressure of reactor outlet to the operating pressure of flash separator. Since the energy to be recovered is unknown at this stage, the tube side outlet temperature is first assumed as 90 °C. Once these steps are completed, it can be seen from Fig. 14.10 that the F_t correction factor of the heat exchanger is determined as 0.675, which is lower than the minimum acceptable threshold of 0.75 (the F_t value is preferred to be 0.85 or higher) (Sinnott & Towler, 2020). Alternative ways to resolve this is by increasing the number of shell passes,[3] or by increasing the tube side outlet temperature. The latter is demonstrated here,

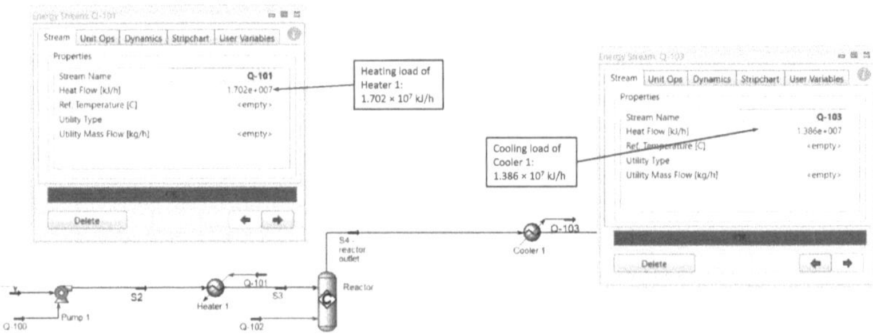

FIGURE 14.9 Energy requirements of Heater 1 and Cooler 1.

3. See Chapter 7 for the use of more shell to improve the F_t value of shell-and-tube heat exchanger.

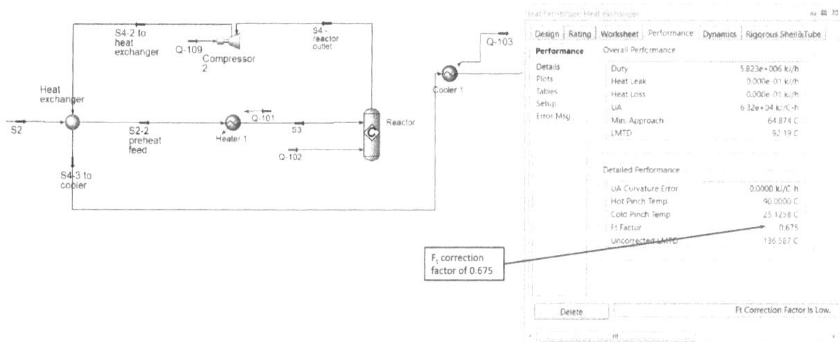

FIGURE 14.10 Streams connection specification of heat exchanger E-101.

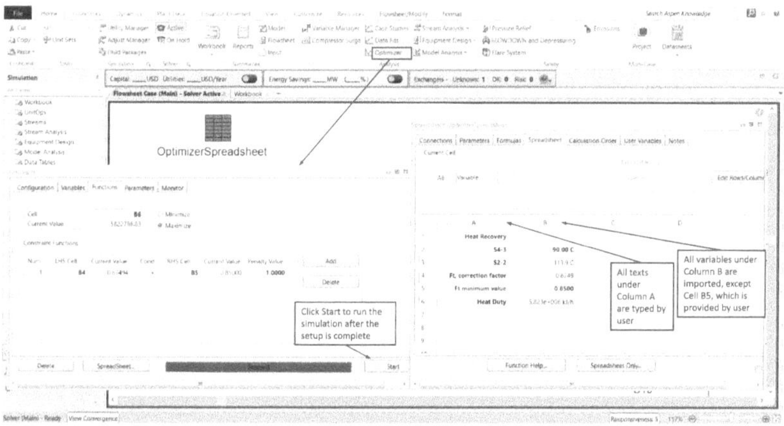

FIGURE 14.11 Specification of Optimizer for energy recovery.

where the energy to be recovered is maximized while fulfilling the constraint of the F_t value.

In this case, the outlet temperature of stream S4-3 (tube side of heat exchanger) is to be increased with the built-in function of Optimizer model. Detailed steps and specification for the Optimizer are shown in Fig. 14.11.

Simulation result shows that the maximum energy to be recovered from the heat exchanger is identified as 5.087×10^6 kJ/h, with an increased temperature of stream S4-3 at 128.5°C (from the assumed temperature of 90°C), while fulfilling the F_t value of 0.85 (see Fig. 14.11). Heater 1 now raises the temperature of stream S2-2 from 113.8 to 340°C (with heating load of 1.702×10^7 kJ/h), while Cooler 1 lowers the temperature of stream S4-3 from 128.5 to 2°C (with cooling load of 9.296×10^7 kJ/h). It is worth noting that the condenser and reboiler of distillation column are excluded from heat integration so to allow better temperature control in these units.

14.3.2 Operating temperature of flash separator

In addition to energy recovery, it is worth exploring the operating temperature of flash separator, in order to reduce its utility cost. In Fig. 14.8, high operating cost is expected for Cooler 1 as it is required to lower the flash separator inlet stream (S6) to 2°C using refrigerant as utility. Its utility cost can be reduced if it is possible to operate the cooler at higher temperature while maintaining an acceptable separation efficiency for the flash separator. A built-in function of Aspen HYSYS, i.e., Case Studies can be utilized for this purpose. As shown in Fig. 14.12, the flash separator with its inlet and outlet streams are cloned and named as "Flash separator (case study)," with streams S6-2, S7-2 and S8-2.

In order to separate hydrogen from the other components, it is intended to condense most components other than hydrogen. Hence, relationship between operating temperature of flash separator and uncondensed fraction for each component is analyzed. The built-in spreadsheet of Aspen HYSYS is used for this calculation, with the detailed steps and specification shown in Fig. 14.12.

As shown in Fig. 14.13, independent variable of inlet stream (S6-2), i.e., temperature, and dependent variables of uncondensed fraction for each

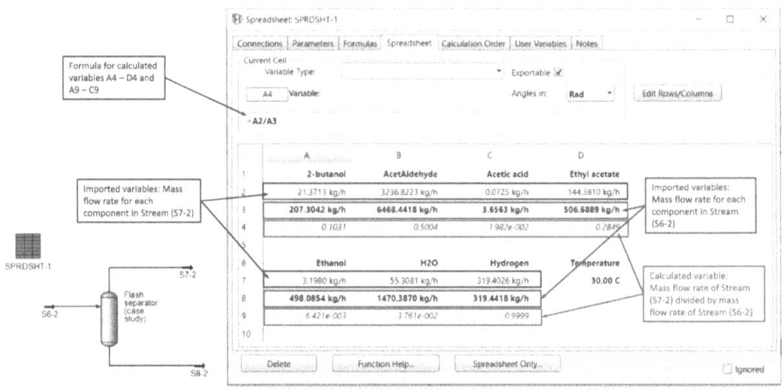

FIGURE 14.12 Specification of Spreadsheet for flash separator case study.

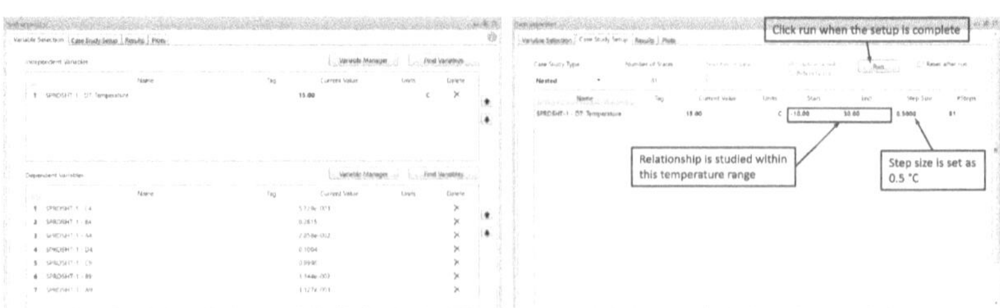

FIGURE 14.13 Specification of flash separator case study.

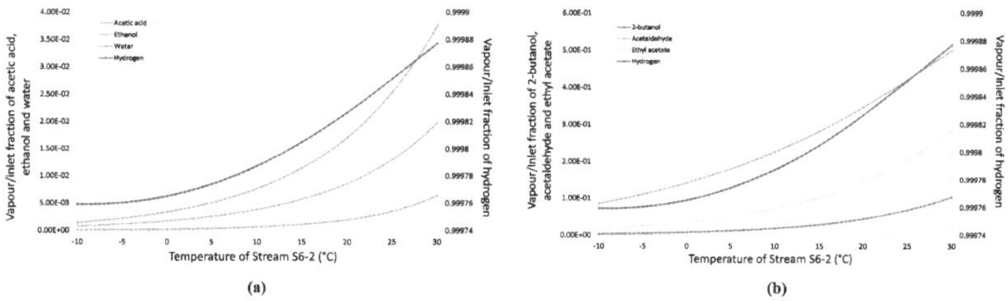

FIGURE 14.14 Graph of uncondensed fraction versus temperature of Stream (S6-2) for (A) acetic acid, ethanol, water, hydrogen; (B) 2-butanol, acetaldehyde, ethyl acetate, hydrogen.

component are added. The relationship is studied within a temperature range of −10−30°C, with a step size of 0.5°C.

Fig. 14.14 shows the uncondensed fraction of all components. As shown, the uncondensed fractions for all components other than acetaldehyde start to level off when the temperature reaches 15°C. Hence, flash separator can be operated at 15°C to reduce the cooling load of Cooler 1 without significantly compromising the separation efficiency. Besides, with the increased operating temperature, chilled water can now be used for Cooler 1, replacing the more expensive refrigeration system.

14.4 Conclusion

This chapter demonstrates the design and simulation of acetaldehyde production through dehydrogenation of ethanol using Aspen HYSYS. Modeling of conversion reactor, absorption column and distillation column are discussed, with particular emphasis on the concept of process scale-up using Adjust logical operation. Detailed process analysis and process enhancement are also illustrated, with elaborative discussion on the use of the built-in tools of Optimizer and Case Studies. Having practiced this chapter, readers are skilled in developing and analyzing an integrated process flowsheet that involves reaction and separation units using Aspen HYSYS.

Exercises

1. The purity of hydrogen by-product can be further enhanced by increasing the amount of scrubbing solution. Find out the highest purity achievable for the hydrogen by-product, and determine the amount of scrubbing solution required.
2. Determine if the condenser and reboiler utilities of distillation column can be reduced by changing the operating pressure without affecting the product quality.

References

Chauvel, A., Lefebvre, G., 1989. Petrochemical Processes: Technical and Economic Characteristics Volume 2. Editions Technip, Paris.

Church, J.M., Joshi, H.K., 1951. Acetaldehyde by dehydrogenation of ethyl alcohol. Industrial and Engineering Chemistry 43 (8), 1804–1811.

Couper, J.R., Penney, W.R., Fair, J.R., Walas, S.M., 2012. Chemical Process Equipment Selection and Design, 3rd ed. Elsevier Inc, Oxford.

Sinnott, R., Towler, G., 2020. Chemical Engineering Design, 6th ed. Butterworth-Heinemann, Oxford.

Speight, J.G., 2019. Handbook of Petrochemical Processes. CRC Press, Boca Raton.

Chapter 15

Dynamic simulation for process control with Aspen HYSYS*

Rafil Elyas
East One-Zero-One Sdn Bhd, Shah Alam, Selangor, Malaysia

Chapter outline

*Simulation and other electronic files of this chapter are made available on a companion website; URL: https://www.elsevier.com/books-and-journals/book-companion/9780323901680.

Chemical Engineering Process Simulation. https://doi.org/10.1016/B978-0-323-90168-0.00015-9

15.1 Introduction

The handling of dynamic behavior and quantification of dynamic process responses for process control design and tuning in chemical engineering has traditionally been based on reducing processes via Laplace transformations into their transfer functions, which is then followed by frequency domain analyses in order to characterize process responses and design control solutions (Stephanopoulos, 1984). This strategy is suitable for mechanical-electrical systems where responses tend to be fast (i.e., little dead time) and linear. Unfortunately, this is not the case for most chemical process systems, as the latter generally process large quantities of material. For example, a crude oil refinery may process between 30,000 and 300,000 barrels of crude oil per day. Furthermore, chemical processes may have residence times ranging from seconds to hours, which introduce a significant amount of dead time; these complicate controller design and tuning. Another challenge in chemical processes is the complexities in fluid behavior. Typical examples include azeotropic behavior and retrograde condensation, where both are highly nonlinear and are functions of composition, temperature, and pressure. Hence, the traditional approach in using Laplace transform-derived proxy models and frequency domain analyses needs to be replaced with more mechanistic, physics-based dynamic models.

Since the 1990s, desktop computers with fast numerical processing capabilities became commonplace and rigorous dynamic process simulation software provided chemical engineers with flexible flowsheet-based environments to model most oil and gas, and chemical processes. This leads to the widespread of commercial dynamic simulation software in the market. Commonly used flowsheet programs for dynamic simulation are roughly grouped into the following "families":

1. **GUI-based interactive flowsheet simulators**[1]
   a. Aspen HYSYS (Aspentech)

1. These are found in various chapter of this book.

   b. UniSim Design (Honeywell)
   c. Petro-SIM (KBC/Yokogawa)
   d. Symmetry (Schlumberger)
2. **Equation-oriented simulators**
   a. Aspen Dynamics (Aspentech)
   b. gPROMS (Siemens/Process Systems Engineering)
3. **Others**
   a. K-Spice (Kongsberg Digital)
   b. DynSim (Aveva)
   c. DWSim (Daniel Wagner Oliveira de Medeiros; open source)

These programs allow chemical engineers to understand process behavior of varying complexity using a time-based approach and take a more intuitive approach to designing and troubleshooting process controls.

15.2 Dynamic model overview

In this section, an overview dynamic modeling will be given for. It should be noted that discussions concerning mathematical structure and framework of dynamic models, as well as the example used are based on the Aspen HYSYS framework.

15.2.1 Steady-state and dynamic models

Before addressing dynamic process modeling, it is beneficial to understand the difference between steady-state and dynamic models and how they are represented in commercial software.

Steady-state models are based on thermodynamic equilibrium models. As there is no accumulation, all material and energy enter and exit a process at the same rate. Information propagates instantaneously across the process, and there is no representation of the process between equilibrium states. Since hold up is not represented, no equipment dimensions are required. In addition, most steady-state models do not consider pressure-flow hydraulic relationships. In other words, changing the inlet pressure of a valve/pipe will not increase its flow. Hence, users usually specify pressure drops across the equipment.

On the other hand, dynamic models utilize thermodynamic equilibrium strategies to compute phase behavior, component partitioning across phases, and the corresponding fluid thermophysical properties. Unlike steady-state models, dynamic models account for material and energy accumulation and hold up. In addition, pressure-flow hydraulic behavior is generally represented.

15.2.2 Dynamic model usage

Dynamic process simulation is used for the following purposes:

1. **As a process design aid or operational troubleshooting.**
   a. Dynamic models allow the incorporation of dynamic considerations into plant design and operational troubleshooting. Steady-state models only provide equilibrium constrained "snapshots" and are used to define idealized operating boundaries and will not be able to predict any excursions from these boundaries during a plant start up, shut down, or emergency.
   b. To determine the impact of material and energy capacitance or hold-up. In some cases, hold-up is useful, when it's used to as a buffer to filter disturbances like a slug catcher on an oil rig. In other cases, hold-up may introduce dead-time or delay, and create controllability problems.
2. **To check the integrity of a plant's distributed control system.** Rigorous dynamic models of a processing facility can be used to test out control strategies before a control system is configured.
3. **To train operators.** Pilots use physics-based models to train. Analogously, plant operators can be trained using rigorous dynamic models that capture the behavior and response of their processes.

15.3 Dynamic modeling concepts[2]

Two main features that separate dynamic from steady-state models are hold-up and pressure-flow hydraulics.

15.3.1 Hold-up

Hold-up can be divided into material and energy hold-up and is typically incorporated in the following equipment:

1. Tanks
2. Separators (two and three phase)
3. Distillation columns, tray liquid, and column vapor space
4. Liquid-liquid contactors
5. Reactors, which include continuously stirred tank reactors and plug flow reactors
6. Valves, pipes, and fittings
7. Heater, cooler, and heat exchangers

To demonstrate the concept of hold-up, a tank model in Fig. 15.1 is used.

15.3.1.1 Material hold-up

The tank in Fig. 15.1 contains a liquid hold-up model. Eq. (15.1) shows the equation that describes the material balance relationship for this

2. Note: These concepts are based on Aspen HYSYS and similar software.

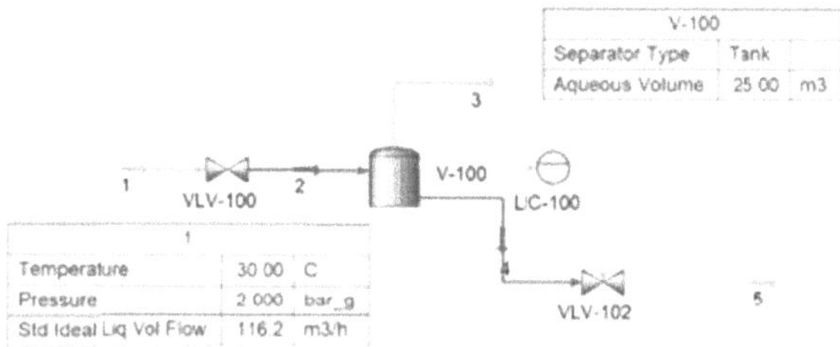

FIGURE 15.1 Tank with liquid level control.

example, which may be interpreted as "liquid flow in − flow out = liquid accumulation":

$$F_{\text{in}}(t) - F_{\text{out}}(t) = \rho \frac{dV(t)}{dt} \tag{15.1}$$

where $F_{\text{in}}(t)$ and $F_{\text{out}}(t)$ are the inlet and outlet fluid flows respectively, ρ is the fluid density and the last term in Eq. (15.1) represents the fluid accumulation in the tank. The fluid inventory is contained in a tank which is assumed to be rigid. Other process variables such as pressure, volume, and temperature effects are also be addressed similarly, for example:

1. Pressure will increase if feed rate is increased while discharge rate is kept constant. Temperature may increase in this case.
2. Pressure will decrease if feed rate is kept constant while discharge rate is increased. Temperature may decrease in this case if liquid expands.
3. Pressure will decrease if vapor is cooled and condenses.

It should be noted that Eq. (15.1) is an initial value problem, an ordinary differential equation which requires a definition of the tank inventory at a given time to be solved. This means that in order to solve this problem (i.e., to run this simulation), an initial value must be supplied. Generally, this initial value will be based on a steady-state model. Thus, prior to constructing a dynamic model, it is often necessary to perform a steady-state simulation in order to generate initial values for the various hold ups and streams in the dynamic simulation. With this material hold-up model come with two other implications, i.e., energy hold-up and pressure-volume effects; these are discussed next.

15.3.1.2 Energy hold-up

The material balance described in (15.1) has a corresponding energy balance, which may be described as "energy in − energy out = energy accumulation":

$$H_{\text{in}}(t) - H_{\text{out}}(t) + Q(t) = \frac{dH(t)}{dt} \tag{15.2}$$

where $H_{in}(t)$ and $H_{out}(t)$ are the respective enthalpies of the inlet and outlet material streams, $Q(t)$ may be heat input (heating) or output (cooling), or heat loss/gain to or from the environment, while the term on the right side of Eq. (15.2) is the heat accumulation of the tank.

15.3.2 Pressure-flow hydraulics

Pressure-flow hydraulics allows the representation of material flow across a process to be related functionally to the pressure gradient. This is necessary in order to model basic phenomena such as opening or closing of valve (to increase/decrease flow rate), as well as more complex phenomena such as compressor surge.

Conservation of energy for a flowing fluid can be summarized by Bernoulli's equation (Eq. 15.3):

$$P_1 + \frac{1}{2}\rho v^2 + \rho g h_1 = P_2 + \frac{1}{2}\rho v^2 + \rho g h_2 \tag{15.3}$$

The flow conductance equation which represents mass flowrate (F) across a valve or nozzle is defined in Eq. (15.4):

$$F = k\sqrt{\rho(P_1 - P_2)} \tag{15.4}$$

where:

P_1 and P_2 are inlet and outlet pressures
v_2 is fluid velocity
h_1 and h_2 are inlet and outlet elevations
k is flow conductance
ρ is fluid mass density

It should be noted that in Aspen HYSYS, the kinetic energy term in Eq. (15.3) $\left(\frac{1}{2}\rho v^2\right)$ is generally set to zero.[3]

When it involves elevation, definition of equipment and nozzle elevations are critical when the liquid phase is being modeled, as these parameters impact fluid head gain and head loss. For gasses, elevations may be neglected if the gas density is low or the resulting static head effects are negligible. Thus, for all intents and purposes, Aspen HYSYS (and other similar software[2]) only capture frictional and elevation pressure contributions. Kinetic energy, fluid acceleration, deceleration, and momentum effects are not represented. The means that these simulators cannot be used to model phenomena like liquid

3. This same principle applies to UniSim Design (Honeywell) and Petro-SIM (KBC/Yokogawa). For Symmetry (Schlumberger), the kinetic energy term is included in nozzle computations.

surge or water hammer. Other simulation platforms that have more comprehensive representation on fluid flow are suitable for these types of analyses.

15.3.2.1 Definition of flow conductance

Practically, all dynamic equipment modules in Aspen HYSYS (and other similar software[2]) require a flow conductance (k) to be defined in some manner.

15.3.2.1.1 Direct flow conductance specification

Direct specification of flow conductance is generally implemented in heat exchangers, distillation column trays (based on dry hole pressure drops), piping, and fittings (where flow conductance may be estimated using diameter and equivalent lengths). The flow conductance is calculated for a known pressure drop and flow rate using Eq. (15.4). It is interesting to note that Eq. (15.4) treats the flow conductance (k) as a constant. However, this is a simplification, as k value is actually the inverse of flow restriction, which can be viewed as a frictional term. The latter is a function of flow regime, which is in turn a function of the Reynold's number (Eq. 15.5):

$$Re = \frac{Dv\rho}{\mu} \tag{15.5}$$

where:

Re is the Reynold's number
D is the equivalent diameter
v is the fluid velocity, which is the volumetric flow divided by the cross-sectional area
μ is the fluid viscosity
ρ is the fluid density

Hence, the value of k may vary depending on the volumetric flow of the fluid. The heat exchanger operation in Aspen HYSYS provides a scaling factor to relate k as a function of flow. In some cases, it may be necessary to perform some steady-state hydraulic computations to determine if there is an appreciable variance of k as a function of flow in the equipment. For example, during the start-up of an equipment, the flow rate may start from zero, then slowly increase to its operational value.

15.3.2.1.2 Valves

The control valve operation in the Aspen HYSYS allows for the input of vendor control parameters (e.g., C_v, C_g), valve equations of Fisher, Mokveld, CCI, and Masoneilan valve vendors. The following equations describe the Fisher Universal Sizing Method (Aspentech Ltd, 2021).

Flowrate (f) through the valve with an inlet (P_1) and outlet pressure (P_2) is described by Eq. (15.6):

$$f = v_{\text{fracfac}}\ 1.06\ C_g\ \sqrt{\rho \times P_1}\ \sin(\arg) \tag{15.6}$$

where:

$$\arg = \frac{56.64}{C_1}\sqrt{1 - \frac{P_2}{P_1}}\ cp_{fac};\ \text{note that arg is bound with upper } \left(\frac{\pi}{2}\right)$$

and lower limits (0). (15.7)

$$C_1 = \frac{C_g}{C_v} \tag{15.8}$$

$$K_m = 0.001334\ C_1 \tag{15.9}$$

$$cp_{\text{fac}} = \sqrt{\frac{0.4839}{1 - \left(\frac{2}{1+\gamma}\right)^{\left(\frac{\gamma}{\gamma-1}\right)}}} \tag{15.10}$$

$$\gamma = \frac{C_p}{C_{vT}} \tag{15.11}$$

where v_{fracfac} is the vapor fraction on a mass basis, $\frac{C_p}{C_{vT}}$ is the ratio of constant pressure and constant volume heat capacities, C_v is the valve flow coefficient, which is defined as the volume of water in US gallons per minute at 60°F that will flow through a fully open valve with a pressure differential of 1 psi across the valve, C_g is the gas flow sizing coefficient and K_m is the valve recovery coefficient. $C_{p\text{fac}}$ is an isentropic correction factor, arg is the flow direction correction factor. Note that C_g and K_m are vendor specific parameters calculated for some design mass flow rate, pressure drop, fluid density, and critical properties.

15.3.2.1.3 Piping hydraulics

Pipe operations in the Aspen HYSYS computes the equivalent flow conductance based on the Darcy or Churchill friction factor relationship (Aspentech Ltd, 2021).

15.3.2.2 Head and energy terms

The head term in Bernoulli equation in Eq. (15.4) may be addressed by either elevation or work from rotating equipment, compressors, pumps, or turbines. This equipment is modeled using their respective performance curves, head and efficiency versus actual volumetric flow. Examples of performance curve

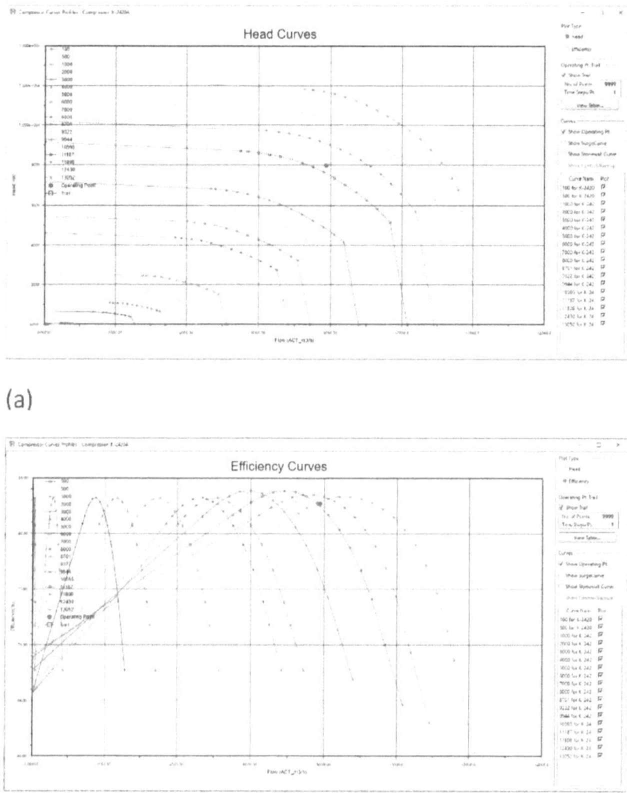

(a)

(b)

FIGURE 15.2 Compressor performance curves: (A) head versus actual volumetric flowrate; (B) polytropic efficiency versus actual volumetric flowrate.

are shown in Fig. 15.2. As shown, these performance curves are plotted for a range of compressor speeds, with head and polytropic efficiency versus its actual volumetric flowrates.

15.3.3 Dynamic model information requirements

It should now be apparent that a dynamic model requires more information as compared to a steady-state model. An example of some information an engineer may require for specific activities are given as follows.

Design and operations troubleshooting:

1. Process flow diagrams (PFD) and the corresponding heat and material balances.
2. Piping and instrumentation diagrams (P&ID).

TABLE 15.1 Example equipment parameters used for dynamic modeling.

Equipment	Typical parameters required for dynamic modeling
Slug catchers, knock out drums, suction scrubbers, flash drums, and tanks	Diameter, height or length, inlet and outlet nozzle locations and sizes, weir location, wall thickness, material of construction heat capacity
Distillation, absorber, stripper, and extraction columns	Diameter, height, weir length and height, downcomer dimensions, feed and product nozzle locations and sizes, dry hole pressure drops for tray flow conductance
Heat exchangers	Thermal and mechanical design data sheets
Control valves	Valve trim characteristic (C_v vs. percent open), actuator response time
Pipes	Diameter, wall thickness and internal roughness, length, insulation details
Compressors	Performance curves (head and polytropic efficiency vs. actual volumetric flow)

3. Isometric drawings and equipment layout.
4. Equipment data sheets, process, and mechanical. An example of required equipment sizes and mechanical geometries are presented in Table 15.1.
5. Control philosophy.

   Operations troubleshooting and operator training simulator:

   In addition to the above, the following would typically be required:
6. Operating manual.
7. Control philosophy, control and alarm settings, cause and effect diagrams, shut down logic.
8. Daily logs, operator feedback for validation.

Table 15.1 summarizes the typical parameters that are required to model commonly encountered process equipment. It should be noted that these parameters are not required for steady-state simulation models as they allow the characterization of volumes, liquid levels, pressure drops, and gains.

15.3.4 Setting up a dynamic model in Aspen HYSYS

Defining a model in HYSYS consists of the following steps.

15.3.4.1 Creating a steady-state model

It is highly recommended to create a steady-state model before constructing a dynamic model for the following reasons:

1. **Model initialization.** The dynamic model needs to be initialized (see Section 15.3.1). The steady-state model will provide these initial values for streams and material and energy hold-ups. It is recommended to save this steady-state model in a separate file to be referred to when necessary.
2. **Preliminary equipment sizing or estimation of missing equipment parameters.** In some cases, equipment data sheets may not be available. It is possible that the equipment of interest may not have been designed or sized yet, but the engineer needs to perform some preliminary dynamic studies for the overall process. It could also be a situation that the engineer may be working with an old process where engineering drawings and data sheets are no longer available.

   In this case, the steady-state model may be used to estimate approximate equipment sizes. Some typical examples include the following:
   a. If a control valve data sheet is unavailable, it is a standard practice to size the valve based on a design flow rate (Eq. 15.6) assuming a 50% valve opening.
   b. A vessel may be sized using the steady-state design flow rate by defining a desired residence time and estimating the corresponding volume.
   c. The value of heat exchanger heat transfer coefficient may be estimated using steady-state heat and energy balances.
3. **Insight on fluid behavior.** It is recommended to generate phase envelopes in steady-state simulation to get a clear understanding of the fluid behavior. This helps to flag potential issues before dynamic analysis is performed. For example, phase envelope of for a compression train should be plotted, where the operating envelope can be overlaid to determine if there are situations where liquid may be present in the compressor inlet.

15.3.4.2 Equipment parameter and flowsheet pressure flow configuration

Unlike steady-state simulation, equipment parameters that influence hold-up and pressure drop are required; these details were summarized in Section 15.3.3.

All inlet and outlet material streams of a process within the battery limit must have either a fixed pressure or fixed flow specification defined. For Aspen HYSYS (and other similar software), pressure specifications are normally used and consistent with a physical process, where fluids flow is based on a pressure gradient. Fixing the material stream flow specification (instead of pressure) may be done for specific analyses; however, caution must be exercised since this may yield unrealistic pressure gradients.

15.3.4.3 Numerical solver configuration

Note that Aspen HYSYS utilizes an implicit Euler integration strategy. Fig. 15.3 shows the parameters and recommended settings.

As shown in Fig. 15.3, the default integrator time step is 0.5 s and should be modified based on transient behavior of the process. The time step should be two or more times smaller than the fastest transient phenomena in the process. Examples of fast transients in a process are controller sampling and action, valve actuator stroke time and compressor surge. An anti-surge valve may open in 1–2 s, a compressor may enter the surge region in 0.1 s, and a controller sampling time may be around 40 ms.

In Aspen HYSYS, the integrator may be accelerated or slowed down (the effect would be analogous to slowing down or speeding up a movie). Accelerating the simulator is done by increasing the integration time step and this may result in a loss of resolution or instability. Acceleration is used when the model is being used as an operator training system engine to speed up calculation. For this type of application, high resolution behavior may not be required. The integrator execution rate definition screen is shown in Fig. 15.4.

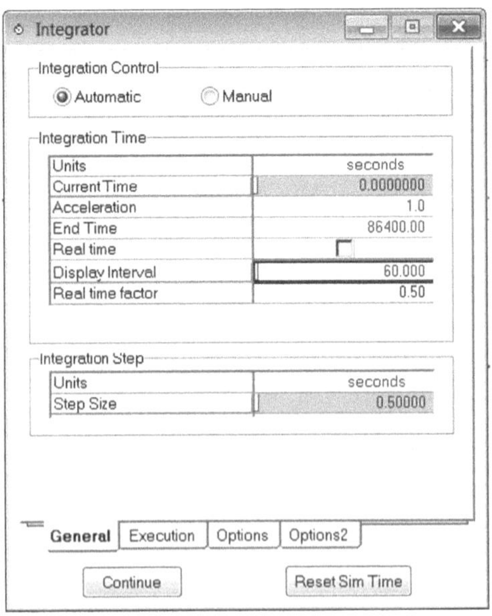

FIGURE 15.3 Integrator settings time step.

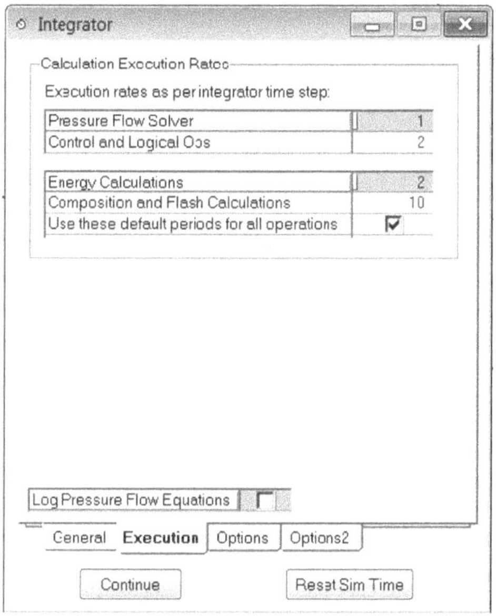

FIGURE 15.4 Execution rates.

This setting determines how often specific equations are solved. The default settings are:

1. Pressure-flow is solved every time step. It is recommended to maintain this setting.
2. Controllers are solved every two time steps. This should be set based on controller sampling time.
3. Energy balances are performed every two time steps. This depends on how fast heat or work may be changing.
4. Flash calculations are performed every 10 time steps. This setting needs to be changed depending on how fast compositions, pressures and temperatures are may change.

From this, it is important to note that the engineer needs to have an idea of how fast the various phenomena change within the process of interest in order.

In Aspen HYSYS, it is possible to ignore static head contributions. This reduces the number of equations to be solved, which speeds up the computation. However, caution must be exercised if static head were to be ignored. The latter is feasible if the process fluid was purely gas, since its density is low. On the other hand, if static head were ignored when a liquid system is modeled, this would lead to an unrealistic result. Without static head computation enabled in a two phase vessel, the pressure of its liquid and vapor nozzles would be treated as identical, which would yield an erroneous

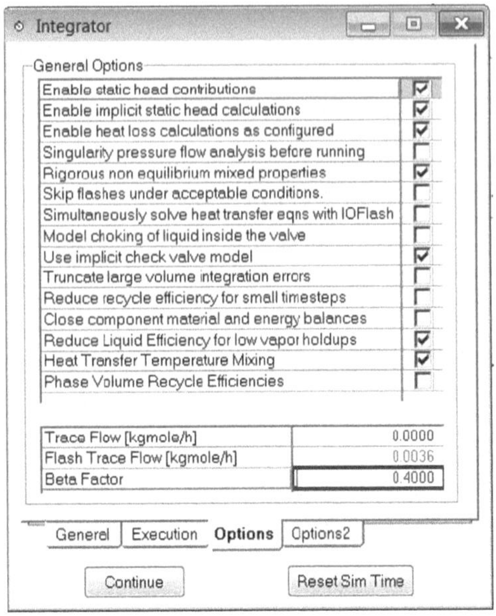

FIGURE 15.5 Static head definition and other transient phenomena.

hydraulic gradient across the process. Static head computations and other transient pressure-volume parameters can be defined in the General Options screen, shown in Fig. 15.5.

15.4 Constructing a dynamic model in HYSYS (Aspentech Ltd, 2021)

Aspen HYSYS has two simulation environments: steady state (default) and dynamic. For small cases such as this example, a steady-state model is first constructed. The model is then transited into dynamic environment. For larger and more complex cases, steady-state and dynamic models are constructed separately. The steady-state model is used as a reference model, where initial values (e.g., compositions, pressure, temperature) may be used for the dynamic model.

The concepts in Section 15.3 are demonstrated with a heated crude oil separator example. As shown in Fig. 15.6, a fluid is heated from 50 to 65°C by heat exchanger E-100 before entering inlet separator V-100.

The steps for setting up a dynamic model as outlined in Section 15.3.4 is used as guideline here.

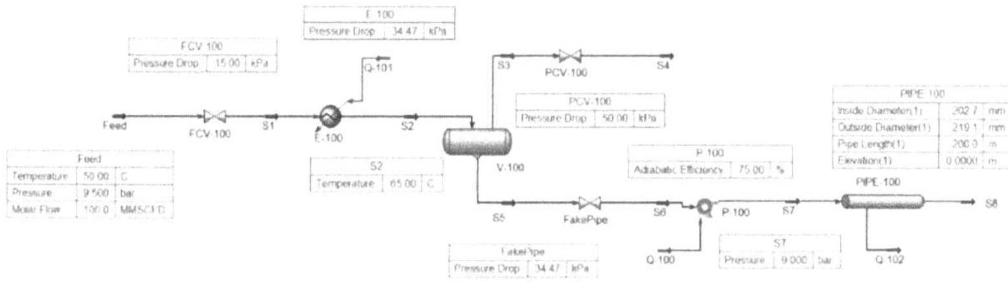

FIGURE 15.6 Simulation flowsheet for heated tank model.

15.4.1 Steady-state process modeling[4]

A steady-state model is first built in Aspen HYSYS. The thermodynamic model used in this example is Peng-Robinson package. Condition of the feed stream is presented in Table 15.2. Note component C7* in is hypothetical, with properties shown in Table 15.3.

TABLE 15.2 Feed flow rate, pressure, temperature, and composition.

Temperature (°C)	50.0
Pressure (bar abs)	9.5
Flow rate (MMscfd)	100.0
Mole percent	
H_2O	1.1%
Nitrogen	0.3%
CO_2	0.1%
Methane	43.7%
Ethane	13.2%
Propane	13.2%
i-Butane	2.6%
n-Butane	5.0%
i-Pentane	1.3%
n-Pentane	1.3%
n-Hexane	1.3%
C7*	16.8%

4. Refer to Topside-SS.hsc simulation file for this section. Readers may refer to Chapter 13 for detailed guide on how to perform steady-state process simulation with Aspen HYSYS.

TABLE 15.3 Properties of component C7*.

Parameter	Value
Normal boiling point (°C)	167.0
Molecular weight (kg/kgmole)	235.0
Ideal liquid density (572.0 kg/m^3)	572.0

TABLE 15.4 Equipment parameters.

Equipment	Tag	Specifications
Flow control valve	FCV-100	Pressure drop = 15 kPa
Production heater	E-100	Pressure drop = 35 kPa Outlet temperature 65°C (specification on stream S2)
Production separators	V-100	Adiabatic No pressure drop
Pressure control valve	PCV-100	Pressure drop = 50 kPa
Separator piping	Fake pipe	Pressure drop = 35 kPa
Production liquid pump	P-100	Adiabatic efficiency = 75% Outlet pressure = 9 bar ab
Facility piping	Pipe-100	Size = 8″ schedule 40 Length = 1000 m

Stream conditions and some important parameters for the equipment are presented in the simulation flowsheet in Fig. 15.6. The production heater E-100 (modeled using the heater model) raises the feed temperature to 65°C prior to it entering the production separator V-100. Pressure of the latter is controlled by pressure control valve PCV-100, while its level is controlled by a level control valve, which will be added in Section 15.4.2.1 in the dynamic model. The bottom liquid stream is then pumped by P-100 to a pipe with length of 1000 m (Pipe-100). The valve connecting V-100 and P-100 acts as a fake pipe which introduces a small pressure drop. The equipment specifications are summarized in Table 15.4.

Once the model is converged, the pump icon shows yellow sign, which indicate a warning. This is because the liquid product of V-100 exits the separator as a saturated liquid, it then gets flashed across the fake pipe and resulted in a small amount of vapor.[5] The pump model flags this as a warning, since vapor entering a pump may result in cavitation.

Note that in this steady-state model, the pressure of vapor and liquid products exiting V-100 are identical. That is because static head contributions of the liquid in the separator are not accounted for. In reality, the liquid product would have higher pressure than the vapor stream, which would put it in the subcooled region. This phenomena can be observed in the dynamic model (see later section).

15.4.2 Setting up dynamic parameters in the steady-state environment

In this section, dynamic parameters will be added to the steady-state model constructed in Section 15.4.1.[6] Equipment parameters that characterize process hold-ups need to be added. These include the following:

1. Valve trim, C_v, and actuator stroke time
2. Heater volume and K
3. Separator volume (length/diameter)
4. Pump curves
5. Pipe pressure drop method and volume

15.4.2.1 Valve

In this case, valve sizes will have to be estimated as they are not provided. With the steady-state model taken as the design basis, it is assumed that the valves would be 50% open at the design conditions. An assumption of 12.5% per second shall be made for the valve's actuator stroke time. A rule of thumb typically used for control valve stroke times is 1″/second. Hence, a 4″ valve would open from 0% to 100% in 4 s. Detailed procedure to set up the valve are given in Figs. 15.7 and 15.8.

Table 15.5 summarizes the parameters for other valves used in the heated tank example. Note that fake pipe is represented by a valve with negligible flow resistance.

5. Verify this from the simulation model.
6. Refer to Topside-SS-WithDynParams.hsc simulation file for this section. This file has a steady-state model but with dynamic parameters. Readers are also encouraged to build the model using Topside-SS.hsc simulation file.

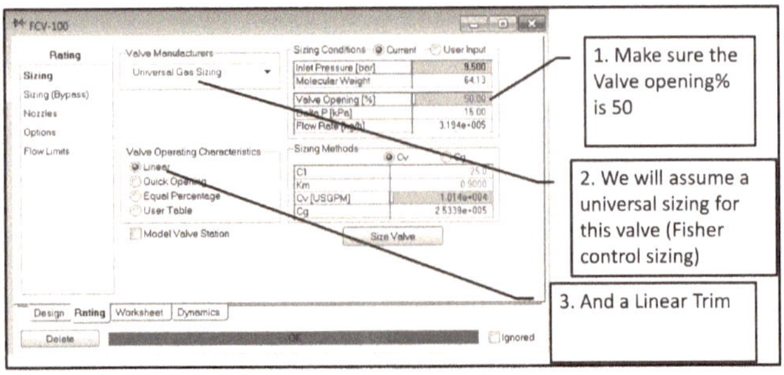

FIGURE 15.7 Valve sizing trim sizing.

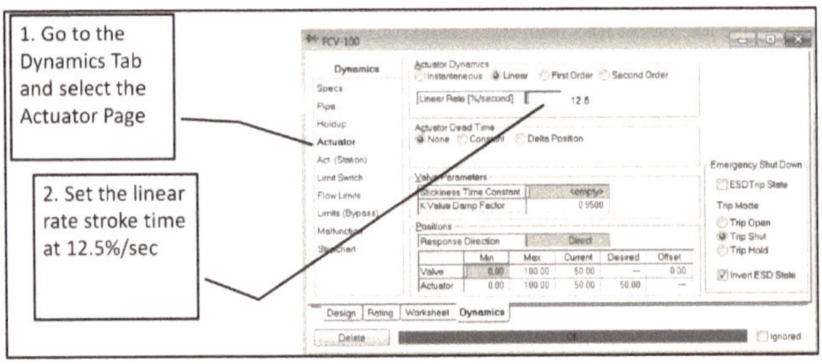

FIGURE 15.8 Actuator definition.

TABLE 15.5 Specifications for all valves.

Valve	Function	Trim	C_v	Stroke (%/s)
PCV-100	Pressure control	Equal percentage	3662	5
VLV-100	Level control	Linear	1200	12.5
Fake pipe	Pipe	Linear	1.00E+05	NA

15.4.2.2 Separator

For this case, the separator is modeled as a horizontal separator with a diameter of 3.5 m and length of 10.0 m. Its geometry is simplified as a flat head cylinder (see Fig. 15.9).

15.4.2.3 Pump

For a steady-state heat and material balance model, it would suffice to characterize the pump by its efficiency and pressure/work. For a dynamic model, this would make the model extremely inflexible since there is no pressure-flow behavior represented. It is necessary to model a pump (and other rotating

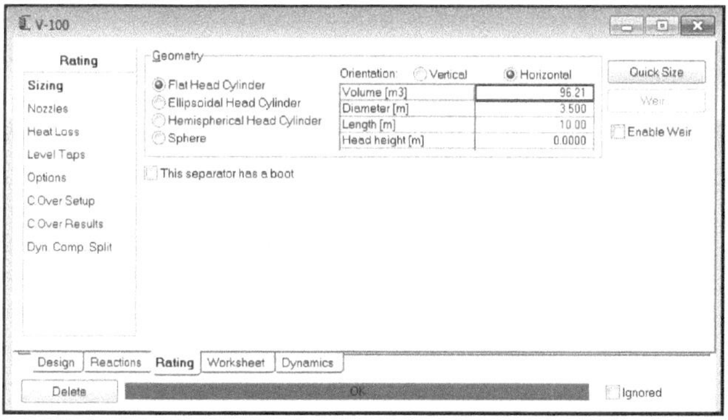

FIGURE 15.9 Separator definition.

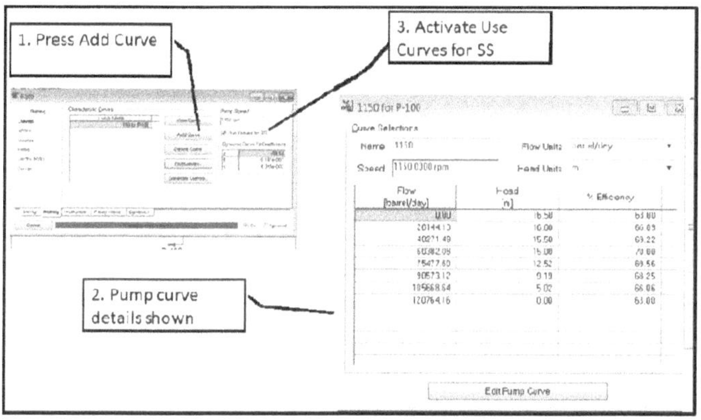

FIGURE 15.10 Pump performance curve definition.

equipment) using performance curves which relate the pump's head and efficiency to the volumetric throughput.

The definitions of pump head, efficiency curves, and dynamic specifications are presented in Figs. 15.10, 15.11, and 15.12 respectively.

15.4.2.4 Heat exchanger

Three parameters are needed to be defined for heat exchanger model, i.e., duty, volume, and pressure-flow hydraulics. These are described as follows.

15.4.2.4.1 Duty

In reality, a shell-and-tube exchanger is used for this type of application, while the common heating fluid is steam or thermal oil. For this example, a heater with an energy stream is used. The latter mimic a utility fluid, in which later will be modulated with a controller.

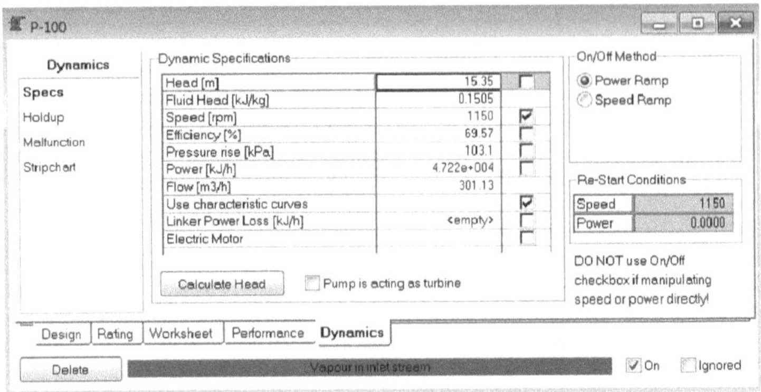

FIGURE 15.11 Pump dynamic specifications.

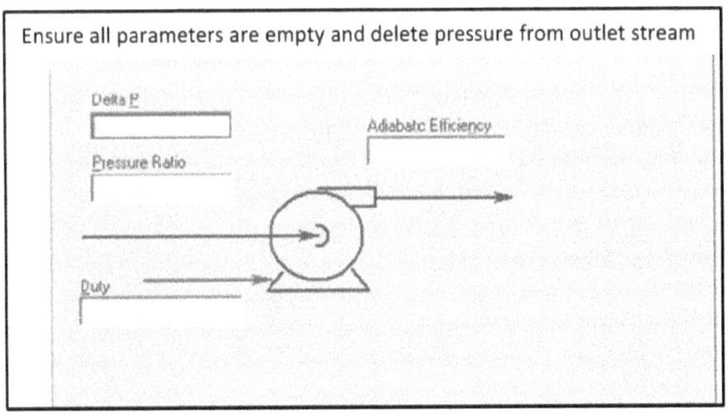

FIGURE 15.12 Pump efficiency specifications (ensure that all specification are cleared).

15.4.2.4.2 Volume

In this example, the volume of the process side is assumed as 0.1 m^3.

15.4.2.4.3 Pressure-flow hydraulics

Since the pressure drop and flow rate are known for this heat exchanger, its flow conductance (k) can be calculated using Eq. (15.4). Detailed steps for specifying the heat exchanger are shown in Fig. 15.13.

15.4.2.5 Pipe

The pipe dynamic parameters need to be specified and its heat transfer coefficient defined, as illustrated in Figs. 15.14 and 15.15.

15.4.2.6 Controllers

Controllers will be set up, starting with the liquid level controller, LIC-100 which controls the liquid level of V-100. This is a PI (proportional, integral) controller with set point of 50% liquid level, a gain of 0.1 and reset time

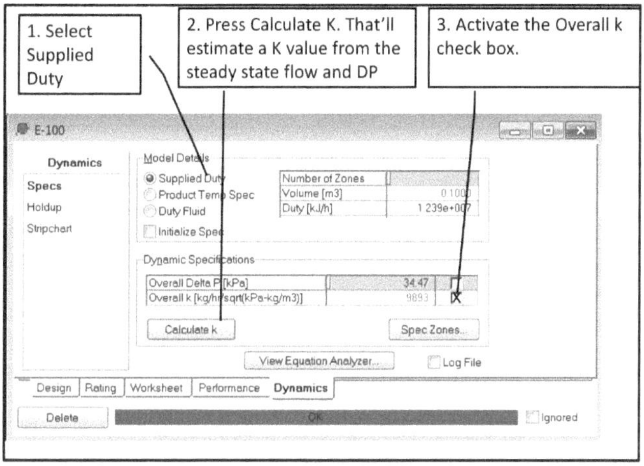

FIGURE 15.13 Heat exchanger dynamic specifications.

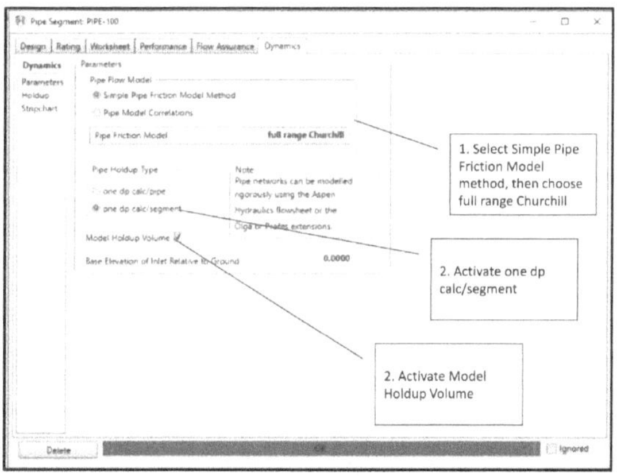

FIGURE 15.14 Pipe pressure drop and volume definitions.

of 6.43 min. Figs. 15.16 and 15.17 illustrate the configuration of LIC-100. Table 15.6 summarizes the specifications of all controllers used in this model.

Note that unlike FIC-100 and PIC-100, the final control element for TIC-100 is not a valve, but an energy stream instead. This simplification is made since the steam system is not being modeled; this is illustrated in Fig. 15.18.

15.4.2.7 Stream pressure boundaries within the battery limit

In this example, all battery limit stream pressures will be specified and fixed. This will define the pressure gradient across the process. Aspen HYSYS has a useful way to track stream pressure flow specifications, and it will color its streams based on the pressure-flow specification. Detailed steps to enable this are shown in Figs. 15.19 and 15.20.

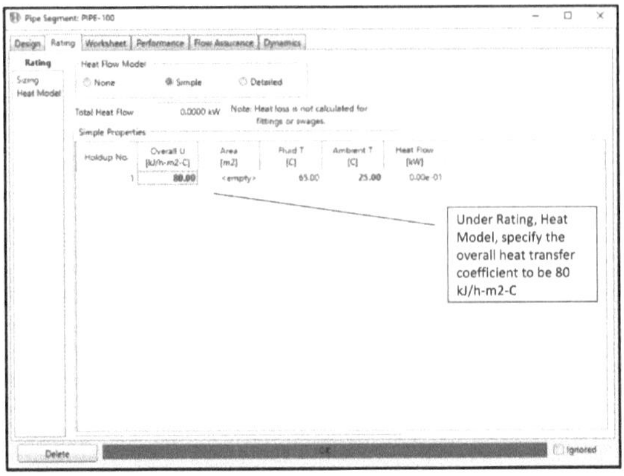

FIGURE 15.15 Pipe heat transfer definition.

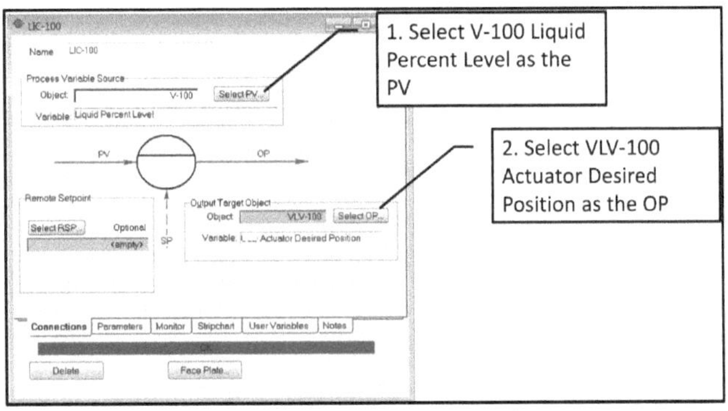

FIGURE 15.16 Controller connections.

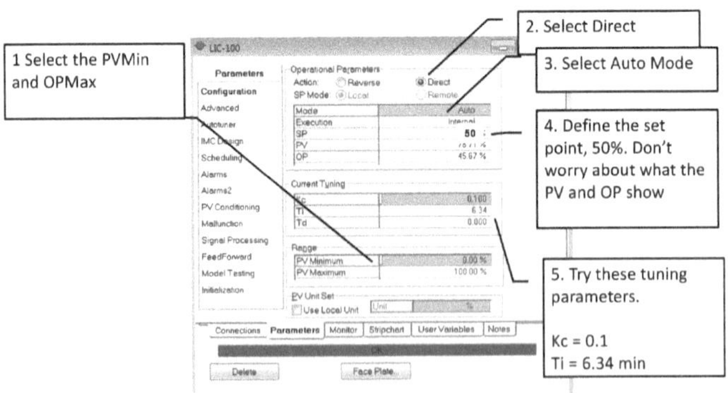

FIGURE 15.17 Controller tuning parameters.

TABLE 15.6 FIC, PIC, and TIC controller definitions.

Tag	Process/ measured variable (PV)	Units	Operating/ manipulated variable (OP)	Set point (SP)	Action	Minimum process variable (PVMin)	Maximum process variable (PVMax)	K_c	T_i
LIC-100	V-100 liquid present level	%	VLV-100 Act desired Pos	50	Reverse	0	100	0.1	6.34 min
FIC-100	Feed mass flow	kg/h	FCV-100 Actuator desired Position	319,411	Reverse	0	600,000	0.05	2 s
PIC-100	Stream 3 pressure	Bar abs	PCV-100 Actuator desired position	9	Direct	20	20	1	5 min
TIC-100	Stream 8 temperature	°C	Q-100 Control valve	65	Reverse	80	80	2	10 min

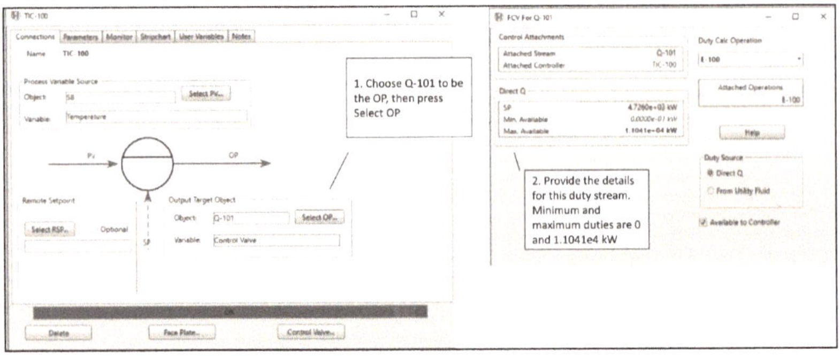

FIGURE 15.18 Definition of duty stream as final control element of TIC-100.

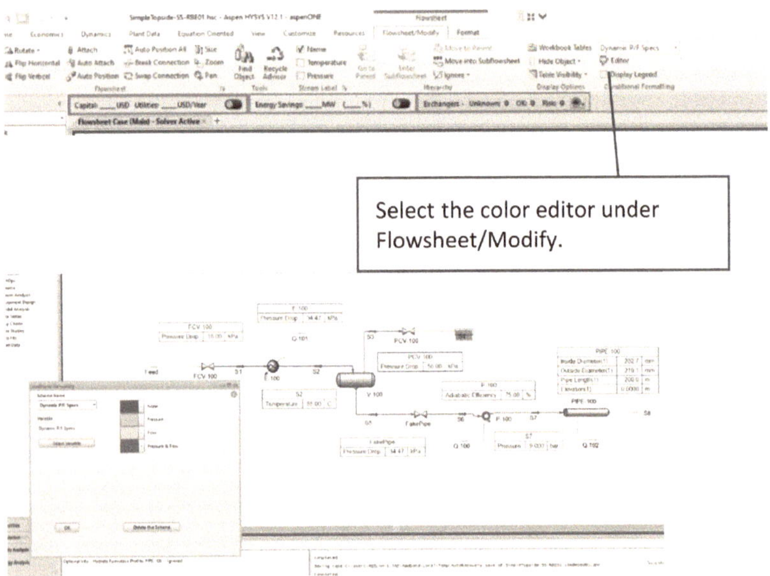

FIGURE 15.19 Color editor.

15.4.2.8 Integrator settings

Specifications of the integrator is then set according to Fig. 15.21.[7] Details of these settings were discussed in the numerical solver configuration description outlined in Section 13.3.4.3. With the completion of integrator specification, this completes the definition of dynamic parameters.[8]

7. Press Control-I to bring up the integrator and dynamic solver settings.

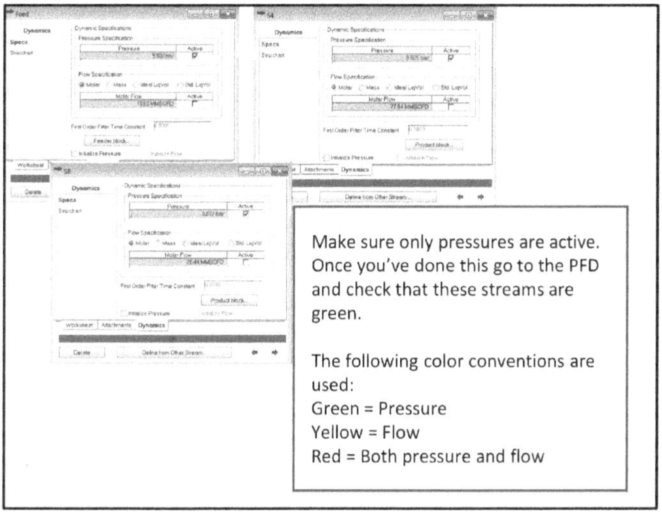

FIGURE 15.20 Defining pressure boundaries on battery limit streams.

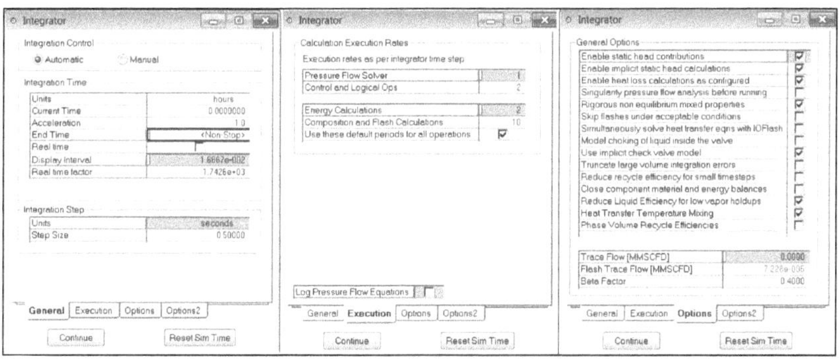

FIGURE 15.21 Integrator and dynamic solver settings.

15.4.3 Transitioning to dynamics[8]

Aspen HYSYS provides strip charts which allows the visualization of process data, temperature, pressure, flow rates, compositions, etc., similar to the strip charts in a plant control room. Fig. 15.22 illustrates the installation and setup of a strip chart.

8. Refer to TopsideDyn.hsc simulation file. Reader may also construct the model using Topside-SS-WithDynParams.hsc simulation file. Note that it is necessary to ensure the steady-state file is preserved, as once the file is transited to the dynamic environment, it cannot be seamlessly converted back to its steady-state model.

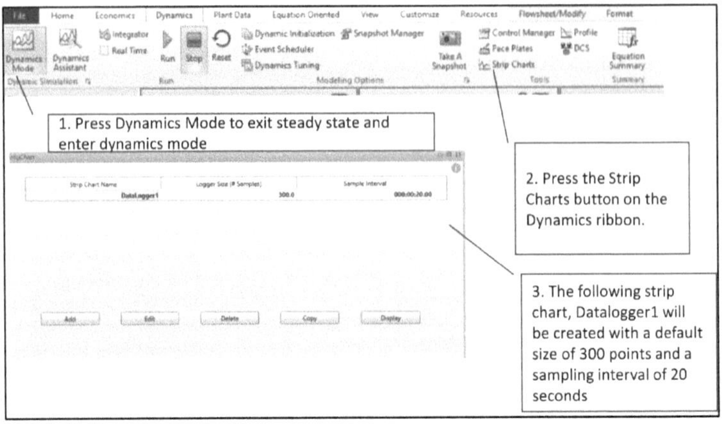

FIGURE 15.22 Switching from steady state to dynamics and setting up the strip chart.

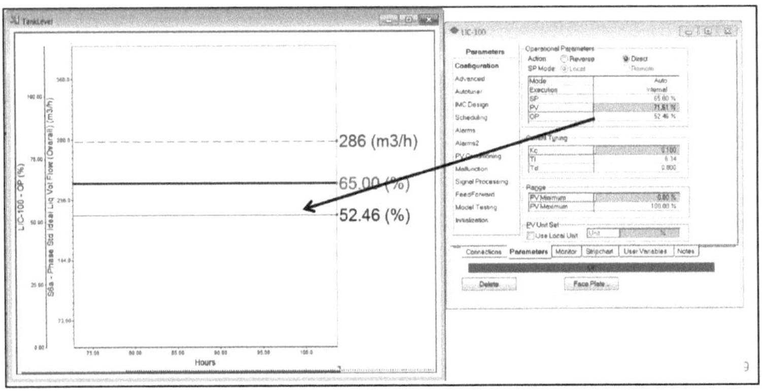

FIGURE 15.23 Adding a variable to a strip chart.

By default, the data will be collected every 20 s. A total of 300 points will be kept. This is generally insufficient, and often defined by the user to meet the requirements of the simulation.

There are several ways to add a variable to the strip chart. An intuitive way for doing so is shown in Fig. 15.23, where the object of interest is placed side by side, where its desired variable is dragged into the strip chart.

The simulation model is now complete and can be used for temperature controller tuning. This is discussed in the following section.

15.5 Using a dynamic model for process control tuning

Now that a dynamic model has been constructed, it may be used to tune the temperature controller.

Controller TIC-100 in the heated tank example is meant to ensure that the temperature of stream S8 to be maintained at 65°C, by manipulating the energy stream Q-101 in heat exchanger E-100. This is a challenge because the temperature sensor is located at stream S8, which has some distance away from E-100. The volume of vessel V-100 combined by the pressure drop, hold up and heat loss at Pipe-100 will introduce some dead-time to the system.

Various controls strategies have been deployed in the industry. In this example, a simple trial-and-error methodology will be used to determine the viable controller tuning constants for TIC-100.

15.5.1 Single loop feedback control overview

15.5.1.1 Definition of feedback control

This chapter focuses on single closed loop controls. TIC-100 in the heated tank example is a typical example for a single loop proportional, integral, and derivative (PID) controller.

For the heated tank example, the feedback control loop does the following mechanism:

1. A process variable (PV) of stream S8, i.e., temperature is measured
2. It is compared to the set point defined in TIC-100, which has been set to 65°C. Its deviation variable then calculated, i.e., ε = (SP-PV), where ε is referred to as the deviation variable
3. The controller will then send a signal based on the magnitude and sign of the deviation variable to the operating variable (OP). The latter corresponds to the "valve" in E-100 in this case, which controls the utility flowrate.

The goal of the control loop is to get the PV to equal to, or as close as possible to the SP. In other words, the control loop drives the deviation variable (ε) to zero as close as possible.

15.5.1.2 PID control

A PID control algorithm consists of three types of actions:

- **i.** Proportional—Signal is linearly proportional to the present value of ε, following Eq. (15.12).
- **ii.** Proportional and integral—An integral term is used to "kick" the output based on the accumulated errors, following Eq. (15.13).
- **iii.** Proportional, integral, and derivative—The derivative term provides some preemptive correction "future errors," following Eq. (15.14).

$$\mathrm{OP}(t) = B + K_c\varepsilon(t) \tag{15.12}$$

$$\mathrm{OP}(t) = B + K_c\varepsilon(t) + \frac{K_c}{\tau_i}\int_0^T \varepsilon(t)\,dt \tag{15.13}$$

$$\mathrm{OP}(t) = B + K_c\varepsilon(t) + \frac{K_c}{\tau_i}\int_0^T \varepsilon(t)\,dt + K_c\,\tau_d\frac{d\varepsilon(t)}{dt} \tag{15.14}$$

where:

OP(t) is the controller output at time t
B is the controller output at steady state (no error)
$\varepsilon(t)$ is the error at time t ($SP(t) - PV(t)$)
SP(t) is the setpoint at time t
PV(t) is the Process Variable at time t
K_c is the controller gain
τ_i is the controller integral time (min/repeat)
τ_d is the controller derivative time (min)

In general, an "aggressively" tuned controller has a large gain, short integral time and large derivative time.

15.5.2 Setting up the tuning scenario[9]

In this case, the set point of TIC-100 is changed from 65 to 75°C. It is necessary to determine the tuning parameters that would provide the quickest and most stable trajectory to the new set point. To facilitate this, an event will be set up to change the set point of TIC-100 from 65 to 75°C automatically at 10 min simulation time. This case will be used to investigate sets of controller gain (K_p) and reset time (τ_i) by performing a sensitivity analysis.

In the dynamics ribbon, go to the Event Scheduler and follow the instructions illustrated in Fig. 15.24.

The integrator start time needs to be set to zero. This is shown in Fig. 15.25. Upon this setting, the dynamic model is ready to be executed.

15.5.3 Running the case studies

The K_c value is varied for 0.1, 2, and 4–10 while keeping the reset time T_i at 1 min. The result of this analysis is presented in Fig. 15.26.

The result shows that with $K_c = 0.1$, the controller response is very sluggish and takes almost 1 h bring the system to the new temperature set point (not shown), the other values K_c are more aggressive, in some cases some second-order behavior or overshoot is observed (see Fig. 15.26). Table 15.7 presents the time constant for each response:

9. Refer to BaseTTuning.hsc simulation file. Readers are also encouraged to construct the file from TopsideDyn.hsc simulation file.

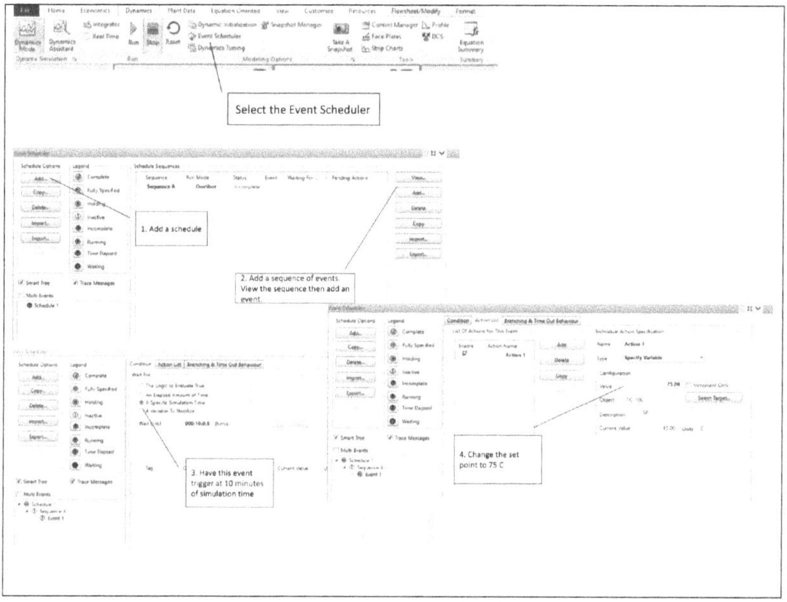

FIGURE 15.24 Event scheduler set up.

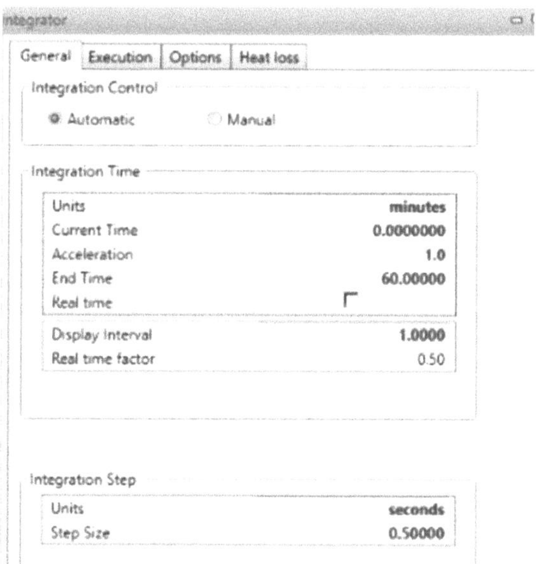

FIGURE 15.25 Integrator time setting.

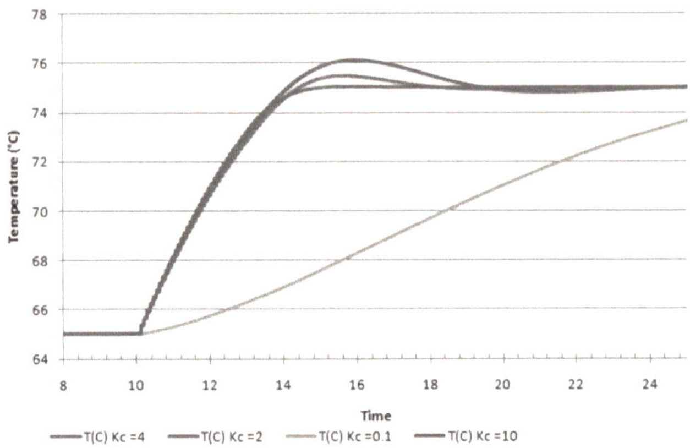

FIGURE 15.26 Results of varying controller gain on temperature response.

TABLE 15.7 Time constants for various values of K_C.

K_C	Time constant (min) (Time taken to reach 63% of target)
10	2.1
4	2.2
2	2.2
0.1	10.5

15.5.4 Other tuning strategies

Section 15.5.3 presented a simple and intuitive way to tune a controller utilizing a dynamic model. It is possible to use these simulation models to determine a system's ultimate period and gain by introducing limit cycle into the closed loop system.

15.5.4.1 Ziegler-Nichols

The Ziegler-Nichols strategy for the closed-loop method is given as follows:

1. Switch the controller to automatic with some low gain (no integral or derivative action)
2. Increase the controller gain until a constant amplitude limit cycle is established. As shown in Fig. 15.27, the limit cycle yields an *ultimate gain* (K_u), while the period between successive oscillation peaks is referred to as the *ultimate period* (P_u).

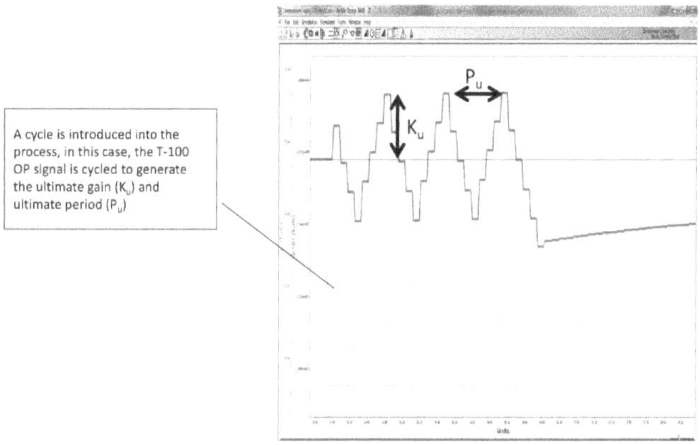

FIGURE 15.27 Limit cycle method to determine ultimate gain (K_u) and period (P_u).

TABLE 15.8 Zeigler-Nichols tuning parameters.

	K_c	τ_i (min/repeat)	τ_d (min)
Proportional	$K_u/2$		
Proportional-integral	$K_u/2.2$	$P_u/1.2$	
Proportional-integral-derivative	$K_u/1.7$	$P_u/2$	$P_u/8$

Once the ultimate period and gain for the system are determined, controller gain, reset time, and integral time can be defined as per Table 15.8.

15.5.4.2 Auto-tune variation (ATV) technique

The auto-tune variation (ATV) technique of Åström and Hagglund (1984) is similar except a perturbation is introduced to the output signal (OP). For the heated tank example, TIC-100 could be put on manual, while the valve opening is stepped up and down by a given percentage (h), in which some magnitude sufficient to elicit an appreciable response from the process (a), while not destabilizing it. For example, the OP could be increased and decreased by 5%, respectively.

The cycle may be done manually, using an event scheduler or a periodic transfer function. The ultimate gain and period for the ATV method are given by Eq. (15.15), while corresponding controller gain and reset time are given by Eq. (15.16).

$$K_u = \frac{4h}{\pi a} \tag{15.15}$$

$$K_p = \frac{K_u}{3.2} \text{ and } \tau_i = \frac{2.2}{P_u} \tag{15.16}$$

where:

h is the normalized amplitude of output signal (OP) change
a is the normalized amplitude of process or measured variable (PV) change
P_u is the period of limit cycle

The tuning parameters determined by the ATV method are significantly less aggressive (have less tendency to overshoot) than those determined by the Ziegler-Nichols strategy. This is because the latter is better suited for electromechanical systems, while the ATV method was developed for minimizing controller overshoot in fluid processing systems, where material and thermal hold-up is common.

15.6 Conclusion

While the use of transfer functions and frequency domain analysis have utility in industry, these systems oversimplify chemical processes and often do not provide adequate resolution on system responses. The current commercial dynamic simulation software are powerful tools in the chemical engineer's arsenal. They enable chemical engineers to model rigorous complex process interactions and responses in user friendly and intuitive environments. This chapter should give the reader a starting point to begin constructing dynamic models and using them to design robust control systems and troubleshoot operational issues.

Exercises

1. Pump warning
   It's likely your pump is throwing a warning (it may be yellow) in dynamics, despite static head being enabled. It is detecting vapor in the suction. Why is this the case and what do you need to do?
2. Integral time constant
   Repeat the tuning case in Section 15.5.3. However, the K_c value is kept at 2. Investigate the T_i value of 0.5, 1, 5, and 10 min.
3. Level control
   Apart from vapor liquid separation, separation vessels (slug catchers, buffer tanks) also serve to provide process buffering and to prevent disturbances from propagating downstream. In Fig. 15.6, vessel V-100 is used for this purpose. Its volume is used to attenuate process disturbances. The tuning of LIC-101 in this case will be based on an Averaging Level Control strategy, where the controller gain and reset time are defined by Eq. (15.17).

$$K_p T_i = 4\tau_{hu} \tag{15.17}$$

where:

K_p is the controller gain
T_i is the integral time constant (min)
τ_{hu} is the residence time (Vol/Max Flow (min))

Use the model in Section 15.5.2 [10] with a liquid level of 45% as a start. The level is then increased to 55%. The reaction of the controller is observed. Find the best K_p and T_i values that will allow the movement from one steady state to another without disrupting the downstream processes.

HINT: Residence time is first calculated. A large gain (e.g., 10) is attempted. The integral time is calculated using the averaging level control criteria and the system response is observed. Next, a small gain (e.g., 0.1) is attempted. The corresponding integral time is calculated and the system response is observed.

References

Aspentech Ltd, 2021. Aspen HYSYS user manual. www.aspentech.com.
Åström, K.J., Hagglund, T., 1984. Automatic tuning of simple regulators with specifications on phase and amplified margins. Automatica 20, 645.
Stephanopoulos, G., 1984. USA, chemical process control: An introductory to theory and practice. Prentice-Hall Inc.

Further reading

Foo, D.C.Y., Chemmangattuvalappil, N., Ng, D.K.S., Elyas, R., Chen, C.-L., Elms, R.D., Lee, H.-Y., Chien, I.-L., Chong, S., Chong, C.H., 2017. Chemical engineering process simulation. Elsevier, Netherlands.

10. Start with the BaseTTuning.hsc simulation file.

Chapter 16

Basics of process simulation with Aspen Plus*

John Frederick D. Tapia
De La Salle University, Manila, Philippines

Chapter outline

This chapter provides a step-by-step guide to perform process simulation of an integrated process flowsheets using Aspen Plus. The concept of simulation applied here is based on sequential modular approach and follows the onion model for flowsheet synthesis (see Chapter 1 for details). The simulation is illustrated using a case study on *n*-octane production.

16.1 Example on *n*-octane production

The production of *n*-octane (C_8H_{18}) (Foo et al., 2005) is summarized in Fig. 16.1. The process is like that of the process given in Example 1.1 with slight modification in the information provided for the distillation column. The simulation steps include the component setup and thermodynamic model in the Aspen properties environment, and the reactor, separation, and recycle

*Simulation and other electronic files of this chapter are made available on a companion website; URL: https://www.elsevier.com/books-and-journals/book-companion/9780323901680.

Chemical Engineering Process Simulation. https://doi.org/10.1016/B978-0-323-90168-0.00016-0

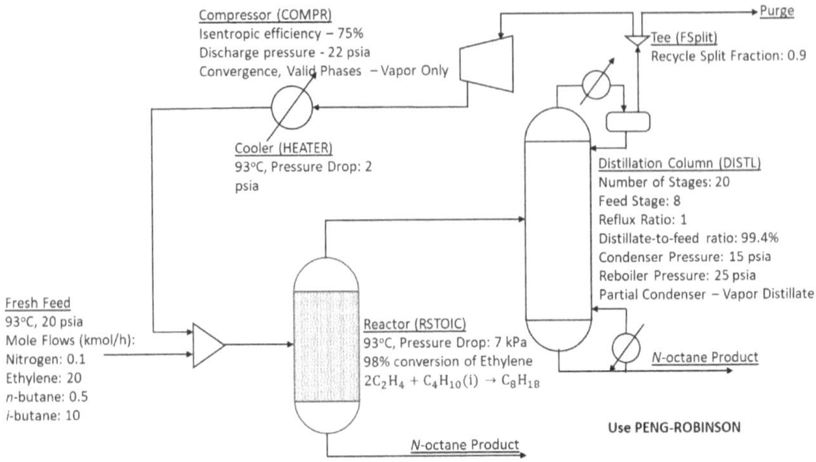

FIGURE 16.1 *n*-Octane production, with important process specifications.

system modeling in Aspen simulation environment. Each step in the simulation is discussed in the following subsections.

16.1.1 Stage 1: simulation setup in properties environment

To set up a process simulation flowsheet, the first two steps are to select components and the thermodynamic properties within Aspen properties environment. There are two ways where components may be entered, i.e., enter component names directly (option A), or search for the intended component from the database (option B). This is illustrated in Fig. 16.2. Aspen Plus automatically populates the information in the rows once the component ID is typed and matches the compounds in the database (option A). On the other hand, the *Find Compounds* dialogue (option B) can be used to allow searching of the component in the Aspen Plus database.

Once the components are specified, the thermodynamic model is then chosen.[1] It can be specified by navigating to the *Specifications* under the *Methods* folder. For the simulation of *n*-octane production, the model that will be used is the Peng-Robinson equation of state since the components involved are hydrocarbons. Steps for specifying the models are shown in Fig. 16.3. After specifying the thermodynamic model, its binary parameters can be viewed. Detailed steps for doing so are shown in Fig. 16.3, with the binary parameters shown in Fig. 16.4.

16.1.2 Stage 2: modeling of reactor in Simulation environment

The next step is to proceed in the *Simulation* environment. Here, the process flowsheet will be implemented using the unit operations model. For the

1. See Chapter 3 for detailed discussion on thermodynamic selection.

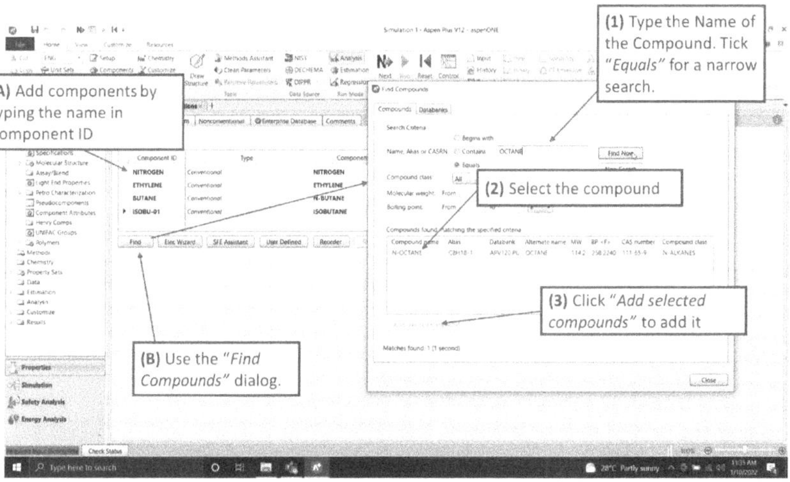

FIGURE 16.2 Adding compounds in Aspen properties environment.

reactor, the model to be used is *RStoic*, where the basis of the output stream is the fractional conversion and for the reactor the model to be used is *Distl*, a shortcut distillation column model which uses the distillate-to-feed ratio as basis. Fig. 16.4 shows the step to enter the *Simulation* environment. The specifications for the reactor and the separator is shown in Table 16.1.

Based on the onion model (see Chapter 1 for details), the synthesis of a process starts with design of the reactor. In Aspen Plus, different types of reactor models may be used, depending on the modeling requirements. These

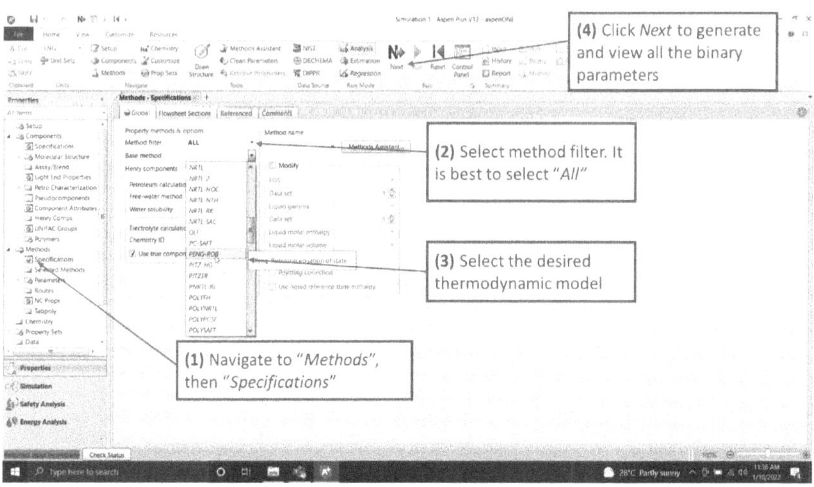

FIGURE 16.3 Selecting thermodynamic models in Aspen properties environment.

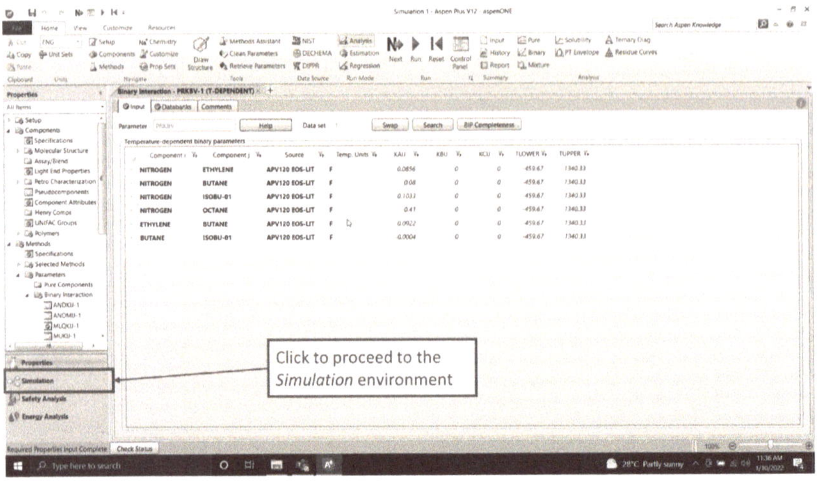

FIGURE 16.4 Summary of binary parameters and switching to *Simulation* environment.

TABLE 16.1 Specifications of the reactor and separation unit for the *n*-octane production.

Unit (*Model*)	Specifications
Reactor (*RStoic*)	$T = 93°C$, $\Delta P = -7$ kPa
	Conversion: 98% of C_2H_4
	Reaction: $2C_2H_4 + C_4H_{10}$ (i) → C_8H_{18}
Separation unit (*Distl*)	Number of stages: 20
	Feed stage: 8
	Reflux ratio: 1
	Distillate-to-feed ratio: 99.4%
	Condenser pressure: 15 psia
	Reboiler pressure: 25 psia
	Partial condenser—Vapor distillate

reactor models can be classified into three main types: (1) simple reactors such as *RStoic* and *RYield*, (2) equilibrium-based reactors such as *RGibbs* and *REquili*, and (3) rate-based reactors such as *RCSTR*, *RBatch*, and *RPlug*. The first type of reactor models requires only mole- or mass-related information, such as fractional conversion (*RStoic*) and product mass yield (*RYield*). On the other hand, information about the Gibbs free energy is used for equilibrium-based reactors. The earlier-mentioned rate-based reactors requires kinetic data to determine the output yield. Certain reactor models such as *RGibbs* and *RYield* do not require a stoichiometric relation to be specified, while other models require it.

For this simulation, *RStoic* model is used. Once the feed stream is connected to the reactor, its specifications are provided. The feed stream is supplied at 93°C and 20 psia (see Fig. 16.1) for component flowrates. Detailed steps for setting up the feed stream are illustrated in Fig. 16.5 and the specifications for the feed is shown in Table 16.2. Two product streams are then added, i.e., a liquid and a vapor stream (see detailed steps in Fig. 16.6A. The operating conditions for the reactor is then specified as illustrated in Fig. 16.6B). The reactor is operating at 93°C with a pressure drop of 7 kPa. Note that phases of the product streams are also specified in the reactor, especially when two or more product streams are involved.

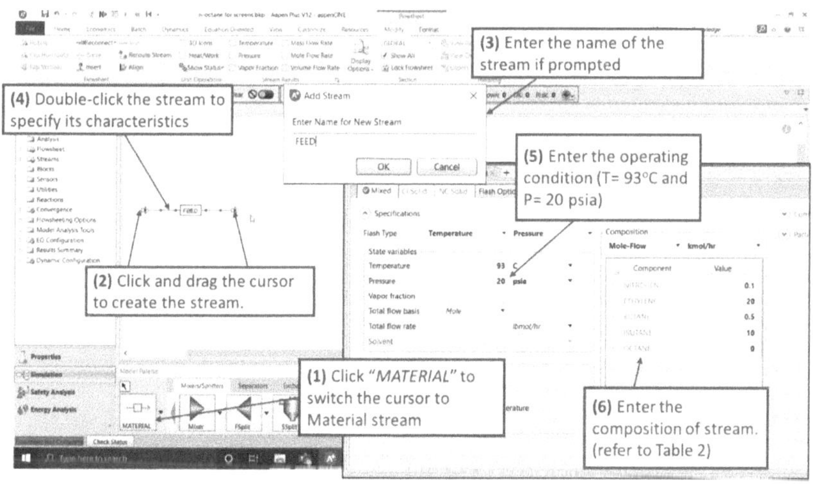

FIGURE 16.5 Adding stream and specifying the feed stream in *Simulation* environment.

TABLE 16.2 Specifications of the feed stream for *n*-octane production.

Component flowrates (kmol/h)	Pressure	Temperature
$N_2 = 0.1$	20 psia	93°C
$C_2H_4 = 20$		
$C_4H_{10} = 0.5$		
C_4H_{10} (*i*) $= 0.5$		

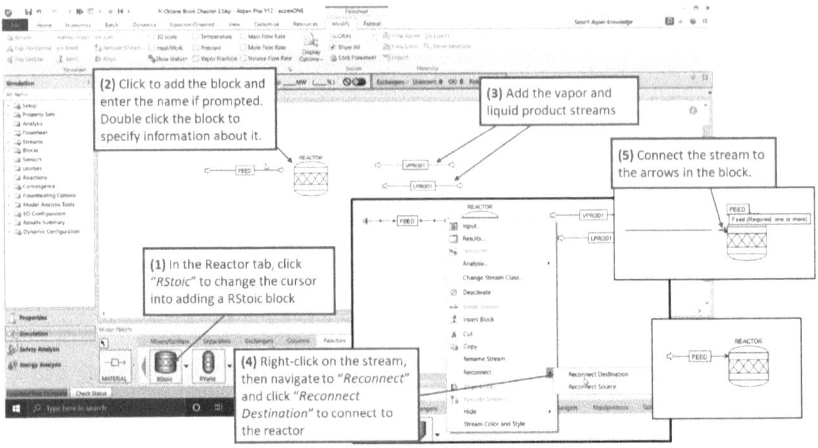

(a)

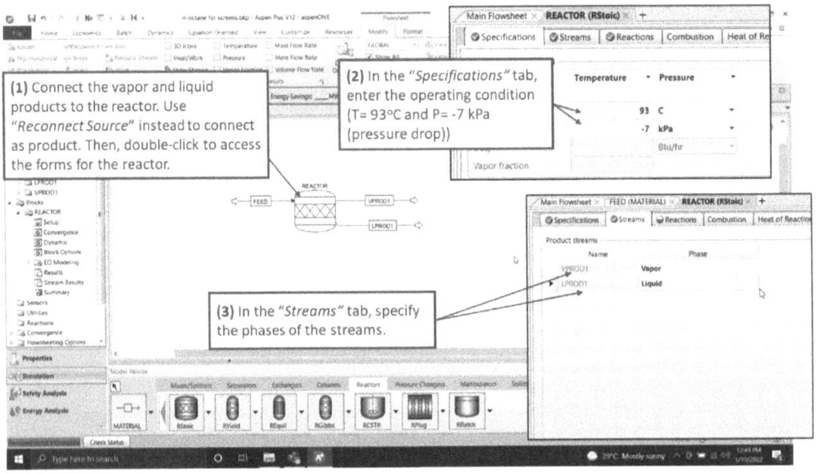

(b)

FIGURE 16.6 Setting-up the reactor model: (A) adding reactor and its products streams in the flowsheet; (B) specifying the operating conditions of the reactor.

The *RStoic* model in Aspen Plus requires the reaction stoichiometry to be specified (detailed steps are shown in Fig. 16.7). For this case, 2 moles of ethylene (C_2H_4) will react to 1 mole of isobutane (i-C_4H_{10}) to produce 1 mole of *n*-octane (C_8H_{18}). As specified in Fig. 16.1, the output will be based on 98% conversion of ethylene.

16.1.3 Stage 3: modeling of separator in Simulation environment

Upon the completion of reactor modeling, simulation is next performed for separation and recycle systems, i.e., layer 2 of the onion model (see Chapter 1

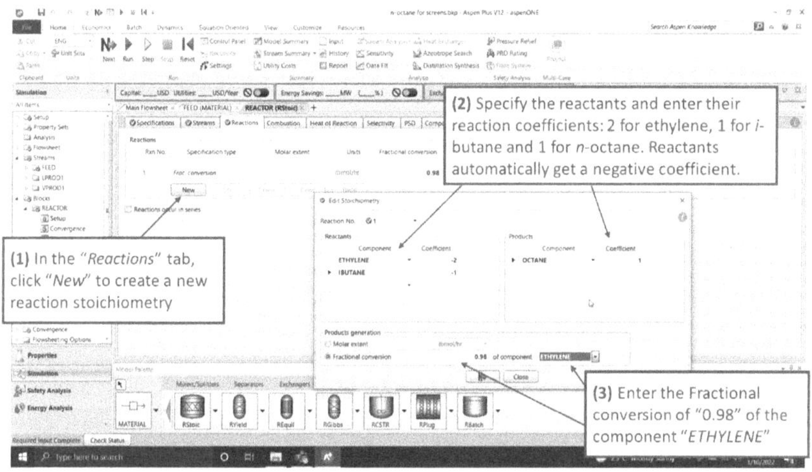

FIGURE 16.7 Specifying reaction stoichiometry for the reactor.

for details). Hence, simulation is next performed for product separation for the reactor effluent. The simulation model reveals that the desired product, i.e., *n*-octane is found to be more concentrated in the liquid phase than that in the vapor phase.[2] Aspen Plus performs flash calculation whenever two products streams of liquid and vapor phases are indicated in the simulation. It is expected that the concentration of component with lower boiling point (i.e., *n*-octane) is present in higher concentration in the liquid phase. A separator is required not only to recover the remaining product but also to allow further reaction of the remaining reactants. In this case, the *Distl* model in Aspen Plus is used to simulate the separator unit.

The *Distl* model uses the Edmister approach for shortcut distillation (Edmister, 1957). The input information for the separation is based on the number of stages, feed location, and the distillate-to-feed ratio. The phase of the distillate is also configured. Important parameters for the *Distl* model are given in Table 16.1. Detailed steps for setting up the *Distl* model is illustrated in Fig. 16.8, while Fig. 16.9 shows detailed steps for its configuration. The pressure in the column decreases from the reboiler at 25 psia to the condenser at 15 psia. For this simulation, the number of trays is set to 20, with feed tray located at the eighth stage. The simulation intends to further recover more *n*-octane in the reactor product and recycle the rest, thus, the distillate-to-feed ratio is set at 0.994.

The bottom stream of the column contains the *n*-octane product while the distillate contains most of the unreacted ethylene. The latter is recycled so to increase the conversion of *n*-octane. However, due to the presence of inert

2. Readers should verify this with their own simulation model.

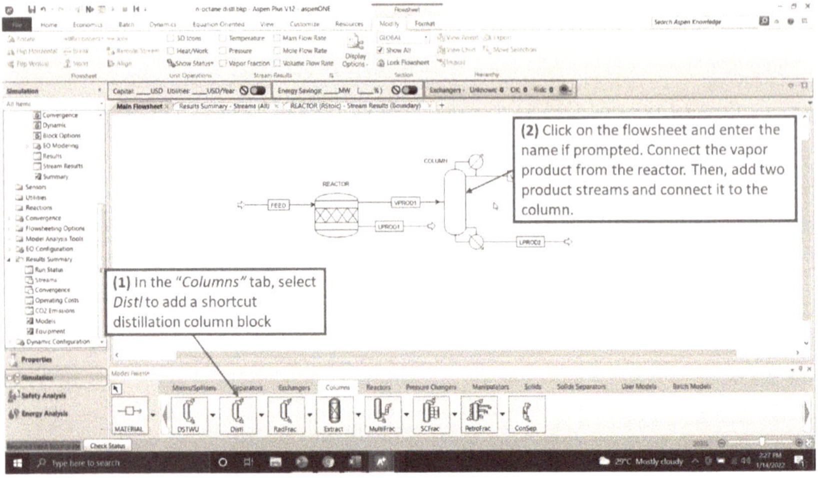

FIGURE 16.8 Adding distillation column and connecting the streams.

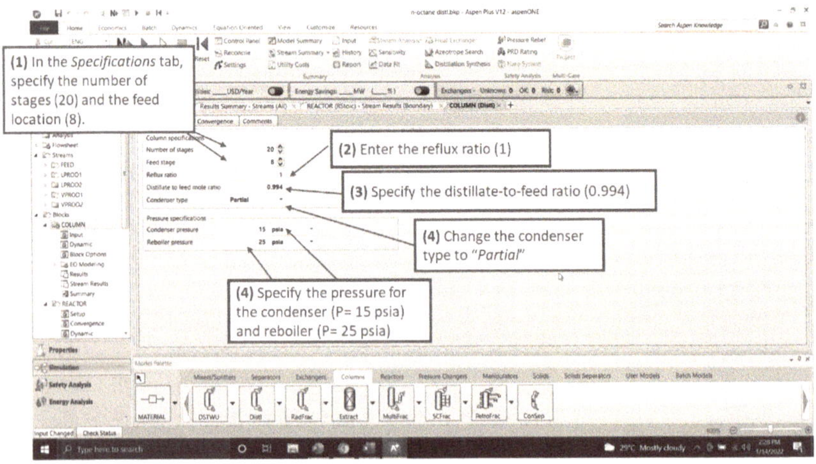

FIGURE 16.9 Specifying information for the distillation column.

components such as nitrogen and n-butane, a portion of the distillate should be purged to prevent accumulation in the system. Next section illustrates how recycling in Aspen Plus is implemented.

16.1.4 Stage 4: modeling of recycling in the Simulation environment

The recycle stream from distillation will be mixed with the fresh feed before entering the reactor. Hence, the recycle stream must have the same operating condition as that of the fresh feed. Thus, its temperature and pressure must be changed to match those of the fresh feed. To do this, the following steps are done:

1. The recycle stream is compressed to 22 psia using a compressor. Energy in the form of mechanical work is used to achieve the desired pressure.
2. The compressed stream is cooled to 93°C, while the stream experiences a pressure drop of 2 psia in the heat exchanger. Energy in the form of heat is involved in the temperature change.

Splitting of the column distillate stream to recycle and purge streams is illustrated in Fig. 16.10. Here a split fraction of 0.9 in the recycle is used,

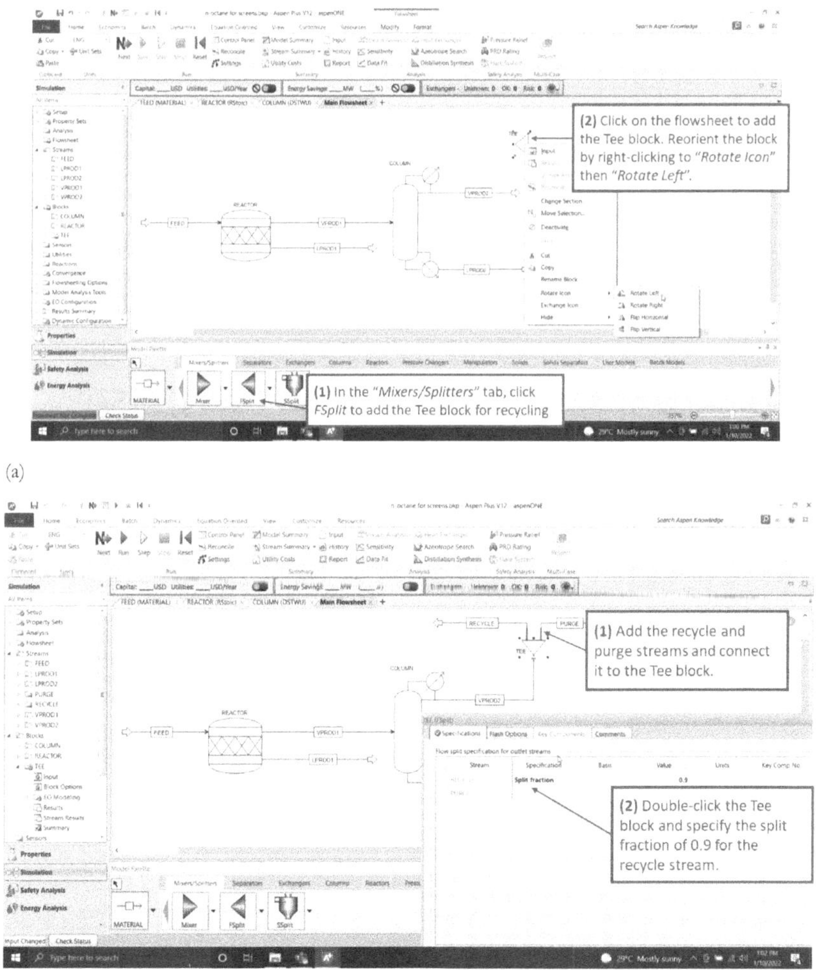

FIGURE 16.10 Setting up the recycle and purge streams using the *FSplit* block: (A) Adding and orienting the tee (*FSplit*) block; (B) Specifying the split fraction in the tee.

TABLE 16.3 Specifications of the splitter, compressor, and cooler for *n*-octane production.

Unit (*Model*)	Specifications
Tee (*FSplit*)	Recycle split fraction: 0.9
Compressor (*Compr*)	Compressor type: Isentropic
	Discharge pressure: 22 psia
	Isentropic efficiency—0.75
	Valid phases—Vapor only
Cooler (*Heater*)	$T = 93°C$, $\Delta P = -2$ psia

which signifies that 90% of the distillate is recycled (see Table 16.3). The *FSplit* block used in Aspen Plus assumes that the separation is homogenous, that is, the temperature, pressure, and composition in the two product streams are the same. For the compressor, which is implemented using the *Compr* model in Aspen Plus, the compressor type is isentropic with an efficiency equal to 0.75 (see Table 16.3). For the cooler, which is implemented using the *Heater* block in Aspen, the absolute pressure can be negative, where it is interpreted as pressure drop in the simulation. The output of the heater can then be mixed with the fresh feed. Detailed steps for simulating the compressor and cooler blocks are illustrated in Figs. 16.11 and 16.12, respectively. Table 16.3 shows the specifications of the units used for the recycle stream.

Once the conditions of the recycle stream are met, it is mixed with the fresh feed through a mixer block. Setting up the mixer block is illustrated in Fig. 16.13. The steps included disconnecting the fresh feed to the reactor and then, a new stream will be created that is connected from the mixer to the reactor.

After finishing the connection, the flowsheet can now be executed in Aspen Plus. The complete flowsheet diagram with the instruction on running the simulation is shown in Fig. 16.14. The status bar in Aspen Plus will indicate that the input is complete and can now be solved. Solving the simulation will generate the characteristics of the output stream as well as the energy requirements of each block. The Stream Results can be accessed by navigating in Results Summary then to Streams in the navigation pane. The result for the

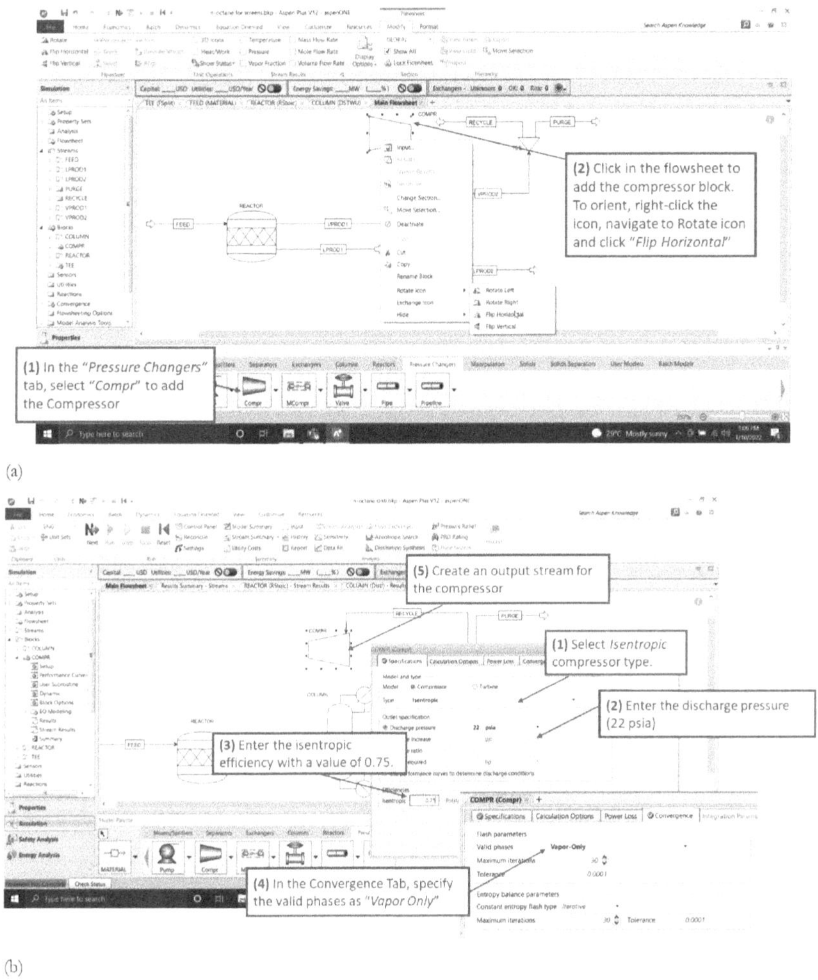

FIGURE 16.11 Setting up the compressor (*Compr*) block: (A) Adding and orienting the compressor (*Compr*) block; (B) Specifying the operating parameters of the compressor.

n-octane production is shown in Fig. 16.15 where the overall stream properties is shown in Fig. 16.15A and the mass composition is shown Fig. 16.15B.

The final flowsheet is illustrated in Fig. 16.16, where the temperature and pressure of all stream are shown, as well as the molar flow rates of *n*-octane and ethylene (other components are not shown for brevity). The resulting overall conversion of ethylene to *n*-octane is 99.7% (calculated based on

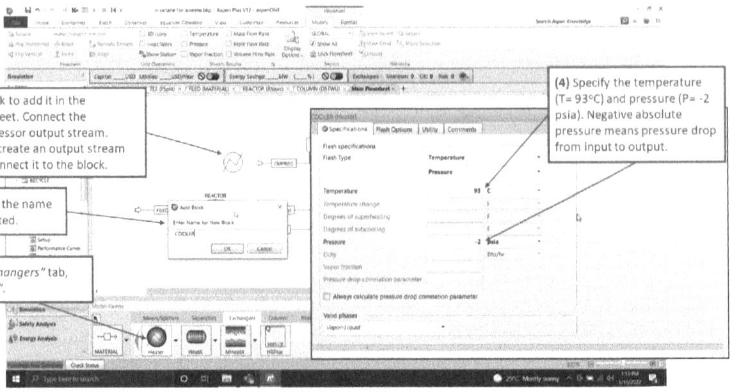

FIGURE 16.12 Setting up the cooler (*Heater*) block.

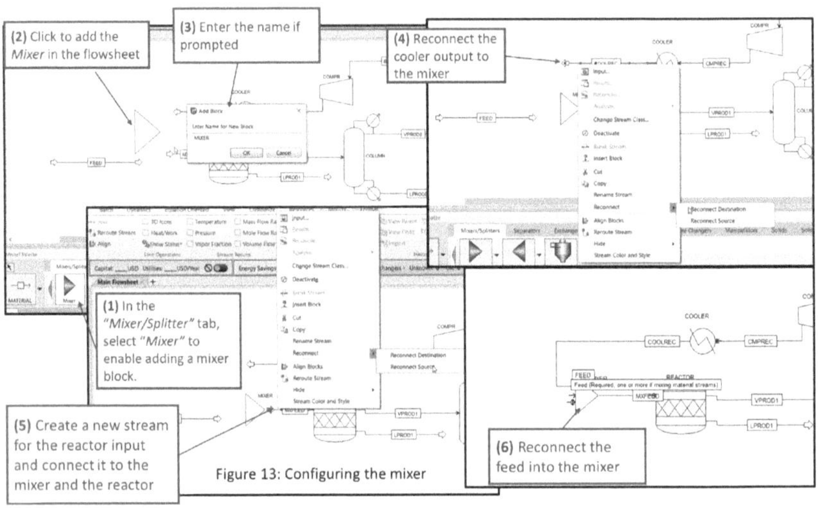

FIGURE 16.13 Setting up the mixer for the fresh feed and the recycle stream.

19.94 kmol/h *n*-octane product form from 20 kmol/h of ethylene in the feed stream).The cooler removes 3579 Btu/h of heat from the recycle stream while cooling it from 99.53 to 93°C.

Now suppose that the available feed is at 30°C and 24 psia instead of the original 93°C and 20 psia. A heater with a pressure drop of 4 psia can be added to meet the specifications of the original feed (see Table 16.4). For the simulation, the original stream to the mixer stream will be maintained and a new feed stream

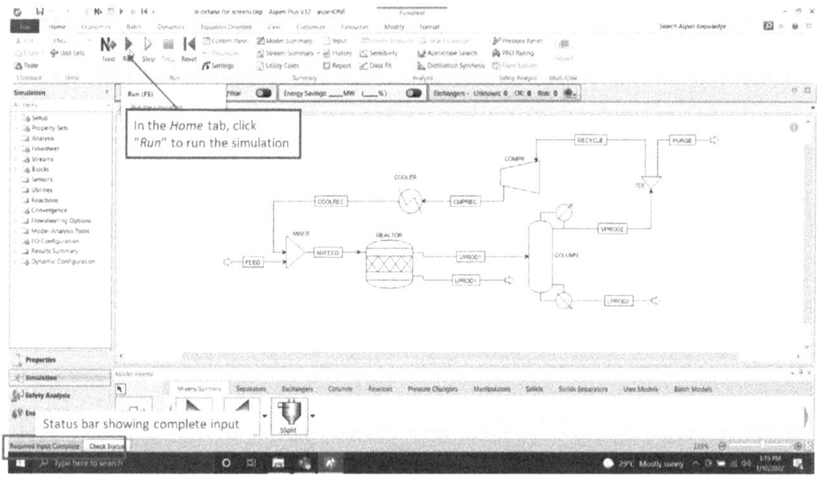

FIGURE 16.14 Flowsheet configuration and running the simulation.

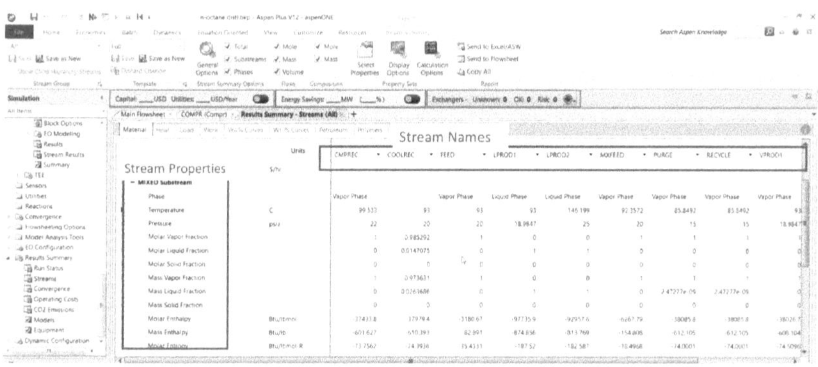

(a)

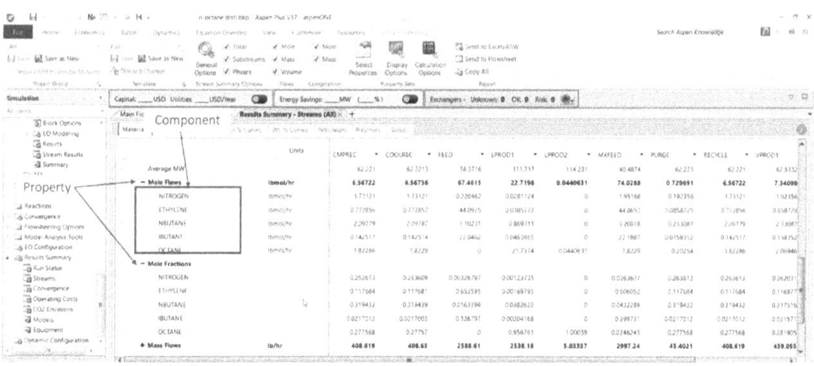

(b)

FIGURE 16.15 Stream results of the *n*-octane production implemented in Aspen Plus: (A) Resulting stream properties generated by the simulation; (B) Resulting stream properties generated by the simulation.

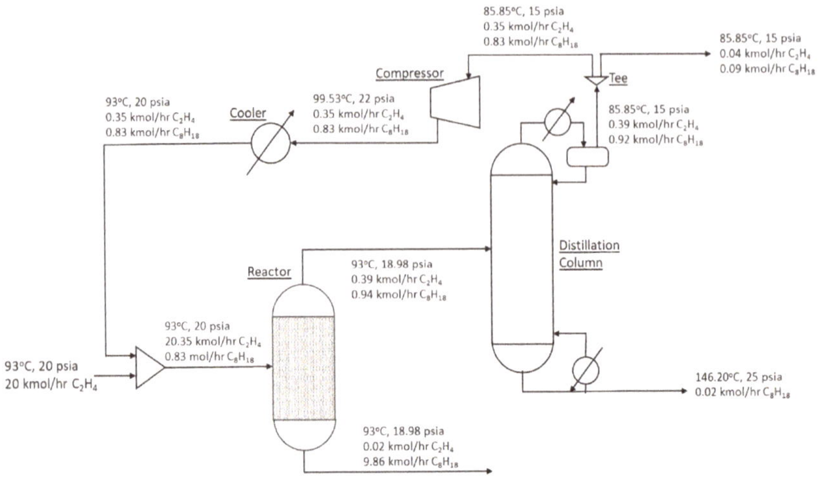

FIGURE 16.16 Final flowsheet of the *n*-octane production simulated in Aspen Plus.

TABLE 16.4 Specifications of the feed heater for *n*-octane production.

Unit (*Model*)	Specifications
Heater (*Heater*)	$T = 93°C$
	$\Delta P = -4$ psia
Heat exchanger (*HeateX*)	Model fidelity—Shortcut
	Hot stream outlet temperature = 93°C
	Hot side pressure drop = −2 psia

will be made as the input to the heater. The new stream will be based on the original feed stream. Fig. 16.17 shows the steps on how the new feed stream is created and the new feed characteristics are specified, while Fig. 16.18 shows how the heater is configured. For the cooler and the heater, the heat involved can be facilitated by using utilities (e.g., cooling water, steam). The utilities used in the process can be reduced through heat integration, which is discussed in the following section. Running the simulation again and the heating duty for the heater is determined as 123,800 Btu/h.

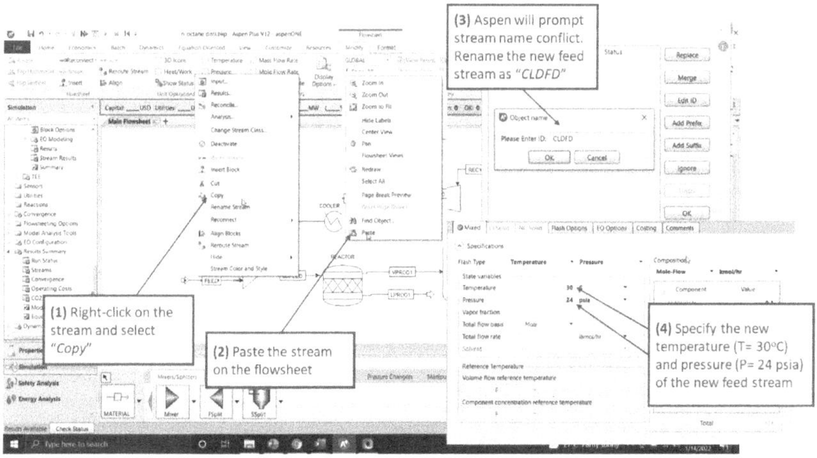

FIGURE 16.17 Steps for duplicating streams in Aspen Plus.

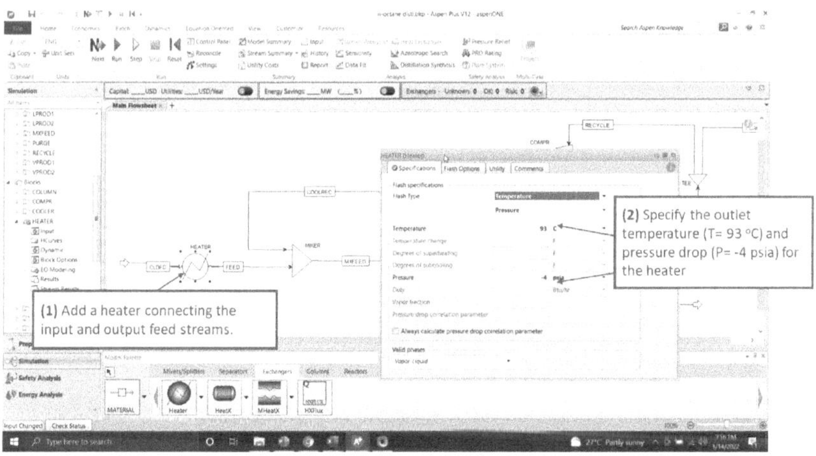

FIGURE 16.18 Adding heater for the feed stream.

16.1.5 Stage 5: simulation of heat integration scheme

Once the simulation of separation and recycle systems are completed, the simulation exercise next moves to layer 3 of the onion model, i.e., heat recovery system. The change in feed condition requires additional heat to be added in the process. This heat will be added to the feed stream to achieve the desired operating condition before being mixed with the recycle stream. Note that 3579 Btu/h of heat have been removed from recycle stream (see Fig. 16.19). The heat removed from the recycle stream can be recovered to the heater to reduce its heating duty of in raising the feed stream temperature.

Fig. 16.20A shows the detailed steps on how the cooler is replaced by the heat exchanger. For this simulation, a shortcut heat exchanger model, i.e., *Heatx* is used; the latter only performs heat balances. Note that the connections for the heat exchanger is important as it will affect the result of the simulation. The hot side of the heat exchanger should be allocated with the recycle stream as it will be cooled. On the other hand, the feed stream should take the cold side as it will be heated using the recycle stream. To adapt the specifications of the cooler used previously, the hot outlet stream has a temperature of 93°C, and a pressure drop of 2 psia. Fig. 16.20B shows the detailed steps (see Table 16.4 for the specifications for the heat exchanger).

The final flowsheet for *n*-octane production is shown in Fig. 16.21. After the heat exchanger, the feed stream has its temperature rise to 32°C (from 30°C, see Fig. 16.21). It is then heated to 93°C using the heater. The latter has lower heating duty of 120,300 Btu/h, instead of the original duty of 123,800 Btu/h (see Fig. 16.21), i.e., a saving of 2.8% is achieved through heat integration. Besides, the use of cooler (Fig. 16.16) is removed. For larger processes where cooling and heating are extensive, heat integration will result in larger utility and cost savings.

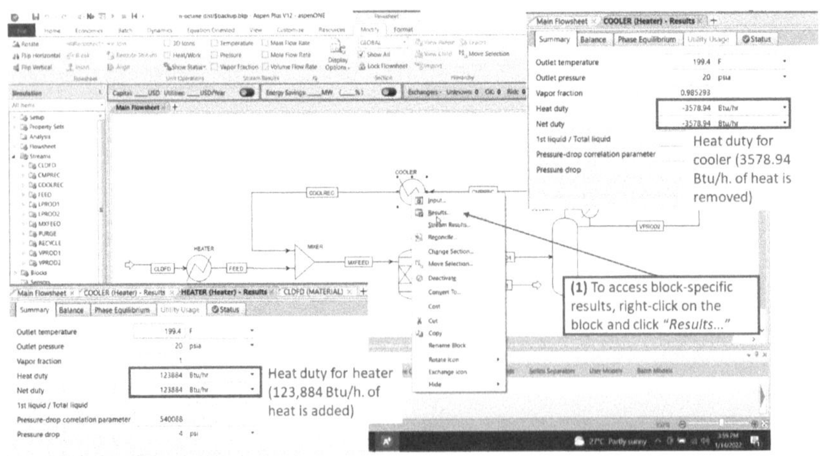

FIGURE 16.19 Results for the duties of heater and the cooler.

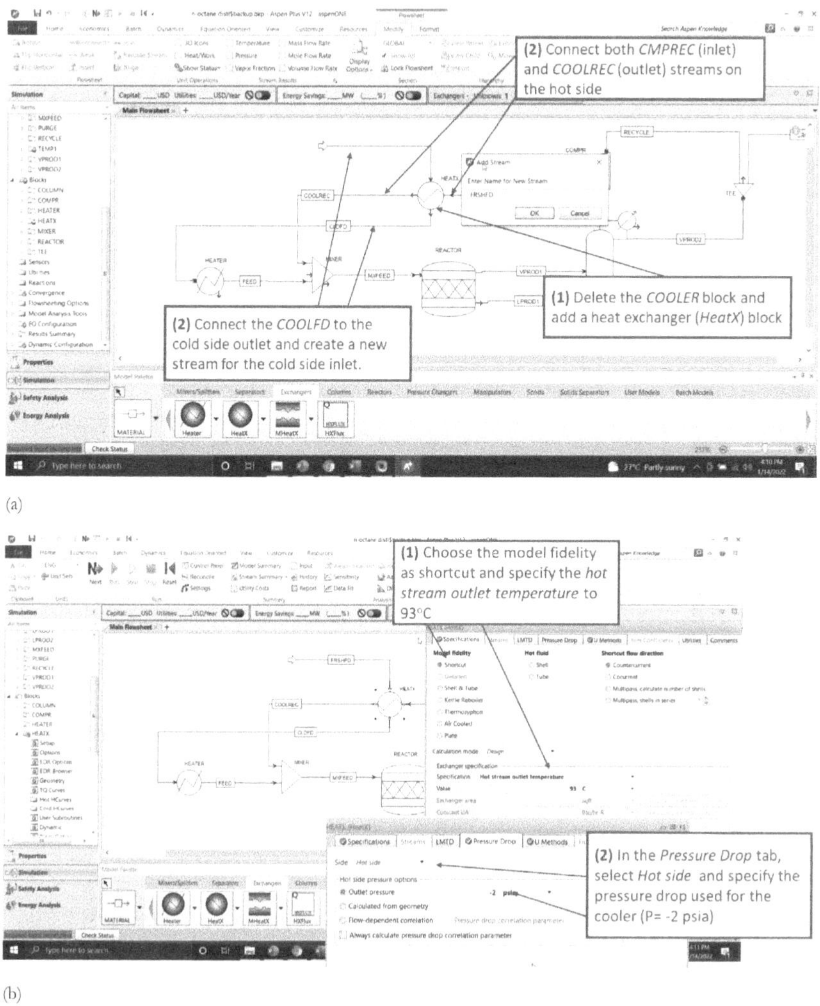

FIGURE 16.20 Configuring the heat exchanger as a replacement for the cooler: (A) Replacing the cooler with heat exchanger; (B) Specifying the operating conditions for the heat exchanger.

16.2 Summary of the *n*-octane simulation

The use of Aspen Plus for process synthesis and design is performed by applying it to *n*-octane production. The concept of onion-model is used wherein the following aspects are considered:

1. The conversion of ethylene and isobutane is performed using a reactor modeled using *RStoic* model. The specified extent of reaction is based on reaction stoichiometry with the reactant conversion

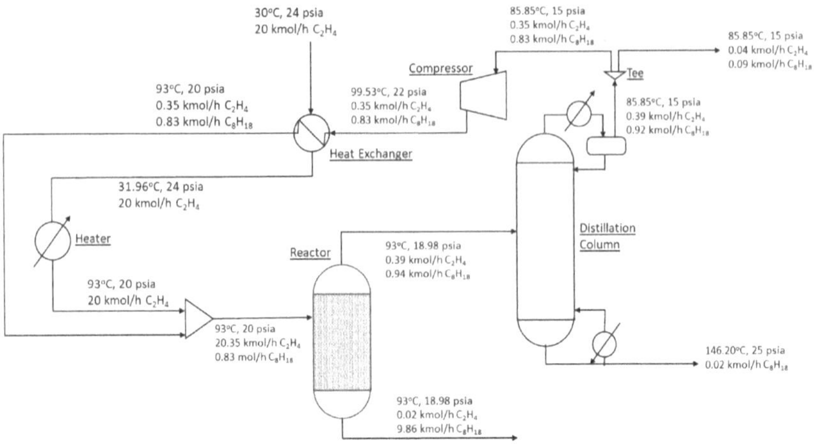

FIGURE 16.21 Final flowsheet of *n*-octane production considering heat integration.

2. The separation of the desired product is performed using a distillation column modeled by *Distl* model. The separation is based on a specified column design (i.e., number of stages, feed location, and reflux ratio) and product yield (i.e., distillate-to-feed ratio).
3. A recycle stream is added to further convert the unreacted components. The operating condition of the recycle is modified to match with the feed. A compressor and a cooler is involved in this step.
4. Heat integration is illustrated using Aspen Plus' *Heatx* model. Utility savings is achieved through heat integration.

References

Edmister, W.C., 1957. Absorption and stripping-factor functions for distillation calculations by manual- and digital-computer methods. AIChE Journal 3 (2), 165–171.

Foo, D.C.Y., Manan, Z.A., Selvan, M., McGuire, M.L., 2005. Integrate process simulation and process synthesis. Chemical Engineering Progress 101 (10), 25–29.

Further readings

Adams, T.A., 2018. Learn Aspen Plus® in 24 hours. McGraw-Hill education, ISBN 978-1-260-11646-5.

Schefflan, R., 2011. Teach yourself the basics of Aspen Plus™. John Wiley & Sons, 978-1118980590.

Chapter 17

Design and evaluation of alternative processes for the manufacturing of bio-jet fuel (BJF) intermediate*

Bor-Yih Yu
Department of Chemical Engineering, National Taiwan University, Taipei, Taiwan

Chapter outline

*Simulation and other electronic files of this chapter are made available on a companion website; URL: https://www.elsevier.com/books-and-journals/book-companion/9780323901680

Chemical Engineering Process Simulation. https://doi.org/10.1016/B978-0-323-90168-0.00017-2

17.1 Introduction

Biomass is one of the most important elements in renewable energy sector. It is an excellent alternative energy source in the era of depleting fossil fuels and global concern toward anthropogenic CO_2 emissions. According to a latest data reported by REN21 (Renewables, 2021), the modern renewables have been estimated to account for 11.2% of the total final energy consumption (TFEC) in year 2019, in which some 5.1% comes from the modern biomass technologies. Among them, 3.7% is used for heating (e.g., in industry or buildings), 1.0% as transportation fuels (e.g., bio-ethanol or bio-diesel), and the remaining for generating electricity (Renewables, 2021).

Technologies converting biomass into various bio-fuels and bio-sourced chemicals has attracted keen interests, especially those utilization routes involving lignocellulosic biomass. The latter comprising polysaccharides (cellulose, hemi-cellulose) and aromatic polymers (lignin) is perhaps the most abundant biomass around the world. Many kinds of agricultural wastes (i.e., bagasse, cereal straws), forest residues (i.e., pine wood), energy crops, are included in this category.

Producing jet-fuel range hydrocarbons from biomass has been considered promising for reducing CO_2 emission from the aviation industry. There are some widely known techniques for this purpose. One of the most established pathway is the gasification of biomass into syngas, which then undergoes Fischer-Tropsch (F-T) reaction to become liquid fuels (Bond et al., 2014; Olcay et al., 2018). As the process is very capitally intensive, it is only economically viable when it is operated in large scale. Its main difficulty lies with the low heating value of biomass, and their inconsistent amount and quality. Another important method is the hydro-processing of ester and fatty acid (HEFA), which uses the waste oil or algae as the source material. The HEFA process is able to produce liquid fuels of good quality (Bond et al., 2014; Olcay et al., 2018). However, it is also hindered by the unstable availability of source materials, and the lack of economic attractiveness (Bond et al., 2014; Olcay et al., 2018). Hence, this method has not been commercialized.

In a novel process scheme, the jet-fuel range hydrocarbons were successfully synthesized through aldol condensation of furfural and acetone (Huber et al., 2005). Furfural has been a popular platform chemical with growing demand, which can be commercially derived from biomass (Zeitsch, 2000). However, the relevant technologies have been slowly improved since its first development in 1920s. Using the steam-stripping reactor, the process faces the bottleneck with low furfural yield (59.5%), and difficulty of separating furfural from a very diluted solution (refers to the ROSENLEW process) (Zeitsch, 2000; Nhien et al., 2016; Silva et al., 2017). This is also the major cause of high furfural price (i.e., 1200−1600 USD/t). To improve the process economics, various studies have been reported, which include the purification of reactor effluent in a dividing-wall column (Nhien et al., 2016) or a hybrid extraction-distillation configuration (Nhien et al., 2017), as well as using a gaseous stream (e.g., nitrogen, air, carbon dioxide) for stripping (Agirrezabal-Telleria et al., 2013; Krzelj et al., 2019).

On the other hand, aldol condensation is a base-catalyzed reaction which occurs between two ketones, two aldehydes, or one aldehyde and one ketone. As reported, the products of the aldol condensation between furfural and acetone are mainly 4-(2-furyl)-s-buten-2-one (FAc) and 1,4-pentadien-3-one,1,5,-di-2-furanyl (F_2Ac), which is could be used as the intermediate for jet-fuel hydrocarbons (O'Neill et al., 2014; Huber et al., 2006; Huber and Dumesic, 2006). Many kinds of catalyst, such as sodium hydroxide (NaOH) solution(O'Neill et al., 2014), activate dolomite(O'Neill et al., 2014), mixed metal oxides(Desai and Yadav, 2019), graphite, MOFs, and zeolites(Kikhtyanin et al., 2015), or the homogeneous liquid amidine (1,8-Diazabicyclo[5.4.0]undec-7-ene, DBU) (JIang et al., 2018) have been reported for use.

This chapter aims to develop and evaluate alternate process schemes for producing bio-jet fuel (BJF) intermediate. The whole process has three steps, i.e., conversion of biomass to furfural (step 1), purification of furfural (step 2) and condensation of furfural/acetone (step 3). The differences in between the schemes are mainly in Step 1 (i.e., steam- or air-stripping) and Step 2 (i.e., separation based on distillation or hybrid extraction-distillation). Each scheme of the process is rigorously simulated with consideration on their physical properties and intersectional heat integration. Finally, these proposed schemes are compared by yearly unit manufacturing cost (YUMC) of the BJF intermediate, and a throughput-independent CO_2 emission rate (CO_2-e).

17.2 Overview

17.2.1 Components and physical properties

The simulation model in this chapter was performed using Aspen Plus V11. The simulation model consists of 12 components, namely furfural (F), acetone (Ac), water, methanol (MeOH), furfural-acetone (4-(2-furyl)-3-buten-2-one, FAc), furfural-acetone-furfural (1,4-pantadien-3-one,1,5-di-2-furanyl, F_2Ac), 4-(furan-2-yl)-4-hydroxybutan-2-one (FAcOH), xylose (XYL), carbon-graphite (C), benzene (BEN), oxygen (O_2), and nitrogen (N_2). Among them, only FAcOH is not found in the Aspen built-in component database, and is added as a user-defined component. The pure-component physical properties of FAcOH are provided in Tables A.1 and A.2 in the appendix. On the other hand, properties of other pure-component are found in the built-in database, and are not listed here.

To describe the phase equilibrium in a mixture comprised of polar components, an activity coefficient model has to be selected in simulation. In this work, NRTL model is used, for its wide capability of describing both vapor-liquid and liquid-liquid equilibrium. Herein, all the Aspen built-in parameters are used, while COMSO-based calculations are performed to estimate the missing parameters (Lin and Sandler, 2002; Liang et al., 2019). All binary interaction coefficients used for simulation are provided in Table A.3 in the appendix.

17.2.2 Reaction kinetics of the aldol condensation reaction

Herein, the kinetic expressions proposed by O'Neill et al. (2014) and the modified version by Yu and Tsai (2020) were used for simulation. The reactions included are listed in Eqs. (17.1)–(17.3):

$$F + Ac \overset{k_1/k_{-1}}{\leftrightarrow} FAcOH \quad \Delta H(298\ K) = -47.2\ (kJ/mol) \tag{17.1}$$

$$FAcOH \overset{k_2}{\rightarrow} FAc + H_2O \quad \Delta H(298\ K) = 19.9\ (kJ/mol) \tag{17.2}$$

$$FAc + F \overset{k_3}{\rightarrow} F_2Ac + H_2O \quad \Delta H(298\ K) = -65.4\ (kJ/mol) \tag{17.3}$$

More specifically, Eq. (17.1) is a reversible reaction between furfural (F) and acetone (Ac) in forming the intermediate (FAcOH); Eq. (17.2) represents the dissociation of FAcOH into FAc and water; Eq. (17.3) describes the reaction between FAc furfural to become F_2Ac and water. The overall reaction (17.1)–(17.3) is exothermic.

The kinetic expressions in power law are listed in Eqs. (17.4)–(17.7), with the parameters provided in Table 17.1. In these equations, the concentration is given in unit of kmol/m^3, and the activation energy in kJ/kmol.

$$r_{1f} = k_1 \exp\left(\frac{-E_1}{RT}\right) C_F C_{Ac} \tag{17.4}$$

$$r_{1r} = k_{-1} \exp\left(\frac{-E_1}{RT}\right) C_{FAcOH} \tag{17.5}$$

$$r_2 = k_2 \exp\left(\frac{-E_2}{RT}\right) C_{FAcOH} \tag{17.6}$$

$$r_3 = k_3 \exp\left(\frac{-E_3}{RT}\right) C_F C_{FAc} \tag{17.7}$$

TABLE 17.1 The kinetic parameters (Yu and Tsai, 2020).

	Pre-exponential factor[a]	Activation energy (kJ/kmol)
k_{1f}	2.657 E3	41,300
k_{1r}	1.422 E10	79,900
k_2	2.922 E5	50,200
k_3	9.353 E−1	18,900

[a] *The unit of new k_1 and k_3 are $\frac{m^6}{kmol \cdot s}$, for k_{1r} and k_2 it are dimensionless.*

17.2.3 Economic evaluation and CO_2 emission analysis

The economics of each process scheme is compared using the yearly unit manufacturing cost (YUMC), defined as in Eqs. (17.8) and (17.9).

$$\mathrm{YUMC}\left(\frac{\mathrm{USD}}{\mathrm{kg}}\right)=\left(\frac{\mathrm{TCC}}{3}+\mathrm{TOC}\right)\times\frac{1}{\mathrm{PR}} \tag{17.8}$$

$$\mathrm{TOC}=\mathrm{TUC}-\mathrm{TEP}+\mathrm{TRMC}+\mathrm{TWWTC} \tag{17.9}$$

where TCC is the total capital cost; TOC is the total operating cost; TUC is the total utility cost; TEP is the total extra profit; TRMC is the total raw material cost, and TWWTC is the total waste water treatment cost. All cost terms are in 1000 US dollar (kUSD). Besides, PR is the production rate of BJF intermediate. Detailed description of each term is provided in Tables 17.2, 17.3 and 17.4. Note that in this work, the cost equations are manually input to an Excel

TABLE 17.2 Items included in economic evaluation.

		Calculation
Total capital cost (TCC)		TCC of vessels, heat exchangers, compressors were taken from Luyben (2011) TCC of furnace was taken from Turton et al. (2009) See Table 17.3 for details
Total operating cost (TOC)	Total utility cost (TUC)	General utilities were taken from Luyben (2011) Fuel: bituminous coal See Table 17.3 for details
	Total extra profit (TEP)	1. The recovered steam generates revenue equal to 90% of its selling price 2. Methanol: 0.276 USD/kg (Nhien et al., 2017)
	Total raw material cost (TRMC)	1. Bagasse: 0.05 USD/kg (Xing et al., 2010) 2. Acetone: 0.95 USD/kg (ICIS.com, 2017) 3. Benzene: 1.33 USD/kg (ICIS.com, 2014)
	Total waste water treatment cost (TWWTC)	Set to 56 USD/1000 m^3 for the waste water stream with purity less than 99.99 wt%
Production rate of the BJF intermediate (PR)		In kg/year, assuming 8000 h of annual operation

TABLE 17.3 The correlations for calculating capital costs (Luyben, 2011; Turton et al., 2009; Douglas, 1988).

Equipment	Capital cost correlation
Vessels (reactor, flash, decanter, column)	$TCC = 17,640 \times D^{1.066} \times L^{0.802} \times \frac{(1+F'_p)}{3.18}$ $F'_p = \frac{\frac{(P+1)D}{2 \times 0.9 \times S - 1.2(P+1)} + CA}{0.0015}$ where P: pressure of vessel (bar); D: vessel diameter (m); CA: corrosion tolerance (=0.00315 m); S: maximum allowed working pressure for the carbon steel (=944 bar)
Heat exchangers (including condenser and reboiler)	$TCC = 7296 \times A^{0.65}$ where A: heat exchange area (m^2)
Compressors	$TCC = 7429.2 \times hp^{0.82}$ where hp: compressor work in horsepower
Furnace	$TCC = 10^\wedge(K_1 - K_2 LOG10(W_f) + K_3 (LOG10(W_f))^\wedge 2)$ where W_f: energy rate of fuel consumption (kW); while $K_1 = 7.3488$, $K_2 = 1.1666$, $K_3 = 0.2028$

TABLE 17.4 The unit price and unit CO_2 emission of utilities employed in this work (Luyben, 2011; Turton et al., 2009; Douglas, 1988).

Utilities	Price (USD/GJ)	CO_2 emission amount(kg-CO_2/GJ)
2-bar steam (120°C)[a]	7.05	66.68 (Gadalla et al., 2005)
6-bar steam (160°C)	7.78	72.86 (Gadalla et al., 2005)
11-bar steam (184°C)	8.22	76.60 (Gadalla et al., 2005)
42-bar steam (254°C)	9.88	91.14 (Gadalla et al., 2005)
Cooling water	0.354	–
Chilled water	4.43	–
Refrigerant (–20°C)	7.88	–
Electricity	16.9	120.06 (Schmidt et al., 2001)
Coal[b]	3.143	88.98 (Yu and Chien, 2015)[a]

[a]*The saturation temperature.*
[b]*Calculated based on the Kaltim Prima Coal.*

file for economic evaluation. However, the readers are encouraged to explore the use of MATLAB/Aspen package for economic evaluation.[1]

Similar as in Chapter 18, the indirect CO_2 emissions generated from consuming heating utilities is considered. A throughput independent term, denoted as "CO_2-e", is defined as the kg of CO_2 emission per kg of BJF intermediate produced. The contribution of CO_2 emission from different grades of utility is calculated based on the data in Table 17.4.

17.3 Process development

The overall process to produce BJF intermediate from biomass comprises of three steps, as shown in Fig. 17.1. In the first step, biomass is converted into a raw furfural product in a stripping reactor. In the following step, furfural is purified from the reactor effluent. In the final step, furfural is reacted with acetone through aldol condensation reaction to form the BJF intermediate.

In the following, three combinations of different techniques used in step 1 (steam stripping or air stripping) and step 2 (distillation-based or HED based) are demonstrated. The specifications of techniques are outlined in Table 17.5. Note that for each scheme, the solvent-less aldol condensation process is used in step 3 in order to reduce energy consumption.

17.3.1 Scheme 1[2]

This scheme represents the conventional production route by incorporating steam-stripping reactor for furfural production, and distillation-based furfural separation. The simulated flowsheet is depicted in Fig. 17.2. The detailed simulation settings are provided as follows.

17.3.1.1 Steam-stripping

The investigations on the steam-stripping reactor is based on the operation of the ROSENLEW reactor (Zeitsch, 2000), which processes bagasse as the feed. In this operation, the pretreated bagasse, which is rich in pentosan and contains

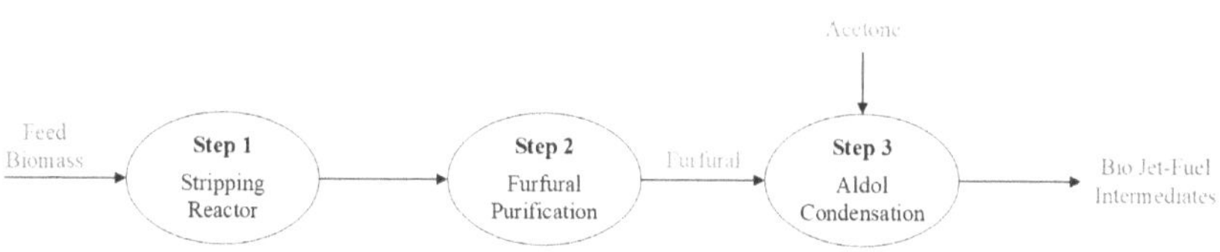

FIGURE 17.1 Steps for producing BJF from biomass.

1. See details in Chapter 18.
2. Please refer to the simulation files in the "*Scheme-1*" folder.

TABLE 17.5 Combination of different techniques.

Scheme	Step 1: stripping reactor	Step 2: furfural purification	Step 3: aldol condensation
1	Steam-stripping	Distillation	Solvent-less
2	Steam-stripping	HED	Solvent-less
3	Air-stripping	HED	Solvent-less

large amount of moisture, is fed into the reactor from the top. On the other hand, a great amount of superheated steam at 265°C and 10 bar is fed from the bottom. During the reaction, pentosan is hydrolyzed to monomeric pentose, and then dehydrated to furfural. Apart from acting as a reactant, the superheated steam also strips the formed furfural to the vapor phase in order to prevent it from deactivation. Some organic acids (i.e., acetic acid, formic acid, etc.) are also formed during the reaction, which serve as homogeneous catalyst for the acid-catalyzed hydrolysis reactions. After the reaction, the overhead vapor is condensed and purified, where furfural product is obtained. Note that owing to massive amount of steam used in the stripping reactor, this vapor stream typically contains only 2–6 wt% furfural. This poses challenges in purification.

Due to the lack of reaction kinetics, the reactor is modeled based on operation data reported by Zeitsch (2000). As reported, the bagasse is fed at 6290 kg/h (25.3% pentosane and 49.1 wt% moisture). The production rate of furfural was reported as 350 kg/h, corresponding to a yield 59.5% (Zeitsch, 2000). The steam consumption was estimated as 30 kg per kg of furfural produced. As for the side products, acetic acid, formic acid, and the light components are formed at the yield of 0.489, 0.050, 0.043 kg per kg furfural, respectively (Silva et al., 2017).

In modeling, the ROSENLEW reactor was scaled up to produce 8094.0 kg/h furfural in the stripping reactor. Some assumptions were used to simplify the modeling on the reactor. Firstly, the reactor was modeled using a 17-tray RADFRAC module in Aspen Plus. The tray number was determined from the original reactor height (12 m) (Zeitsch, 2000) with assumption of 2 ft (0.6096 m) tray spacing. The column pressure is set at 10 bar, and each stage encountered a pressure drop of 0.0068 atm. Secondly, due to the lack of kinetics, the reactor was modeled as a column separating the post-reaction mixture, with its composition described based on the aforementioned yields. The superheated steam (10 bar and 265°C), and those species in the liquid phase in the real reactor (i.e., moisture in the feed bagasse, formed furfural, and unreacted pentose) were assumed to enter from the column bottom. On the other hand, the unreacted bagasse, and all the side products formed during the

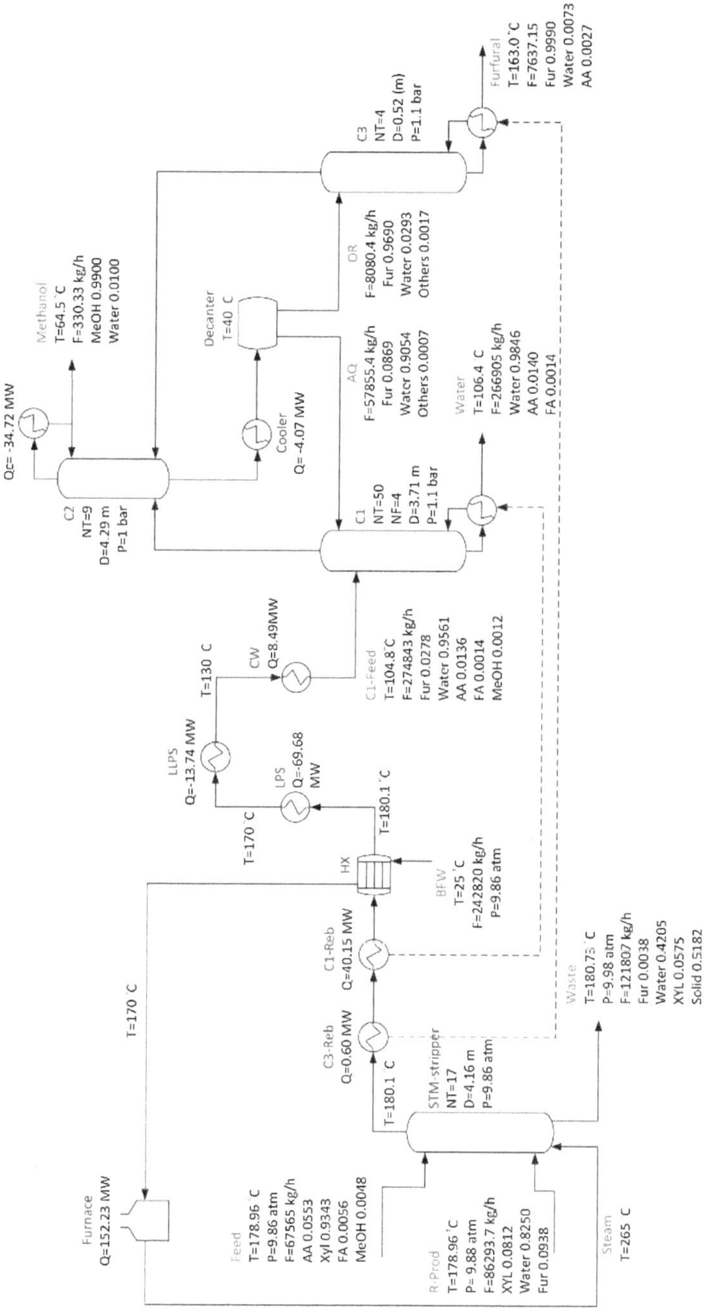

FIGURE 17.2 Flowsheet of Scheme 1, which consists of steps 1 and 2.

reaction were assumed to enter from the top. For simplicity, the unreacted bagasse was modeled using solid carbon (C), pentose was represented by xylose, and the light components was assumed to be methanol. Under this operation, the overhead vapor contains the stripped furfural and large amount of steam, while the bottom liquid is consisted of the residue bagasse (simulated as carbon), unreacted pentose, and liquid water.

As shown in Fig. 17.2, the overhead vapor contains 2.78 wt% furfural and 95.61 wt% water, which is very close to the original data (2.80 wt% furfural and 95.57 wt% water) (Zeitsch, 2000). It can be concluded that the model describes the reactor performance well, despite of the various assumptions made.

17.3.1.2 Distillation-based furfural separation

The separation of the condensed reactor effluent ("C1-Feed" in Fig. 17.2) could be very energy-intensive. This is evidenced from the T-x-y diagram of furfural/water as depicted in Fig. 17.3, that the composition of "C1-Feed" lies in the very narrow region at the left of the heterogeneous azeotrope. For separating this kind of mixture, distillation is performed firstly in this narrow region to get the high purity water and a near-azeotropic mixture of furfural/water. Next, this near-azeotropic mixture is condensed and decanted. The high purity furfural product is then obtainable by processing the furfural-rich phase in another distillation column.

The flowsheet of the furfural purification section is depicted together with the upstream reaction section as in Fig. 17.2. Herein, three columns (C1, C2, C3) are used for furfural purification. C1 and C3 are designed as strippers,

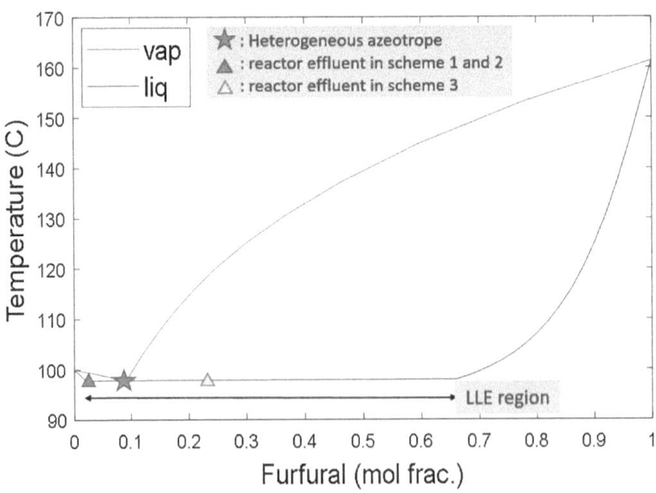

FIGURE 17.3 Txy diagram of furfural and water.

while C2 as a rectifier. The stream "C1-feed" is firstly sent to C1 to perform regular distillation to get the high purity water from the bottom, and heterogeneous azeotrope from the top. This stream as well as the overhead vapor stream from C3 are sent to C2, in which high purity methanol (99 wt%) is separated from the top. The bottom of C2 is then cooled to 47°C, and sent into a decanter to perform liquid-liquid separation. The aqueous and organic phase are then sent to the first stage of C1 and C3, respectively, to serve as liquid reflux. Note that a small amount of acid does not go to the organic phase in the decanter. Thus, it comes out with the high purity water from the C1 bottom. Further purification of this water stream is beyond the scope of this work.

As the reactor effluent contains a great amount of steam, condensing it requires the removal of a huge amount of latent heat. A good alternative for energy conservation is to carry out inter-sectional heat integration. As depicted in Fig. 17.2, condensation is firstly performed by supplying heat to C3 and C1 reboilers. Next, the partly condensed overhead vapor is used to preheat the boiler feed water from 25 to 170°C, which saves the energy required in producing superheated steam. The remaining heat is used to generate the 6-bar steam and the 2-bar steam (see Table 17.4), and is finally removed by cooling water to become saturated liquid before entering the distillation section. Hence, for a very energy-intensive process like the current one, inter-sectional heat integration is important, as it reduces the net energy demand of the whole process significantly.

The design parameters, i.e., total number of stages of each column and feed location (NF) of the stream "C1-feed", could be determined from optimization. The optimal results based on sequential iterative method could be found from a recent paper by Yu and Tsai (2020).

17.3.2 Scheme 2[3]

This scheme incorporates hybrid extraction-distillation (HED) technique for furfural separation. In general, HED utilizes the naturally occurred liquid-liquid existence region for separation. It has been widely considered as being more energy-saving as compared with distillation-based processes (Shen and Chien, 2019; Yu et al., 2017). Herein, benzene is used as extraction agent for the HED process (Nhien et al., 2017).

The proposed flowsheet containing steam-stripping reaction and furfural separation based on HED is depicted in Fig. 17.4. In this process, stream C1-feed (same as that in Fig. 17.2) is cooled to 43.5°C before it is sent to the extractor, in which 99.9% of furfural is recovered. The extract phase consisted of benzene and furfural, which is further purified to 99.9 wt% in the downstream column (COL). Prior to entering this column, the feed underwent

3. Please refer to the simulation files in the "*Scheme-2*" folder.

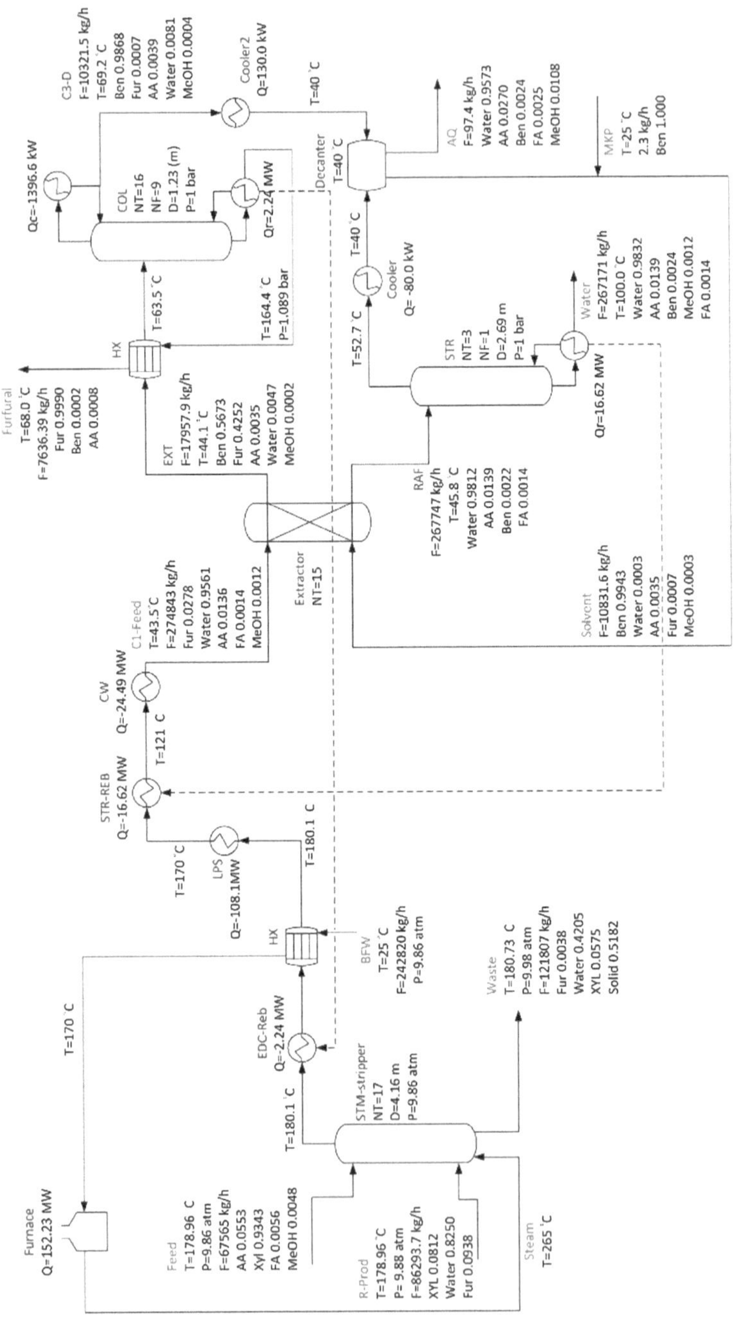

FIGURE 17.4 Flowsheet of Scheme 2, which consists of steps 1 and 2.

heat-exchange with the column bottom for pre-heating. On the other hand, the raffinate stream that contains majorly water and slight amount of benzene, and is sent to a downstream stripper (STR) for benzene recovery. Note that benzene and water forms a minimum boiling azeotrope. Hence, recovering benzene inevitably brings some water to the overhead vapor. To this end, this vapor is cooled to 40°C, and decanted along with the cooled distillate stream from COL. The organic phase from the decanter is recovered to the extraction tower. The aqueous phase and the bottom stream of STR are combined and sent to the waste water treatment section.

Similar to scheme 1, a massive amount of latent heat released via overhear vapor condensation from reactor effluent provides opportunities for heat integration. In the proposed heat integration strategy depicted in Fig. 17.4, the overhead vapor is first condensed by exchanging heat with the reboiler heat of COL. It is then cooled by heating the BFW, followed by generating 6-bar steam, and as energy source for the reboiler of STR in sequence. Finally, this overhead vapor is cooled using cooling water prior to entering the extractor.

As in scheme 1, the energy demand in step 2 (i.e., furfural purification) can be supplied using the latent heat from overhead vapor condensation. However, as the HED-based separation is less energy intensive than the distillation-based process (i.e., scheme 1: 40.75 MW; scheme 2: 18.87 MW), greater amount of steam can thus be generated, which enhances the process economics and helps to reduce CO_2 emission.

17.3.3 Scheme 3[4]

This scheme incorporates air-stripping reactor and HED-based purification for furfural production. The simulation flowsheet of the air-stripping process is illustrated in Fig. 17.5A, while that of the HED-based separation is depicted in Fig. 17.5B.

In simulating the air-stripping reactor, the basic assumptions (i.e., column configuration, operating conditions, feed locations, surrogate species used, etc.) were the same as those in the steam-stripping reactor, except for two differences. First, air replaces steam as the stripping agent, with air-to-furfural ratio set to 34.47 (Agirrezabal-Telleria et al., 2013). Second, the furfural yield was set at 70% based on experimental results reported by Krzelj et al. (2019). Due to the enhanced furfural yield, less amount of bagasse, acetic acid, formic acid, methanol, and pentose present in the reactor at the same production rate of furfural.

The process starts from pressurization of fresh air by a two-staged compression steps with intermediate cooling. The compression ratio in each

4. Please refer to the simulation files in the "*Scheme-3*" folder.

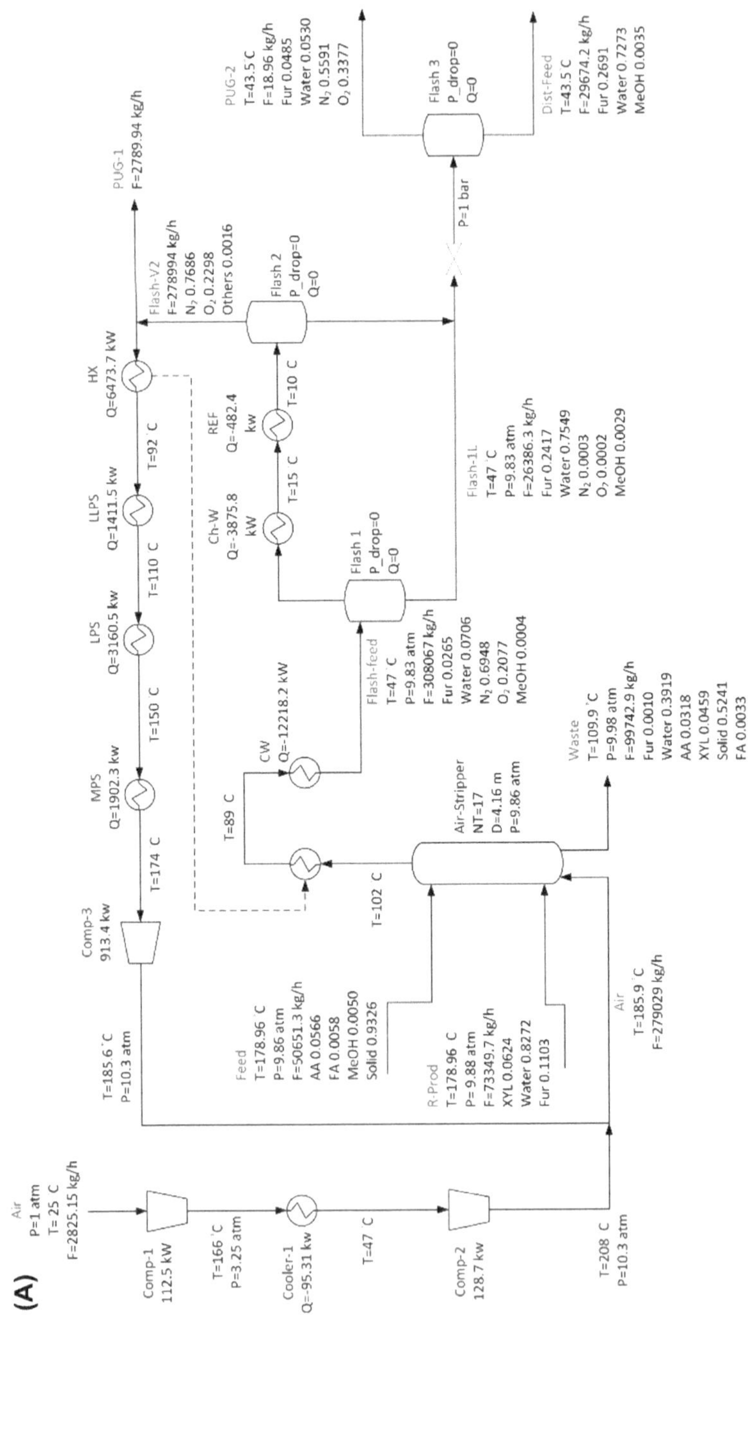

FIGURE 17.5 Simulation flowsheets of (A) air-stripping reaction section. (B) HED-based furfural separation section.

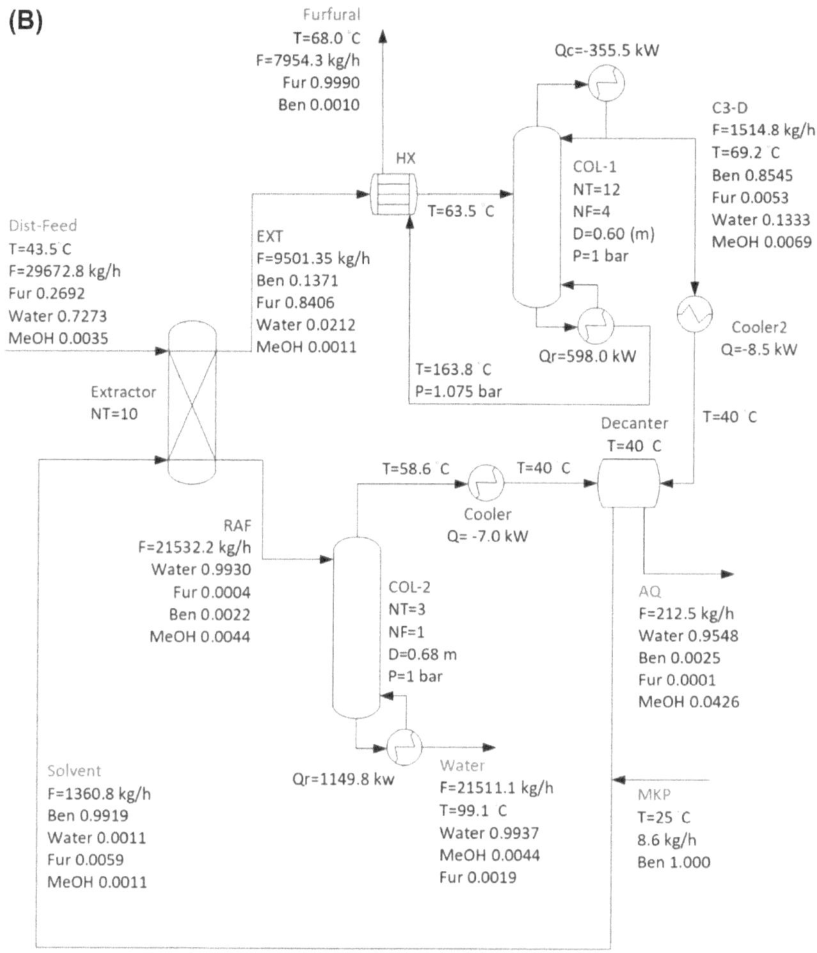

FIGURE 17.5 Con't.

stage is set to 3. The compressed air is combined with recycled air, and fed at column bottom. After reaction, the overhead vapor is condensed, and underwent vapor-liquid separation in two consecutive flash separators for the recovery of furfural into liquid phase. The vapor stream from the second flash separator is mostly air. It is heated by a series of steps using different grades of heating utility in an ascending order, which is then compressed and recycled. The reason for compressing this recycle stream is that it requires much less energy as compared to compressing fresh ambient air. Besides, note that organic acids were not stripped to the vapor phase in the air-stripping case. This could be a benefit, as stringent specification on water purity downstream may be reached without further waste water treatment steps.

The configuration and separation targets for the HED section in this scheme is identical to those in scheme 2, yet the feed composition is different (stream "Dist-Feed" in Fig. 17.5). Note that by incorporating an air-stripping reactor, the raw product now contains 26.9 wt% furfural, which is significantly higher than that of the steam-stripping case (2.78 wt%). More importantly, this composition now is at the right side of the heterogeneous azeotrope (see Fig. 17.3), and in the LLE separable region. This leads to a much smaller amount of benzene (i.e., 1360.8 kg/h in scheme 3, and 10,831.6 kg/h in scheme 2) required to recover 99.9% furfural.

Note that using air as the stripping agent leads to a lower temperature (i.e., 102°C) of the overhead vapor comparing with the steam-stripping case (scheme 1). Hence, energy released by overhead vapor condensation cannot be used as the reboiler heat source as that in scheme 1, hindering the intersectional heat integration. As demonstrated in Fig. 17.5A, this condensing energy is only sufficient to be partially used as heating source for the vapor stream from flash separator. After this heat exchange, the partly-condensed overhead vapor is cooled to 47°C using cooling water, before it enters the flash separators.

Developing a distillation-based process (i.e., as in scheme 1) to purify the cooled air-stripping reactor effluent (i.e., stream "Dist-Feed" in Fig. 17.5A) would also be noteworthy. As this composition now lies in the LLE region (see Fig. 17.3), this stream can be decanted first, and the organic and aqueous streams are processes using two columns in order to obtain high purity furfural and water, respectively (Yu and Tsai, 2020).

17.3.4 Aldol condensation process

In the final step of jet-fuel hydrocarbon production, the purified furfural obtained from step 2 is reacted with acetone through aldol condensation reaction. The kinetic expressions listed in Eqs. (17.4)–(17.7) were derived from a series of experiment using aqueous methanol as solvent. However, this leads to heavy tasks in the post-reaction sections. Also, a bulk amount of water is previously removed from furfural in step 2. Hence, adding water in the downstream process after it is removed should be avoided. To this end, the solvent-less aldol condensation process is a good alternative to enhance the process sustainability. This idea was experimentally investigated recently, revealing the interests toward operating this reaction in a solvent-less condition (Desai and Yadav, 2019). In modeling the reactor, an assumption was made that the reaction performances do not change under the solvent-less condition.

The optimal flowsheet of this solvent-less process is illustrated in Fig. 17.6. The fresh feed containing furfural and acetone is firstly mixed with three recycled streams to become an equal-molar combined feed. Before entering the reactor, the combined feed undergoes a two-staged heating, heated to 110 and 130°C using the 2-bar, and the 6-bar steam, respectively.

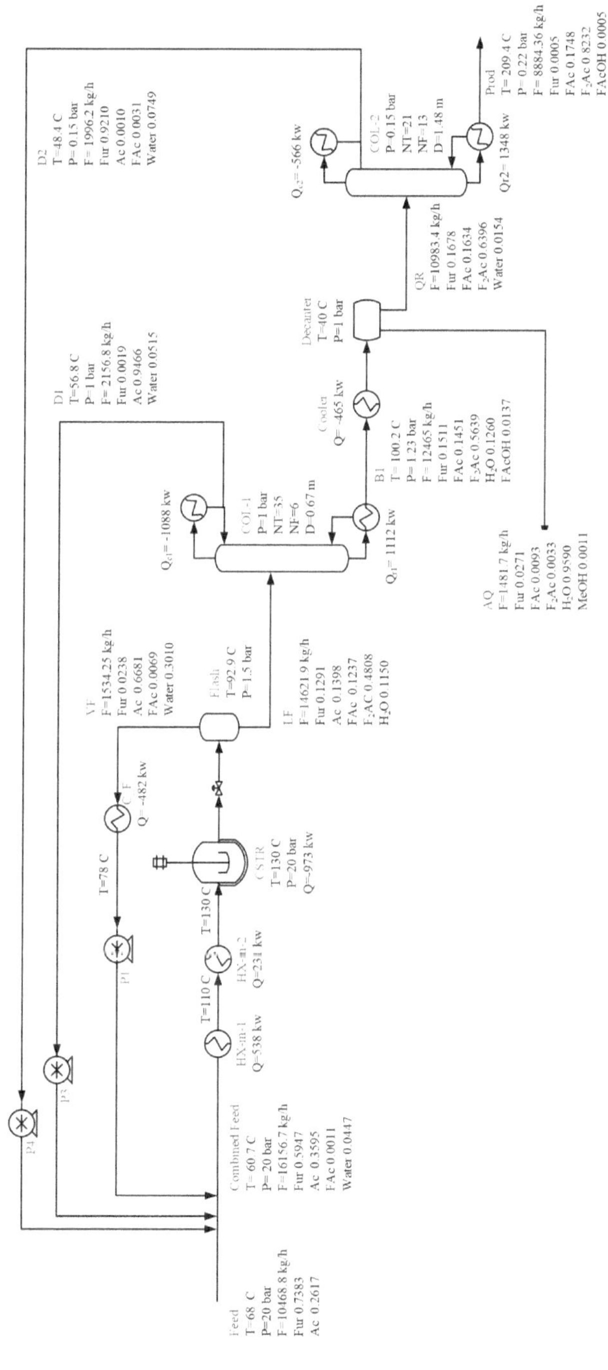

FIGURE 17.6 Flowsheet of the solvent-less aldol condensation process.

The reactor is modeled using the built-in RCSTR module. This reactor is set to operate at 20 bar and 130°C (O'Neill et al., 2014). Volume of this reactor is set to 2.57 m^3, which allows furfural conversion of 80%, the highest value as reported by O'neill et al. (2014). High pressure effluent of the reactor is adiabatically depressurized to 1.5 bar, before it is sent for flash separation. The flashed vapor which is rich in acetone is then condensed, and recycled to the reactor inlet. On the other hand, the liquid stream from the flash is sent to downstream separation.

In the downstream separation section, 35 trays, feed at tray 6 is used to separate the remaining light components from the distillate. It is operated under normal pressure, with the pressure drop between the stages set constant at 0.0068 atm. In this column, its distillate stream is set to recover 99.9% of methanol, while water recovery of 99.9% is set for its bottom stream. Reflux ratio and reboiler duty were adjusted to reach these two targets.

The bottom product from COL-1 (containing furfural, FAc, F_2Ac, and water) is cooled to 40°C before it is sent to a decanter, where it is separated by its naturally occurred liquid-liquid equilibrium. The aqueous stream from the decanter is 95.9 wt% rich in water. Further recovery of its organic species could be performed, but has not been investigated here. On the other hand, the organic stream of the decanter that contains mainly furfural, FAc, F_2Ac is sent to another distillation column (COL-2), in which the jet-fuel intermediate is obtained from the bottom. As FAc and F_2Ac has high boiling point, COL-2 should be operated under vacuum to allow the use of the 42-bar steam as the reboiler heat source. To this end, the pressure of COL-2 is set at 0.15 bar, which is the lowest operating condition where cooling water may still be used in its condenser with distillate temperature no less than 47° C. In this column, 99.99% recovery of FAc and F_2AC is set at the bottom, and the purity of combined FAc and F_2Ac is set to 99.9 mol%. Similar to COL-1, reflux ratio and reboiler duty of this column were also adjusted to reach these specifications. In order to avoid large pressure drop in this vacuum column, pressure drop over each stage was set constant at 0.005 atm. This can be actually achieved by setting lower weir height in the distillation column. Note that the liquid distillates from COL-1 and COL-2 are pumped and recycled to the reaction section.

As FAc and F_2Ac are to be used as liquid fuels together. Hence, the bottom stream of COL-2 that contains high quantity of these components will not be further separated. Note that the mixture of FAc and F_2Ac still requires an upgrading treatment to reduce its oxygen content before it can become the final product. This is the main reason why this mixture is known as BJF intermediate. To date, the mechanism of upgrading reaction is still under investigation. Once the mechanism is well-established, the overall biomass-to-BJF process will be further evaluated.

Note also that the furfural production rates vary among three alternative schemes, i.e., 7637.1 kg/h (scheme 1), 7636.4 kg/h (scheme 2) and 7954.3 kg/h (scheme 3). Hence, the production scale for the aldol condensation process is adjusted for consistency. This is considered in economic evaluation section.

17.4 Process analysis

17.4.1 Economic evaluation

A comparison of YUMC for each alternative scheme is given in Table 17.6, with their itemized results. The major observations are given as follows. First, scheme 3 has the lowest TCC, due to its great saving from utilizing an air-stripping reactor. Although compressors are equipped, the high pressure recycled air stream effectively reduces the compressor work. This supports the idea of air recycling. From the table, it is also clear that the stripping strategy is more influential on TCC, as compared to than in distillation and HED.

For TUC, much higher values are calculated for schemes 1 and 2. The reason is that these schemes require superheated steam, and thus higher amount of fuel is required in the furnace. However, the TEPs in these schemes are also much greater, as most of the energy input to the furnace can be recovered through steam generation. From Table 17.6, it is noted that TUC is nearly identical to TEP in scheme 1, indicating that the benefit from intersectional heat-integration is significant. When switching from distillation to HED (scheme 2), TEP becomes higher than TUC, further revealing that HED is less energy intensive. As mentioned in the previous section, intersectional heat integration for scheme 3 is not feasible. Given that the steam value is high enough to overcome the expense of fuel, applying steam-stripping may be more favorable.

As for the TRMC, scheme 3 is lower than the other two alternatives. The reason is that the use of air-stripping reactor ensures high furfural yield (70%). This reduces the cost of the bagasse feed. Note that for schemes 2 and 3, the cost of benzene makeup is also considered. With proper recovery of benzene from the raffinate phase, the makeup cost is almost negligible. On the other hand, the TWWTC in all case are relatively small, as purification of water has been considered in each process scheme. Scheme 3 has the highest PR among all schemes, benefited from higher furfural recovery in the air-stripper (98.7% in scheme 3, and 94.3% in the other two).

Finally, the YUMC of all scenarios are compared. It is found that Scheme 2 has the lowest YUMC of 0.698 USD/kg, i.e., 9.2% lower than that of scheme 1 (with YUMC of 0.769 USD/kg). Scheme 3 has a YUMC value that is slightly higher than scheme 2, i.e., 0.716 USD/kg. As there is no intersectional heat integration, this scheme could be superior to scheme 2 for cases with lower steam price.

Moreover, it is found that these YUMC results are comparable with the recent jet fuel data. According to International Air Transport Association (IATA) (International Air Transport Association (IATA)), the commercial jet fuel price fluctuated in between 0.586–0.693 USD/kg in year 2018–2020, owing to the variation of crude oil price. The average jet fuel price was calculated as 0.63 USD/kg, while its highest was reported at 0.789 USD/kg in September 2018 (International Air Transport Association (IATA)). Although the analysis stands on an optimistic side (i.e., good energy recovery, feasible of solvent-less aldol condensation process, etc.) by only focusing on the production, it still indicates that producing the BJF could be economically viable starting from the commercial-available techniques.

TABLE 17.6 Detailed YUMC results for each scheme.

	Scheme 1	Scheme 2	Scheme 3
Capital cost (×1000 USD)			
Stripping	14,695.93	12,609.50	7361.50
Aldol	1324.82	1324.47	1353.50
Distillation	–	–	–
Extraction	–	1674.45	718.37
TCC (×1000 USD)	**16,020.75**	**15,608.42**	**9433.37**
Utility cost (×1000 USD)			
Stripping	15,752.27	15,561.21	2667.64
Aldol	794.80	794.28	832.18
Distillation	-	-	-
Extraction	-	17.36	375.74
TUC (×1000 USD)	**16,547.07**	**16,372.85**	**3875.56**
Extra profit (×1000 USD)			
2-bar steam	14,052.50	175.35	183.31
6-bar steam	2511.63	21,797.4	-
Methanol	729.37	-	-
TEP (×1000 USD)	**17,293.50**	**21,972.75**	**183.31**
Raw material cost (×1000 USD)			
Bagasse	29,927.51	29,927.51	25,438.38
Acetone	19,900.19	19,817.02	20,726.42
Benzene	0.00	26.60	90.76
TRMC (×1000 USD)	49,827.69	49,771.13	46,255.56
Waste water treatment (×1000 USD)			
Stripping	30.63	30.63	23.05
Aldol	0.67	0.66	0.69
Extraction	131.38	130.56	10.48
TWWTC (×1000 USD)	**162.68**	**161.85**	**34.23**
TOC (×1000 USD)	**44,268.90**	**39,367.62**	**44,762.22**
TMC (×1000 kUSD)	**49,609.15**	**44,570.42**	**47,906.68**
PR (kg/h)	**8870.74**	**8869.77**	**9276.10**
YUMC (USD/kg)	0.769	0.698	0.716

17.4.2 CO_2 emission analysis

The CO_2 emission of each scheme is provided in Table 17.7, primarily based on unit CO_2 emission data provided in Table 17.4. Besides, some basic assumptions were made. First, steam generated from waste heat recovery is considered as net reduction in CO_2, with the reduction rate set identical to the emission rate from consuming the same grade of steam. Second, the use of bagasse as raw material is considered as CO_2 consumption. The calculation is explained as follows. Based on the formula of furfural (i.e., $C_5H_4O_2$), it is known that carbon accounts for 62.5% of its total mass. This portion of carbon is originated from bagasse, which has been avoided due to the use of bagasse for furfural production. According to the reaction stoichiometry, combusting 1 kg of carbon (MW = 12) generates 3.67 kg of CO_2 (MW = 44). Hence, once the furfural production rate is determined, the corresponding CO_2 reduction rate can be evaluated. Note that any item leading to CO_2 emission is represented by a positive value, whilst that resulting in CO_2 consumption or saving is given in a negative value. Similarly, a positive CO_2-e means the net CO_2 generation from the plant, and vice versa.

Among three schemes, scheme 1 has the most CO_2 emission, with CO_2-e of 1.22 kg/kg. The main hotspot of CO_2 emission for this scheme is its furnace, which uses great amount of fuel (i.e., coal) for superheated steam production. Scheme 2 reveals a near carbon-neutral feature (CO_2-e = 0.28 kg/kg), showing great improvement comparing with scheme 1. This result is supported by the fact that HED is less energy-intensive, as greater amount of waste heat is recovered from overhead vapor condensation through steam generation. Finally, scheme 3 is carbon-negative (CO_2-e = −1.65). The main reason for this significant improvement is due to the use of an air-stripping reactor, which removes the need of superheated steam production; this results in lower fuel consumption.

17.4.3 Future prospects in BJF production

The current biomass-to-BJF route still has unfavorable process economics. Hence, two aspects of improvement are suggested.

First, note that furfural production only uses hemi-cellulose content in the bagasse. If the remaining portions (i.e., cellulose and lignin) can be converted to other valuable chemicals, its process economics could be enhanced. This could be realized through the concept of a "bio-refinery," which has been proposed in the literature (Bond et al., 2014; Olcay et al., 2018; Xing et al., 2010). To date, most relevant investigations were focused on the overall economics of a bio-refinery. For in-depth analyses, clarifying the details (i.e., thermodynamics, kinetics, etc.) of other reaction routes in a bio-refinery is highly recommended.

Secondly, products from the proposed process (i.e., FAc, F_2Ac) are actually the BJF intermediate, which have to be upgraded (i.e., with removal of oxygen) before it becomes the final product. Hence, a relatively simple index, YUMC, was used for the current economic analysis, which was considered adequate for showing the merits and drawbacks for different integration

TABLE 17.7 Detailed CO_2 emission analysis.

		Scheme 1		Scheme 2		Scheme 3	
		kw	CO_2 (ton/y)	kw	CO_2 (ton/y)	kw	CO_2 (ton/y)
Step 1 (stripping reactor)	2-bar steam	−13,740	−26,387	0	0	1112	2135
	6-bar steam	−63,870	−134,023	−108,200	−227,043	3161	6632
	11-bar steam	0	0	0	0	1902	4197
	Electricity	0	0	0	0	1155	3992
	Fuel	152,240	390,134	152,240	390,134	0	0
	Sum	**229,724**		**163,091**		**16,956**	
Step 2 (furfural separation)	2-bar steam	0	0	0	0	1150	2208
	11-bar steam	0	0	0	0	577	1273
	Sum	**0**		**0**		**3482**	
Step 3 (aldol condensation)	2-bar steam	6,68	1282	664	1276	695	1336
	6-bar steam	228	478	227	476	237	498
	42-bar steam	1329	3489	1323	3473	1384	3634
	Electricity	18	63	18	63	19	65
	Sum	**5313**		**5288**		**5533**	
Raw material		−148,534		−148,534		−148,534	
Product (kg/h)		8871		8870		9284	
CO_2-e (kg CO_2/kg prod)		1.22		0.28		−1.65	

strategies. Given that the details for upgrading are well-defined, a detailed economic analysis could be performed.

17.5 Conclusion

In this work, three alternative schemes of biomass-to-bio-jet fuel (BJF) intermediate process were simulated, and compared in terms of their economic (using yearly unit manufacturing cost, or "YUMC") and environmental performances (based on CO_2 emission per unit kg of product, or "CO_2-e"). Scheme 1 was proposed based on current technologies (i.e., steam-stripping reaction, distillation-based furfural separation, and aldol condensation). The results showed that BJF intermediate can be produced with acceptable YUMC (0.769 USD/kg) if intersectional heat-integration was carried out. However, the overall process emits a great amount of net CO_2 (CO_2-e = 1.22 kg/kg). For process improvement, utilizing the HED configuration for furfural separation was effective (Scheme 2, with YUMC of 0.698 USD/kg and CO_2-e of 0.28 kg/kg). Besides, further reduction of CO_2 emission without deterioating the economics (Scheme 3, with YUMC of 0.716 USD/kg and CO_2-e of −1.65 kg/kg) was achieved by incorporating air-stripping technology in furfural production.

Exercise

1. Starting from the stream "Dist-Feed" in Fig. 17.5A, please design a distillation-based process for deep removal of furfural, water, and methanol (each specified at 99.9 wt%). Note that the composition of furfural and water now lies in the liquid-liquid split region (see Fig. 17.3). This alternative can be started from the decantation of this feed.
2. From Fig. 17.5B, it is apparent that recovering benzene from the raffinate phase is energy intensive, as evidenced from much larger reboiler duty (1149.8 kw) of COL-2 than that of COL-1 (598.0 kW). One alternative is not to recover benzene, and uses the makeup flow to maintain the overall mass balance. Based on the benzene price of 1.33 USD/kg, please find whether the alternative is better. If not, please also indicate the benzene price, under which the break-even would occurs (i.e., where the benefit of recovering benzene would overcome the expenses of consuming steam).
3. Starting from the aldol condensation process in Fig. 17.6, consider a situation that this process has been operated for a long time that catalyst deactivation occurs. The pre-exponential factor of the reversible FAcOH reaction (i.e., k_1 and k_{-1}) drops by 10%, and those in remaining reaction (i.e., k_2 and k_3) drop by 15%. Assuming that all equipment sizes and separation targets are unchanged. Please show how catalyst deactivation influences the process performance through simulation (note: other operating variables are changeable).

Appendix

TABLE A.1 Scalar pure component properties of FAcOH.

Properties	Units	Values
TC	K	521.415
PC	bar	56.832
VC	cum/kmol	0.23448
ZC	–	0.307401
OMEGA	–	0.239871
MW	–	58.0791

TABLE A.2 Temperature-dependent pure component properties of FAcOH.

Properties	Ideal gas Heat capacity	Vapor Pressure	Heat of vaporization
Correlation	**CPIGDP**	**PLXANT**	**DHVLDP**
T unit	K	K	K
Property units	cal/mol-K	Pa	cal/mol
Parameters			
1	16.2252	116.974	34,929.5
2	90.5454	−12,511.3	2.21496
3	381.373	0	−2.54017
4	−71.5453	0	0.907211
5	448.314	−12.9507	0
6	36.1286	1.18E−31	213.643
7	1035.13	10.8778	640.93
8		213.643	
9		640.93	

TABLE A.3 The binary interaction parameters (NRTL model) used in this work.

Comp. i	Comp. j	Source	AIJ	AJI	BIJ	BJI	CIJ	DIJ	EIJ	EJI	FIJ	FJI
F	AC	NISTV100 NIST-IG	1.37367	−0.61569	32.3214	66.961	0.5	0	0	0	0	0
F	H_2O	APV100 VLE-IG	−5.8732	7.1079	2335.05	−1265.84	0.3	0	0	0	0	0
AC	H_2O	APV100 VLE-IG	6.3981	0.0544	−1808.99	419.972	0.3	0	0	0	0	0
F	FAC	COSMO	−0.0351902	1.2753	5.55888	5.60707	0.3	0	0.127952	−0.32473	−0.0006	0.000671
F	F2AC	COSMO	−5.03926	8.52506	13.1717	−3.02021	0.3	0	0.894847	−1.60527	−0.0034	0.005905
H_2O	FAC	COSMO	−12.7873	−0.73442	35.6307	4.70785	0.3	0	3.5399	0.672859	−0.01109	−0.00569
H_2O	F2AC	COSMO	−3.82689	2.87637	−20.4889	2.94379	0.3	0	2.53381	0.368897	−0.01255	−0.00866
AC	FAC	COSMO	−13.2661	24.9985	15.8559	−33.4563	0.3	0	2.2748	−4.43513	−0.00426	0.008518
AC	F2AC	COSMO	13.7539	−20.9724	−20.5732	47.6951	0.3	0	−2.41335	3.75076	0.004635	−0.0073
FAC	F2AC	COSMO	−9.18675	13.6368	7.19778	−4.04449	0.3	0	1.59038	−2.41817	−0.00306	0.004765
AC	MEOH	APV100 VLE-IG	0	0	101.886	114.135	0.3	0	0	0	0	0
H_2O	MEOH	APV100 VLE-IG	2.7322	−0.693	−617.269	172.987	0.3	0	0	0	0	0
F	MEOH	NISTV100 NIST-IG	−2.80759	3.78784	282.515	25.874	0.1	0	0	0	0	0
FAC	MEOH	COSMO	15.1764	−6.38559	−31.8181	5.94272	0.3	0	−2.65552	1.23017	0.003554	−0.00247
F2AC	MEOH	COSMO	38.3518	−25.1695	−59.7752	69.6245	0.3	0	−7.28292	4.92823	0.015844	−0.01222
H_2O	FAC-OH	COSMO	−14.8381	−9.1741	88.5989	11.8921	0.3	0	3.85895	2.16081	−0.01113	−0.00791

Continued

TABLE A.3 The binary interaction parameters (NRTL model) used in this work.—cont'd

Comp. i	Comp. j	Source	AIJ	AJI	BIJ	BJI	CIJ	DIJ	EIJ	EJI	FIJ	FJI
F	FAC-OH	COSMO	−6.65773	7.2976	8.84145	3.17165	0.3	0	1.50836	−1.5978	−0.00513	0.004849
AC	FAC-OH	COSMO	−9.20627	16.6971	21.0117	−16.8405	0.3	0	1.45726	−2.78249	−0.00247	0.004917
FAC-OH	FAC	COSMO	−2.36735	3.74398	10.7501	1.68357	0.3	0	0.429584	−0.70087	−0.00091	0.001513
FAC-OH	F2AC	COSMO	−14.6732	20.9296	10.5842	−20.2189	0.3	0	2.74453	−3.94721	−0.00601	0.008666
FAC-OH	MEOH	COSMO	4.47585	−0.47013	−12.4972	7.07574	0.3	0	−0.80612	0.222434	0.001618	−0.00183
F	Water	APV100 VLE-HOC	−4.7563	4.2362	1911.42	−262.241	0.3	0	0	0	0	0
F	AA	NISTV100 NIST-HOC	0.22581	0.476171	−0.1263	1.15093	0.5	0	0	0	0	0
Water	AA	NISTV100 NIST-HOC	1.52053	−1.15109	234.527	1.0868	0.1	0	0	0	0	0
F	XYL	R-PCES	0	0	1725.32	−94.2427	0.3	0	0	0	0	0
Water	XYL	R-PCES	0	0	686.547	−474.263	0.3	0	0	0	0	0
AA	XYL	R-PCES	0	0	−558.551	876.383	0.3	0	0	0	0	0
F	FA	APV100 VLE-HOC	0	0	289.216	46.1655	0.3	0	0	0	0	0
Water	FA	NISTV100 NIST-HOC	−0.601879	0.895118	−0.13029	−129.148	0.5	0	0	0	0	0

AA	FA	NISTV100 NIST-HOC	−0.787482	0.815127	21.7323	37.4283	0.101079	0	0	0	0	0
XYL	FA	R-PCES	0	0	1378.19	−741.973	0.3	0	0	0	0	0
Furfural	Benzene	APV100 VLE-HOC	−4.1934	18.6557	1256.5	−6158.1	0.3	0	0	0	0	0
Water	MEOH	APV100 VLE-HOC	4.8683	−2.6311	−1347.53	838.594	0.3	0	0	0	0	0
Water	Benzene	APV100 LLE-ASPEN	140.087	45.1905	−5954.31	591.368	0.2	0	−20.0254	−7.5629	0	0
AA	MEOH	APV100 VLE-HOC	7.4858	−3.8591	−2151.88	975.377	0.3	0	0	0	0	0
FA	Benzene	APV100 VLE-HOC	0	0	771.892	539.855	0.43	0	0	0	0	0
MEOH	Benzene	APV100 VLE-HOC	0.3628	−3.0277	160.549	1657.77	0.4	0	0	0	0	0
F	MEOH	NISTV100 NIST-HOC	−2.37923	3.93443	121.696	9.8658	0.1	0	0	0	0	0
AA	Benzene	APV100 VLE-HOC	1.1158	1.5307	−418.144	104.171	0.3	0	0	0	0	0
XYL	MEOH	R-PCES	0	0	2562.74	−1022.82	0.3	0	0	0	0	0
XYL	Benzene	R-PCES	0	0	532.732	3223.75	0.3	0	0	0	0	0
FA	MEOH	R-PCES	0	0	457.268	−288.11	0.3	0	0	0	0	0

References

Agirrezabal-Telleria, I., Gandarias, I., Arias, P.L., 2013. Production of furfural from pentosan-rich biomass: analysis of process parameters during simultaneous furfural stripping. Bioresource Technology 143, 258–264.

Bond, J.Q., et al., 2014. Production of renewable jet fuel range alkanes and commodity chemicals from integrated catalytic processing of biomass. Energy & Environmental Science 7 (4), 1500–1523.

Desai, D.S., Yadav, G.D., 2019. Green synthesis of furfural acetone by solvent-free aldol condensation of furfural with acetone over La_2O_3-MgO mixed oxide catalyst. Industrial & Engineering Chemistry Research 58 (35), 16096–16105.

Douglas, J.M., 1988. Conceptual Design of Chemical Processes. McGraw-Hill, New York.

Gadalla, M.A., et al., 2005. Reducing CO_2 emissions and energy consumption of heat-integrated distillation systems. Environmental Science & Technology 39 (17), 6860–6870.

Huber, G.W., Dumesic, J.A., 2006. An overview of aqueous-phase catalytic processes for production of hydrogen and alkanes in a biorefinery. Catalysis Today 111 (1–2), 119–132.

Huber, G.W., et al., 2005. Production of liquid alkanes by aqueous-phase processing of biomass-derived carbohydrates. Science 308 (5727), 1446–1450.

Huber, G.W., et al., 2006. Renewable liquid alkanes from aqueous-phase processing of biomass-derived carbohydrates. Abstracts of Papers of the American Chemical Society 231.

ICIS.com, 2014. ICIS Pricing. Benzene, Europe.

ICIS.com, 2017. Acetone Asia Pacific, Chemical Prices, News, Analysis.

International Air Transport Association (IATA), Jet Fuel Price Monitor (Accessed on Jan. 31st, 2020).

JIang, C., Cheng, L., Cheng, G., 2018. Kinetics of aldol condensation of furfural with acetone catalyzed by 1,8-diazabicyclo[5.4.0]undec-7-ene. Journal of Materials Science and Chemical Engineering (6), 65–73.

Kikhtyanin, O., Kubicka, D., Cejka, J., 2015. Toward understanding of the role of Lewis acidity in aldol condensation of acetone and furfural using MOF and zeolite catalysts. Catalysis Today 243, 158–162.

Krzelj, V., et al., 2019. Furfural production by reactive stripping: Process optimization by a combined modeling and experimental approach. Industrial & Engineering Chemistry Research 58 (35), 16126–16137.

Liang, H.H., et al., 2019. Improvement to PR plus COSMOSAC EOS for predicting the vapor pressure of nonelectrolyte organic solids and liquids. Industrial & Engineering Chemistry Research 58 (12), 5030–5040.

Lin, S.T., Sandler, S.I., 2002. A priori phase equilibrium prediction from a segment contribution solvation model. Industrial & Engineering Chemistry Research 41 (5), 899–913.

Luyben, W.L., 2011. Principles and Case Studies of Simultaneous Design (Chapter 5). Wiley, New York.

Nhien, L.C., et al., 2016. Design and optimization of intensified biorefinery process for furfural production through a systematic procedure. Biochemical Engineering Journal 116, 166–175.

Nhien, L.C., et al., 2017. Techno-economic assessment of hybrid extraction and distillation processes for furfural production from lignocellulosic biomass. Biotechnology for Biofuels 10.

O'Neill, R.E., et al., 2014. Aldol-condensation of furfural by activated dolomite catalyst. Applied Catalysis B-Environmental 144, 46–56.

Olcay, H., et al., 2018. Techno-economic and environmental evaluation of producing chemicals and drop-in aviation biofuels via aqueous phase processing. Energy & Environmental Science 11 (8), 2085–2101.

Renewables 2021 Global Status Report: A Comprehensive Annual Overview of the State of Renewable Energy, 2021. Renewables Energy Policy Network For the 21st Century (REN21).

Schmidt, W.P., et al., 2001. Safe design and operation of a cryogenic air separation unit. Process Safety Progress 20 (4), 269–279.

Shen, W.C., Chien, I.L., 2019. Design and control of ethanol/benzene separation by energy saving extraction distillation process using glycerol as an effective heavy solvent. Industrial & Engineering Chemistry Research 58 (31), 14295–14311.

Silva, J.F.L., et al., 2017. Integrated furfural and first generation bioethanol production: process simulation and technoeconomic analysis. Brazilian Journal of Chemical Engineering 34 (3), 623–634.

Turton, R., et al., 2009. Analysis, Synthesis, and Design of Chemical Processes, third ed. Pearson Education, Boston.

Xing, R., et al., 2010. Production of jet and diesel fuel range alkanes from waste hemicellulose-derived aqueous solutions. Green Chemistry 12 (11), 1933–1946.

Yu, B.Y., Chien, I.L., 2015. Design and economic evaluation of a coal-to-synthetic natural gas process. Industrial & Engineering Chemistry Research 54 (8), 2339–2352.

Yu, B.Y., et al., 2017. Energy-efficient extraction distillation process for separating diluted acetonitrile water mixture: rigorous design with experimental verification from ternary liquid liquid equilibrium data. Industrial & Engineering Chemistry Research 56 (51), 15112–15121.

Yu, B.Y., Tsai, C.C., 2020. Rigorous simulation and techno-economic analysis of a bio-jet-fuel intermediate process with various integration strategies. Chemical Engineering Research and Design 159, 47–65.

Zeitsch, K., 2000. The Chemistry and Technology of Furfural and Its Many by-Products. Elsevier Science, Netherlands.

Chapter 18

Production of diethyl carbonate from direct CO_2 conversion*

Bor-Yih Yu[2], Pei-Jhen Wu[1], Chang-Che Tsai[2] and Shiang-Tai Lin[2]
[1]*Department of Chemical Engineering, National Taiwan University of Science and Technology, Taipei, Taiwan;* [2]*Department of Chemical Engineering, National Taiwan University, Taipei, Taiwan*

Chapter outline

18.1 Introduction

The anthropogenic emission of carbon dioxide (CO_2) has led to critical issues in our environment, society, and economics. Many world economics aim to

*Simulation and other electronic files of this chapter are made available on a companion website; URL: https://www.elsevier.com/books-and-journals/book-companion/9780323901680.

Chemical Engineering Process Simulation. https://doi.org/10.1016/B978-0-323-90168-0.00009-3

become an economy with net-zero greenhouse gas emissions by year 2050. To meet these challenges, research topics in CO_2 capture, storage, and utilization have received much attention. Being regarded as an abundant carbon source which is inexpensive and nontoxic, various routes have been reported to convert CO_2 into value-added chemicals, e.g., syngas, methanol, dimethyl carbonate, or polymers (Meylan et al., 2015; Muthuraj and Mekonnen, 2018; Ohno et al., 2018).

Being chemically very stable, conversion of CO_2 has been mostly realized through an indirect route. In this scenario, a highly reactive reactant (e.g., epoxides, hydrogen, etc.) is first reacted with CO_2 to get an intermediate product, which is further reacted with another chemical(s) to become the product. This inevitably complicates the process, and increases their energy demand. On the other hand, direct CO_2 conversion routes are regarded more atom-efficient and sustainable. However, the chemical equilibrium limitation makes it necessary to remove its water content in-situ in order to drive the reaction toward high product yield in direct CO_2 conversion (Gu et al., 2019; Leino et al., 2013; Wang et al., 2017; Tomishige et al., 2019; Tamura et al., 2018; Honda et al., 2013).

In this chapter, a direct conversion process of converting CO_2 to diethyl carbonate (DEC) is demonstrated. DEC is a versatile material which can be used as fuel additive, solvent, or reactive intermediate. Various DEC production routes reported in the literature are listed and compared in Table 18.1. The previously reported routes (i.e., Paths A to F in Table 18.1) suffered from the drawbacks such as the usage of toxic reactants, severe equilibrium limitation, low yield, or complex process configurations. More noteworthy, the heavy loading in the separation section often makes the entire process in net CO_2 emission, albeit CO_2 is used as a reactant (Path B).

In this chapter, Path G in Table 18.1 is investigated rigorously. It is aimed to develop an atomic-efficient direct CO_2 conversion process in reaching net-negative CO_2 emission. In this route, CO_2 is directly reacted with ethanol (EtOH) to produce DEC and water. Besides, 2-cyanopyridine (2-CP), acting as a dehydrant, is added to react with water to form 2-picolinamide (2-PA). With continuous water removal, the conversion is greatly improved to 50%, with more than 95% product selectivity. This greatly reduces the downstream separation load. From the rigorous simulation study, it is found that this novel process contributed a net CO_2 reduction of 0.154 kg per kg of DEC generated, i.e., CO_2-e = −0.154 kg/kg, and is thus considered very attractive in CO_2 utilization.

18.2 Process overview

18.2.1 Physical properties

In this work, simulation was performed in Aspen Plus V11 (Aspen Technology). Ten components were involved in process simulation, i.e., carbon dioxide (CO_2), ethanol (EtOH), diethyl carbonate (DEC), water (H_2O), 2-

TABLE 18.1 Reaction pathways for diethyl carbonate (DEC) manufacture.

Path #	Reaction	Descriptions
A	**Ethanolysis of phosgene** (Muskat and Strain, 1945) $2C_2H_6O(EtOH) + COCl_2(phosgene) \rightarrow C_5H_{10}O_3\ (DEC) + 2HCl$	- Earliest process for DEC production - Inherently toxic
B	**Ethanolysis of CO_2** (Shukla and Srivastava, 2016) $2C_2H_6O(EtOH) + CO_2 \rightarrow C_5H_{10}O_3\ (DEC) + H_2O$	- Cleaner reactants - Thermodynamically non-spontaneous - Severe equilibrium limitation
C	**Oxidative carbonylation** (Shukla and Srivastava, 2016) $2C_2H_6O(EtOH) + CO + 0.5O_2 \rightarrow C_5H_{10}O_3\ (DEC) + H_2O$	- Can be operated in either liquid or vapor phase - Thermodynamically favorable - Equipment corrosion, catalyst deactivation, low yields
D	**Ethanolysis of urea** (Shukla and Srivastava, 2016) $C_2H_6O(EtOH) + CH_4N_2O(urea) \rightarrow C_3H_7NO_2$ (ethyl carbamate) $+ NH_3$ $C_2H_6O(EtOH) + C_3H_7NO_2$ (ethyl carbamate) $\rightarrow C_5H_{10}O_3\ (DEC) + NH_3$	- Cheap and nontoxic raw materials - Thermodynamically non-spontaneous - High operational temperature (>180°C) and pressure (>25 bar)
E	**Trans-esterification** (Shukla and Srivastava, 2018; Zhang et al., 2012) $C_4H_6O_3$ (propylene carbonate) $+ 2C_2H_6O\ (EtOH) \rightarrow C_5H_{10}O_3\ (DEC) + C_3H_8O_2$ (propylene glycol)	- The reactants (i.e., PC, DMC) are derived from CO_2. - High yields under process intensification - Complex process configurations
F	**Trans-esterification** (Wei et al., 2011; Zheng et al., 2017; Yang et al., 2019) $C_3H_6O_3\ (DMC) + C_2H_6O\ (EtOH) \rightarrow C_4H_8O_3\ (EMC) + CH_4O\ (MeOH)$ $C_4H_8O_3\ (EMC) + C_2H_6O\ (EtOH) \rightarrow C_5H_{10}O_3\ (DEC) + CH_4O\ (MeOH)$	
G	**Direct conversion with dehydration** (Yu et al., 2020) $2C_2H_6O(EtOH) + CO_2 \rightarrow C_5H_{10}O_3\ (DEC) + H_2O$ $C_6H_4N_2$ (2-cyanopyridine) $+ H_2O \rightarrow C_6H_6N_2O$ (2-picolinamide)	- Green and atom-efficient - High yields using 2-cyanopyridine as an in situ dehydrant

cyanopyridine (2-CP), 2-picolinamide (2-PA), ethyl picolinimidate (EPI), ethyl picolinate (EP), ammonia (NH_3), and ethyl carbamate (EC). Among them, there is no built-in component data for EPI, and thus it is defined manually. The estimated essential properties for EPI are included in the Appendix.

The required physical properties for performing the overall mass and energy balance through Aspen Plus are summarized in Table 18.2. Both the pure component properties and the mixture properties are required. For the pure component properties, at least the listed scalar (i.e., critical properties, the acentric factor, the Gibbs energy and enthalpy of formation) and temperature-dependent ones (i.e., ideal gas heat capacity, enthalpy of vaporization, vapor pressure) should be specified. More specifically, the Gibbs energy of

TABLE 18.2 The required physical properties for calculating mass and energy balance.

Category	Name	Notation in Aspen Plus
Pure component-(scalar)	Critical properties (i.e., temperature, pressure, volume, compressibility factor)	TC, PC, VC, ZC
	Acentric factor	Omega
	Gibbs energy of formation	DGFORM
	Enthalpy of formation	DHFORM
Pure component-(T dependent)	Ideal gas heat capacity	CPIG (Aspen ideal gas heat capacity polynomial) CPIGDP (DIPPR equation) CPIALE (NIST Aly-Lee equation)
	Enthalpy of vaporization	DHVLDP (DIPPR heat of vaporization equation) DHVLWT (Watson heat of vaporization equation) DHVLTD (NIST TDE Watson heat of vaporization equation)
	Vapor pressure	PLXANT (Extended Antione equation) WAGNER (Wanger vapor pressure equation)
Mixture	Binary interaction parameters	Activity coefficient model (e.g., NRTL, UNIQUAC, Wilson, etc.) Equation of state (e.g., Peng-Robinson, Redlich-Kwong, Soave-Redlich-Kwong, etc.)

formation, the enthalpy of formation, and the ideal gas heat capacity are used to calculate the essential ideal gas properties of pure compounds (i.e., Gibbs energy, enthalpy, entropy). The critical properties and the acentric factor are used in cubic equation of states, which bridges the gap in thermodynamic properties between ideal gas and real fluids. The enthalpy of vaporization and the vapor pressure, on the other hand, enable the calculation of physical properties of pure compounds during phase transition, which are essential for accurate description of pure fluid properties. Note that for each temperature-dependent property, different built-in correlations are available for fitting the experimental data. The identification name of these popularly used correlations in Aspen Plus are also provided in Table 18.2.

For describing the mixture properties, one should first determine whether an activity coefficient model or an equation of state (EOS) should be used.[1] A general guideline is to select an activity coefficient model to describe a system containing primarily polar compounds, and an EOS if otherwise. When an activity coefficient model is used, the binary interaction parameter pairs are required to calculate the activity coefficient and liquid volume (e.g., by using Rackett model) of each compound in the mixture. The corresponding parameters could be regressed from the reported phase equilibrium data (i.e., vapor-liquid equilibrium or liquid-liquid equilibrium). When using EOS, the binary interaction parameters are specified to calculate the mixture properties (i.e., fugacity coefficients, mixture liquid volume, etc.) based on various mixing rules available in Aspen Plus. Note that exceptions to the aforementioned guideline may exist for those slightly-polar systems (i.e., processes involving aromatics or its derivatives, olefins, etc.). Hence, the verification of the binary interaction parameters using the experimental data is considered of significant importance prior to designing the process.

It is noteworthy that a massive amount of experimental data is incorporated in the built-in database (NIST ThermoData Engine (Inc. P.G., 2021)) in Aspen Plus, which could be used for correlating model parameters in order to describe pure component and mixture properties. Yet it is not always possible to find every (or pair of) model parameters needed in process simulation. There are several methods which can be used to generate reliable estimates to these properties, as outlined in Table 18.3. Most commonly, the *group contribution method* (GCM) is applied, which is also the basis for Aspen Plus "Estimation" function. In general, this is a semi-experimental method which divides the component structures into many predefined functional groups, in which their individual contributions to the properties are estimated and summed. Hence, the suitability of GCM to those components owing complex chemical structures become another issue. Another method of estimating the missing properties is through the COSMO-based calculation (i.e.,

1. Refer to Chapter 20 for detailed description of physical property estimation.

TABLE 18.3 Methods for estimating the missing properties.

Category	Name	Aspen built-in GCM	QM/COSMO calculation
Pure component-(scalar)	Critical properties (i.e., temperature, pressure, volume, compressibility factor)	Lyderson, Joback, Fedors, Ambrose, Simple, Gani, Mani, Riedel, Fedors, Definiti	PR + COSMOSAC EOS
	Acentric factor	Definiti, Lee-Kesler	PR + COSMOSAC EOS
	Gibbs energy of formation	Joback, Benson, Gani	Thermal Chemistry Analysis in G09
	Enthalpy of formation	Benson, Joback, Bensonr8, Gani	Thermal Chemistry Analysis in G09
Pure component-(T dependent)	Ideal gas heat capacity	Benson, Joback, Bensonr8	Thermal Chemistry Analysis in G09
	Enthalpy of vaporization	Riedel, Li-Ma, Mani	PR + COSMOSAC EOS
	Vapor pressure	Data, Definiti, Vetere, Gani, Ducros, Li-Ma	PR + COSMOSAC EOS
Mixture	Binary interaction parameters	UNIFAC	COSMO-SAC

PR + COSMOSAC EOS and the COSMO-SAC activity coefficient model). These methods leverage quantum-mechanical (QM) calculations of molecules embedded in a perfect conductor (the COSMO calculation (Lin and Sandler, 2002)) to assess a variety of thermodynamic properties. The computed screening charge density distribution, known as σ-profile, is utilized in the evaluation of activity coefficients of molecules. From these methods, any missing properties could be estimated only from the chemical structure with acceptable accuracy.

In order to describe the dissolution of CO_2 into the liquid phase, the Henry's constants of CO_2 in each other components are also required. This could be realized by setting "Henry's Component" in the directory at "Properties" → "Components," as depicted in Fig. 18.1. Through this function, the selected compounds are considered as sparingly soluble gases in simulation. The built-in parameters of Henry's constant equation will automatically appear

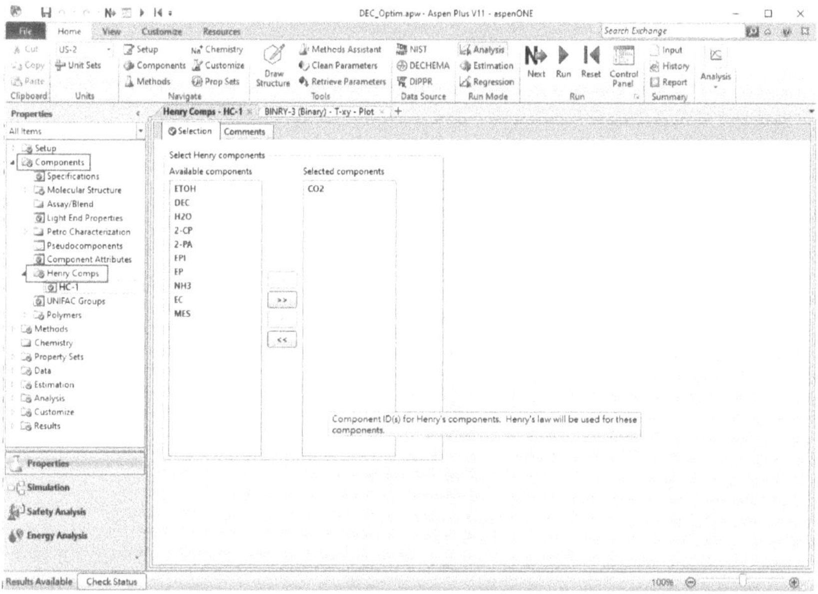

FIGURE 18.1 The snapshot demonstrating how to set the Henry's components in Aspen Plus.

after the sparingly soluble gases are specified, if there is any. Without this setting, the solubility of these gases into the liquid phase could be overly predicted, as vapor-liquid equilibrium in between them are calculated instead. Herein, only CO_2 is selected as the Henry's component.

In modeling the process, all built-in physical property parameters (both for pure components and mixture) are used, with the missing ones estimated by COSMO-based calculation. All parameters employed here are provided in the Appendix of this chapter.

18.2.2 Reaction pathway and kinetic expression

Recently, Giram et al. (2018) reported a series of experiments for direct synthesis of DEC from CO_2 and EtOH using CeO_2 catalyst. 2-CP was added as the chemical dehydrant. The reactions involved are listed in Eqs. (18.1)–(18.9). As reported, the obtained DEC yield was up to 50%, using a stoichiometric molar ratio of 2-CP/EtOH at moderate temperature (120–150°C) and high pressure (20–50 bar)

$$CO_2 + 2(EtOH) \xrightarrow{r1} DEC + H_2O \qquad (18.1)$$

$$DEC + H_2O \xrightarrow{r2} CO_2 + 2(EtOH) \quad (18.2)$$

$$2CP + H_2O \xrightarrow{r3} 2PA \quad (18.3)$$

$$2CP + EtOH \xrightarrow{r4} EPI \quad (18.4)$$

$$2PA + EtOH \xrightarrow{r5} EP + NH_3 \quad (18.5)$$

$$EP + NH_3 \xrightarrow{r6} 2PA + EtOH \quad (18.6)$$

$$DEC + 2PA \xrightarrow{r7} EC + EPI \quad (18.7)$$

$$EC + EPI \xrightarrow{r8} DEC + 2PA \quad (18.8)$$

$$EPI + H_2O \xrightarrow{r9} EP + NH_3 \quad (18.9)$$

Note that Eqs. (18.1) and (18.2) represent the reversible reactions for direct DEC synthesis from CO_2 and EtOH. Eq. (18.3) is the dehydration reaction. The remaining equations represent other side reactions. The power law-based kinetic expression for each reaction are listed in Eqs. (18.10)–(18.18), with parameters provided in Table 18.4 (Yu et al., 2020).

$$r_1 = k_1 \exp\left(-Ea_1/RT\right) \quad (18.10)$$

TABLE 18.4 The regressed kinetic parameters ($i = 1$–9).

Reaction	K_{0i} (kmol/kgcat-s)	Ea_i (kJ/kmol)
r_1	4.608E+14	150,163
r_2	5.550E+17	125,359
r_3	7.048E+00	258
r_4	3.851E−05	15,259
r_5	2.151E−02	6919
r_6	8.726E+00	1937
r_7	7.133E+05	69,837
r_8	3.133E+01	12,915
r_9	1.114E+02	14,592

$$r_2 = k_2 \exp\left(-Ea_2/_{RT}\right) X_{DEC} X_{H_2O} / X_{CO_2} X^2_{EtOH} \tag{18.11}$$

$$r_3 = k_3 \cdot T \cdot \exp\left(-Ea_3/_{RT}\right) X_{2CP} X_{H_2O} \tag{18.12}$$

$$r_4 = k_4 \exp\left(-Ea_4/_{RT}\right) X_{2CP} X_{EtOH} \tag{18.13}$$

$$r_5 = k_5 \exp\left(-Ea_5/_{RT}\right) X_{PA} X_{EtOH} \tag{18.14}$$

$$r_6 = k_6 \exp\left(-Ea_6/_{RT}\right) X_{EP} X_{NH_3} \tag{18.15}$$

$$r_7 = k_7 \exp\left(-Ea_7/_{RT}\right) X_{DEC} X_{2PA} \tag{18.16}$$

$$r_8 = k_8 \exp\left(-Ea_8/_{RT}\right) X_{EC} X_{EPI} \tag{18.17}$$

$$r_9 = k_9 \cdot T \cdot \exp\left(-Ea_9/_{RT}\right) X_{EPI} X_{H_2O} \tag{18.18}$$

18.2.3 Basis for evaluating the process economics and carbon emission

18.2.3.1 Economics

In this work, the cost function defined in Eq. (18.19) is used as the objective function in optimization:

$$\text{Cost(kUSD)} = \frac{\sum TCC}{PBY} + \sum TOC + \sum_i UP_i (F_i - F_{i,b}) \tag{18.19}$$

In this equation, the total capital cost (TCC) is calculated as the summation of the bare module cost of each process equipment, using correlations reported by Turton et al. (2018), given as in Table 18.5. The payback year (PBY) is set as 8 years. Note that the original correlations were developed using USD in year 2001 with Chemical Engineering Plant Cost Index (CEPCI) of 397.0. The CEPCI at year 2018 was reported as 616.5 (Jenkins, 2018), and was used to correct the capital cost of each item.

On the other hand, the total operating cost (TOC) indicates the expense from using utilities. The unit cost of the utilities employed in this work are also referenced from Turton et al. (2018), and are listed in Table 18.6. Also, the

TABLE 18.5 Correlations for calculating capital cost (CEPCI = 616.5).

	Purchase cost (C_{P0}) $\log_{10}C_p^0 = K_1 + K_2\log_{10}A + K_3[\log_{10}A]^2$					Bare module Factor (F_{BM}) $F_{BM} = B_1 + B_2F_pF_M$	
	K_1	K_2	K_3	Capacity	Range	B_1	B_2
Heat exchangers	4.8306	−0.8509	0.3187	Area, m^2	$10 < A < 1000$	1.63	1.66
Reactor Column Flash	3.4974	0.4458	0.1074	Volume, m^3	$0.3 < A < 520$	2.25	1.82
Sieve tray	2.9949	0.4465	0.3961	Area, m^2	$0.07 < A < 12.30$	1.00	0
Compressor	2.2897	1.3604	−0.1027	Work, kW	$450 < A < 3000$	2.80	0
	Pressure factor (F_p) $\log_{10}F_P = C_1 + C_2\log_{10}P + C_3[\log_{10}P]^2$						
	C1		**C2**		**C3**	**Range (barg)**	
Heat exchangers	0		0		0	$P < 5$	
	0.03881		−0.11272		0.08183	$5 < P < 140$	
Vessel	$F_{p,vessel} = \dfrac{\dfrac{PD}{2(850-0.6P)} + 0.00315}{0.0063}$ If $F_{p,vessel} < 1$, set $F_{p,vessel} = 1$. If $P < -0.5$ barg, set $F_{p,vessel} = 1.25$					All	

F_M: material factor, set to 1.0 for all the equipment made of carbon steel.

TABLE 18.6 Cost data and unit amount of carbon emission for the utilities.

Utilities	Price (Turton et al., 2018) (USD/GJ)	Benefit from recovery (USD/GJ)	Latent heat (kg/kJ)	CO_2 emission amount (kg-CO_2/GJ) (Yu et al., 2020)
2-Bar steam (2 bar/120°C)	4.11	3.42	2203.04	66.68
Low pressure steam, LPS (6 bar/160°C)	4.54	3.81	2081.86	72.86
Medium pressure steam, MPS (11 bar/184°C)	4.77	4.01	1999.72	76.60
High pressure steam, HPS (42 bar/254°C)	5.66	4.76	1694.34	91.14
Cooling water	0.378		–	–
Chilled water	4.77			–
Electricity	16.9		–	120.06 (Schmidt et al., 2001)

benefits from recovering waste heat through steam generation are included. They could be calculated from subtracting the selling prices by the cost of boiler feed water (assuming 1.523 USD per 1000 kg used). The amount of boiler feed water used for generating different grades of steam are also calculated based on the corresponding latent enthalpies.

Finally, the last term in Eq. (18.19) represents the stream cost as a difference, in which UPi denotes the unit price of each compound, while F_i and $F_{i,b}$ denote the flowrate of each compound in any given case and the base case, respectively. The unit price of CO_2, EtOH and DEC are set at 0.07, 0.47 and 1.50 (USD/kg), respectively.

18.2.3.2 Carbon emission

Indirect CO_2 emissions are often accompanied by consuming utilities in the chemical plant, especially steam and electricity. The reason is that both generation and recovery of utilities require the use of fossil fuels. Hence, it is important to analyze whether a plant using CO_2 as raw material actually reduces CO_2.

In order to investigate this issue, a correlation equating the unit amount of utility used with the corresponding CO_2 emission rate was reported elsewhere (Gadalla et al., 2005). The emission rates listed in Table 18.6 are derived based on the latent enthalpy of different grades of steam, assuming that natural gas is used as the fuel source. For simplicity, emissions from cooling utilities are not considered. For better demonstration of the result, a throughput-independent term, i.e., kg CO_2 emission per kg DEC formed (or "CO_2-e"), is used for analysis. A positive CO_2-e value indicates net CO_2 emission, while the opposite for net CO_2 reduction. Besides, recovering waste heat through steam generation is regarded as a contribution to CO_2 reduction, as it helps to reduce fuel consumption.

18.3 The direct CO_2-to-DEC process

18.3.1 Process development

The optimized process configuration is presented in Fig. 18.2, with the optimization details addressed in Section 18.4. The process starts from mixing 50 kmol/h of fresh 2-CP, 100 kmol/h fresh EtOH, 50 kmol/h of recycled 2-CP, and 99.7 kmol/h of recycled stream (89.30 mol% EtOH, 9.39% mol% DEC and 1.31 mol% for others) to become a combine liquid mixture. This mixture is pumped to 50 bar and preheated to 130°C (using LPS listed in Table 18.6), before being sent to the reactor. The latter is simulated as an isothermal plug flow reactor with the present of both vapor and liquid phases. The reactor is operated at 50 bar, and has 1 bar pressure drop along the reactor. Besides, the catalyst density is assumed as 1500 kg/m^3, and the void fraction in the reactor

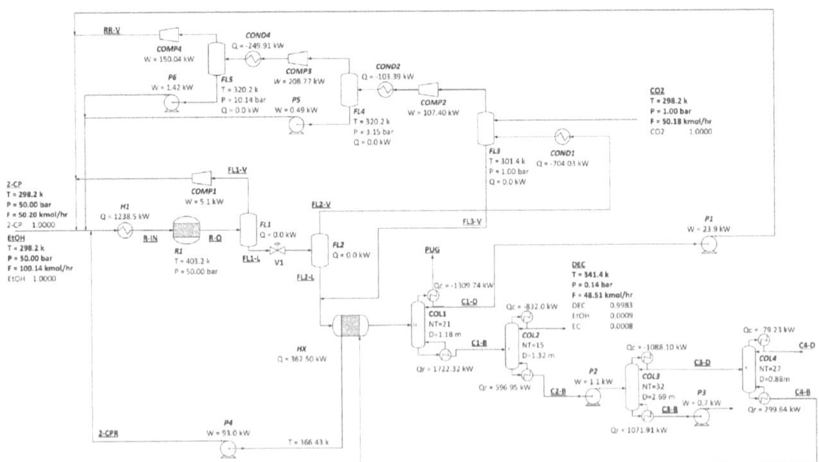

FIGURE 18.2 The optimized flowsheet of the DEC production process.

is set to 80%. The reactor has a length of 4.11 m and a diameter of 1.5 m, leading to 50% conversion of EtOH.

Effluent of the reactor is flashed under the same temperature and pressure (**Flash-1**), which separates the majority of CO_2 from the vapor stream. This stream is compressed to compensate the pressure drop from the reactor. On the other hand, the liquid stream from **Flash-1** is sent to another knock-out drum (**Flash-2**) that is operated at 1.5 atm to remove another portion of the dissolved CO_2 from the generated vapor. As the vapor stream from **Flash-2** contains a certain amount of DEC, it is condensed to 37.1°C, and stripped by 50.2 kmol/h of fresh CO_2 (enters at 25°C) in **Flash-3**. Through this, 90% of DEC in the **Flash-2** vapor is recovered. Next, the **Flash-3** vapor is pressurized via a 3-staged compression, and mixed with **Flash-1** vapor to be recycled to the reactor. Note that in each compression stage, the compression ratio is set at 3.25, and the compressor efficiency is assumed as 85%. The inter-stage coolers lower the temperature to 47°C by using cooling water. A small portion of condensate is separated to avoid the present of liquid phase in the compressors. As the condensate contains EtOH and DEC, they are pumped to 50 bar, in order to be recycled to the reactor.

The liquid streams from Flash-2 and Flash-3 are combined and pre-heated in a heat exchanger, before it is sent to the distillation sequence. For the latter, column **COL1** is used to recover the unreacted EtOH from the distillate. As depicted in Fig. 18.3, close-boiling feature presents between DEC and EtOH, indicating the difficulty to separate EtOH to high purity. Here, the targeted EtOH purity in the distillate is set at 90 mol%. Also, in order to remove the remaining gaseous compounds, a partial vapor-liquid condenser is equipped at the column top, operated at 47°C. The bottom stream from **COL1** is sent to **COL2**, aiming at separating 99.9 wt% DEC from the top. In these two columns, the purity targets were reached by adjusting the corresponding reboiler duties using the built-in "Design Specification/Vary." The reflux ratios of these two columns (i.e., 0.183 for **COL1**, and 0.500 for **COL2**) are determined from

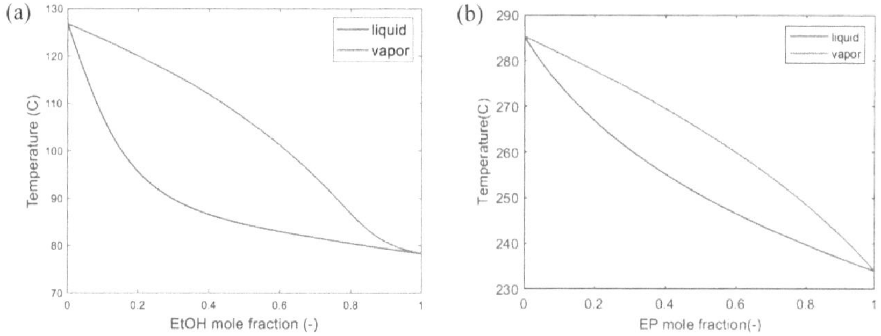

FIGURE 18.3 (A) Txy diagram of EtOH and DEC at normal pressure. (B) Txy diagram of 2-PA and EP at normal pressure.

optimization. The **COL2** bottom contains 2-CP, 2-PA and other side products, and is purified using another two columns. **COL3** performs splitting of 2-CP and 2-PA, whilst **COL4** removes the slight amount of impurity (i.e., EC) in order to recycle 2-CP. Before being recycled, the **COL4** bottom is cooled through exchanging heat with the **COL1** feed. Also note that the heavy components have boiling points up of beyond 300°C. Hence, **COL2** should be operated under vacuum to enable the use of high pressure steam (HPS) (see Table 18.6) as the heat source.

In the current process, the **COL3** bottom is a mixture of 2-PA (95.82 mol %), EP (3.6 mol%) and some trace amount of other species (0.58 mol%). The purpose of this stream poses another environmental issue, as 2-PA has not find its commercial usages so far. However, if 2-PA is useable, further separation of 2-PA from this mixture to the desired purity could be realized via simple distillation, as neither azeotrope or close-boiling behavior is discovered from the Txy diagram in between 2-PA and EP. In view of process sustainability, a better option is to convert 2-PA back to 2-CP. This reduced the net requirement of 2-CP and at the same time reduced the production of 2-PA. However, only one paper by Honda et al. investigated into this reaction. They reported that in order to get a high 2-CP yield, it was required to dissolve 2-PA into a great amount of mesitylene (MES) solvent (i.e., 5 mmol 2-PA in 143 mmol MES), and the reaction was proceeded by an impractically long residence time of 500 h, indicating that the reaction kinetics was unfavorable. Hence, further experimental exploration in the regeneration reaction is required. For simplicity, the current **COL3** bottom stream is regarded as a product in the later economic evaluation.

18.3.2 Optimization

Simulated annealing (SA) algorithm is used for optimization in this work. Its objective is set to minimize the cost function, as defined in Eq. (18.19). The SA method was first proposed by Kirkpatrick et al. (1983) based on the work of Metropolis et al. (1953). SA has become a popular probabilistic method applied for simulator-based optimization problems, as it is capable in finding global solution, and can be programmed easily. In general, SA compares the optimization problem to the annealing process of a metallurgical material, in which the molecules tend to find the lower energy state to settle under the given temperature. With the gradual temperature decrease, the molecules eventually settle at the lowest energy state. Here, the cost function defined in Eq. (18.19) represents the energy level, while the optimization variables resembles the molecules. On the other hand, the temperature is a user-defined variable which indicates the moving ability of the molecules. In this work, the SA algorithm was coded in MATLAB v2021, which is connected with the Aspen files through ActiveX server.

The optimization variables in this process are the total number of trays and the feed location of each column, the operating pressure of the three vacuumed columns (**COL2**, **COL3**, and **COL4**), and the temperature at the partial vapor-liquid condenser of **COL1**. The variables with values fixed during optimization are mentioned as follows. First, the operating pressure of COL1 is set at normal pressure in order to eliminate the use of more expensive heating or cooling utilities at both end of this column, as well as avoid the loss of useful species (i.e., EtOH, DEC) from the purged vapor. Second, all the flash drums are operated adiabatically at the indicated pressures. Third, for safety concerns, the compression ratio of each compressor is set at 3.25. Also, each intermediate cooling temperature is set at 47°C for enabling the use of cooling water. The flowchart for optimization can be found in Fig. 18.4. The general concept is that a new set of variables is randomly generated, based on the rules in Eqs. (18.20) and (18.21):

$$\text{discrete variables}: \quad \theta_{\text{new}} = \theta_{\text{old}} + \text{randn}[-1, 1] * \text{Step} \tag{18.20}$$

$$\text{continuous variables}: \quad \theta_{\text{new}} = \theta_{\text{old}} + 2 * (\text{rand}() - 1) * \text{Step} \tag{18.21}$$

In these equations, the MATLAB built-in command, "randn[−1, 1]," generates a random integer in between −1/1/0, enabling the discrete variables to move one step backward/forward, or no move. On the other hand, "rand()" generates a uniformly distributed random number in between −1 and 1. This enables the continuous variables to move to any value in between one step forward/backward from the current value. After the new variable set is generated, the cost is calculated. If cost from the new run is lower than the previous one, the new set of variables is accepted, and is used as the basis for next run. However, once the cost from the new run is higher, this configuration is conditionally accepted using the metropolis criteria, listed as in Eq. (18.22):

$$\text{Accept probability} = \exp\left(\frac{\text{Cost}_{\text{old}} - \text{Cost}_{\text{new}}}{\text{T}}\right) \tag{18.22}$$

This indicates that a lower cost configuration will always be accepted while a higher cost configuration could be easily accepted under higher temperature. Based on these rules, hundreds or thousands of simulation runs are successively performed, and the process eventually settles at a configuration that has the lowest cost.

The important structural parameters used in the simulated annealing algorithm, including the initial temperature (T_0), equilibrium number of iteration (EqNT), annealing rate (k), and settling temperature (T_f) were set based on literature (Yu et al., 2020; Yu, 2020; Lee et al., 2021), and are described as follows. First, T_0 (in the same unit of energy state) is set to enable the acceptance ratio be higher than 80%. For determining T_0, several hundreds of simulation runs were performed on the base case, in order to observe how

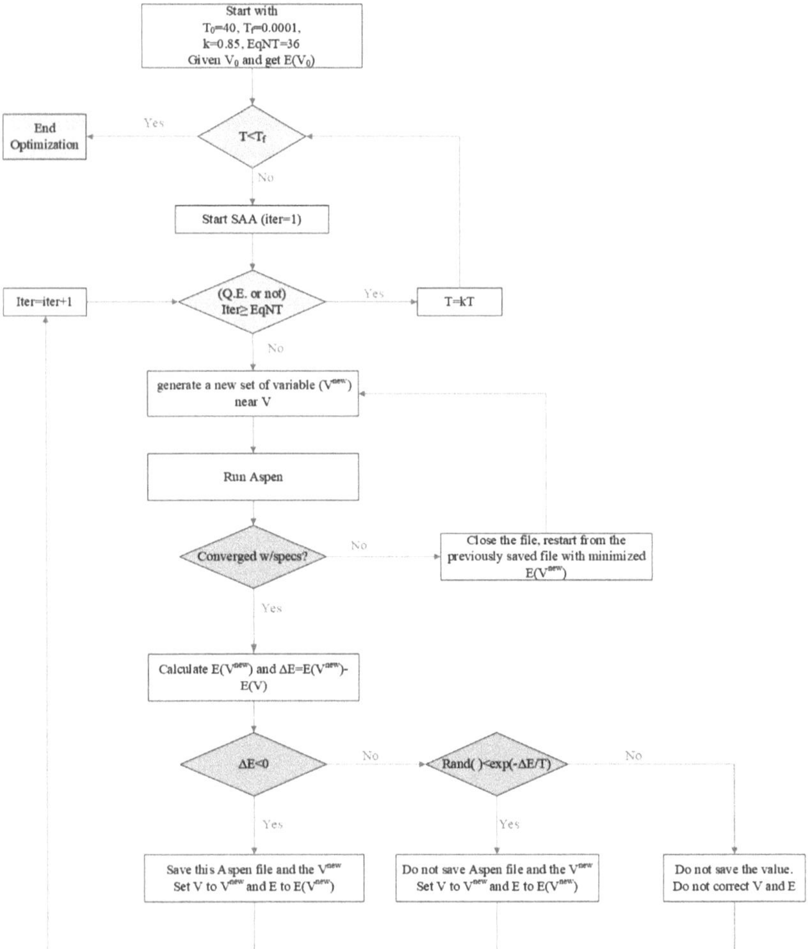

FIGURE 18.4 The flowchart for process variables optimization using simulated annealing.

much percentage of runs is accepted by the algorithm. In this work, T_0 is set to 40. Second, EqNT is typically set at 3 times of the total number of optimization variables. Hence, it is set at 36 here in this work. The annealing rate (k) is suggested to be within 0.8–0.9, and is set at 0.85 here. Note that once the annealing rate is set too small, the optimization problem will not be thoroughly performed. Finally, T_f is set at 10^{-4}. This value should be adequately small to ensure the optimized results within reasonable computation efforts.

As both continuous and discrete variables are optimized, the optimal configurations from different runs may not be identical. Hence, optimization exercises should be performed several times to ensure consistency. For this example, three runs of optimization were performed, starting from different base configurations. The change of cost function is tracked in Fig. 18.5. As depicted, the settling values of cost function from different optimization runs are very similar. This ensures the consistency of the algorithm. The optimal configuration holding the lowest cost among the three runs (i.e., case 3) is depicted in Fig. 18.5.

The comparison of the base and the optimal configurations are provided in Table 18.7. It is noteworthy that TC_1 settles at around 47°C, which is considered as the lowest allowable temperature for using the less expensive cooling water. Moreover, the settling values for P3 and P4 enable the use of HPS at the reboiler, as well as the recovery of condenser duty by generating steam. From the optimization results, it is noticed that NT1 varies the greatest in between cases. For NT_1, it is resulted from using a larger payback period (i.e., 8 years) for evaluating the capital cost, and hence the influence of its variation on the process cost is diluted.

18.4 Techno-economic and CO_2 emission analysis

18.4.1 Techno-economic analysis

The techno-economic evaluation is performed on the optimized processes, using the financial model listed in Table 18.8. Here, the minimum required selling prices (MRSP) leading to an internal rate of return (IRR) of 15% is calculated. The MRSP is compared with the current market price for its

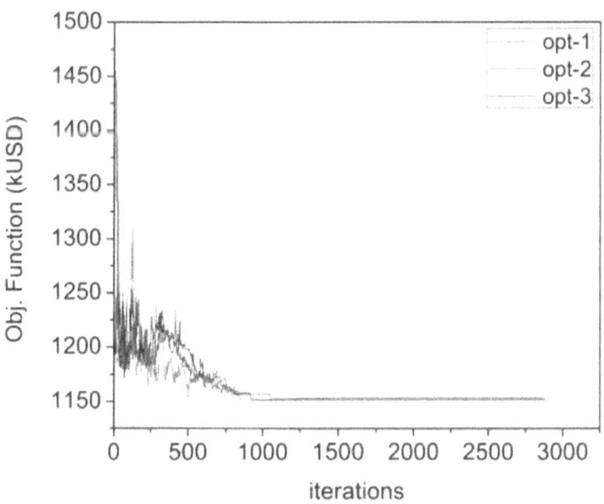

FIGURE 18.5 Tracking of the objective function during optimization.

TABLE 18.7 Comparison of the optimized variable under different conditions.

	Base 1	Optim. 1	Base 2	Optim. 2	Base 3	Optim. 3
NT_1	16	29	19	21	19	21
NF_1	4	19	9	10	9	13
NT_2	22	13	23	15	23	15
NF_2	17	7	18	7	18	8
NT_3	30	33	44	32	36	32
NF_3	15	8	10	8	13	7
TC_1 (°C)	47.03	47.01	47.09	47.00	47.50	47.20
P_2 (barg)	−0.95525	−0.95468	−0.88022	−0.826	−0.9	−0.87575
P_3 (barg)	−0.86325	−0.92192	−0.90634	−0.92934	−0.9	−0.93822
NT_4	20	30	20	25	22	27
NF_4	10	11	13	9	13	9
P_4 (barg)	−0.5	−0.69344	−0.49209	−0.58572	−0.55	−0.49721
TAC (kUSD)	1446.9	1153.1	1252.5	1151.8	1231.2	1151.0
CO_2-e (kg/kg)	−0.101	−0.154	−0.139	−0.150	−0.140	−0.154

TABLE 18.8 The economic model employed in this work.

Items	Descriptions
Total capital cost (TCC)	Based on the bare module cost in Table 18.5
Total operating cost[a] (TOC)	Based on the data in Table 18.6
Fixed capital investment	1.68[a] C_{BM}
Labor cost	14 workers, annual salary = 50,000 USD
Raw material cost[a]	Calculated based on the following cost data: EtOH: 0.47 USD/kg CO_2: 0.07 USD/kg 2-CP: 10 USD/kg
Cost of manufacturing	18%FCI + 2.76C_L + 1.26 (C_{RM} + C_{OPT} + C_{WT}) C_{WT}: 36 USD/1000 ton
Revenues[a]	Calculated based on the following cost data: DEC: MRSP determined based on 15% IRR 2-PA: 8.8 USD/kg Recovered steam: see Table 18.6
Depreciation	MACRS 7 years
Income tax rate	35%
Period of construction	3 years
Project life	12 years

[a]*Assuming 8000 h of annual operation.*

economic feasibility. In the current analysis, 2-CP is regarded as a readily available raw material with unit cost of 10 USD/kg. On the other hand, 2-PA is recovered as a product, with a unit sales price of 8.8 USD/kg (denoted as "$UP_{2\text{-}PA}$").

The itemized results from the techno-economic analysis are provided in Table 18.9. In general, three economic hotspots could be identified. The compression section is the first hotspot, which accounts for about 30% of the TCE and TOC, respectively. The columns used for recovering the reactant (**COL-1**) and performing product purification (**COL-2**) comprises another hotspot, which contribute 23% of TCE and 49% of TOC. The third hotspot refers to the columns (i.e., **COL-3** and **COL-4**) performing the splitting of heavy-boiling components. These two columns combine to contribute 31% TCE and 25% TOC. Note that the condenser temperature of these columns are high enough for generating steam. The values listed in Table 18.9 already consider the benefits from this heat recovery. Based on the aforementioned

TABLE 18.9 Results from the techno-economic analysis.

	Capital cost	Operating cost
Reaction	661.57 (17.21%)	20.16 (2.95%)
Compression	1179.63 (30.68%)	205.03 (30.00%)
COL-1	502.45 (13.07%)	257.47 (37.67%)
COL-2	381.57 (9.92%)	76.99 (11.27%)
COL-3	872.9 (22.70%)	90.31 (13.21%)
COL-4	246.52 (6.41%)	33.47 (4.90%)
Total (×1000 USD)	3844.64	683.43
Raw material cost (×1000 USD)	436,700	
Cost of manufacturing (×1000 USD)	541,080	
Total revenue (×1000 USD)	543,140	
MRSP (USD/kg)	2.38	

information, the MRSP of DEC from this process is calculated as 2.38 USD/kg, which is comparable to the current market price of 1.9 USD/kg.

The influences of unit price of 2-CP and 2-PA on MRSP were investigated via a sensitivity analysis. For convenience, the price difference in between 2-CP and 2-PA was defined for demonstrating the results. In this sensitivity test, the unit price of 2-CP was varied in the range of 2–10 USD/kg, while the price difference was change from 0.4 to 1.6 USD/kg. This enables a thorough investigation of their impacts on the economics. Note that the presumed price differences always give the 2-PA prices lower than that of 2-CP. This assumption avoids an unreasonable situation which tends to compensate the high MRSP of the products using the benefit from 2-PA recovery.

The sensitivity analysis results are provided in Table 18.10. In general, it quantifies the potential extent of improvement on the process economics. The most optimistic combination of these factors is to set 2-CP unit price at 2 USD/kg, and a price difference of 0.4 USD/kg (i.e., the 2-PA unit price at 1.6 USD/kg). Under this circumstance, the MRSP becomes 1.16 USD/kg, which is more competitive as compared with the current/future prices.

18.4.2 CO_2-emission analysis

The result of CO_2 emission analysis on the optimized process is demonstrated through Table 18.11. The contribution to CO_2 emission includes the steam and

TABLE 18.10 Sensitivity analysis on the MRSP of DEC to the price of 2-CP and 2-PA.

	Price differences (USD/kg)			
2-CP unit price (USD/kg)	0.4	0.8	1.2	1.6
10	1.51	1.94	2.38[a]	2.81
8	1.42	1.85	2.29	2.72
6	1.33	1.77	2.20	2.63
4	1.24	1.68	2.11	2.54
2	1.16	1.59	2.02	2.45

[a]*Base case study.*

TABLE 18.11 Results of carbon emission analysis.

		Amount of CO_2 emission (Ton/y)
Emission-	Comp	1295.40 (E)
	QR_1	3907.33 (MPS)
	QR_2	1073.83 (MPS)
	QR_3	3117.33 (HPS)
	QR_4	698.78 (HPS)
	Heater	2599.67 (LPS)
Saving from steam		−2208.17
Consumption		−17,669.18
Purge		137.72
Product		45,812.41
CO_2-e (kg/kg)		**−0.154**

electricity consumption, and the CO_2 purged from the process. On the other hand, those factors leading to CO_2 reduction include its use as reactants, and the recovery of waste by generating steam. As demonstrated, producing 1 kg of DEC leads to 0.154 kg of net CO_2 reduction, indicating that this route is attractive from the view point of CO_2 utilization.

Note that in this work, the economically competitive MRSP for DEC is meant to reach net-negative greenhouse gas emission. Two aspects could be suggested for a deeper exploration into this topic. Firstly, for being more sustainable, regeneration of 2-CP from 2-PA could be considered. This decreases the net usage of 2-CP and avoids the production of 2-PA. Moreover, the economic impacts on the process from these two compounds could be reduced. However, 2-CP regeneration could result in net CO_2 emission in the overall process. More experimental investigation in the regeneration reaction (i.e., in kinetics, behavior of side products, etc.) are necessary, which enable the further investigations on this issue. Second, further improvement on the conversion of EtOH to DEC (currently 50%) would be helpful. Not only that it increases the throughput, it also reduces the separation load downstream. Higher conversion is helpful in improving both economic and environmental performances of this process.

18.5 Conclusions

In this chapter, the design of an economically feasible and carbon negative direct CO_2-to-diethyl carbonate (DEC) process was demonstrated. The detailed investigation (or supplement) on the physical properties, the reaction kinetics, the process configuration, and optimization were systematically presented. Next, the minimum required selling price of DEC (i.e., 2.3 USD/kg, reaching 15% internal rate of return), and the CO_2 emission rate per kg of DEC formed (CO_2-e = −0.154) were calculated to reveal the attractiveness economically and environmentally.

Exercises

1. The entry D in Table 18.1 presents the production path of diethyl carbonate through ethanolysis of urea. For each component involved in this reaction, please find all the necessary properties (listed in Table 18.2) in the Aspen database. Use the group contribution method listed in Table 18.3 for estimation if necessary.
2. In the proposed DEC process, the bottom stream from COL3, obtained at 494.3 (K), 0.23 bar and 50.15 kmol/h, has a composition of 96.56 mol% 2-PA, 2.95% EO, 0.39% EPI and 0.10% 2-CP. Please develop a distillation-based downstream process for 2-PA purification (targeted purity: 99 wt%). [Note: the required physical properties are listed in the appendix]
3. Starting from the simulation/computation files provided with this chapter, please re-investigate the optimization problem encompassing the following changes: (1) The payback period is shortened to 3 years. (2) The cost of all the heating utilities are increased by 10% due to the risen fossil fuel price. (3) The DEC unit price increases to 1.8 USD/kg. Try to compare your optimized process configuration with the one reported in this chapter.

Appendix

A.1 Parameters for pure-component properties

1. Scalar Properties

TABLE A.1 Parameters for pure-component properties.

					1. Scalar properties						
Parameters	Units	CO_2	EtOH	DEC	H_2O	NH_3	EC	2-CP	2-PA	EP	EPI
DGFORM	cal/mol	−94,193.7	−40,090.3	−110,980	−54,593.5	−3917.07	***−64,378.5***	***76,975.3***	***30,245.1***	***−24,942.7***	***72,738.1***
DHFORM	cal/mol	−93,988.2	−56,116.8	−152,646	−57,757.2	−10,962.5	−107,118	67,037.8	−15,104.6	***−62,779.7***	***30,328.7***
OMEGA	–	0.223621	0.643558	0.485692	0.344861	0.252608	0.753264	*0.204198*	1.02125	0.73264	*0.356692*
PC	bar	73.83	61.37	33.9	220.64	112.8	93.3633	47.56	49.5074	36.1421	*36.467*
TC	C	31.06	240.85	302.85	373.946	132.5	407.85	543.85	542.85	427.85	*538.883*
VC	cc/mol	94	168	356	55.9472	72.47	269.578	302.863	331.665	458.998	*569.1*
ZC	–	0.274	0.241	0.252	0.229	0.242	0.474385	0.209064	0.242016	0.284624	*0.307401*

Italic and bold: estimated via Aspen built-in GCM.
Italic: estimated via COSMO-based calculation.

2. Ideal gas heat capacity (units: cal/mol-K)

CPIGDP		1	2	3	4	5	6	7	8	9
CO_2	Aspen	7.014904	8.249737	1428	6.305532	588	−223.15	4726.85		
EtOH	Aspen	11.75122	34.81657	1662.8	22.42763	744.7	0	1226.85		
DEC	Aspen	22.71902	72.45629	1428.6	46.46269	668.2	26.85	1226.85		
H_2O	Aspen	7.968616	6.398682	2610.5	2.124773	1169	−173.15	2000		
NH_3	Aspen	7.983902	11.69867	2036	5.388363	882	−173.15	1226.85		
EPI	*COSMO*	*15.8885*	*44.8894*	*420.434*	*73.9488*	*1389.35*	*30.4291*	*1029.43*		
CPIALE										
2-CP	Aspen	12.60077	51.30959	1080.73	29.58202	487.811	0	8.31447	−73.15	726.85
2-PA	Aspen	18.44492	36.61699	550.816	33.61947	1509.13	0	8.31447	−73.15	726.85
EP	Aspen	17.69046	88.30539	535.418	−44.1533	605.304	0	8.31447	−73.15	726.85
EC	Aspen	14.60934	50.46241	1195.93	28.0171	530.455	0	8.31447	−73.15	726.85

Italic: estimated via COSMO-based calculation.

CPIGDP:

$$C_p^{*,ig} = C_{1i} + C_{2i}\left[\frac{C_{3i}/T}{sinh(C_{3i}/T)}\right]^2 + C_{4i}\left[\frac{C_{5i}/T}{sinh(C_5/T)}\right]^2, \ for \ C_{6i} < T < C_{7i}$$

CPIALE:

$$C_p^{*,ig} = C_{1i} + C_{2i}\left[\frac{C_{3i}/T}{sinh(C_{3i}/T)}\right]^2 + C_{4i}\left[\frac{C_{5i}/T}{sinh(C_5/T)}\right]^2, \ for \ C_{8i} < T < C_{9i}$$

3. Enthalpy of vaporization (unit: cal/mol, Temperature unit: K)

DHVLDP		**1**	**2**	**3**	**4**	**5**	**6**	**7**	**8**
CO_2	Aspen	5190.121	0.382	−0.4339	0.42213	0	−56.57	31.06	
EtOH	Aspen	15,723.46	1.1905	−1.7666	1.0012	0	−114.1	240.85	
DEC	Aspen	15,438.76	0.003765	1.7376	−1.4912	0	−43	302.85	
H_2O	Aspen	13,518.68	0.61204	−0.6257	0.3988	0	0.01	373.95	
NH_3	Aspen	7529.139	0.3914	−0.2289	0.2309	0	−77.74	132.5	
EPI	*COSMO*	*28,248.1*	*1.80763*	*−2.12029*	*0.85059*	*0*	*227.369*	*730.83*	
2-CP	*COSMO*	*22,158.5*	*1.647*	*−2.01009*	*0.88419*	*0*	*194.266*	*728.497*	
DHVLTD									
2-PA	Aspen	10.66176	1.12688	−0.60232	−0.04543	542.85	4	26.85	542.85
EP	Aspen	10.60105	2.5611	−3.1306	1.07653	427.85	4	−48.15	427.85
EC	Aspen	10.38945	2.10666	−1.99665	0.391679	407.85	4	48.248	407.85

Italic: estimated via COSMO-based calculation.

DHVLDP:

$$\Delta_{vap}H_i^* = C_{1i}(1 - T_{ri})^{\left(C_{2i}+C_{3i}T_{ri}+C_{4i}T_{ri}^2+C_{5i}T_{ri}^3\right)},\ for\ C_{6i} \leq T \leq C_{7i}$$

DHVLTD:

$$\ln\left(\Delta_{vap}H_i^*\right) = C_{1i} + \sum_{m=2}^{nTerms} C_{mi}T_r^{m-2}ln(1 - T_r),\ T_r = \frac{T}{T_c}$$

4. Vapor pressure

PLXANT		Units	1	2	3	4	5	6	7	8	9
CO_2	Aspen	bar	35.50407	−2839	0	0	−3.8639	2.81E−16	6	−56.57	31.06
EtOH	Aspen	bar	61.79107	−7122.3	0	0	−7.1424	2.89E−06	2	−114.1	240.85
DEC	Aspen	bar	197.6371	−12,732	0	0	−28.31	2.39E−05	2	−43	302.85
H_2O	Aspen	bar	62.13607	−7258.2	0	0	−7.3037	4.17E−06	2	0.01	373.95
NH_3	Aspen	bar	78.97007	−4669.7	0	0	−11.607	0.017194	1	−77.74	132.5
2-CP	COSMO	Pa	*97.8276*	*−9984.1*	*0*	*0*	*−10.8541*	*3.70E−06*	*2.01878*	*194.266*	*728.497*
EPI	COSMO	Pa	*125.254*	*−12,475.8*	*0*	*0*	*−14.7204*	*2.54E−05*	*1.78115*	*227.369*	*730.83*
WAGNER											
2-PA	Aspen	bar	−11.7923	4.9566	−10.5842	−5.30105	3.902125	542.85	25	542.85	
EP	Aspen	bar	−8.65529	−0.28912	−2.61643	−8.0715	3.587465	427.85	−53.15	427.85	
EC	Aspen	bar	−10.0981	1.23846	0	0	4.536505	407.85	48.248	407.85	

Italic: estimated via COSMO-based calculation.

PLXANT:

$$\ln P_i^{*,l} = C_{1i} + \frac{C_{2i}}{T + C_{3i}} + C_{4i}T + C_{5i}\ln T + C_{6i}T^{C_{7i}},\ for\ C_{8i} \leq T \leq C_{9i}$$

WAGNER:

$$\ln P_{ri}^{*,l} = \frac{C_{1i}(1 - T_{ri}) + C_{2i}(1 - T_{ri})^{1.5} + C_{3i}(1 - T_{ri})^{3} + C_{4i}(1 - T_{ri})^{6}}{T_{ri}},$$

$$where\ T_{ri} = \frac{T}{T_c},\ P_{ri}^{*,l} = \frac{P_i^{*,l}}{P_{ci}}$$

A.2 Binary interaction parameters for the NRTL model

TABLE A.2 Binary interaction parameters for the NRTL model.

Comp i	Comp j	Source	AIJ	AJI	BIJ	BJI	CIJ	DIJ	EIJ	EJI	FIJ	FJI
EtOH	H_2O	Aspen	−0.8009	3.4578	246.18	−586.081	0.3	0	0	0	0	0
H_2O	NH_3	Aspen	−6.2684	9.6121	1525.454	−3232.82	0.3	0	0	0	0	0
EtOH	DEC	Aspen	0.560888	−6.98708	−18.8314	2754.07	0.298159	0	0	0	0	0
EtOH	NH_3	Aspen	0	0	13.1784	−311.188	0.3	0	0	0	0	0
DEC	H_2O	Aspen	−0.60219	1.49444	911.975	1012.22	0.341099	0	0	0	0	0
CO_2	EtOH	Aspen	−0.02476	1.19821	0	0	0.48223	0	0	0	0	0
CO_2	H_2O	Aspen	10.064	10.064	−3268.14	−3268.14	0.2	0	0	0	0	0
H_2O	MES	Aspen	10.9461	−7.63942	−398.12	3551.77	0.154977	0	0	0	0	0
DEC	NH_3	COSMO	54.5857	−22.5422	−0.33369	24.669	0.3	0	−10.3445	4.25675	0.0233	−0.01049
EtOH	2-CP	COSMO	6.0941	7.31675	0.682505	5.31285	0.3	0	−1.17298	−1.13315	0.002596	0.000259
EtOH	2-PA	COSMO	−6.02542	−14.1011	23.3619	18.3212	0.3	0	0.92899	2.75705	0.000386	−0.00676
EtOH	EPI	COSMO	22.6273	−6.85973	−12.9908	0.043679	0.3	0	−3.7258	0.854017	0.002318	0.001873
EtOH	EP	COSMO	5.7103	−3.26784	6.6474	10.2462	0.3	0	−1.04099	0.680079	0.001336	−0.00147
EtOH	EC	COSMO	−9.97006	−14.1598	29.4171	23.1938	0.3	0	1.81754	2.51088	−0.00266	−0.00415
DEC	2-CP	COSMO	−10.5387	9.51712	16.9828	9.1747	0.3	0	2.0287	−1.81319	−0.00468	0.004103
DEC	2-PA	COSMO	29.5649	−21.7333	−24.1206	23.7607	0.3	0	−5.57448	4.11553	0.010803	−0.00793
DEC	EPI	COSMO	13.7215	−8.48593	−1.38305	10.4625	0.3	0	−2.51106	1.5396	0.004662	−0.00287
DEC	EP	COSMO	5.88906	−4.87852	2.47798	10.1285	0.3	0	−1.0768	0.88505	0.002162	−0.00177
DEC	EC	COSMO	19.5978	−16.0008	−9.40707	29.9245	0.3	0	−3.68009	3.05903	0.006708	−0.00578
H_2O	2-CP	COSMO	−11.2773	17.3986	0.768179	−34.9217	0.3	0	2.78253	−2.78882	−0.007	0.001652
H_2O	2-PA	COSMO	22.7392	−19.4311	18.8086	43.8431	0.3	0	−3.55213	3.54481	0.004524	−0.0065

H_2O	EPI	COSMO	11.168	−25.9749	2.17188	72.7837	0.3	0	−1.09271	4.9693	−0.00094	−0.01085
H_2O	EP	COSMO	−17.2504	−18.5295	24.188	32.5917	0.3	0	4.36693	3.85491	−0.01231	−0.01071
H_2O	EC	COSMO	33.411	−17.0919	−81.9747	8.53046	0.3	0	−5.54964	2.94963	0.007657	−0.00424
2-CP	2-PA	COSMO	4.49761	−6.10019	−2.92534	167.967	0.3	0	−0.7362	0.962009	0.000858	−0.00096
2-CP	EPI	COSMO	3.441	−0.55656	2.76489	7.44532	0.3	0	−0.44746	−0.08714	−0.00093	0.0017
2-CP	EP	COSMO	11.0961	−8.14258	0.702333	16.7224	0.3	0	−1.96357	1.40765	0.003077	−0.00211
2-CP	NH_3	COSMO	35.975	−14.3032	−48.1324	20.4251	0.3	0	−6.4414	2.47161	0.012823	−0.00516
2-CP	EC	COSMO	2.92083	1.34492	4.44401	6.52229	0.3	0	−0.45238	−0.31648	1.83E−05	0.001257
2-PA	EPI	COSMO	21.5464	−13.4564	−3.73274	16.5104	0.3	0	−3.8172	2.29396	0.005781	−0.00292
2-PA	EP	COSMO	28.8387	−19.7583	−21.7851	29.8059	0.3	0	−5.32428	3.57764	0.009607	−0.00592
2-PA	NH_3	COSMO	−20.9733	12.9576	37.5336	−1.05332	0.3	0	3.66748	−2.2101	−0.00665	0.00394
2-PA	EC	COSMO	−4.29928	2.96516	10.9356	2.94679	0.3	0	0.771308	−0.50687	−0.00162	0.001054
EPI	EP	COSMO	−4.59721	5.24143	8.34879	2.38563	0.3	0	0.826996	−0.94912	−0.00139	0.001618
EPI	NH_3	COSMO	−18.2992	13.9677	2.71174	−24.207	0.3	0	3.39461	−2.54708	−0.00822	0.006025
EPI	EC	COSMO	−17.9067	27.3455	100.861	7.34323	0.3	0	3.04467	−4.85946	−0.00402	0.007639
EP	NH_3	COSMO	46.0047	−16.9369	−53.8443	6.75312	0.3	0	−8.51571	3.06628	0.019526	−0.00768
EP	EC	COSMO	−19.7268	30.328	48.5866	−26.8201	0.3	0	3.53836	−5.54481	−0.00593	0.009988
NH_3	EC	COSMO	13.9825	−21.8987	3.85281	35.2546	0.3	0	−2.37434	3.78514	0.003971	−0.00633
CO_2	DEC	COSMO	8.91508	−13.2638	−0.19137	16.1946	0.3	0	−1.49759	2.28418	0.002947	−0.00449
CO_2	2-CP	COSMO	−4.98956	10.1909	−9.35352	−4.83108	0.3	0	0.688589	−1.57188	−0.00153	0.003419
CO_2	2-PA	COSMO	12.9657	1.2028	−16.531	5.67592	0.3	0	−1.92478	−0.59062	6.85E−05	0.002973
CO_2	EPI	COSMO	3.44419	−5.26142	4.61738	9.17245	0.3	0	−0.43393	0.725866	0.000828	−0.00135
CO_2	EP	COSMO	6.12512	−8.89839	−0.72378	5.57034	0.3	0	−0.93321	1.41134	0.001737	−0.00264
CO_2	NH_3	COSMO	32.5621	−15.5746	3.57801	49.2769	0.3	0	−5.73054	2.67361	0.008222	−0.00323
CO_2	EC	COSMO	22.085	−1.63484	−16.3114	9.98911	0.3	0	−3.74261	0.055755	0.005278	0.000556

A.3 Parameters for Henry's constant equation (temperature in °C)

TABLE A.3 Parameters for Henry's constant equation (temperature in °C).

Comp i	Comp j	Source	Units	AIJ	BIJ	CIJ	EIJ	TLOWER	TUPPER	EJI
CO_2	EtOH	Aspen	bar	89.58308	−5018.8	−11.891	0	10	40	0
CO_2	H_2O	Aspen	bar	159.8651	−8741.55	−21.669	0.001103	−0.15	79.85	0
CO_2	DEC	COSMO	bar	−64.9414	−310.468	13.74381	−0.02909	300	500	0
CO_2	2-CP	COSMO	bar	33.82405	−3062.01	−3.2636	−0.00251	300	500	0
CO_2	2-PA	COSMO	bar	34.57416	−2957.5	−3.41034	−0.0024	300	500	0
CO_2	EPI	COSMO	bar	34.74492	−3082.61	−3.4644	−0.0022	300	500	0
CO_2	EP	COSMO	bar	32.72526	−3127.01	−3.05382	−0.00341	300	500	0
CO_2	NH_3	COSMO	bar	−468.711	10,919.49	83.33925	−0.13306	300	450	0
CO_2	EC	COSMO	bar	15.99874	−2390.53	−0.16617	−0.00789	300	500	0

Supplementary materials

Supplementary data related to this article can be found online at doi:10.1016/B978-0-323-90168-0.00009-3

References

Aspen Technology, Aspen Plus V11 User Guide.

Gadalla, M.A., et al., 2005. Reducing CO_2 emissions and energy consumption of heat-integrated distillation systems. Environmental Science & Technology 39 (17), 6860–6870.

Giram, G.G., Bokade, V.V., Darbha, S., 2018. Direct synthesis of diethyl carbonate from ethanol and carbon dioxide over ceria catalysts. New Journal of Chemistry 42 (21), 17546–17552.

Gu, Y., et al., 2019. Direct synthesis of alternating polycarbonates from CO_2 and diols by using a catalyst system of CeO_2 and 2-furonitrile. ACS Sustainable Chemistry & Engineering 7 (6), 6304–6315.

Honda, M., et al., 2013. Ceria-catalyzed conversion of carbon dioxide into dimethyl carbonate with 2-cyanopyridine. Chemsuschem 6 (8), 1341–1344.

Inc, P.G, 2021. On-Site Nitrogen Generation. Available from: https://puritygas.ca/.

Jenkins, S., 2018. Chemical Engineering Magazine-Chemical Engineering Plant Cost Index: 2018 Annual Value.

Kirkpatrick, S., Gelatt, C.D., Vecchi, M.P., 1983. Optimization by simulated annealing. Science 220.

Lee, C.T., et al., 2021. Screening of CO_2 utilization routes from process simulation: design, optimization, environmental and techno-economic analysis. Journal of CO_2 Utilization 53.

Leino, E., et al., 2013. The influence of various synthesis methods on the catalytic activity of cerium oxide in one-pot synthesis of diethyl carbonate starting from CO_2, ethanol and butylene oxide. Catalysis Today 210, 47–54.

Lin, S.T., Sandler, S.I., 2002. A priori phase equilibrium prediction from a segment contribution solvation model. Industrial & Engineering Chemistry Research 41 (5), 899–913.

Metropolis, N., et al., 1953. Equation of state calculations by fast computing machines. The Journal of Chemical Physics 21.

Meylan, F.D., Moreau, V., Erkman, S., 2015. CO_2 utilization in the perspective of industrial ecology, an overview. Journal of CO_2 Utilization 12, 101–108.

Muskat, I.E., Strain, F., 1945. Preparation of Carbonic Acid Esters.

Muthuraj, R., Mekonnen, T., 2018. Recent progress in carbon dioxide (CO_2) as feedstock for sustainable materials development: co-polymers and polymer blends. Polymer 145, 348–373.

Ohno, H., Sato, H., Fukushima, Y., 2018. Configuration of materially retained carbon in our society: a WIO-MFA-based approach for Japan. Environmental Science & Technology 52 (7), 3899–3907.

Schmidt, W.P., et al., 2001. Safe design and operation of a cryogenic air separation unit. Process Safety Progress 20 (4), 269–279.

Shukla, K., Srivastava, V.C., 2016. Diethyl carbonate: critical review of synthesis routes, catalysts used and engineering aspects. Rsc Advances 6 (39), 32624–32645.

Shukla, K., Srivastava, V.C., 2018. Efficient synthesis of diethyl carbonate from propylene carbonate and ethanol using Mg-La catalysts: characterization, parametric, and thermodynamic analysis. Industrial & Engineering Chemistry Research 57 (38), 12726–12735.

Tamura, M., et al., 2018. Direct catalytic synthesis of N-arylcarbamates from CO_2, anilines and alcohols. Chemcatchem 10 (21), 4835–4839.

Tomishige, K., Tamura, M., Nakagawa, Y., 2019. CO_2 conversion with alcohols and amines into carbonates, ureas, and carbamates over CeO_2 catalyst in the presence and absence of 2-cyanopyridine. Chemical Record 19 (7), 1354–1379.

Turton, R., et al., 2018. Analysis, Synthesis, and Design of Chemical Processes, fifth ed. Pearson Education, Boston.

Wang, J.H., Hao, Z.P., Wohlrab, S., 2017. Continuous CO_2 esterification to diethyl carbonate (DEC) at atmospheric pressure: application of porous membranes for in situ H_2O removal. Green Chemistry 19 (15), 3595–3600.

Wei, H.Y., et al., 2011. Design and control of reactive-distillation process for the production of diethyl carbonate via two consecutive trans-esterification reactions. Journal of Process Control 21 (8), 1193–1207.

Yang, A., et al., 2019. Energy-saving investigation for diethyl carbonate synthesis through the reactive dividing wall column combining the vapor recompression heat pump or different pressure thermally coupled technique. Energy 172, 320–332.

Yu, B.Y., 2020. Development of two plant-wide glycerol carbonate production processes: design, optimization and environmental analysis. Journal of the Taiwan Institute of Chemical Engineers 117, 19–25.

Yu, B.Y., et al., 2020. Evaluating the direct CO_2 to diethyl carbonate (DEC) process: rigorous simulation, techno-economical and environmental evaluation. Journal of CO_2 Utilization 41, 101254.

Zhang, Y., et al., 2012. Experimental and kinetic studies on a homogeneous system for diethyl carbonate synthesis by transesterification. Chemical Engineering & Technology 35 (4), 693–699.

Zheng, L., et al., 2017. Design and control of reactive dividing-wall column for the synthesis of diethyl carbonate. Chemical Engineering and Processing-Process Intensification 111, 127–140.

Chapter 19

Multiplatform optimization on unit operation and process designs*

Vincentius Surya Kurnia Adi
Department of Chemical Engineering, National Chung Hsing University, Taichung, Taiwan

Chapter outline

19.1 Introduction

Many optimization problems on unit operation and process designs involve multiple conflicting objectives that should be considered simultaneously. For example, to design a reactor, one may want to optimize the reaction temperature for optimal conversion. However, its economic benefit should also be optimized simultaneously. This problem can be formulated as a multiobjective optimization problem (also known as multiobjective mathematical

*Simulation and other electronic files of this chapter are made available on a companion website; URL: https://www.elsevier.com/books-and-journals/book-companion/9780323901680.

Chemical Engineering Process Simulation. https://doi.org/10.1016/B978-0-323-90168-0.00019-6

programming). The set of feasible solutions is not explicitly known in advance but is restricted by constraint functions.

Moreover, there is a hierarchy of decision-makers in many decision processes, where decisions are made at different levels. One way to handle such hierarchies is to focus on one level and include properties of other levels as assumptions. *Multilevel programming* is a research area that focuses on the entire hierarchical structure. In terms of modeling, the constraint domain associated with a multilevel programming problem is implicitly determined by a series of optimization problems that must be solved in a predetermined sequence. If a two-level optimization problem is considered, the upper level is called the leader and the lower level as the follower. Note that most of the available tools are prepared for single-level optimization. Hence, it is necessary to implement multiobjective and multilevel optimization as a multiplatform optimization problem to achieve optimal solutions.

In multiobjective optimization problems, there exists no unique solution. In most cases, several equally good solutions exist, because multiobjective optimization task considers several conflicting objectives simultaneously. The solutions are a set of alternatives with different trade-offs, called *Pareto optimal solutions* or *non-dominated* solutions. Despite multiple Pareto optimal solutions, in practice, usually only one of these solutions is to be chosen. In a multiobjective optimization problem, there are at least two equally important tasks: an optimization task for finding Pareto optimal solutions (involving a computer-based procedure) and a decision-making task for choosing a single most preferred solution. The latter typically necessitates preference information from the user (Branke et al., 2008). Multiobjective optimization problems have been intensively studied, for example, in Pareto (1896, 1906) and Kuhn and Tucker (1951). Many ideas and approaches have their foundation in the theory of mathematical programming. Kuhn and Tucker (1951) also developed multiobjective optimization problems while formulating optimality conditions of nonlinear programming.

In general, solving a multiobjective optimization problem will help users make decisions by considering the multiple objectives simultaneously and finding an optimal solution that is the most acceptable. Thus, the solution process needs some involvement of the user in specifying preference information. The final solution can then be chosen by the users in one way or the other. Therefore, the user may use different tools to specify the multiobjective optimization problem and use several tools and/or platforms to find the respective solutions.

Multilevel programming was first defined and studied by Bracken and McGill (1973, 1974) as a generalization of mathematical programming. A variety of special cases were later reported. In the last decade, multilevel programming problems have been extensively studied in their general setting. A bibliography review on multilevel programming is reported by Vicente and Calamai (1994). Moreover, a much detailed discussion on multilevel optimization can be found in the work by Migdalas et al. (1998).

Several tools and platforms can be used to construct the desired multi-objective and multilevel optimization methods. For example, Microsoft Excel can be exploited through Aspen Simulation Workbook in Aspen Plus software as an add-in that links simulation with process data. Besides, the programming software of MATLAB can also be made interacting with Aspen Plus through a COM Automation Server. Hence, the structure of the multiplatform optimization depends on the feature of the simulation software to allow interaction with external mathematical tools through its automation interface (see Fig. 19.1).

19.2 Aspen Plus automation interface

The Aspen Plus Windows user interface is an ActiveX Automation Server. The ActiveX technology enables an external Windows application to interact with Aspen Plus through a programming interface such as Microsoft's Visual Basic. The server exposes objects through the Microsoft Component Object Model (COM). With the ActiveX Automation Server, users can explore the following features:

- Connect both inputs and results of Aspen Plus simulations to other applications such as design programs or databases.
- Write own user interface to an Aspen Plus plant model. This interface allows others to run the Aspen Plus model without learning how to use its user interface.

Note, however, that to use Aspen Plus Automation Server, one must have Aspen Plus installed on the computer and have its user license.

19.3 COM objects in MATLAB

The Microsoft Component Object Model (COM) provides a framework for integrating reusable, binary software components into an application. As program components are implemented with compiled code, the source code can be written in programming languages that support COM. Upgrades to applications are simplified, as program components can simply be swapped without recompiling the entire application. In addition, a program

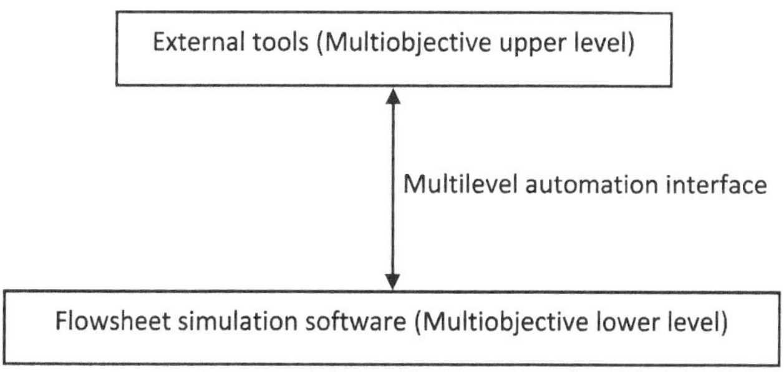

FIGURE 19.1 Multiobjective and multilevel problem in multiplatform optimization structure.

component's location is transparent to the application, so program components can be relocated to a separate process or even a remote system without having to modify the application.

As mentioned above, the Aspen Plus user interface is a COM Automation Server. The automation technology enables an external Windows application to interact with Aspen Plus through a programming interface using a language such as Microsoft's Visual Basic or C#. The following code shows how to make the connection from MATLAB.

```
AspenVersion = 'apwn.document'; % Programmatic identifier (ProgID) of
Aspen Plus document class.
% Also 'apwn.document.34' to specify V8.8 as the version to be used.
AspenPath = 'C:\Users\vska\Downloads\opt.bkp'; % Replace with the
simulation path
AspenVisible = 0; % 0 for not visible
AspenDialogs = 1;

% Get pointer and load Aspen Plus comserver:
Aspen = actxserver(AspenVersion);
% Open the Aspen-Simulation
invoke(Aspen,'InitFromFile2',AspenPath);
Aspen.visible = AspenVisible; % Make it visible
Aspen.SuppressDialogs = AspenDialogs;

%Reinit and run simulation
Aspen.Reinit();
Aspen.Run2();
% Wait to update GUI
pause(1);

%Release COM object
Aspen.Quit();
Aspen.Parent.Quit();
delete(Aspen);
```

19.4 Aspen Simulation Workbook (ASW)

Aspen Simulation Workbook (ASW) can assist the user in creating a Microsoft Excel interface for their simulations. With ASW, the user can run their simulation, obtain and set variable values in Microsoft Excel cells. In some cases, the user might want to add additional custom automation of their simulation. Three types of automation are possible with ASW, which include:

1. User Defined Functions (UDFs) are custom Microsoft Excel functions to retrieve information from their simulation. UDFs are the simplest form of Microsoft Excel automation that require no programming.
2. Action Macros: These are a series of macros delivered with ASW for completing common tasks. The macros are most commonly used and run from a workbook without writing code. Moreover, it can also be called from within custom VBA code written by the user.

3. ASW COM Automation: ASW exposes an interface that allows users to write custom VBA scripts to further control their simulation and thus is the most complicated method of automation. It is only accessible via VBA code. This feature is used for tasks not covered by other automation methods (e.g., setting simulation path).

The UDFs and Action Macros are delivered in the form of XLA files. The XLA files are automatically opened when the ASW is enabled by the user. On the other hand, the ASW COM automation is facilitated by a series of COM libraries installed and registered on the user system by ASW. Before understanding the different methods of automating ASW, it is helpful to know what files are delivered with ASW that support automation. There are two types of files that ASW uses for automation: Excel XLA files and Active X COM DLL files.

When user enables the ASW Excel add-in, the ASW custom ribbon tab and two "XLA" files will be added into Excel. These files can be seen if the user enters the Excel Visual Basic for Applications Editor (Developer Ribbon Tab → Visual Basic). Fig. 19.2 shows steps to enable the tab. The names of the two files are:

- AspenOSEWorkbook.xla
- AspenSimulationWorkbook.xla

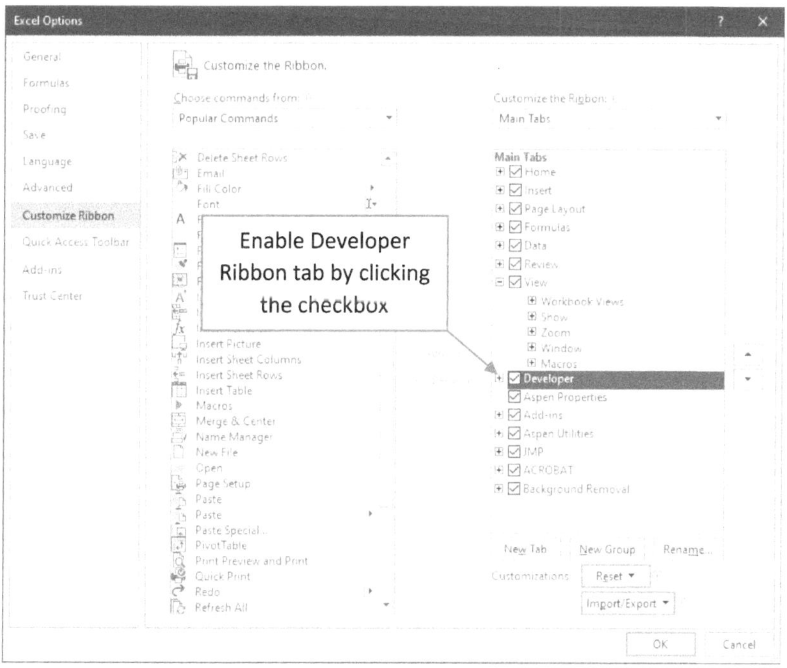

FIGURE 19.2 Steps to enable Developer Ribbon tab in Excel.

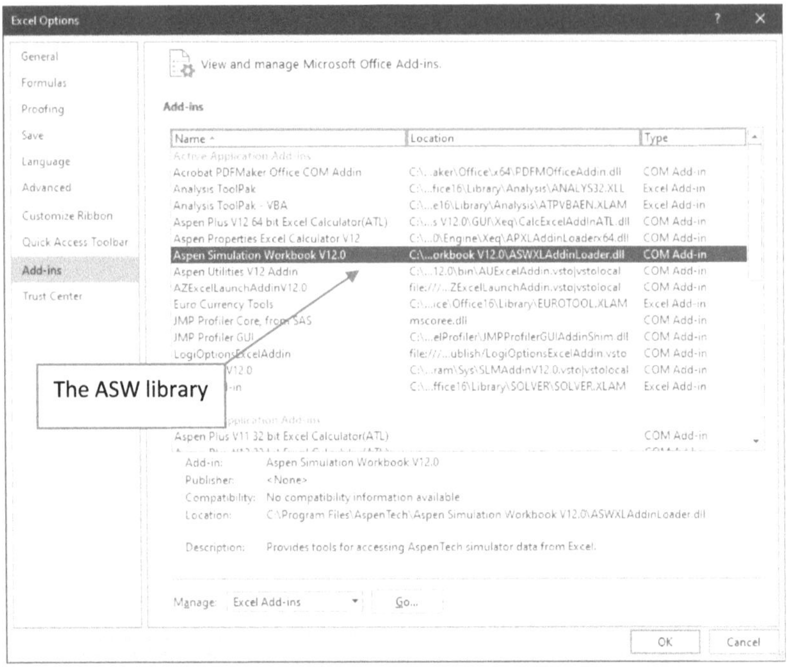

FIGURE 19.3 Aspen Simulation Workbook Excel add-ins.

These two XLA files are VBA code files. When the user enables ASW UDFs or Action Macros, the functions that the user may be calling are within these XLA files.

The second type of automation file installed with ASW is an ActiveX COM DLL. These are automatically registered during the install process (see Fig. 19.3). The DLL file can directly control ASW objects such as workbooks via custom code that the user writes in VBA or other languages that support ActiveX. The VBA code within the previously mentioned XLA files (AspenOSEWorkbook.xla and AspenSimulationWorkbook.xla) all use the DLL file to control ASW.

If the user is writing custom ASW VBA automation code, then the user must add a reference to these classes. The display name of the library is "Aspen Simulation Workbook VX.X" (where X.X is replaced by the version of ASW).

19.5 Multiplatform optimization

A good understanding in multiobjective optimization is crucial as a starting point in multiplatform optimization concept. A multiobjective optimization problem may be presented as follows:

$$\begin{aligned} &\text{minimize } \{f_1(\mathbf{x}), f_2(\mathbf{x}), \ldots, f_k(\mathbf{x})\} \\ &\text{subject to } \mathbf{x} \in S \end{aligned} \tag{19.1}$$

This form of multiobjective optimization problem involves $k(\geq 2)$ conflicting *objective functions* $f_i : (\mathbf{R}^n \rightarrow \mathbf{R})$ that is minimized simultaneously. The *decision (variable) vectors* $\mathbf{x} = (x_1, x_2, ..., x_n)^T$ belong to the nonempty *feasible region* $S \subset \mathbf{R}^n$. In this general problem formulation, the types of constraints forming the feasible region are not fixed. *Objective vectors* are images of decision vectors and consist of *objective (function) values* $\mathbf{z} = \mathbf{f}(\mathbf{x}) = \{f_1(\mathbf{x}), f_2(\mathbf{x}), \ldots, f_k(\mathbf{x})\}^T$. Furthermore, the image of the feasible region in the objective space is called a *feasible objective region* $Z = \mathbf{f}(S)$.

Following the aforementioned definition of multiobjective optimization, multilevel optimization is also dominant in multiplatform optimization. For the general multilevel optimization problem $(\mathbf{P})$ can be formulated as follows:

$$
(\mathbf{P}) \quad \begin{array}{c}
(P_1) \min\limits_{x_1} f_1(x_1, x_2, \ldots, x_k) \\
\text{s.t. } g_1(x_1, x_2, \ldots, x_k) \leq 0 \\
\text{where } x_2 \text{ solves} \\
(P_2) \min\limits_{x_2} f_2(x_1, x_2, \ldots, x_k) \\
\text{s.t. } g_2(x_1, x_2, \ldots, x_k) \leq 0 \\
\vdots \\
\text{where } x_k \text{ solves} \\
(P_k) \min\limits_{x_k} f_k(x_1, x_2, \ldots, x_k) \\
\text{s.t. } g_k(x_1, x_2, \ldots, x_k) \leq 0
\end{array} \tag{19.2}
$$

P_1 is called the first level problem and corresponds to the highest level in the hierarchy. The decision-maker at this level controls the decision variables x_1, and the objective is to minimize the function f_1. Consequently, P_k is the kth level problem and corresponds to the lowest level in the hierarchy. If only two levels are considered, the two decision-makers are referred to as *leader* and *follower*, respectively.

The following is a typical procedure for optimization problems in Aspen Plus. The user usually wants to optimize the operating margin for a simple flowsheet. Consequently, they may use optimization to maximize or minimize a user-specified objective function by manipulating decision variables (feed stream, block input, or other input variables). In Aspen Plus, the objective function can be any valid Fortran expression involving one or more flowsheet quantities, which can form a multiobjective optimization. The tolerance of the objective function is the tolerance of the convergence block associated with the optimization problem. The user has the option of imposing equality or inequality constraints on the optimization. Equality constraints within an optimization are similar to design specifications. The constraints can be any function of flowsheet variables computed using Fortran expressions or in-line Fortran statements. The user must specify the tolerance of the constraint.

For larger flowsheets, tear streams and the optimization problem can be converged simultaneously or separately. For simultaneous convergence, the tear stream is treated as an additional constraint.

Aspen Plus solves optimization problems iteratively. By default, Aspen Plus generates and sequences a convergence block for the optimization problem. The user can override the convergence defaults by entering convergence specifications on Convergence forms. SQP and Complex methods may be chosen to converge the optimization problems.

The manipulated variable's value provided in the Stream or Block input is used as the initial estimate. Providing a reasonable estimate for the manipulated variable helps the optimization problem converge in fewer iterations. This is especially important for optimization problems with a large number of various variables and constraints. Optimization problems can be challenging to formulate and converge. It is vital to have a good understanding of the simulation problem before adding the complexity of optimization.

The recommended procedure for creating an optimization problem is given as follows:

1. Start with a simulation (instead of starting with optimization). There are several reasons for this approach:
   - It is easier to detect flowsheet errors in a simulation.
   - The user can get a good feel for what reasonable specifications are.
   - The user can get a good feel for a reasonable range of decision variables.
   - The user can get a good estimate for the tear streams.
2. Perform sensitivity analysis before optimization to find appropriate decision variables and their ranges.
3. Evaluate the solution using sensitivity analysis to find out if the optimum is broad or narrow.

For more information about optimization, press F1 to see the Aspen Plus Help topic Simulation and Analysis Tools → Sequential Modular Flowsheeting Tools → Optimization.

19.5.1 Case study—dichloro-methane solvent recovery system

In a dichloro-methane solvent recovery system (see Fig. 19.4), two adiabatic flash units, i.e., TOWER1 (operated at 19.7 psia) and TOWER2 (18.7 psia) are fed with saturated steam STEAM1 and STEAM2, respectively. The two steam feeds provide heat and separation to the FEED mixture; the latter consists of water and dichloro-methane (MECL2). The details of the inlet streams of the flash units can be found in Table 19.1. In this case, it is desired to limit the maximum allowable concentration of dicloro-methane in the effluent stream from TOWER2. Hence, optimization is used to find the minimum steam flow

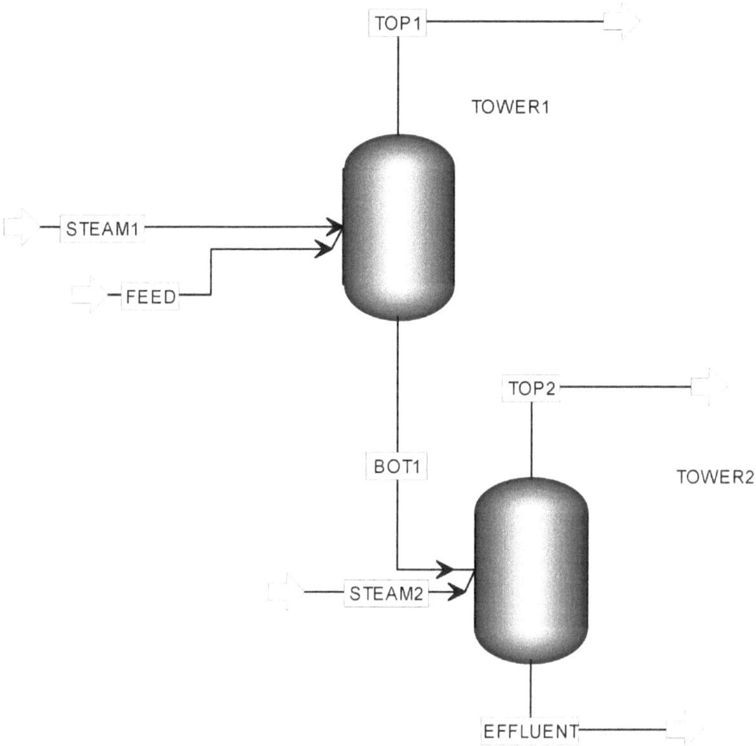

FIGURE 19.4 Dichloro-methane solvent recovery system.

TABLE 19.1 Feed stream conditions.

Stream	FEED	STEAM1	STEAM2
Temperature (F)	100	381.981 (saturated)	381.981 (saturated)
Pressure (psia)	24	200	200
Mass flows (lb/hr)	100,000	18,100 (initial)	18,100 (initial)
Mass fractions			
WATER	0.986	1	1
MECL2	0.014	0	0

rates while satisfying the maximum concentration of dicloro-methane in TOWER2 effluent stream.

The mass flow rate of stream STEAM1 and STEAM2 are set as manipulated variables for the Optimization block O-1 as shown in Fig. 19.5. The user

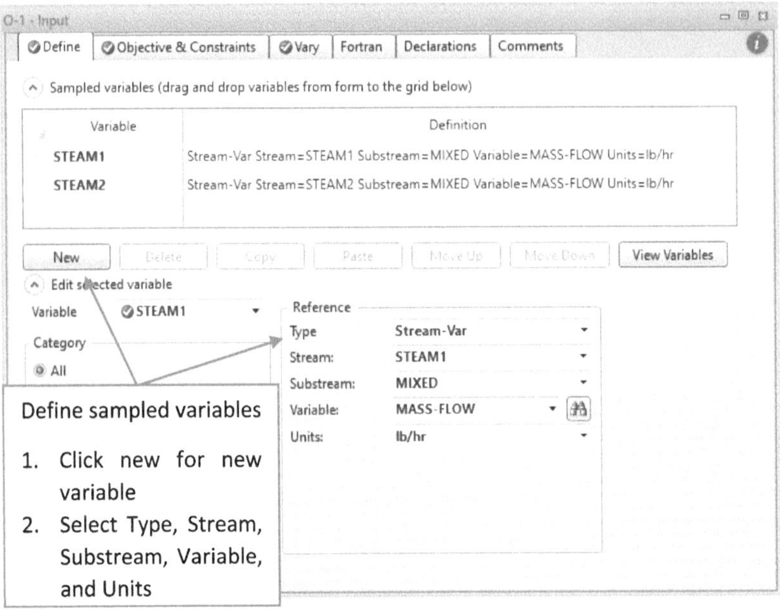

FIGURE 19.5 Steps to define sampled variables for STEAM1 and STEAM2.

can define the optimization objective function, i.e., TOTSTEAM as a summation of flowrates of STEAM1 and STEAM2 as described in Eq. (19.3). Note that the two variables STEAM1 and STEAM2 lead the problem into multi-objective function. The optimization problem is converged when the total sum of STEAM1 and STEAM2 reaches its minimum value. This can be coded as a Fortran expression, such as that shown in Fig. 19.6. The specified mass flow values for STEAM1 and STEAM2 in Table 19.1 will be used as the initial estimate by the optimization convergence block. The user does not have to specify convergence of the design specification. Aspen Plus automatically generates a convergence block in meeting the specification. User may set a convergence block OPT for a proper output variable to address the said block in the multilevel optimization problem.

Only one constraint is associated with this optimization problem, i.e., EFFL. The latter is satisfied when the concentration of dichloro-methane in TOWER2 effluent stream is less than 150 ppm.

Mathematically, this case study can be expressed as in Eq. (19.3):

$$\begin{aligned} \text{TOTSTEAM} &= \min_{\text{STEAM1,STEAM2}} \text{STEAM1} + \text{STEAM2} \\ \text{s.t. EFFL} &\leq 150\ \text{ppm} \end{aligned} \tag{19.3}$$

while simultaneously satisfy all mass and energy balances in the simulation flowsheet in Fig. 19.4.

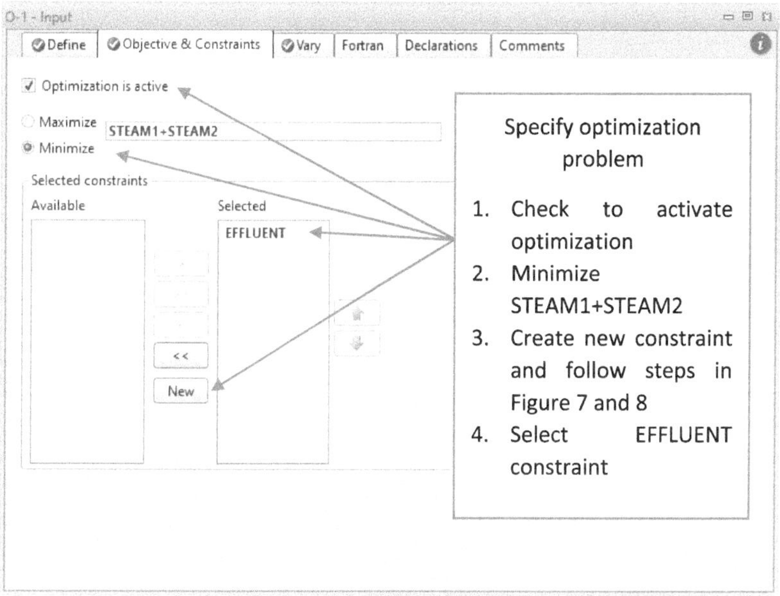

FIGURE 19.6 Steps to specify optimization objective and constraint.

User can enter the necessary input parameters (e.g., STEAM1, STEAM2, objective function STEAM1+STEAM2, EFFL constraint) to set up the one level multiobjective optimization problem (follow Figs. 19.5–19.9 for detailed steps). The user can find these features in the Model Analysis Tools in Aspen Plus. In Fig. 19.9, it can be observed that the STEAM1 and STEAM2 are varied independently and simultaneously in a multiobjective problem. The Optimization O-1 in Aspen Plus can then be run to determine the minimum total steam TOTSTEAM. As shown in Fig. 19.10, the TOTSTEAM flowrate is determined as 14,198.15 lb/hr, which is contributed by STEAM1 (12,190.9 lb/hr) and STEAM2 (2007.25 lb/hr).

A sensitivity analysis can be carried out to explore the minimum value of TOTSTEAM for different feed compositions. Multiple tasks must be implemented simultaneously, i.e., vary feed composition while minimizing TOTSTEAM as multiobjective optimization, which renders the embedded sensitivity analysis module under the Model Analysis Tools in Aspen Plus challenging to use. The automation interface through MATLAB could be helpful on this occasion. This will be discussed in the following section.

19.5.2 Sensitivity analysis with automation interface in MATLAB

To facilitate the corresponding sensitivity analysis, user needs to create an evaluation function optaspen.m for each feed composition with the following MATLAB code:

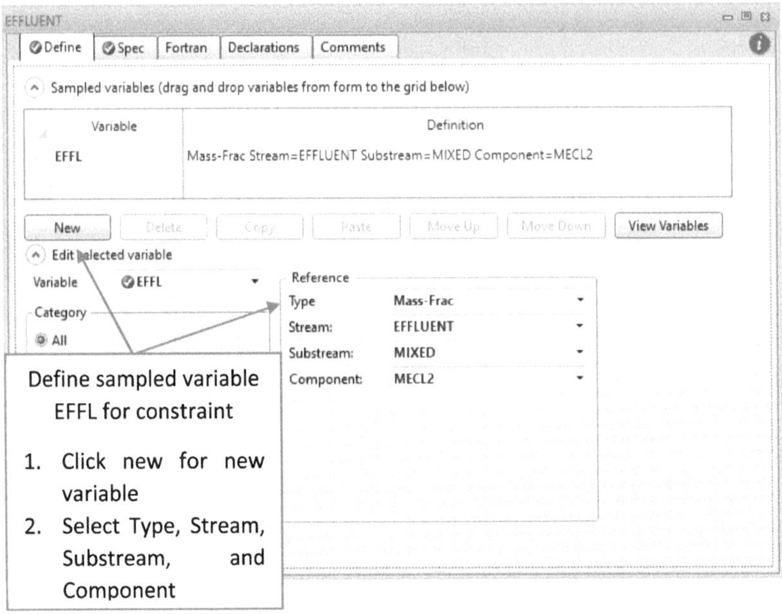

FIGURE 19.7 Steps to define EFFLUENT constraint for Optimization block O-1.

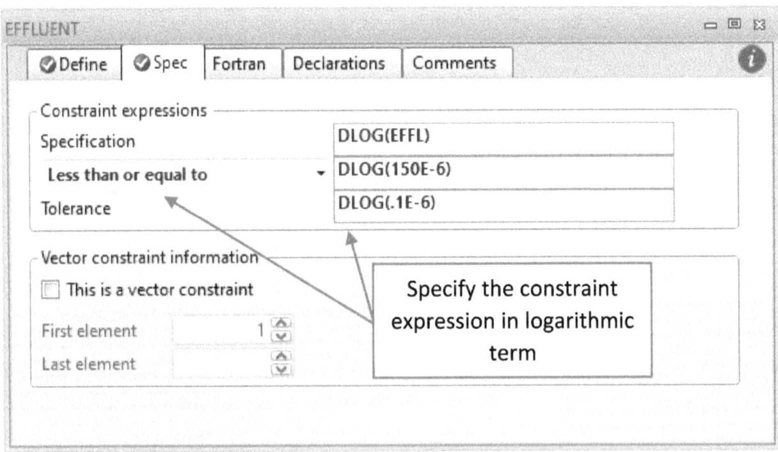

FIGURE 19.8 Specification for EFFLUENT constraint.

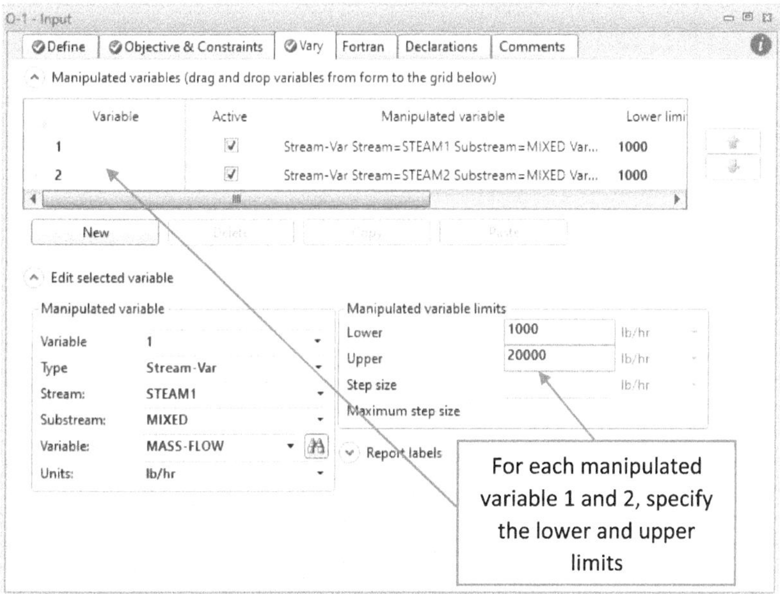

FIGURE 19.9 Steps to specify manipulated variables STEAM1 and STEAM2.

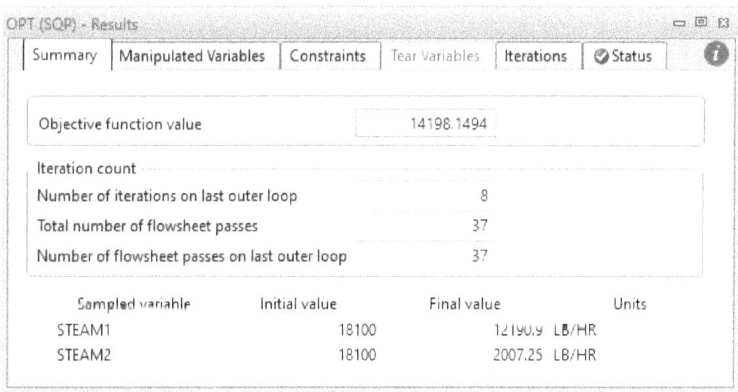

FIGURE 19.10 Optimization result of O-1 Optimization block shown under OPT Convergence block results for the minimum steam flow rates STEAM1 and STEAM2.

```
function TOTSTEAM = optaspen(WATER)

AspenVersion = 'apwn.document';
AspenPath = 'C:\Users\vska\Downloads\opt.bkp'; % Replace with the
simulation path
AspenVisible = 0; % 0 for not visible
AspenDialogs = 1;
% Get pointer and load Aspen Plus comserver:
```

```
Aspen = actxserver(AspenVersion);
% Open the Aspen-Simulation
invoke(Aspen,'InitFromFile2',AspenPath);
Aspen.visible = AspenVisible; % Make it visible
Aspen.SuppressDialogs = AspenDialogs;
%Input parameter
MECL2=1-WATER; %FEED MECL2 mass fraction
Aspen.Tree.FindNode('\Data\Streams\FEED\Input\TOTFLOW\MIXED').
Value=(100000);
Aspen.Tree.FindNode('\Data\Streams\FEED\Input\FLOW\MIXED\
WATER').Value=WATER;
Aspen.Tree.FindNode('\Data\Streams\FEED\Input\FLOW\MIXED\
MECL2').Value=MECL2;

Aspen.Reinit();
Aspen.Run2();

%Output parameter
TOTSTEAM = Aspen.Tree.FindNode('\Data\Convergence\Convergence\
OPT\Output\VAR_VAL2').Value;

%Release COM object
Aspen.Quit();
end
```

Note that the structure of the evaluation function is quite straightforward. The input parameters are first defined, and the output parameter from convergence block OPT is collected after the simulation. As the MATLAB code calls the simulation file opt.bkp with Aspen.Run2() command, the embedded single level optimization procedure, i.e., minimizing TOTSTEAM, is simultaneously computed (as the multiobjective lower level problem shown in Fig. 19.1). The FEED MECL2 mass fraction is computed as MECL2=1-WATER according to mass balance. The whole evaluation function optaspen.m acts as a wrapper to the multiobjective lower level TOTSTEAM minimization problem. Sensitivity analysis can be done for different WATER mass fractions from 0.8 to 0.99 to check if there is any difference in the minimum TOTSTEAM with the following MATLAB code:

```
i=1;
for x=0.8:0.01:0.99
   TOTSTEAM(i)=optaspen(x);
   WATER(i)=x;
   i=i+1;
end
plot (WATER,TOTSTEAM)
xlabel('WATER mass fraction')
ylabel('TOTSTEAM (lb/hr)')
```

From Fig. 19.11, it can be seen that the TOTSTEAM is peaked at 15,851.3 lb/hr at 0.88 WATER mass fraction. This sensitivity analysis shows that there is a maximum TOTSTEAM. This gives the user insight to avoid

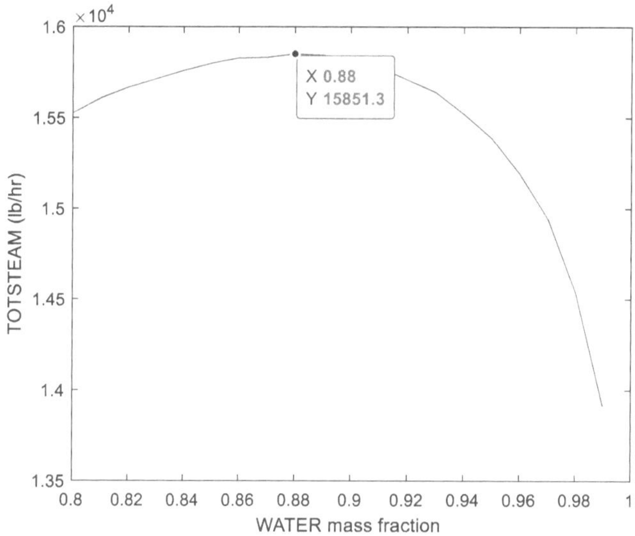

FIGURE 19.11 Sensitivity analysis using automation interface of Aspen Plus and MATLAB.

FEED stream around the WATER mass fraction value, which may be costly in steam consumption.

19.5.3 Multiobjective and multilevel problem under multiplatform optimization with automation interface in MATLAB

As shown in Fig. 19.11, it can be seen that there is a peak where TOTSTEAM is at maximum (roughly at 15,851.3 lb/hr) at a certain WATER mass fraction (about 0.88). An optimization problem can be formulated to find the accurate value of TOTSTEAM as follows:

$$\max_{\text{WATER mass fraction}} \quad \text{TOTSTEAM} \tag{19.4}$$

It is evident that this optimization problem is a multilevel optimization problem as TOTSTEAM can be elaborated as

$$\begin{aligned} &\max_{\text{WATER mass fraction}} \; \min_{\text{STEAM1,STEAM2}} \quad \text{STEAM1} + \text{STEAM2} \\ &\text{s.t. EFFL} \leq 150 \text{ ppm} \end{aligned} \tag{19.5}$$

Under the MATLAB platform, the user can exploit essential optimization tools (fminbnd) as the upper level optimization. The user can then rely on Aspen Plus Optimization wrapped under the evaluation function optaspen.m for the lower level problem. Consequently, the multilevel optimization problem, i.e., Eq. (19.5), can be solved with the following code:

```
WATER = fminbnd(@(WATER) -optaspen(WATER),0.8,0.99)
```

Note that "fminbnd" is a minimization tool. As the upper level problem in Eq. (19.5) is a maximization problem, the negative sign is introduced to the optaspen function to modify the fminbnd minimization procedure into a maximization one. The accurate solution is 0.8793 for WATER mass fraction, which corresponds to 15,851.4 lb/hr of TOTSTEAM. In other words, if the FEED entering the TOWER1 contains 0.8793 mass fraction of WATER, steam consumption will increase to 15,851.4 lb/hr substantially. The user may want to avoid such conditions by maintaining the desired FEED WATER mass fraction.

In this case study, the fminbnd function is utilized to find the minimum single-variable function on fixed intervals. It is mainly chosen to showcase the concept of multiobjective and multilevel problem on two different platform of optimization. It is always possible to use more advanced optimization constraints and models, i.e., fmincon (find minimum of constrained nonlinear multivariable function), fminsearch (find minimum of unconstrained multivariable function using derivative-free method), fminunc (find minimum of unconstrained multivariable function), etc.

19.5.4 Sensitivity analysis with automation interface in excel using ASW

Similar to the previous approach in MATLAB, the ASW interface can also be exploited for the same optimization purpose (see Figs. 19.12–19.13).

Tables can then be created in ASW to allow users in manipulating variables directly in Microsoft Excel without finding them in the Aspen Plus model. Once a table has been created, the user can change the cell values in the workbook to adjust the model directly. Four variables (Table 19.2) should be entered into the Organizer pane using the Browse for Variables button (see details in Figs. 19.14–19.15). These variables are inserted into a table in Microsoft Excel. The user can select four variables in Table 19.2 and export them to Microsoft Excel by clicking on the Create Table button and select the desired location in Microsoft Excel (Fig. 19.16). The user can customize the corresponding table using the Simulation Workbook Table Wizard (Fig. 19.17).

User may modify the "Simulation Workbook Tables" following the steps shown in Fig. 19.18.

In ASW, scenario tables allow users to conduct case studies or sensitivity studies. The user can manipulate a variety of inputs and see the influence of

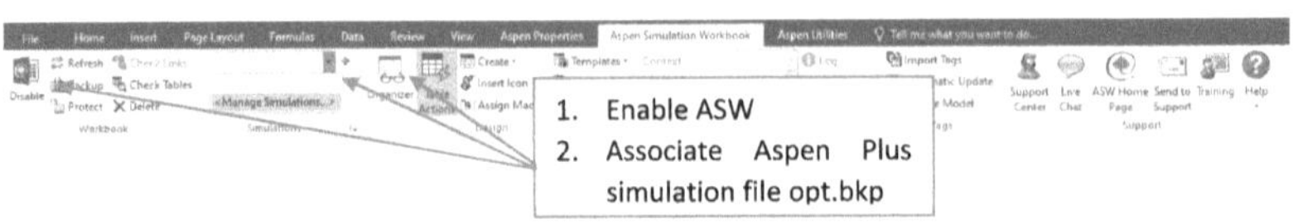

FIGURE 19.12 Steps to enable ASW and associate Aspen Plus simulation file in Microsoft Excel.

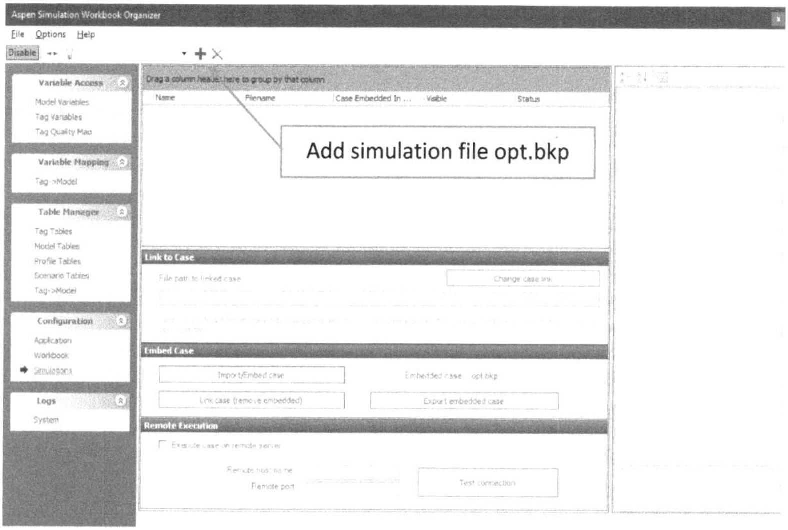

FIGURE 19.13 Step to add new simulations in the organizer pane.

TABLE 19.2 ASW variable paths from Aspen Plus simulation file opt.bkp.

Variable name	Variable paths
TOTFLOW	Top.appModel.Streams.FEED.Input.TOTFLOW.MIXED
WATER	Top.appModel.Streams.FEED.Input.FLOW.MIXED.WATER
MECL2	Top.appModel.Streams.FEED.Input.FLOW.MIXED.MECL2
TOTSTEAM	Top.appModel.Convergence.Convergence.OPT. Output.VAR_VAL2

these changes on a number of model results. Detailed steps to create a scenario table are shown in Fig. 19.19.

The sensitivity analysis can be done using the scenario table for different WATER mass fraction from 0.8 to 0.99. A total of 20 cases should be checked if there is any difference on the minimum TOTSTEAM assuming a 0.01 step size (Fig. 19.20). Detailed steps for doing so are found in Figs. 19.21 and 19.22, with sensitivity analysis results found in Fig. 19.23. The exact correlation of WATER mass fraction and TOTSTEAM can be made in a Microsoft Excel chart, as shown in Fig. 19.23.

Similar to previous sections, the objective of TOTSTEAM reaches its maximum value, i.e., 15,851.33 lb/hr at the WATER mass fraction of 0.88 (Case 9). In other words, if FEED entering TOWER1 contains 0.88 WATER

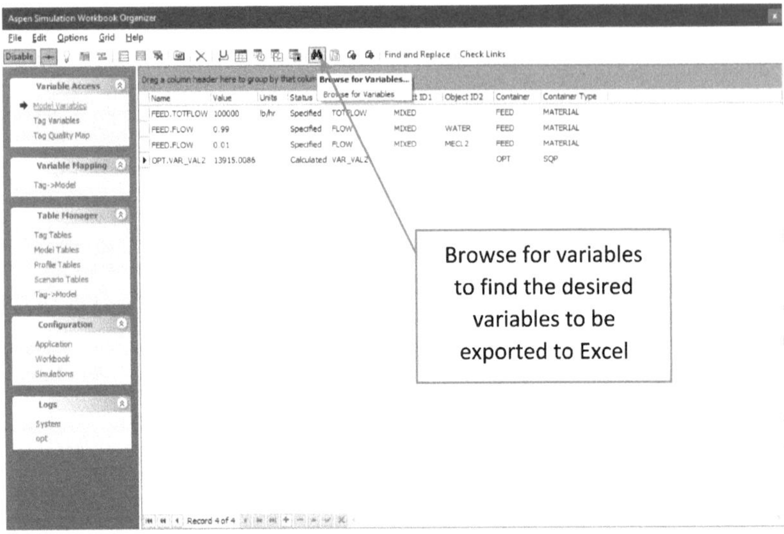

FIGURE 19.14 Step to browse for desired variables to be exported to Microsoft Excel.

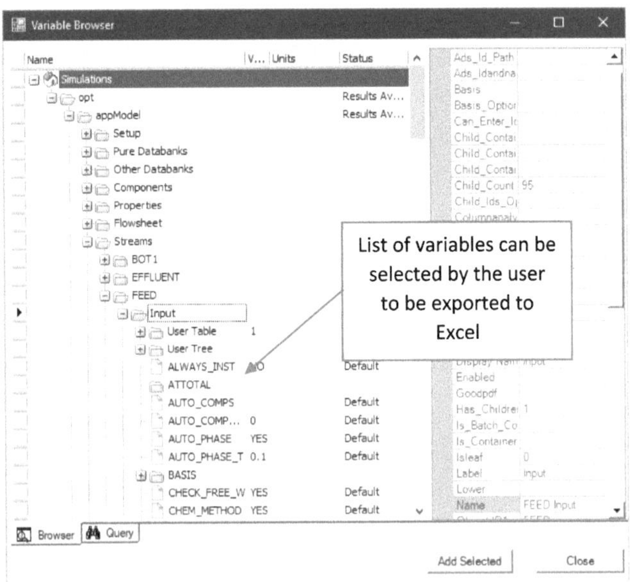

FIGURE 19.15 Variable browser window to find the list of variables from the simulation file.

mass fraction, the steam consumption will be peaked at 15,851.33 lb/hr (see Fig. 19.4). The result is slightly different from the results obtained using multiplatform optimization in MATLAB, but the conclusion is identical.

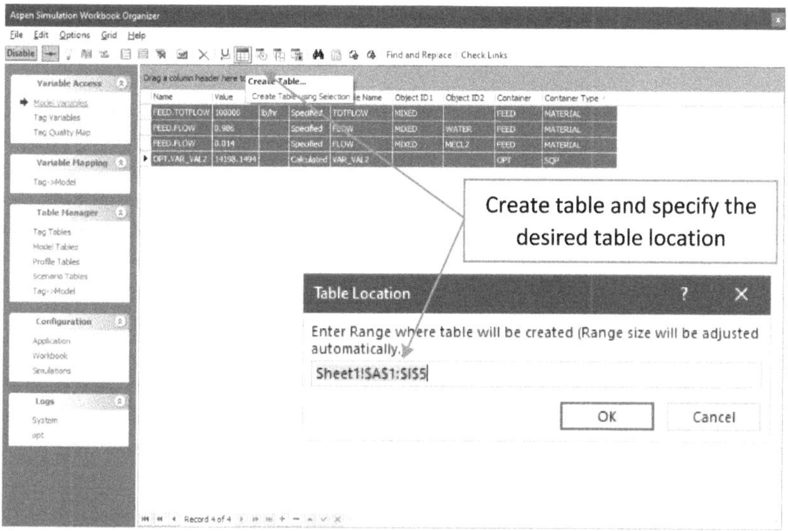

FIGURE 19.16 Steps to create table and select table location.

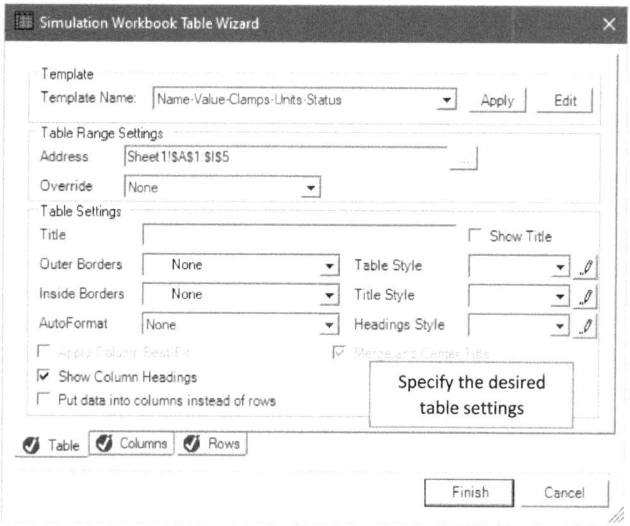

FIGURE 19.17 Simulation Workbook Table wizard in ASW.

Unfortunately, ASW is a one-directional interface to Aspen Plus. Adjusting variables in the table will change their simulation values but do not automatically run the simulation. To see the effect of changing a variable, the user must click the play button in the "Run" group or in the actual simulation.

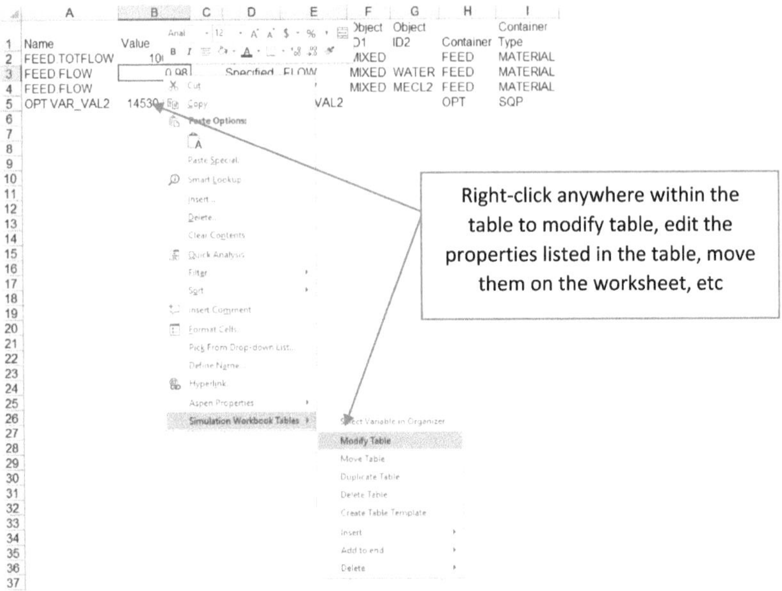

FIGURE 19.18 Steps to modify table in Microsoft Excel using right-click.

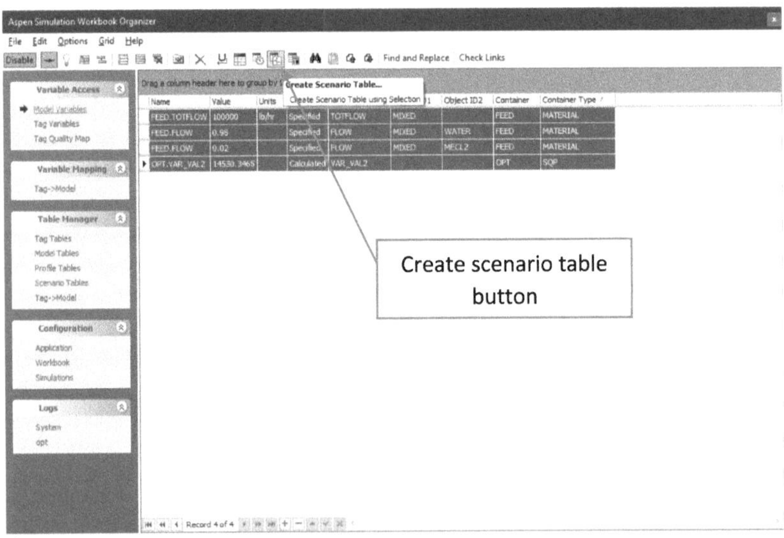

FIGURE 19.19 Create scenario table button.

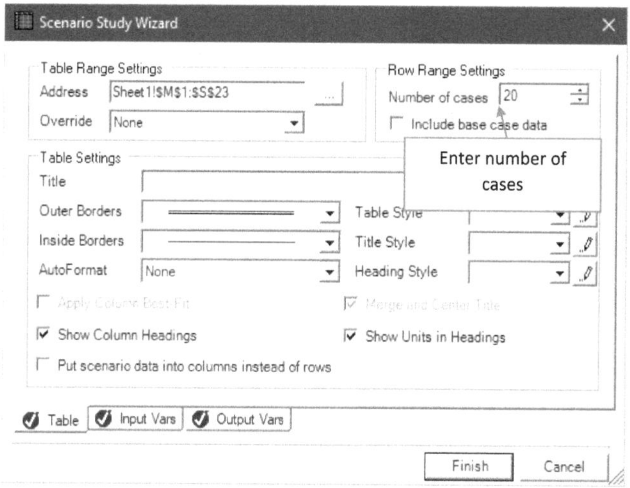

FIGURE 19.20 Scenario study wizard in ASW.

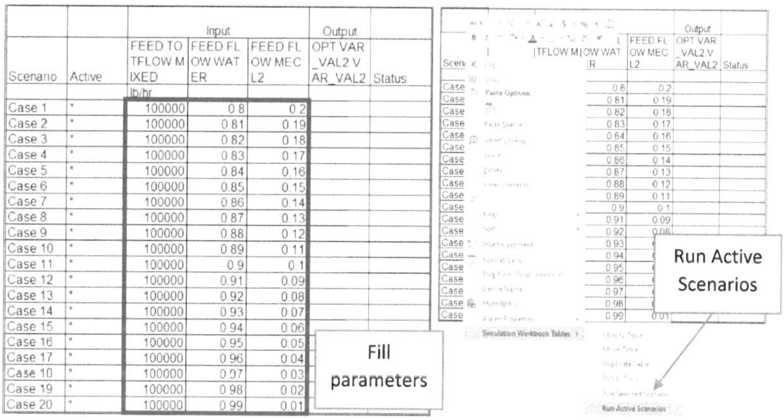

FIGURE 19.21 Steps to input parameters and to run active scenarios in Microsoft Excel.

Consequently, this adds difficulty to perform similar automatic upper level optimization like in MATLAB. The user may use the macros feature or other approaches to achieve the multilevel optimization found elsewhere. This case study shows that software interface is essential for multilevel optimization. However, the lower and upper level software's capabilities play a more significant role in the multiobjective and multilevel optimization, limiting the possibility of problem complexity the user may desire.

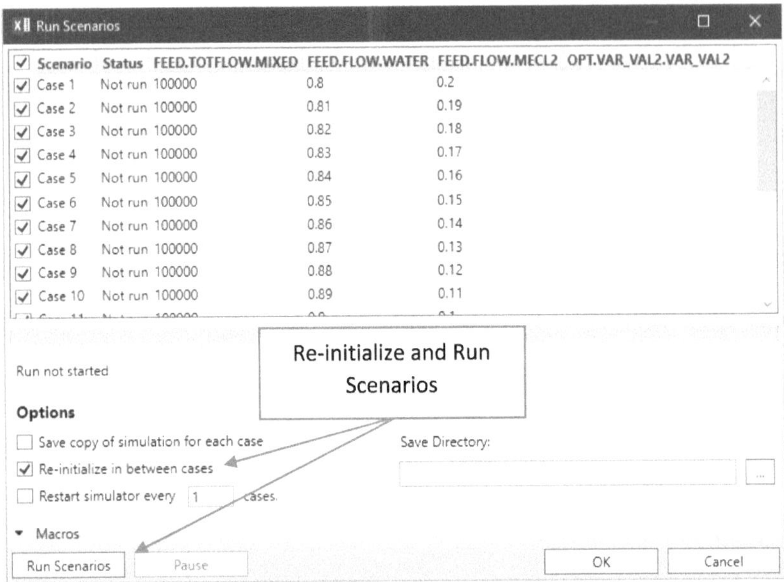

FIGURE 19.22 Steps to run scenarios in Microsoft Excel.

		Input			Output	
Scenario	Active	FEED.TOTFLOW.MIXED lb/hr	FEED.FLOW.WATER	FEED.FLOW.MECL2	OPT.VAR_VAL2.VAR_VAL2	Status
Case 1	•	100000	0.8	0.2	15525.11	Results Available
Case 2	•	100000	0.81	0.19	15607.32	Results Available
Case 3	•	100000	0.82	0.18	15667.44	Results Available
Case 4	•	100000	0.83	0.17	15712.51	Results Available
Case 5	•	100000	0.84	0.16	15759.9	Results Available
Case 6	•	100000	0.85	0.15	15799.59	Results Available
Case 7	•	100000	0.86	0.14	15827.94	Results Available
Case 8	•	100000	0.87	0.13	15832.1	Results Available
Case 9	•	100000	0.88	0.12	15851.33	Results Available
Case 10	•	100000	0.89	0.11	15843.73	Results Available
Case 11	•	100000	0.9	0.1	15820.73	Results Available
Case 12	•	100000	0.91	0.09	15779.11	Results Available
Case 13	•	100000	0.92	0.08	15709.05	Results Available
Case 14	•	100000	0.93	0.07	15640.63	Results Available
Case 15	•	100000	0.94	0.06	15521.05	Results Available
Case 16	•	100000	0.95	0.05	15387.01	Results Available
Case 17	•	100000	0.96	0.04	15192.01	Results Available
Case 18	•	100000	0.97	0.03	14944.08	Results Available
Case 19	•	100000	0.98	0.02	14541.59	Results Available
Case 20	•	100000	0.99	0.01	13915.01	Results Available

Highest TOTSTEAM value 15851.33 lb/hr at 0.88 WATER mass fraction

FIGURE 19.23 Results available shown in scenario table.

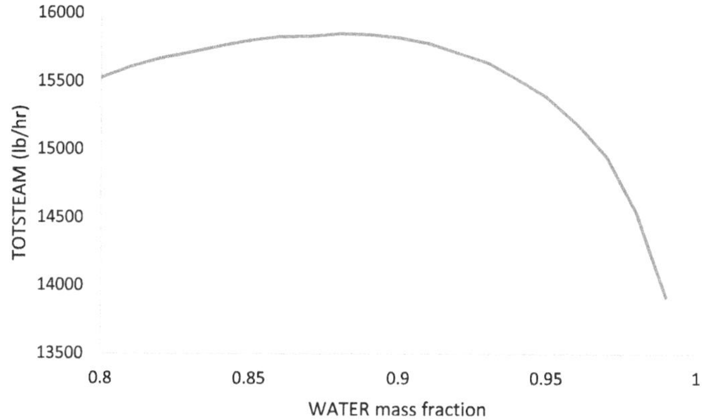

FIGURE 19.24 Scatter with straight lines chart of water mass fraction versus TOTSTEAM (lb/hr).

19.6 Conclusion

This chapter elaborates on multiobjective and multilevel optimization using different platform to facilitate more advanced optimization procedure. It is expected that the same technique can be implemented using different mathematical and programming software by exploiting the automation interface available in commercial process simulation software. The user can explore more advanced features depending on the capability of the upper and lower level software to achieve the desired optimization hierarchy and strategy.

Exercises

1. In Eq. (19.3), objective function in the multiobjective optimization problem is defined with TOTSTEAM as a summation of STEAM1 and STEAM2 flowrates. It can be observed that STEAM1 flowrate is sixfold of that in STEAM2 for such case (see Fig. 19.10). In a different scenario, the objective function is changed to the costs of steam production, i.e., COSTSTEAM ($/hr). Assuming that the steam generation cost of STEAM1 is 20% more expensive than STEAM2 (due to its higher consumption), the optimization objective is re-written as COSTSTEAM ($/hr) = 1.2 ($/lb)*STEAM1 (lb/hr) + 1 ($/lb)*STEAM2 (lb/hr). Determine the resulting flowrates of STEAM1 and STEAM2 for the same feed conditions (see Table 19.1). What can be concluded from this new result?
2. With the same strategy of the multiplatform optimization, what would be the maximum COSTSTEAM and the corresponding feed concentration if the same conditions in Question 1 is considered?

References

Bracken, J., McGill, J.M., 1973. Mathematical programs with optimization problems in the constraints. Operations Research 21, 37–44.

Bracken, J., McGill, J.M., 1974. A method for solving mathematical problems with nonlinear programs in the constraints. Operations Research 22, 1097–1101.

Branke, J., Deb, K., Miettinen, K., Slowiński, R., 2008. Multiobjective Optimization: Interactive and Evolutionary Approaches. Springer-Verlag, Berlin Heidelberg.

Kuhn, H., Tucker, A., 1951. Nonlinear programming. In: Neyman, J. (Ed.), Proceedings of the Second Berkeley Symposium on Mathematical Statistics and Probability. University of California Press, Berkeley, pp. 481–492.

Migdalas, A., Pardalos, P.M., Värbrand, P., 1998. Multilevel Optimization: Algorithms and Applications. Springer, Boston MA.

Pareto, V., 1896. Cours d'Economie Politique. Rouge, Lausanne.

Pareto, V., 1906. Manuale di Economia Politica. Piccola Biblioteca Scientifica, Milan. Translated into English by A.S. Schwier (1971), Manual of Political Economy, MacMillan, London.

Vicente, L.N., Calamai, P.H., 1994. Bilevel and multilevel programming: A bibliographic review. Journal of Global Optimization 5, 291–306.

Chapter 20

Flexible design strategy for process controllability*

Vincentius Surya Kurnia Adi
Department of Chemical Engineering, National Chung Hsing University, Taichung, Taiwan

Chapter outline

20.1 Introduction

Traditionally, design and control decisions are made at sequential stages over the process plant design. In the design phase, the optimal operating conditions and the related material and energy balance data are established mainly based on economic considerations. The control systems are configured to maintain the key process conditions at the fixed nominal values in the subsequent step. It is desirable to address the operability issues at the earliest possible stage before stipulation of control schemes. Moreover, dealing with uncertainties is one of the practical issues that must be addressed in designing and operating any production process. A unit operation or a series of unit operations in a production process is expected to be fully functional in the presence of uncertain feed flow rate, feed concentration, or ambient temperature, to name a few. Faulty upstream operations can eventually affect the feed flow rate and feed concentration entering the unit operation as a process disturbance. The ability of a system to maintain feasible operation despite unexpected disturbances is referred to as its operational flexibility (Halemane and Grossmann, 1983), which is clearly a feature of critical importance that must be

*Simulation and other electronic files of this chapter are made available on a companion website; URL: https://www.elsevier.com/books-and-journals/book-companion/9780323901680.

Chemical Engineering Process Simulation. https://doi.org/10.1016/B978-0-323-90168-0.00020-2

incorporated into the design considerations. Grossmann and his coworkers first proposed a formal definition of *operational flexibility* and developed a quantitative performance measure accordingly to facilitate process analysis (Grossmann and Halemane, 1982; Grossmann and Floudas, 1987). More specifically, the *flexibility index* (denoted as FI) was computed numerically by solving a multilevel optimization problem. Many researchers later successfully integrated such a metric in their design procedure, e.g., Riyanto and Chang (2010) and Adi and Chang (2011). A dedicated book on operational flexibility is also found (Chang and Adi, 2018).

In the usual practice of process design, the operational flexibility of a process is ensured in an ad hoc fashion by choosing conservative operating conditions, applying empirical overdesign factors, and introducing additional or redundant units. Several drawbacks of this approach can be summarized as follows:

1. As the interactions among units are not considered, the actual flexibility level of the entire process cannot be accurately determined.
2. As the economic penalties of the heuristic design practices are not evaluated, their financial implications cannot be adequately assessed.

Note that the systematic incorporation of flexibility analysis in process synthesis and design has received considerable attention in recent years. Several mathematical programming models have been developed to facilitate quantitative flexibility analyses to provide the designers with the capabilities to (1) determine the performance index of any given design concerning the standard requirements in actual operation, (2) identify the bottleneck conditions that limit the operational feasibility, and (3) compare alternative designs on an objective basis (Swaney and Grossmann, 1985a, b). More details on flexibility analyses can be found in the work by Chang and Adi (2018).

This FI which was first defined by Swaney and Grossmann (1985a, b) can be used as a precise gauge of the feasible region in the parameter space. Specifically, it is associated with the maximum allowable deviations of the uncertain parameters from their nominal values, by which viable operation can be assured with proper manipulation of the control variables. Swaney and Grossmann (1985b) also showed that under certain convexity assumptions, critical points that limit feasibility and/or flexibility must lie on the vertices of the uncertain parameter space. This property will be exploited in the following case study to show that a proper design needs to incorporate the uncertainty consideration. Therefore, the end result will be resilient to uncertainty and thus controllable.

20.2 Flexibility index model

The basic framework (Biegler et al., 1997) for computing such an index is presented to provide an accurate explanation of the FI model by first introducing two label sets:

$$\mathbb{I} = \{i | i \text{ is the label of an equality constraint}\} \tag{20.1}$$

$$\mathbb{J} = \{j | j \text{ is the label of an inequality constraint}\} \quad (20.2)$$

The general design model of the continuous processes can be expressed accordingly as

$$h_i(\mathbf{d}, \mathbf{z}, \mathbf{x}, \boldsymbol{\theta}) = 0 \quad \forall i \in \mathbb{I} \quad (20.3)$$

$$g_j(\mathbf{d}, \mathbf{z}, \mathbf{x}, \boldsymbol{\theta}) \leq 0 \quad \forall j \in \mathbb{J} \quad (20.4)$$

where h_i is the ith equality constraint in the design model (e.g., the steady-state mass or energy balance equation of a processing unit); g_j is the jth inequality constraint (e.g., a capacity limit); **d** represents a vector in which all design specifications are stored; **z** denotes the vector of adjustable control variables; **x** is the vector of state variables; $\boldsymbol{\theta}$ denotes the vector of uncertain parameters which are assumed to be bounded within a hypercube in the parameter space, i.e.,

$$\boldsymbol{\theta}^L \leq \boldsymbol{\theta} \leq \boldsymbol{\theta}^U \quad (20.5)$$

Note that $\boldsymbol{\theta}^U$ and $\boldsymbol{\theta}^L$ represent the vectors of expected upper and lower bounds of the uncertain parameters, respectively. Note also that although the uncertain parameters may function as design, control, and/or state variables, these possibilities are excluded for simplicity in this case.

The following mathematical program can be utilized to determine a so-called *feasibility function* $\psi(\mathbf{d}, \boldsymbol{\theta})$, i.e.,

$$\psi(\mathbf{d}, \boldsymbol{\theta}) = \min_{\mathbf{x}, \mathbf{z}} \max_{j \in \mathbb{J}} g_j(\mathbf{d}, \mathbf{z}, \mathbf{x}, \boldsymbol{\theta}) \quad (20.6)$$

subject to the equality constraints given in Eq. (20.3). Notice that this formulation means that, for a *fixed* design defined by **d** and the *fixed* values of parameters $\boldsymbol{\theta}$, the largest $g_j(\forall j \in \mathbb{J})$ is minimized by adjusting the control variables **z** while keeping $h_i = 0$ $(\forall i \in \mathbb{I})$. The given system is clearly operable if $\psi(\mathbf{d}, \boldsymbol{\theta}) \leq 0$, while infeasible if otherwise (see Fig. 20.1).

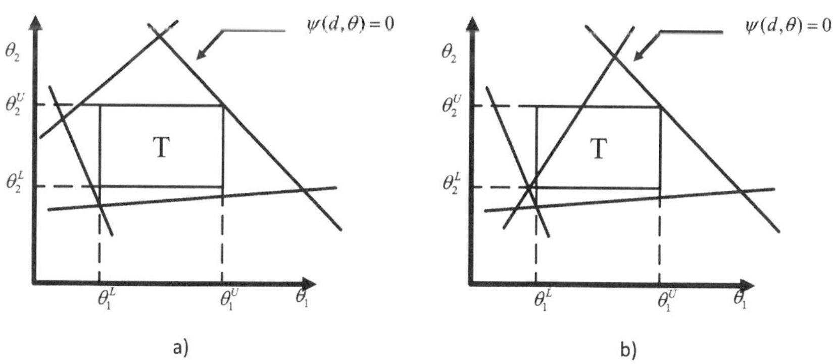

FIGURE 20.1 Feasible and infeasible designs in the parameter space. (A) Feasible design, $\chi(d) \leq 0$ (B) Infeasible design, $\chi(d) > 0$. *Adapted from Biegler, L.T., Grossmann, I.E., Westerberg, A.W., 1997. Systematic methods of chemical process design. Prentice Hall PTR, Upper Saddle River, NJ.*

On the other hand, the above optimization problem can be defined alternatively with a different formulation by introducing an extra scalar variable u, i.e.,

$$\psi(\mathbf{d},\boldsymbol{\theta}) = \min_{\mathbf{x},\mathbf{z},u} u \tag{20.7}$$

subject to Eq. (20.3) and

$$g_j(\mathbf{d},\mathbf{z},\mathbf{x},\boldsymbol{\theta}) \leq u \quad \forall j \in \mathbb{J} \tag{20.8}$$

Notice also that, if $\psi(\mathbf{d},\boldsymbol{\theta}) = 0$, at least one of the inequality constraints should be active, i.e., $g_j = 0(\exists j \in \mathbb{J})$.

Since the feasibility function is evaluated according to a deterministic model with *constant* $\boldsymbol{\theta}$, it is necessary to perform the feasibility check on a more comprehensive basis by considering all possible values of the uncertain parameters. A permissible hypercube $\mathbf{T}$ can be defined in the parameter space, i.e.,

$$\mathbf{T} = \left\{\boldsymbol{\theta} \middle| \boldsymbol{\theta}^N - \Delta\boldsymbol{\theta}^- \leq \boldsymbol{\theta} \leq \boldsymbol{\theta}^N + \Delta\boldsymbol{\theta}^+\right\} \tag{20.9}$$

where $\boldsymbol{\theta}^N$ denotes a vector of given nominal parameter values and $\Delta\boldsymbol{\theta}^+$ and $\Delta\boldsymbol{\theta}^-$ represent vectors of the expected deviations in the positive and negative directions, respectively. Hence, an additional optimization problem can be formulated to facilitate this more rigorous test:

$$\chi(\mathbf{d}) = \max_{\boldsymbol{\theta} \in \mathbf{T}} \psi(\mathbf{d},\boldsymbol{\theta}) \tag{20.10}$$

where $\chi(\mathbf{d})$ denotes the feasibility function of a *fixed* design defined by $\mathbf{d}$ over $\mathbf{T}$. The given system should therefore be feasible if $\chi(\mathbf{d}) \leq 0$, while infeasible if otherwise (Fig. 20.2).

The permissible hypercube $\mathbf{T}$ is expanded/contracted with another scalar variable δ to provide a unified measure of the maximum tolerable range of variation in every uncertain parameter (Swaney and Grossmann, 1985a, b):

$$\mathbf{T}(\delta) = \left\{\boldsymbol{\theta} \middle| \boldsymbol{\theta}^N - \delta\Delta\boldsymbol{\theta}^- \leq \boldsymbol{\theta} \leq \boldsymbol{\theta}^N + \delta\Delta\boldsymbol{\theta}^+\right\} \tag{20.11}$$

where δ can be determined by solving the FI model given below:

$$FI = \max \delta \tag{20.12}$$

subject to

$$\chi(\mathbf{d}) \leq 0 \tag{20.13}$$

The maximized objective value FI is the steady-state *flexibility index*, representing the most significant value of δ that guarantees $g_j \leq 0(\forall j \in \mathbb{J})$, i.e., $\chi(\mathbf{d}) \leq 0$, in the parameter hypercube. Note also that $\delta \geq 1$ essentially implies that the system is feasible under the original constraint of Eq. (20.11). Several effective strategies are available for solving the optimization problem defined by Eqs. (20.12) and (20.13). The details can be found elsewhere (Chang and Adi, 2018).

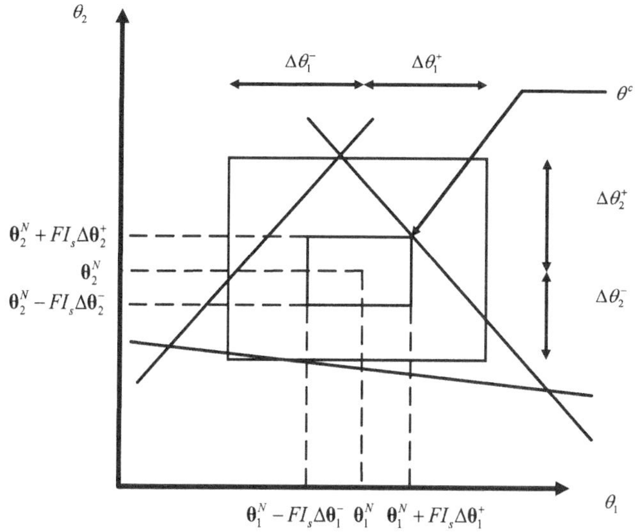

FIGURE 20.2 Geometrical interpretation of steady-state *flexibility index. Adapted from Biegler, L.T., Grossmann, I.E., Westerberg, A.W., 1997. Systematic methods of chemical process design. Prentice Hall PTR, Upper Saddle River, NJ.*

The vertex method (Swaney and Grossmann, 1985a, b) is elaborated explicitly for proof of concept and to greatly simplify the problem. It is assumed that the optimal solution is always associated with one of the vertices of the feasible hypercube the parameter space. Let $\Delta\boldsymbol{\theta}^k(\forall k \in V)$ denotes the kth vertex and V is the set of all vertices. It is possible to determine the largest possible value δ along a specific vertex direction, i.e., $\Delta\boldsymbol{\theta}^k$, by solving the following programming model:

$$\delta^k = \max_{\mathbf{x},\mathbf{z},\delta} \delta \tag{20.14}$$

subject to Eqs. (20.3) and (20.4), and

$$\boldsymbol{\theta} = \boldsymbol{\theta}^N + \delta\Delta\boldsymbol{\theta}^k \tag{20.15}$$

Among all resulting parameter hypercubes, i.e., $\mathbf{T}(\delta^k)$ and $\forall k \in V$, it is clear that only the smallest one can be totally inscribed within the feasible region defined by Eqs. (20.3) and (20.4). Hence,

$$FI = \min_{k \in V}\{\delta^k\} \tag{20.16}$$

Thus, the following simple procedure applies:

Step 1: Solve the optimization problem defined by Eqs. (20.3), (20.4), (20.14), and (20.15) for each vertex $k \in V$.

Step 2: Select FI according to Eq. (20.16).

Swaney and Grossmann (1985a, b) showed that under certain convexity conditions, the optimal solution of *FI* is associated with one of the vertices. However, even when these conditions are not met, it is often found that this approach is still applicable in practice. Note also that the vertex method may be computational demanding due to dimension explosion. For example, $2^{10} = 1024$ optimization runs are needed for 10 uncertain parameters and, if the number of parameters is raised to 20, the computation load for the required $2^{20} = 1{,}048{,}576$ runs can be very overwhelming. However, in specific, realistic applications, many of these runs may be omitted based on physical insights (Li and Chang, 2011).

20.3 Aspen Plus RCSTR module case study

A case study is provided for a continuous stirred tank reactor (CSTR) using the Aspen Plus RCSTR module. Notably, it is assumed that a CSTR is to be designed for a liquid phase esterification reaction of acetic acid (AA) and ethanol (ETHANOL), where produce water (WATER) and ethyl acetate (ACETATE) is to be produced. This reaction is reversible, exothermic, and the equilibrium composition is a weak function of temperature. A reactor model is created in Aspen Plus, with the details found in Figs. 20.3–20.4.

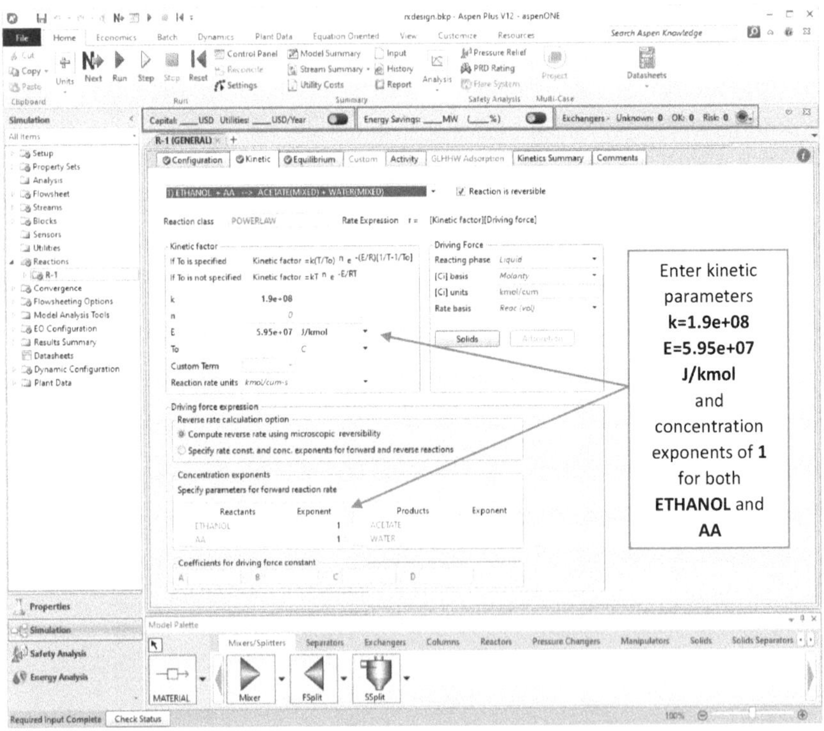

FIGURE 20.3 Steps to define R-1 kinetic details.

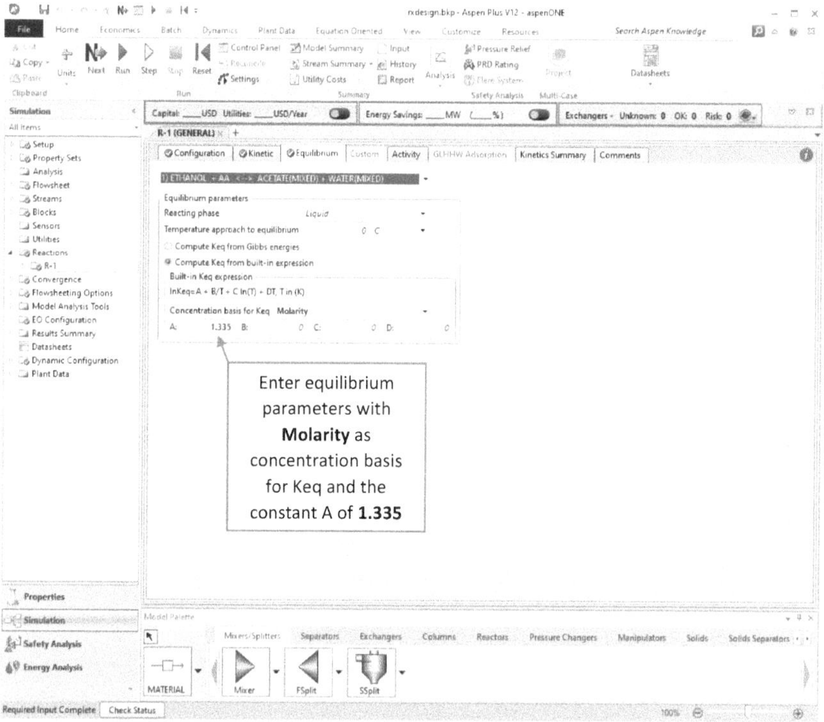

FIGURE 20.4 Steps to define R-1 equilibrium details.

First feed stream of the RCSTR module is an ethanol-water mixture with 0.95 mol fraction ethanol. This ETHANOL stream is fed at 200 kmol/h, 70°C and atmospheric pressure. Second feed stream to the RCSTR reactor contains of 200 kmol/h pure acetic acid (AA) at the same condition. Users can specify the RCSTR module with an initial reactor volume of 0.1 m^3 and associate the R-1 reaction to the RCSTR module (see Fig. 20.5). The design specification feature of Aspen Plus can be utilized by the user to determine the proper reactor volume in achieving 30% mole fraction of ethyl acetate at the reactor outlet. By adding a Design-Spec block into the flowsheet, the user can specify the targeted ethyl acetate mole fraction by varying the RCSTR volume (see detailed steps in Figs. 20.6–20.8). The Design-Spec block can be found in the Model Pallete under Manipulators tab. The simulation is then executed to obtain the RCSTR volume at 0.077 m^3, where 0.3 mol fraction of ethyl acetate is obtained at the reactor outlet. The overall strategy is typically done for the process design stage of the RCSTR module. The corresponding Aspen Plus simulation file is provided (rxdesign.bkp).

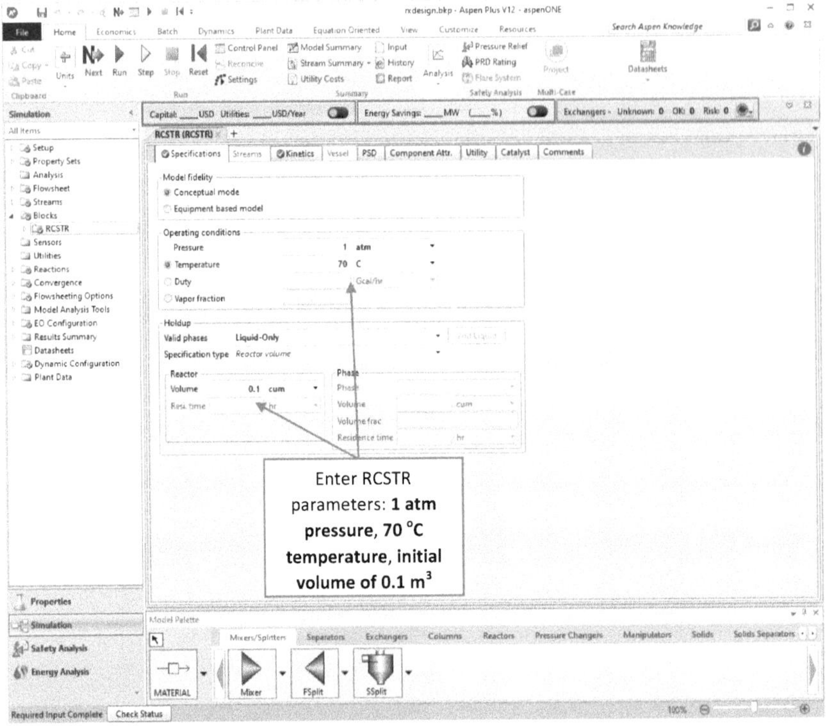

FIGURE 20.5 Steps to define RCSTR specifications.

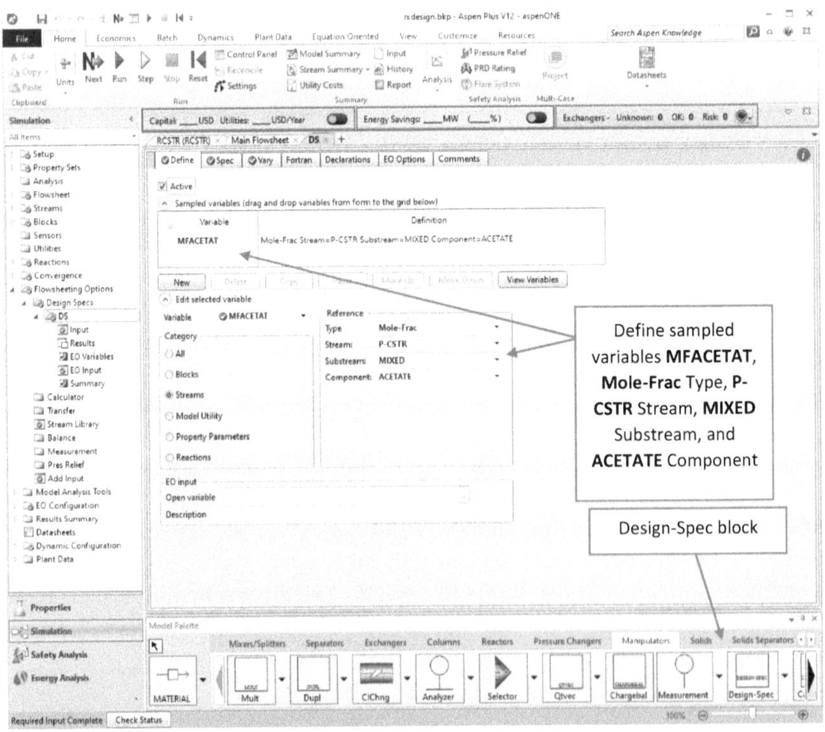

FIGURE 20.6 Steps to define the sampled variable of ACETATE mole fraction (MFACETAT).

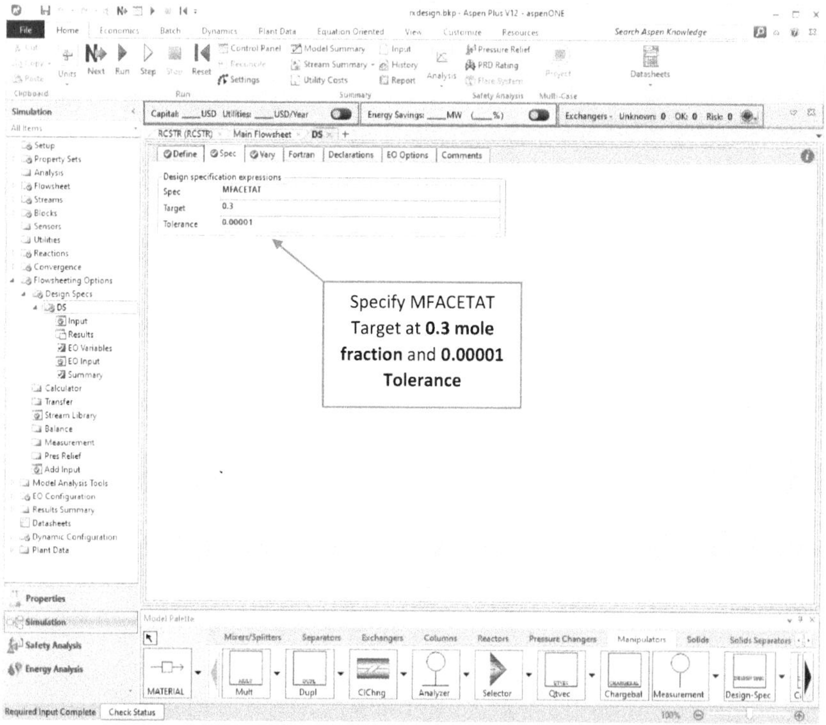

FIGURE 20.7 Steps to define MFACETAT mole fraction target variable.

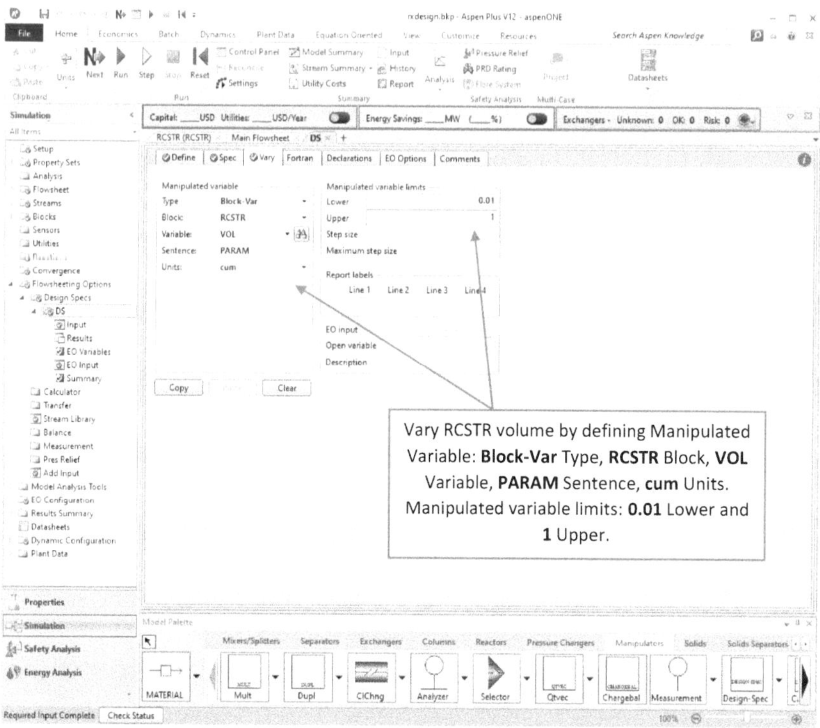

FIGURE 20.8 Steps to define RCSTR volume as manipulated variable.

20.4 Vertex methods for calculating FI of RCSTR

For the RCSTR design above, flexibility analysis can be carried out for the case when the ETHANOL stream has an expected uncertainty with vertex direction of ± 10 kmol/h. In other words, the ETHANOL flow rate may fluctuate between 190 and 210 kmol/h from its nominal value of 200 kmol/h (see Eqs. 20.5 and 20.15). On the other hand, the molar flow rate of AA is available from 190 to 210 kmol/h. In this case, the molar flow rate of AA should be defined as the compensating variable (i.e., control output) to meet the targeted ethyl acetate mole fraction of 0.3. The upper and lower bounds of the control output represent one of the inequality constraints of the RCSTR (see Eq. 20.4). The original Design-Spec block is modified to mimic such control constraints by varying the AA molar flow rate (see Fig. 20.9), while the reactor volume is fixed at 0.077 m^3 (Fig. 20.10). This ensures that the ACETATE mole fraction is perfectly controlled at 0.3 while fluctuation occurs at the ETHANOL flow rate.

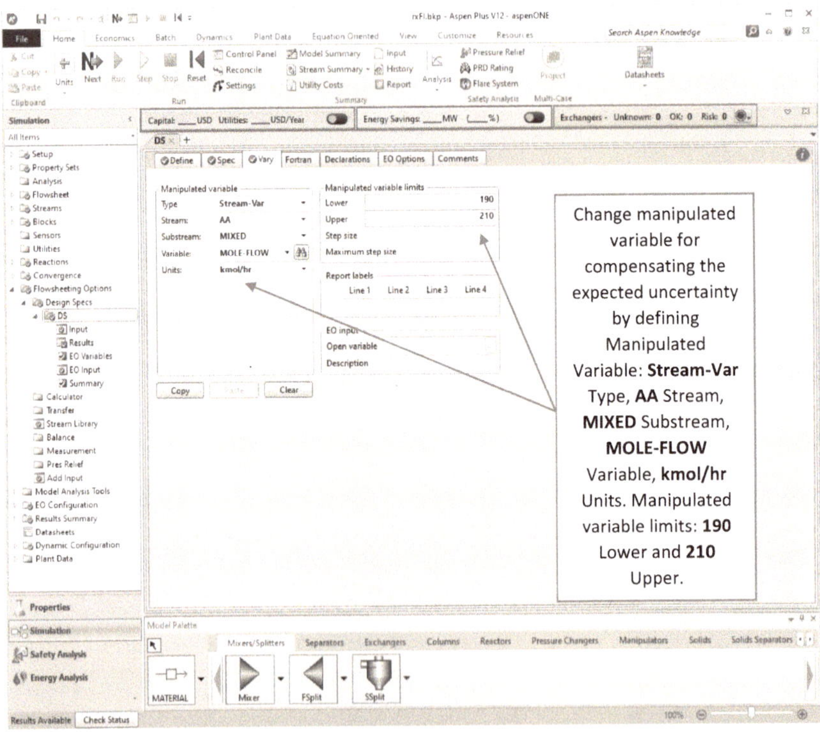

FIGURE 20.9 Steps to specify the modified Design-Spec block manipulated variable (AA molar flow rate).

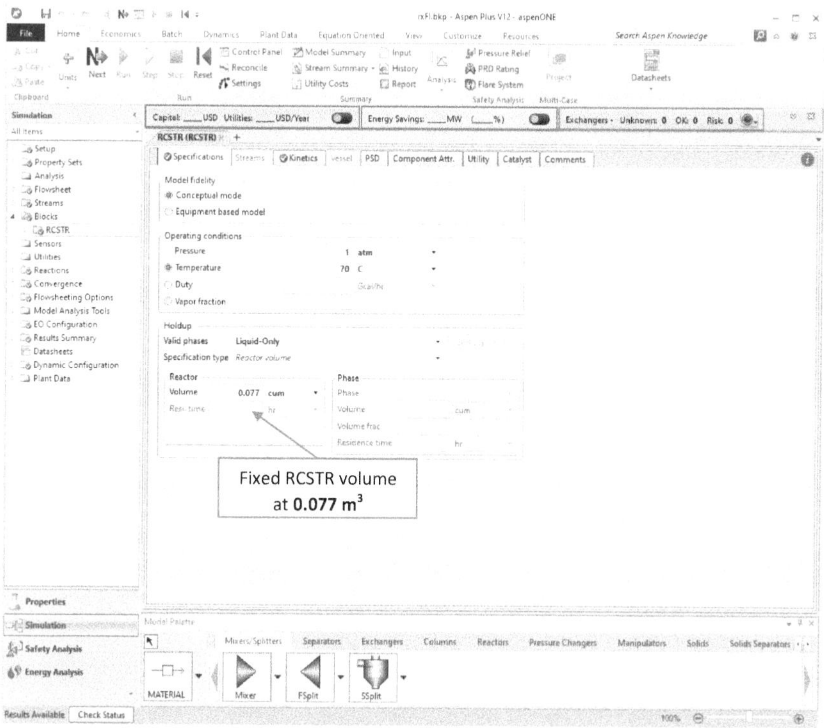

FIGURE 20.10 Steps to fix reactor volume.

To evaluate the FI of the RCSTR reactor, the automation interface of Aspen Plus and MATLAB is exploited.[1] An evaluation function (evalrx.m) is created in MATLAB with the following code to run the newly created Aspen Plus simulation file (rxFI.bkp).

```
function [ACETATE, AA, STATUS] = evalrx(TOTFLOW, VOL)
AspenVersion = 'apwn.document'; % A programmatic identifier (ProgID)
of Aspen Plus document class.
AspenPath = 'C:\Users\vska\Downloads\rxFI.bkp'; % Replace with the
simulation path
AspenVisible = 0; % 0 for not visible
AspenDialogs = 1;
% Get pointer and load Aspen Plus comserver:
Aspen = actxserver(AspenVersion);
% Open the Aspen-Simulation
invoke(Aspen,'InitFromFile2',AspenPath);
```

1. The details of Aspen Plus and MATLAB interface can be found in Chapter 2 - Multiplatform optimization on unit operation and process designs chapter.

```
Aspen.visible = AspenVisible; % Make it visible
Aspen.SuppressDialogs = AspenDialogs;
%Reinit and run simulation

%ETHANOL mole flow rate
Aspen.Tree.FindNode('\Data\Streams\ETHANOL\Input\
TOTFLOW\MIXED').Value=TOTFLOW;
%RCSTR volume
Aspen.Tree.FindNode('\Data\Blocks\RCSTR\Input\VOL').Value=VOL;

Aspen.Reinit();
Aspen.Run2();
% Wait to update GUI

ACETATE = Aspen.Tree.FindNode('\Data\Streams\P-CSTR\Output\
MOLEFRAC\MIXED\ACETATE').Value;
AA = Aspen.Tree.FindNode('\Data\Streams\AA\Output\TOT_FLOW').Value;
DSBLKSTAT = Aspen.Tree.FindNode('\Data\Flowsheeting Options\Design-
Spec\DS\Output\BLKSTAT').Value;
RCSTRBLKSTAT = Aspen.Tree.FindNode('\Data\Blocks\RCSTR\Output\
BLKSTAT').Value;
if DSBLKSTAT == 0 && RCSTRBLKSTAT == 0
   STATUS = 1;
else
   STATUS = 0;
end

%Release COM object
Aspen.Quit();
Aspen.Parent.Quit();
delete(Aspen);
end
```

The above code describes that within the MATLAB platform, Aspen Plus simulation file (rxFI.bkp) is executed for a given ETHANOL flow rate (TOTFLOW) at a given RCSTR volume (VOL). The Aspen Plus simulation file (rxFI.bkp) is already embedded within the Design-Spec block to achieve a 0.3 mol fraction of ethyl acetate by varying AA molar flow rate for a perfect control scenario. The corresponding ACETATE mole fraction, AA molar flow rate, RCSTR block status, and Design-Spec block status (BLKSTAT) are extracted to MATLAB. Both RCSTR block and Design-Spec block will report the BLKSAT variables which value of 0 indicate that the simulation has been executed succesfully. Therefore, a conditional statement is added that it requires both RCSTR block RCSTRBLKSTAT and Design-Spec block DSBLKSTAT to be 0 so that the STATUS of the evaluation function (evalrx.m) is 1, and 0 otherwise. The feasibility STATUS value in the evaluation function (evalrx.m) represents the feasibility Eq. (20.6).

A MATLAB code (FIbisect.m) is written to implement the vertex method mentioned earlier to compute the FI of the RCSTR. A bisection algorithm is embedded in the code to mimic the maximization model of Eq. (20.14).

Basically, the feasible RCSTR volume (VOL) value boundary on each given vertex, i.e., ETHANOL molar flow rate 190−210 kmol/h is computed. A multiplier variable (Mul) in the FIbisect.m is introduced for subsequent sensitivity analysis if user wants to check the FI of different RCSTR volumes.

```
function [FI] = FIbisect(Mul)
VOL=0.077*Mul;
TOTFLOWNom = 200;
TOTFLOWUp = 210;
TOTFLOWLo = 190;

%Simple bisection method for Flexibility Index computation at each
vertex
a = TOTFLOWNom;
b = TOTFLOWUp;
k = 0;
while b-a > 0.1
   x = (a+b)/2;
   [ACETATE, AA, STATUS] = evalrx(x,VOL);
   if STATUS == 1
     a = x;
   else
     b = x;
   end
   k = k+1;
end
FIRight = (x-TOTFLOWNom)/(TOTFLOWUp-TOTFLOWNom);

a = TOTFLOWLo;
b = TOTFLOWNom;
k = 0;
while b-a > 0.1
   x = (a+b)/2;
   [ACETATE, AA, STATUS] = evalrx(x,VOL);
   if STATUS==1
     b = x;
   else
     a = x;
   end
   k = k+1;
end
FILeft = (TOTFLOWNom-x)/(TOTFLOWNom-TOTFLOWLo);
FI = min(FIRight,FILeft);
end
```

By running the corresponding code in MATLAB command window by typing FIbisect(1), the FI of the given RCSTR volume at 0.077 m^3 (or Mul = 1) is determined as 0.52. The RCSTR can only tolerate 52% of the incoming uncertainty range. Practically, the ETHANOL molar flow rate that the RCSTR can handle is reduced to only ±5.2 kmol/h from the original expected range ±10 kmol/h. The reduced range is correlated to the ETHANOL flow rates of 194.8–205.2 kmol/h for the RCSTR to perform normally. The same MATLAB code (FIbisect.m) can also be executed to do a sensitivity analysis of the volume multiplier (Mul) from 0.9 to 1.1 so that the user can check the FI of different RCSTR volumes (see sensanFI.m).

```
i=1;
for x=0.9:0.02:1.1
   FI(i) = FIbisect(x);
   Mul(i)=x;
   i=i+1;
end
plot (Mul,FI)
xlabel('Volume multiplier')
ylabel('FI')
```

An FI profile from the sensitivity analysis is shown in Fig. 20.12. It is evident that when the reactor volume is too small (i.e., low volume multiplier), the residence time is too fast that the esterification reaction cannot reach the expected conversion target of 0.3 ethyl acetate molar fraction. When the reactor is too big (i.e., high volume multiplier), the retention time is too long that the esterification reaction has reached equilibrium. Note that the compensating variable AA molar flow rate range is already restricted at 190–210 kmol/h. Hence, the RCSTR cannot achieve the 0.3 ethyl acetate molar fraction when the volume is beyond the appropriate values. A careful design strategy with flexibility analysis can provide higher FI, as shown in Fig. 20.11. At volume multiplier of 1.04, the design can tolerate the expected range of the ETHANOL molar flow rate uncertainty (±10 kmol/h). Dynamic simulation with appropriate control scenarios to verify this conclusion is shown in the following section.

20.5 Aspen Plus Dynamics for RCSTR controllability verification

Dynamic simulation is an extension of steady-state simulation whereby time-dependence is built into the models via derivative terms, i.e., accumulation of mass and energy. This includes the definition of start-up and shutting down a plant, changes of conditions during a reaction, holdups, thermal changes, among others. Dynamic simulations require prolonged calculation of time and are mathematically more complex than steady-state simulations. It can be seen as repeatedly calculated steady-state simulations (based on a fixed time step) with constantly changing parameters. Dynamic simulation can be used in both online and offline fashion. The formal can be used for model predictive

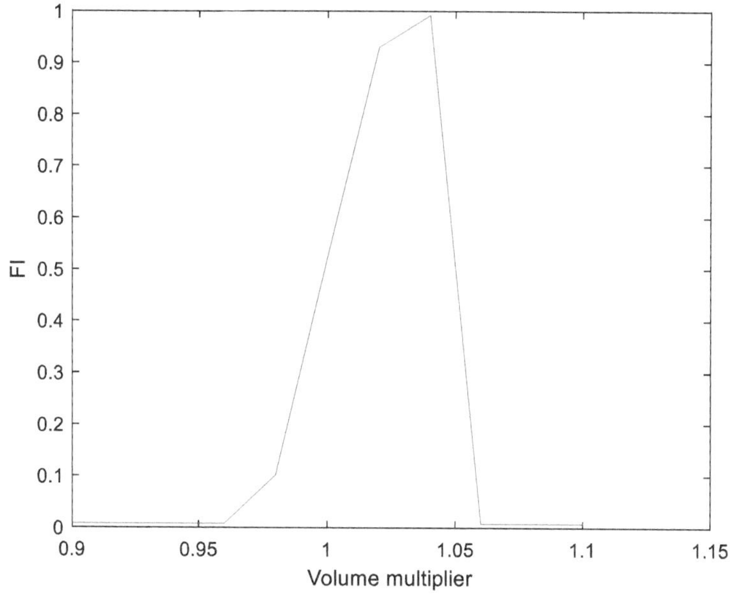

FIGURE 20.11 Sensitivity analysis of volume multiplier versus FI.

control. The real-time simulation results are used to predict the changes for a control input change, and the control parameters are optimized based on the results. On the other hand, offline dynamic simulation can be used in the design, troubleshooting, and optimization of process plants and the conduction of case studies to assess the impacts of process modifications. In this case, Aspen Plus Dynamics is used for verification purpose to show that an RCSTR with flexibility analysis will encounter controllable condition under an expected uncertainty.

The Aspen Plus basic workflow for dynamic simulation study is depicted in Fig. 20.12. First, a steady-state simulation should be built. Dynamic information is then added into the simulation file and exported to Aspen Plus Dynamics. Dynamic simulation can then be executed using Aspen Plus Dynamics platform. User can evaluate several modifications on the control configuration and other features.

The user can utilize the same steady-state Aspen Plus simulation file (rxFI.bkp) as a basis to verify whether the given design could tolerate the expected disturbance or otherwise. In this case study, two different volume multipliers are to be verified for process controllability as shown in Table 20.1, i.e., 1 and 1.04, which are correlated to FI of 0.52 and 1, respectively. There are three scenarios with positive and negative vertices to be studied (see Table 20.1), along with the expected process controllability.

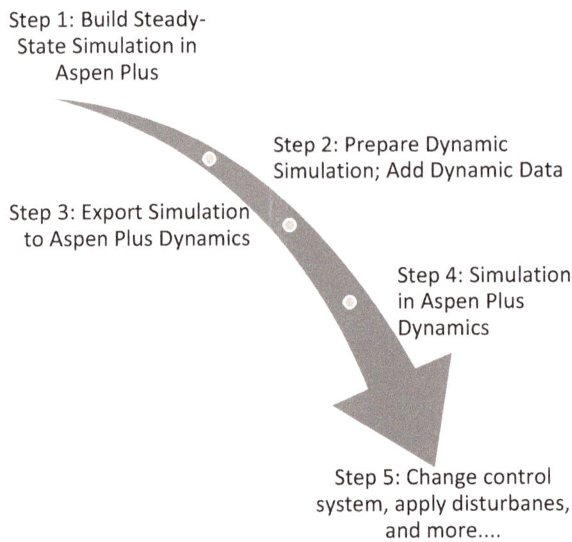

FIGURE 20.12 Aspen Plus Dynamics workflow.

TABLE 20.1 Verification matrix for RCSTR controllability.

RCSTR volume multiplier	FI	Disturbance vertices	Expected controllability
1	0.52	±5.2 kmol/h	Normal
1	0.52	±10 kmol/h	Abnormal
1.04	1	±10 kmol/h	Normal

Following the step 2 in Fig. 20.12, the vessel geometry should be defined. In this case, a flat head type with length of 0.5 m and an initial liquid volume fraction of 0.5 are arbitrarily selected for a vertical RCSTR vessel (Fig. 20.13). Since the operation is carried out at low temperature and liquid phase, flow-driven dynamic simulation is chosen.

An Aspen Plus Dynamics simulation can then be created (rxFI.dynf). Note that all control parameters are at their default values, and no further parameter tuning is implemented to simplify the problem. Arguably, control parameter tuning is necessary for actual operation. The dynamic simulations in this study are meant to distinctly verify the process controllability instead of determining the optimal control strategy. Two PI controllers are created by default for temperature and liquid levels (see Fig. 20.14). As defined in the steady-state Aspen Plus simulation, the RCSTR temperature is controlled at 70°C by

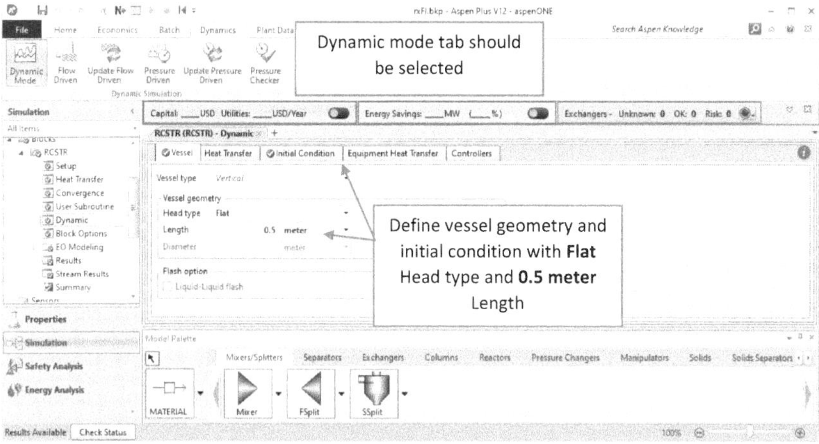

FIGURE 20.13 Steps to define vessel geometry for dynamic mode.

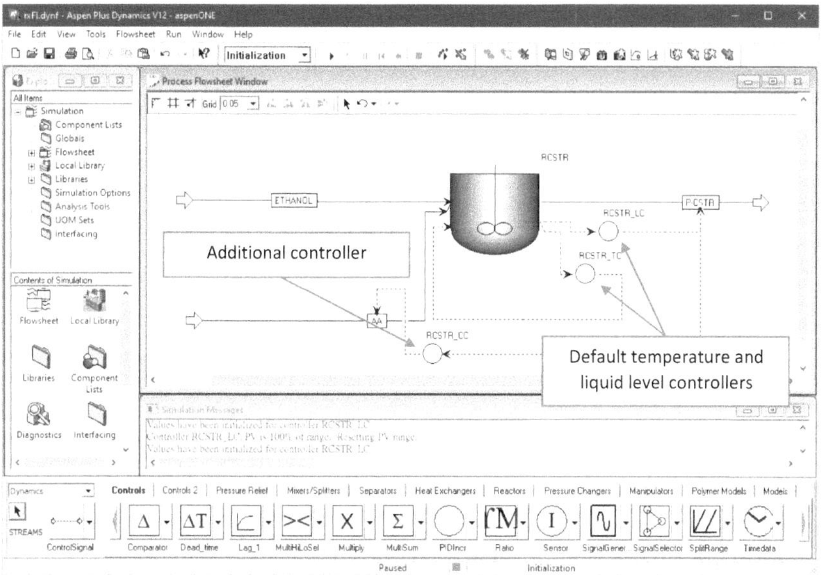

FIGURE 20.14 Aspen Plus Dynamics flowsheet of RCSTR module with PI controller for manipulating AA molar flow rate to achieve 0.3 ACETATE mole fraction.

manipulating the heat duty. The liquid level is maintained at 0.25 m (50% volume) by manipulating the RCSTR outlet molar flow rate. As defined previously, the initial liquid volume fraction is set as 0.5 (see Fig. 20.14). Due to the initialization in Aspen Plus Dynamics, the RCSTR volume is now twice as large, i.e., 0.154 m^3 (vertical flat head vessel with 0.5 m height and 0.626 m diameter). Therefore, at 0.25 m, the liquid volume is determined as 0.077 m^3, as specified in the original problem.

An additional PI controller is added accordingly to mimic the behavior of the modified Design-Spec block in the vertex method (Fig. 20.9), as shown in Fig. 20.14. The AA molar flow rate is set as manipulated variable to compensate the uncertain ETHANOL molar flow rate, in order to achieve 0.3 ACETATE molar fraction at the RCSTR outlet. The particular PI controller set point is 0.3 kmol/kmol and the initial output is 200 kmol/h. For fast action, the gain is arbitrary set at 100%/% and integral time at 20 min. The process variable and set point range is given as 0–1 kmol/kmol, while the output range is set to 190–210 kmol/h. A 12 h of simulation is carried out to show a steady-state operation, and disturbances are introduced for the next 48 h to follow the verification matrix of Table 20.1.

It can be observed that the RCSTR volume obtained from the traditional design strategy could not satisfy the controllability requirement for the expected ETHANOL molar flow rate uncertainty, as shown in the verification results of Aspen Plus Dynamics in Figs. 20.15 and 20.16. The RCSTR is only operable at a limited uncertainty range (±5.2 kmol/h). Fig. 20.15 obviously shows that the process variable (ACETATE molar fraction) can return back to 0.3.

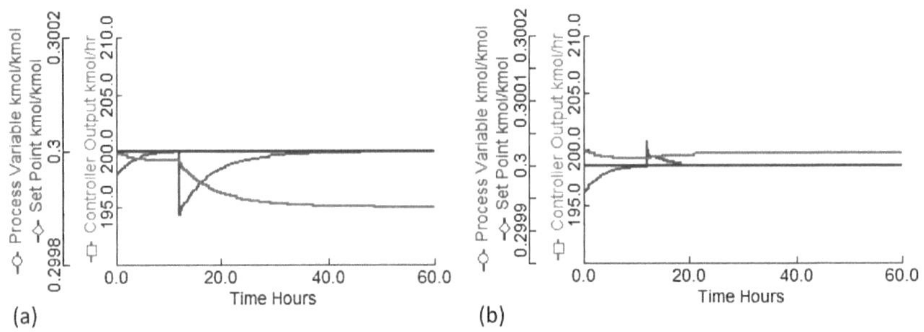

FIGURE 20.15 The dynamic simulation of the original RCSTR volume (0.077 m^3 or liquid level 0.25 m) with ETHANOL molar flow rate: (A) 205.2 kmol/h; (B) 194.8 kmol/h.

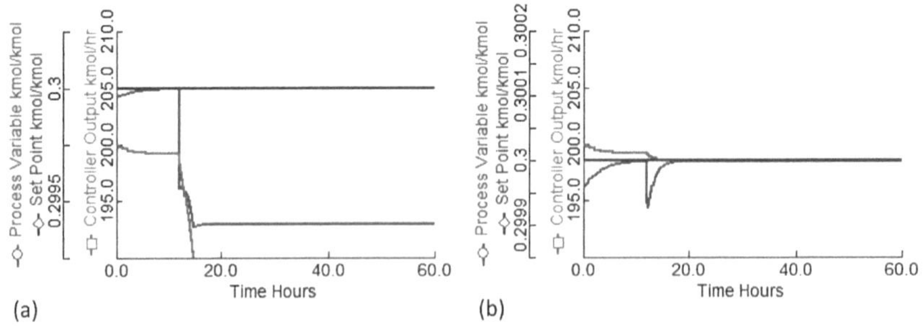

FIGURE 20.16 The dynamic simulation for the original RCSTR volume (0.077 m^3 or liquid level 0.25 m) with ETHANOL molar flow rate: (A) 210 kmol/h; (B) 190 kmol/h.

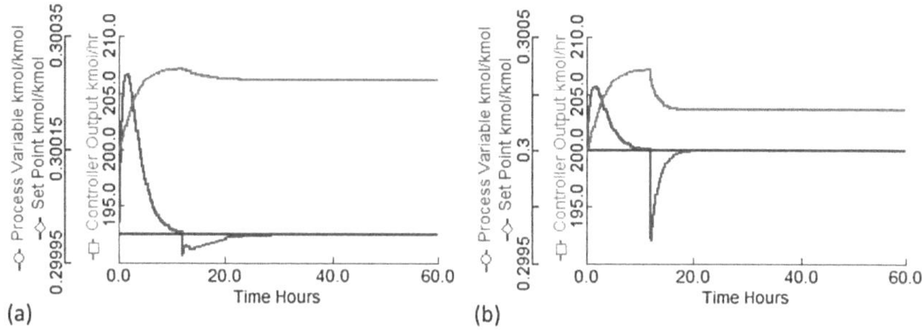

FIGURE 20.17 The dynamic simulation for RCSTR (liquid level $1.04 \times 0.25\ \text{m} = 0.26\ \text{m}$) with ETHANOL molar flow rate: (A) 210 kmol/h; (B) 190 kmol/h.

A full range test of ETHANOL molar flow rate (190−210 kmol/h shown in Fig. 20.16) demonstrates that the Controller Output AA molar flow rate (the green profile in Fig. 20.16A) is eventually at its lowest value (190 kmol/h) for ETHANOL molar flow rate of 210 kmol/h. This means that the control system is unable to properly maintain the molar fraction of ACETATE in the RCSTR outlet stream lfto 0.3 after a positive disturbance (+10 kmol/h) is introduced. In Figs. 20.15B and 20.16B, the 0.3 ACETATE molar fraction can be achieved for negative disturbance upset (−5.2 and −10 kmol/h). These results clearly describe that with an FI of 0.52, the RCSTR is only controllable within a fraction of the expected uncertainty range.

On the opposite side, a slightly larger RCSTR volume (1.04 volume multiplier with liquid level setpoint of 0.26 m), shows better process controllability and system flexibility. In Fig. 20.17, it is shown in the dynamic profiles where the RCSTR is now controllable to achieve the desired ACETATE molar fraction for both directions of ETHANOL molar flow rate range (190−210 kmol/h). This case study obviously shows that an appropriate design strategy should incorporate the process uncertainty consideration. A careful flexibility analysis will help the user to determine the desired design size that is resilient and controllable to disturbance.

20.6 Conclusion

This chapter elaborates on the concept of incorporating *flexibility index* in the traditional design strategy. The flexibility analysis is beneficial in enhancing resiliency and controllability of the final process. The results from the case study show that the design strategy with simultaneous consideration of process uncertainty will be more controllable than the conventional approach. The dynamic simulations also reinforce the concept of flexibility index implementation in improving the controllability of the final design. By utilizing the flexibility measures to improve the controllability and resiliency of the desired design, the unnecessary equipment oversizing can be avoided and eventually save costs.

Exercises

1. In the case study, the RCSTR module design target is specified at 0.3 mol fraction for ACETATE product. Please design a new RCSTR module with a 0.2 ACETATE mole fraction target. As the new specification is lower than the original value, the RCSTR volume is expected to be smaller. What about the FI of the new RCSTR module? What can be concluded from this new result?
2. Please verify the controllability of the new RCSTR module with Aspen Plus Dynamics. What do you suggest to improve the flexibility of the new RCSTR module?

References

Adi, V.S.K., Chang, C.T., 2011. Two-tier search strategy to identify nominal operating conditions for maximum flexibility. Industrial and Engineering Chemistry Research 50, 10707–10716.

Biegler, L.T., Grossmann, I.E., Westerberg, A.W., 1997. Systematic Methods of Chemical Process Design. Prentice Hall PTR, Upper Saddle River, NJ.

Chang, C.T., Adi, V.S.K., 2018. Deterministic Flexibility Analysis: Theory, Design, and Applications. CRC Press, Boca Raton, FL.

Grossmann, I.E., Floudas, C.A., 1987. Active constraint strategy for flexibility analysis in chemical processes. Computers and Chemical Engineering 11, 675–693.

Grossmann, I.E., Halemane, K.P., 1982. Decomposition strategy for designing flexible chemical-plants. AIChE Journal 28, 686–694.

Halemane, K.P., Grossmann, I.E., 1983. Optimal process design under uncertainty. AIChE Journal 29, 425–433.

Li, B.H., Chang, C.T., 2011. Efficient flexibility assessment procedure for water network designs. Industrial and Engineering Chemistry Research 50, 3763–3774.

Riyanto, E., Chang, C.T., 2010. A heuristic revamp strategy to improve operational flexibility of water networks based on active constraints. Chemical Engineering Science 65, 2758–2770.

Swaney, R.E., Grossmann, I.E., 1985a. An index for operational flexibility in chemical process design.1. Formulation and theory. AIChE Journal 31, 621–630.

Swaney, R.E., Grossmann, I.E., 1985b. An index for operational flexibility in chemical process design.2. Computational algorithms. AIChE Journal 31, 631–641.

Index

'*Note:* Page numbers followed by "f" indicate figures and "t" indicate tables.'

H

I

M

O

P

R

S

T

U

V

W

FSC
www.fsc.org
MIX
Papier | Fördert
gute Waldnutzung
FSC® C083411

Kolibri 360 GmbH
Ferdinand-Jühlke-Straße 7
99095 Erfurt, Deutschland
produktsicherheit@kolibri360.de